Handbook of Computational Social Science
for Policy

Eleonora Bertoni • Matteo Fontana •
Lorenzo Gabrielli • Serena Signorelli •
Michele Vespe

Editors

Handbook of Computational Social Science for Policy

 Springer

Editors

Eleonora Bertoni
Scientific Development Unit
Centre for Advanced Studies, Science and
Art, European Commission - Joint Research
Centre
Ispra, Italy

Matteo Fontana
Scientific Development Unit
Centre for Advanced Studies, Science and
Art, European Commission - Joint Research
Centre
Ispra, Italy

Lorenzo Gabrielli
Scientific Development Unit
Centre for Advanced Studies, Science and
Art, European Commission - Joint Research
Centre
Ispra, Italy

Serena Signorelli
Scientific Development Unit
Centre for Advanced Studies, Science and
Art, European Commission - Joint Research
Centre
Ispra, Italy

Michele Vespe
Digital Economy Unit
European Commission - Joint Research
Centre
Ispra, Italy

ISBN 978-3-031-16626-6   ISBN 978-3-031-16624-2   (eBook)
https://doi.org/10.1007/978-3-031-16624-2

This Springer imprint is published by the registered company Springer Nature Switzerland AG
The registered company address is: Gewerbestrasse 11, 6330 Cham, Switzerland

Preface[1]

Without a doubt, we live in an era in which data production is ubiquitous and data storage is cheap and widely available. Since the first decade of the new millennium, we are increasingly witnessing the widespread availability of smartphones, connected devices and sensor arrays able to provide all sorts of data that carry information about human activity and behaviour, in the form of "digital traces".

Combined with improvements in data storage and processing capabilities, it was just a matter of time before researchers started to explore such datasets for scientific purposes. Computational and analytical techniques have also evolved in order to deal with these new forms of data, including unstructured data. Among the many fields in which this "data revolution" has provided valuable input, one in which the contribution has been most disruptive is that of the social sciences (Einav & Levin, 2013; González-Bailón, 2013; Lazer & Radford, 2017).

In this context we can see the birth of a new discipline, Computational Social Science (CSS), which can be defined as "the development and application of computational methods to complex, typically large-scale, human (sometimes simulated) behavioural data" (Lazer et al., 2009). For the purpose of this handbook, our wish is to propose and to interpret the "behavioural data" in the broadest sense possible. Indeed we interpret as being "CSS-grade" data all pieces of information that, to some extent, provide information about humans: from survey data analysed with advanced computational methods to mobility data from network operators, from news articles to administrative data from municipalities. Since the pioneering contribution of Lazer and co-authors, CSS has reached an advanced degree of maturity, with academic journals completely devoted to the issue (such as the *Journal of Computational Social Science*[2] or *EPJ Data Science*[3]), as well as special issues of highly regarded publications covering different aspects of the discipline

---

[1] The views expressed are purely those of the authors and may not in any circumstances be regarded as stating an official position of the European Commission.

[2] https://www.springer.com/journal/42001

[3] https://epjdatascience.springeropen.com/

(such as one published on *Nature* in July 2021[4]) and academic handbooks devoted to the subject (such as the two volumes of the *Handbook of Computational Social Science* (Engel et al., 2022).

Being a scientific discipline that explicitly aims at understanding and modelling human behaviour and the interaction between humans and the environment they live in, the potential of CSS as a policymaking tool is self-evident. Despite this potential, a systematic approach towards mainstreaming the use of advanced computational methods and nontraditional data (either as the main source of information or in combination with more traditional ones) in policymaking has yet to be identified. A first step is to map the so-called demand side of CSS across several areas of policymaking by sourcing thematic questions at the interface between policy and research that can be addressed using CSS methods.

Based on the investigative/exploratory approach pioneered by the New York University GovLab with the 100 Questions Initiative,[5] the editors of the present book have performed a similar yet more extensive and EU-oriented exercise in Bertoni et al. (2022), where a list of thematic questions, drafted by scientists of the European Commission's Joint Research Centre, is presented. The policy relevance of the questions is ensured by the specific role of JRC scientists at the frontier between science and policy, as well as by a mapping of the specific questions onto the political priorities of the European Commission headed by Ursula Von der Leyen (Von der Leyen, 2019), as well as onto the UN Sustainable Development Goals.[6] These questions have represented an enabling factor for the editors to design and produce this handbook, as well as a starting point for the chapter authors to develop their work. We provide a table showing the correspondence between the questions in "Mapping the Demand Side of CSS for Policy" and the chapters of the present book in the front matter of the book.

This focus on specific policy-relevant issues sets the present book apart with respect to the previous literature on CSS. Another relevant difference is its ambition to shed light on the role of CSS techniques in all the phases of the policy cycle (as described in Jann & Wegrich, 2007). CSS methods can be used to help governments and supranational organisations at different stages, from the formulation of policy proposals to their adoption, implementation and evaluation. This is achieved by providing insights on foundational issues and on methodological aspects, as well as by direct applications to policy-relevant fields. This ambition to encompass all steps of the policy cycle represents a distinguishing point with respect to the book edited by Paruolo and Crato (2019), which is aimed at describing the state of the art with respect to ex-post policy evaluation.

The *Handbook of Computational Social Science for Policy* (CSS4P) is thus divided into three parts: foundational issues, methodological aspects and thematic application of CSS4P respectively. The first set of chapters on foundational issues in

---

[4] https://www.nature.com/collections/cadaddgige/

[5] https://the100questions.org/

[6] https://sdgs.un.org/

CSS4P opens with an exposition and description of what the key policymaking tasks are in which CSS can provide insights and information. Despite the recent COVID-19 emergency highlighting the need for access to nontraditional data sources, obstacles still remain to systematic use of CSS in government (Chap. 1). The widespread adoption of CSS in policymaking is still hindered by the presence of limiting factors in terms of access to key data sources, as well as the availability of analytical capabilities; Chap. 2 goes into the details of this issue, providing a taxonomy of governance and policy challenges. Of particular relevance to the policymaking setting are the social justice implications of the use of computational methods, which should be taken into account every time a public sector body decides to implement CSS solutions (Chap. 3). Ethical considerations of Computational Social Science approaches should be factored in not only when it comes to their implementation phase, but from the outset of the definition of the problem and possible solution, following an ethics-by-design approach; Chap. 4 provides an extended view of this issue from a researcher's perspective and gives some guidelines in the form of a framework that could help managing this particularly sensitive and important topic.

One of the aims of the present book is also to provide critical reviews of the current methodological literature, to better place Computational Social Science studies and their policy applications in the right technical context. Among the most important issues to be tackled, a prominent position is covered by complex systems, which require specific empirical and simulation methodologies such as agent-based models or machine learning techniques (Chap. 5). Moreover, digital trace data processed in CSS applications—such as large observational data, textual data or behavioural data gathered from large-scale online experiments—requires specific models, methods and modelling approaches, such as text mining techniques, large-scale behavioural experiments, causal inference and statistical techniques aimed at the reproducibility of science. A discussion around these issues is developed in Chaps. 6, 7 and 8. The use of CSS allows also a systematic improvement of more traditional tasks that involve data gathering and data processing, namely, the territorial impact assessment of policy measures (Chap. 9) and the production of statistics by official statistics offices (Chap. 10)

The remaining part of the handbook is devoted to critically surveying those scientific fields in which the potential impact of digital trace data and advanced computational methods is significant. CSS has proven to be an effective solution to address current gaps in economic policymaking (Chap. 12), by also providing insights in terms of labour market analysis (Chap. 13) and education economics (Chap. 16), as well as on the economics of social interactions and the related issue of access to economic opportunities (Chap. 21). Another area in which CSS has shown much promise relates to migration topics (Chap. 18) and more generally demography (Chap. 17), as well as the empirical study of human mobility where, in particular, the access to digital trace data can help describe dynamics of our society not captured by traditional sources of data (Chap. 23). Many themes related to the climate crisis, environmental sustainability and climate change mitigation or adaptation strategies are topical areas of interest for policy to which CSS can provide a substantial contribution: the socioeconomic consequences of climate change can be modelled using advanced computational methods, both by using simulation techniques such as Integrated Assessment Models, but also via statistical techniques (Chap. 14), while mitigation strategies such as more sustainable transport systems can also be explored (Chap. 24), and crisis management strategies can be improved (Chap. 22). Regional policy can be greatly aided by CSS methods, for example, in terms of understanding the regional variations of food security and nutrition (Chap. 11), but also by analysing what the problems are with the sustainability of tourism economies, e.g. via the use of data coming from online booking and short-term rental platforms (Chap. 19). The recent COVID-19 pandemic and the widespread presence of disinformation and misinformation on vaccines have put a strong attention on epidemiology (Chap. 15) as well as on understanding the information environment connected to such important and critical issues through the scanning and analysis of traditional and nontraditional media sources using neural embeddings, classification algorithm and network models (Chap. 20).

Ispra, Italy   Eleonora Bertoni
August 2022   Matteo Fontana
                                                                                       Lorenzo Gabrielli
                                                                                       Serena Signorelli
                                                                                       Michele Vespe

References

Bertoni, E., Fontana, M., Gabrielli, L., Signorelli, S., & Vespe, M. (Eds). (2022). *Mapping the demand side of computational social science for policy*. EUR 31017 EN, Luxembourg, Publication Office of the European Union. ISBN 978-92-76-49358-7, https://doi.org/10.2760/901622

Crato, N., & Paruolo, P. (Eds.). (2019). *Data-driven policy impact evaluation: How access to microdata is transforming policy design* (1st ed). Springer. https://doi.org/10.1007/978-3-319-78461-8

Einav, L., & Levin, J. D. (2013). *The data revolution and economic analysis*. National Bureau of Economic Research. Retrieved from http://www.nber.org/papers/w19035

Engel, U., Quan-Haase, A., Liu, S. X., & Lyberg, L. (Eds.). (2022). *Handbook of computational social science*. Routledge. https://doi.org/10.4324/9781003024583

González-Bailón, S. (2013). Social science in the era of big data: Social science in the era of big data. *Policy & Internet, 5*(2), 147–160. https://doi.org/10.1002/1944-2866.POI328

Jann, W., & Wegrich, K. (2007). Theories of the policy cycle. In F. Fischer, G. Miller, & M. S. Sidney (Eds.), *Handbook of public policy analysis: Theory, politics, and methods*. CRC Press. https://doi.org/10.4324/9781315093192

Lazer, D., & Radford, J. (2017). Data ex Machina: Introduction to big data. *Annual Review of Sociology, 43*(1), 19–39. https://doi.org/10.1146/annurev-soc-060116-053457

Lazer, D., Pentland, A., Adamic, L., Aral, S., Barabási, A.-L., Brewer, D., Christakis, N., Contractor, N., Fowler, J., Gutmann, M., Jebara, T., King, G., Macy, M., Roy, D., & Van Alstyne, M. (2009). Computational social science. *Science, 323*(5915), 721–723. https://doi.org/10.1126/science.1167742

Von der Leyen, U. (2019). *A union that strives for more: My agenda for Europe : political guidelines for the next European Commission 2019 2024*. European Commission. Directorate General for Communication (Ed.). Publications Office of the European Union. Retrieved from https://data.europa.eu/doi/10.2775/018127

Chapter Correspondence Between *Mapping the Demand Side of Computational Social Science for Policy* and *Handbook of Computational Social Science for Policy*

Chapter of *Mapping the Demand Side of Computational Social Science for Policy*	Chapter of *Handbook of Computational Social Science for Policy*
1. Data Governance	2. Computational Social Science for the Public Good: Toward a Taxonomy of Governance and Policy Challenges
2. Data Innovation for Official Statistics	10. Opportunities and Challenges of Computational Social Science for Official Statistics
3. Data Justice	3. Data Justice, Computational Social Science and Policy
4. Enablers of CSS4P	1. Computational Social Science for Public Policy
5. Ethics in CSS4P	4. The Ethics of Computational Social Science
6. Statistical Issues in CSS4P	6. From Lack of Data to Data Unlocking
7. Systems Modelling Using Advanced Techniques and/or Nonconventional Data Sources	5. Modelling Complexity with Unconventional Data: Foundational Issues in Computational Social Science
8. Analysis of Mobility due to COVID-19	23. The Empirical Study of Human Mobility: Potentials and Pitfalls of Using Traditional and Digital Data
9. Citizen Attitudes on Policy Areas and Related Societal Issues	7. Natural Language Processing for Policymaking
10. Climate Change Modelling and Policy	14. Computational Climate Change: How Data Science and Numerical Models Can Help Build Good Climate Policies and Practices
11. Data and Modelling for Territorial Impact Assessment of Policies	9. Data and Modelling for the Territorial Impact Assessment (TIA) of Policies
12. Demography	17. Leveraging Digital and Computational Demography for Policy Insights

(continued)

Chapter of *Mapping the Demand Side of Computational Social Science for Policy*	Chapter of *Handbook of Computational Social Science for Policy*
13. Describe Social Behaviour Through Computational Social Science to Inform Policy	8. Describing Human Behaviour Through Computational Social Science
14. Education	16. Learning Analytics in Education for the Twenty-First Century
15. Emergency Response and Disaster Risk Management	22. Social Media Contribution to the Crisis Management Processes: Towards a More Accurate Response Integrating Citizen-Generated Content and Citizen-Led Activities
16. Employment and Labour	13. Changing Job Skills in a Changing World
17. Epidemiology	15. Digital Epidemiology
18. Food Security and Development Policies	11. Agriculture, Food and Nutrition Security: Concept, Datasets and Opportunities for Computational Social Science Applications
19. Inclusive Rural Development	
20. International Migration	18. New Migration Data: Challenges and Opportunities
21. Macroeconomic Policy	12. Big Data and Computational Social Science for Economic Analysis and Policy
22. Political Analysis, Misinformation and Democracy	20. Computational Social Science for Policy and Quality of Democracy: Public Opinion, Hate Speech, Misinformation and Foreign Influence Campaigns
23. Social Interactions and Inequalities	21. Social Interactions, Resilience, and Access to Economic Opportunity: A Research Agenda for the Field of Computational Social Science
24. Sustainable Transports	24. Towards a More Sustainable Mobility
25. Tourism	19. New Data and Computational Methods Opportunities to Enhance the Knowledge Base of Tourism

Acknowledgments

This idea for this book was conceived during the first months of the "Computational Social Science for Policy" project, incubated at the Centre of Advanced Studies (CAS) of the Joint Research Centre (JRC) of the European Commission (EC), located in Ispra (VA), Italy. The role of the CAS inside the JRC is to enhance the capabilities of the European Commission to better understand and address complex and long-term societal challenges faced by the EU.

The road to its completion has been a long one. We started with the project kick-off workshop in May 2021 and then spanned into an exercise of mapping future policy questions to be tackled using computational social science methods. This led to the publication of a science for policy report "Mapping the Demand Side of Computational Social Science for Policy"[7] that acts as the inspiration guide for this handbook, where we collected valuable contributions from a community of experts about how those policy questions could be addressed. We are extremely grateful to all the chapter authors for their insightful and wholehearted collaboration.

We are also extremely thankful to all the external experts, academics and EC colleagues who believed in the CSS4P project from the start, participated in the kick-off workshop, as well as to the colleagues of the JRC and of the rest of the EC who acted as reviewers, provided suggestions and engaged in active and frank discussions about the book. We would like to thank Adalbert Wilhelm, Albrecht Wirthmann, Alexander Kotsev, Alexandra Balahur, Anna Berti Suman, Anne Goujon, Béatrice d'Hombres, Biagio Ciuffo, Carlo Lavalle, Carolina Perpiña, Charles MacMillan, Daniela Ghio, Domenico Perrotta, Eimear Farrell, Emanuele Ciriolo, Emanuele Ferrari, Emilia Gómez Gutiérrez, Enrico Pisoni, Enrique Fernández-Macías, Fabio Ricciato, Fabrizio Natale, Federico Biagi, Filipe Batista, François J. Dessart, Gary King, Ginevra Marandola, Giulia Listorti, Guido Tintori, Hannah Nohlen, Haoyi Chen, Helen Johnson, Hendrik Bruns, Jens Linge, Juan Carlos Císcar-Martínez, Julia Le Blanc, Kristina Potapova, Laurenz Scheunemann, Luc Feyen, Luca Barbaglia, Luca Onorante, Lucia Vesnić-Alujević, Marco Colagrossi,

---

[7] https://publications.jrc.ec.europa.eu/repository/handle/JRC126781

Marco Ratto, Marco Scipioni, María Alonso Raposo, Marianna Baggio, Marina Micheli, Marta Sienkiewicz, Matteo Sostero, Myrto Pantazi, Néstor Duch-Brown, Nikolaos Stilianakis, Panayotis Christidis, Paola Rufolo, Paolo Paruolo, Pascal Tillie, Paul Smits, Peter Salamon, Pieter Kempeneers, René Van Bavel, Ricardo Barranco, Sara Grubanov-Boskovic, Sergio Consoli, Sergio Gomez y Paloma, Solomon Messing, Stefano Maria Iacus, Tom De Groeve, Valerio Lorini, Victor Nechifor and Zsuzsa Blaskó. The book could not have been possible without the constant support and encouragement of Jutta Thielen-Del Pozo, Head of the Scientific Development Unit, Shane Sutherland, Project Leader of the Centre of Advanced Studies, as well as Desislava Stoyanova and Carolina Oliveira for the administrative and legal support. We are also very thankful to the Joint Research Centre of the European Commission for the financial support to offer this book as open access.

Finally, we would like to thank our Editor, Ralf Gerstner, and Ramya Prakash of Springer for having believed in this book since the beginning and for the support.

Ispra, Italy   Eleonora Bertoni
August 2022   Matteo Fontana
                                                                            Lorenzo Gabrielli
                                                                            Serena Signorelli
                                                                               Michele Vespe

Contents

Abbreviations

3D	Three-dimensional
ABM	Agent-based model
AI	Artificial intelligence
AME	Adult male equivalent weights
ANN	Artificial neural networks
AoIR	Association of Internet Researchers
APEX	Agricultural Policy/Environmental eXtender
API	Application Programming Interface
AREA	Anticipate, Reflect, Engage, Act
ARMAX	Autoregressive Moving Average with Exogenous variables
ARS	Airline reservation systems
ASD	Adaptive survey design
AV	Autonomous vehicle
BART	Bayesian Additive Regression Trees
BDI	Beck Depression Inventory
BDAaaS	Big Data Analytics as a Service
BERT	Bidirectional Encoder Representations from Transformers
BLS	Bureau of Labour Statistics
BSA	British Sociological Association
CARE & Act	Consider context, Anticipate impacts, Reflect on purposes, positionality, and power, Engage inclusively, Act responsibly and transparently
CART	Classification and Regression Tree
CAV	Connected autonomous vehicles
CBS	Dutch Central Bureau of Statistics
CES	Current Employment Statistics
CDC	Centers for Disease Control and Prevention
CCTV	Closed-circuit television
CDR	Call detail record
Cedefop	The European Centre for the Development of Vocational Training
CES-D	Center for Epidemiologic Studies Depression Scale

CFIO	Coordinated Foreign Influence Operation
CGE	Computable General Equilibrium
CIMMYT	International Maize and Wheat Improvement Center
CNG	Compressed natural gas vehicles
CO2	Carbon dioxide
COMPAS	Correctional Offender Management Profiling for Alternative Sanctions
COVID-19	Coronavirus disease 2019
CMCC	Centro Euro-Mediterraneo per i Cambiamenti Climatici
CPI	Consumer Price Index
CSS	Computational Social Science
DEA	Data Envelopment Analysis
DHS	Demographic Health Surveys
DICE	Dynamic Integrated Climate-Economy
Doc2Vec	Document to vector
DSA	Digital Services Act
DSGE	Dynamic stochastic general equilibrium
E3ME	Energy-Environment-Economy Global Macro-Economic
EASO	European Asylum Support Office
ECDC	European Centre for Disease Prevention and Control
EDPS	European Data Protection Supervisor
EERP	Experimental Economics Replication Project
EFSA	European Food Safety Authority
EMS	Emergency medical services
EPSRC	Engineering and Physical Sciences Research Council
EPU	Economic and political uncertainty
E&T	Education and training
ESCO	The multilingual classification of European Skills, Competences, Qualifications and Occupations
EU	European Union
EVs	Electric vehicles
FAIR	Findable, Accessible, Interoperable, and Reusable
FAO	Food and Agriculture Organization of the United Nations
FARMSIM	Farming Simulator
FCTs	Food composition tables
FSMS	Food Security Monitoring System
GASC	Generality, Adequacy Specificity, and Comparability
GBM	Gradient boosting machine
GCM	Global Compact for Safe, Orderly and Regular Migration
GDELT	Global Database of Events, Language, and Tone
GDP	Gross domestic product
GDPR	General Data Protection Regulation
GEM-E3	General Equilibrium Model for Economy-Energy-Environment
GHG	Greenhouse gas
GLEaM	Global Epidemic and Mobility

GMDAC	Global Migration Data Analysis Centre
GO	Governmental organization
GovLab	Governance Lab
GPS	Global positioning system
GPU	Graphics processing unit
GTU	Google Trends Uncertainty
HEVs	Hybrid electric vehicles
HICs	High-income countries
IAM	Integrated Assessment Model
IaaS	Infrastructure as a Service
IASC	Inter-Agency Standing Committee
ICE	Internal combustion engine
ILI	Influenza-like illness
ILO	International Labour Organization
IMP	Impact Mitigation Plan
IoT	Internet of Things
IOM	International Organization for Migration
IP	Intellectual property
IPCC	Intergovernmental Panel on Climate Change
IPR	Intellectual property rights
IRA	Internet Research Agency
IRB	Institutional Review Board
IRE	Internet Research Ethics
IRS	Internal Revenue Service
ISCO	International Standard Classification of Occupations
ISIS	Islamic State of Iraq and Syria
IT	Information technology
JOAPS	Job-Oriented Asymmetrical Pairing System
JRC	Joint Research Centre
KPI	Key performance indicator
KCMD	EU Commission Knowledge Centre on Migration and Demography
KNN	K-nearest neighbour
LASSO	Least Absolute Shrinkage and Selection Operator
LDA	Latent Dirichlet Allocation
LMI	Labour market intelligence
LMICs	Low-/middle-income countries
LSMS	Living Standard Measurement Studies
MaaS	Mobility as a Service
MAD	Minimum acceptable diet for infants and young children
MDD-W	Minimum dietary diversity for women
ML	Machine learning
MICS	Multi-indicator cluster surveys
MOOC	Massive open online course

NESH	National Committee for Research Ethics in the Social Sciences and the Humanities
NeurIPS	Neural Information Processing Systems
NGO	Nongovernmental organization
NLP	Natural language processing
NMEW	National Mission for Empowerment of Women
N-NCD	Nutrition-related noncommunicable diseases
OECD	Organisation for Economic Co-operation and Development
OJAs	Online job advertisements
OPAL	OPen ALgorithm project
OSS	Open-source software
PBSEs	Population-based survey experiments
PET	Privacy-enhancing technologies
PHEV	Plug-in hybrid electric vehicles
PISA	Programme for International Student Assessment
POI	Point of interest
POS	Points of sale
PPP	Public-private partnership
PTSD	Post-traumatic stress disorder
PUE	Pop-up experiment
QCA	Qualitative comparative analysis
RCA	Revealed comparative advantage
RCP	Representative Concentration Pathway
RCT	Randomised controlled trial
RF	Random forest
RFF	Resources for the Future
RFID	Radio-frequency identification
RPP	Reproducibility Project – Psychology
RRI	Responsible Research and Innovation
SIR	Susceptible, Infected, Recovered
SCI	Social Connectedness Index
SCC	Social cost of carbon
SDG	Sustainable Development Goal
seff	Effective Mesh Density
SEM	Structural Equation Models
Skills-OVATE	Skills Online Vacancy Analysis Tool for Europe
SMEs	Small and medium enterprises
SMM	Social media monitoring
SNA	Social network analysis
SSA	Social Security Administration
SSP	Shared Socioeconomic Pathway
SSRP	Social Sciences Replication Project
STATPRO	Shared, Transparent, Auditable, Trusted, Participative, Reproducible, and Open
STEM	Science, technology, engineering and mathematics

STeMA	A Sustainable Territorial Economic/Environmental Management Approach
SVM	Support vector machines
TEU	Twitter Economic Uncertainty
TIA	Territorial Impact Assessment
TRIPOD	Transparent Reporting of a multivariable prediction model for Individual Prognosis or Diagnosis
TSA	Tourism Satellite Account
UAV	Unmanned aerial vehicles
UGC	User generated content
UK	United Kingdom
UN	United Nations
UNDP	United Nations Development Programme
US	United States
UNHCR	The UN Refugee Agency
UNSD	United Nations Statistics Division
UNWTO	United Nations World Tourism Organization
VA	Value added
VOST	Virtual Operations Support Team
WHO	World Health Organization
WiFi	Wireless Fidelity
WMA	World Medical Association

Part I
Foundational Issues

Chapter 1
Computational Social Science for Public Policy

Helen Margetts and Cosmina Dorobantu

Abstract Computational Social Science (CSS), which brings together the power of computational methods and the analytical rigour of the social sciences, has the potential to revolutionise policymaking. This growing field of research can help governments take advantage of large-scale data on human behaviour and provide policymakers with insights into where policy interventions are needed, which interventions are most likely to be effective, and how to avoid unintended consequences. In this chapter, we show how Computational Social Science can improve policymaking by detecting, measuring, predicting, explaining, and simulating human behaviour. We argue that the improvements that CSS can bring to government are conditional on making ethical considerations an integral part of the process of scientific discovery. CSS has an opportunity to reveal bias and inequalities in public administration and a responsibility to tackle them by taking advantage of research advancements in ethics and responsible innovation. Finally, we identify the primary factors that prevented Computational Social Science from realising its full potential during the Covid-19 pandemic and posit that overcoming challenges linked to limited data flows, siloed models, and rigid organisational structures within government can usher in a new era of policymaking.

1.1 Introduction

These are exciting times for social science. Large-scale data was formerly the province of the physical and life sciences, while social science relied mostly on qualitative data or survey data to understand human behaviour. The data revolution from the 2010s onwards, where huge quantities of transactional data are

H. Margetts (✉) · C. Dorobantu
The Alan Turing Institute, London, UK

Oxford Internet Institute, University of Oxford, Oxford, UK
e-mail: helen.margetts@oii.ox.ac.uk; cdorobantu@turing.ac.uk

© The Author(s) 2023
E. Bertoni et al. (eds.), *Handbook of Computational Social Science for Policy*,
https://doi.org/10.1007/978-3-031-16624-2_1

generated by people's online actions and interactions, means that for the first time, social scientists have access to large-scale, real-time transactional data on human behaviour. With this influx of data, social scientists can and need to develop and adapt computational methods for analysis of large-scale social data. Computational Social Science—the marriage of computational methods and social sciences—can transform how we detect, measure, predict, explain, and simulate human behaviour. Given that public policy is about understanding and potentially changing the world outside—society and the economy—Computational Social Science is well placed to help policymakers with a wide range of tasks, combining as it does computational methods with social scientific lines of enquiry and theoretical frameworks. Given the struggle that social science often has to demonstrate or receive recognition for policy impact (Bastow et al., 2014), CSS might act as a channel for social science to be appreciated in a policy context. This chapter examines how CSS might assume an increasingly central role in policymaking, bringing social science insight and modes of exploration to the heart of it.

The seminal article on CSS (Lazer et al., 2009) laid out how the capacity to analyse massive amounts of data would transform social science into Computational Social Science, just as data-driven models and technologies had transformed biology and physics. 'We define CSS as the development and application of computational methods to complex, typically large-scale, human (sometimes simulated) behavioural data [. . .] Whereas traditional quantitative social science has focused on rows of cases and columns of variables, typically with assumptions of independence among observations, CSS encompasses language, location and movement, networks, images, and video, with the application of statistical models that capture multifarious dependencies within data' (Lazer et al., 2009, p. 1060). Although there is no definitive list of the methodologies that would fall into the category of CSS, it is clear that agent computing, microsimulation, machine learning (ML), complex network analysis, and statistical modelling would all fall into the field. We might also add large-scale online experimental methods and some of the ethical thinking that should accompany the handling of large-scale data about human behaviour.

Lazer et al.'s early article did not discuss how policymaking or government might also be transformed, although the second article 12 years on (Lazer et al., 2021) emphasised the need to articulate how CSS could tackle societal problems. Given CSS's emphasis on data and data analysis, the transformative potential of Computational Social Science for policymaking is huge. Traditionally, governments have made little use of transactional data for policymaking (Margetts & Dorobantu, 2019). That is not surprising, given that bureaucratic organisation from the earliest forms of the state relied on 'the files' for information (Muellerleile & Robertson, 2018). Paper-based files offer the capability to find individual pieces of data but generate no usable data for analysis. Likewise, the large-scale computer systems which gradually replaced these files from the 1950s onwards in the largest governments also had no capacity to generate usable data (Margetts, 1999). For decades, governments' transactional data resulting from their interactions with citizens languished in 'legacy systems', unavailable to policymakers. During this period, data and modelling existed in government but relied on custom built 'official statistics' or performance indicators, or long-running annual surveys, such as the UK 'British Crime Survey'. Only with the internet and the latest generation of data-driven models and technologies has there been the possibility for policymakers to use large-scale transactional data to inform decision-making.

This chapter outlines key policymaking tasks for which CSS can be used: detection, measurement, prediction, etiology,[1] and simulation. It discusses how CSS needs to be 'ethics-driven,' revealing bias and inequalities and tackling them by taking advantage of research advancements in ethics and responsible innovation. Then, the chapter examines how the potential of CSS tools has been highlighted in the pandemic crisis but also how CSS failed to realise this potential due to weaknesses in data flows, models, and organisational structures. Finally, the chapter considers how CSS might be used to tackle future crisis situations, renewing the policy toolkit for more resilient policymaking.

1.2 Detection

Detection is one of the 'essential capabilities that any system of control must possess at the point where it comes into contact with the world outside' (Hood & Margetts, 2008). Government is no exception, needing to understand societal and economic behaviour, trends, and patterns and to calibrate policy accordingly. That includes detection of unwanted (or less often, wanted) behaviour of citizens and firms to inform policy responses.

Data-intensive technologies, such as machine learning, lend themselves very well to the performance of detection tasks. Advances in machine learning over the past decade make it a powerful tool in the analysis of both structured and unstructured data. Structured data refers to data points that are stored in a machine-readable format. ML is well suited to the performance of detection tasks that rely on structured data, such as pinpointing fraudulent transactions in large-scale financial data. The progress made by researchers and practitioners in the fields of natural language processing (NLP) and computer vision now also makes ML well suited to the analysis of unstructured data, such as human language and visual data (Ostmann & Dorobantu, 2021). ML can perform detection tasks that were outside the realm of possibilities in earlier decades due to our inability, in the past, to process large quantities of structured and unstructured data.

A good illustration of where Computational Social Science and policymakers can work side-by-side to detect unwanted behaviour relates to online harm. Online harm is a growing problem in most countries, including (but not limited to) the gen-

---

[1] Etiology is a term often used in the medical sciences, meaning 'the cause, set of causes, or manner of causation of a disease or condition'. (Oxford Languages from Oxford University Press). Here we use it more broadly, to refer to the cause, set of causes, or manner of causation of phenomena of interest within the social sciences.

eration, organisation, and dissemination of hate speech, misinformation, misleading advertising, financial scams, radicalisation, extremism, terrorist networks, sexual exploitation, and sexual abuse. Nearly all governments are tackling at least some of these harms via a range of public agencies. Criminal justice agencies need to track and monitor the perpetrators of harm; intelligence agencies need to scrutinise security threats, while regulators need to detect and monitor the behaviour of a huge array of data-powered platforms, particularly social media firms.

How can Computational Social Science help policymakers? A growing number of computational social scientists are focusing on the detection of harmful behaviour online, seeking to understand the dissemination and impact of such behaviour, which is a social as well as a computational task. Machine learning classifiers need to be built, and this is a highly technical task, requiring cutting edge computer science expertise and facing huge challenges (see Röttger et al., 2021). But it is those with social science training that are comfortable dealing with the normative questions of defining terms such as 'hate'. And it is social scientists who are able to explore the motivations behind harmful online behaviour; to understand the differential impacts of different kinds of harm (e.g., misinformation has different dynamics from hate speech, see Taylor et al., 2021); and to explore how we can build distinct classifiers for different kinds of online harm or different targets of harm, such as misogyny (Guest et al., 2021) or sinophobia (Vidgen et al., 2020). By bringing together the development of technical tools and the rigour and normative stance of the social sciences, Computational Social Science offers a holistic and methodologically sound solution to policymakers interested in tackling online harm.

Regulation for online safety is a key area where CSS is uniquely qualified to help. Regulators need to develop methodological expertise but often struggle to keep ahead of the perpetrators of unwanted online behaviour and the massive platforms where these harms play out. While CSS expertise is growing in this area, the platforms themselves have incubated parallel streams of in-house research, with different motivations, confidentiality, secrecy, and lack of data sharing preventing knowledge transfer between the two. This leaves an important role for academic researchers, working directly with regulators to help them understand the 'state-of-the-art' research in promoting online safety.

1.3 Measurement

Another key capability of government is measurement. Policymakers need to be able to monitor and track societal and economic trends and patterns in order to understand when interventions are needed.

The technologies that were available to us prior to the data revolution limited our ability to collect, store, and analyse data. These technological limitations meant that in the past, policymakers and academic researchers alike were at best able to measure socio-economic phenomena imprecisely and at worst unable to measure them at all. For example, policymakers and researchers have been trying for decades to understand visitation rates at public parks (see, e.g., Cheung, 1972). This understanding is needed for a range of policy interventions, from protecting green spaces and increasing investment in parks to driving up community usage. But what seems like a simple metric, the number of visitors to a park, has been difficult to produce in practice. The solution preferred by many local authorities has been to hire contractors and ask them to stand at the entrance of a park and count the number of people going in. This solution has obvious limitations: it is costly, it can only measure park attendance for limited periods of time, it is prone to measurement error, and it fails to capture characteristics of the people visiting the park—to name only a few.

Complex socio-economic phenomena are even more difficult to measure. Firms, consumers, and policymakers are increasingly worried about inflation, a phenomenon that threatens the post-pandemic economic recovery. Yet despite the fact that so many eyes and newspaper headlines focus on the consumer price index, few know the difficulties of collecting and generating it. In the UK, for example, the Office for National Statistics calculates the Consumer Prices Index. The index largely rests on the physical collection of data in stores across 141 locations in the UK. At a time when we needed precise inflation measures the most, during the Covid-19 crisis, the data collection efforts for the Consumer Prices Index were severely affected by store closures and social distancing measures. Furthermore, the labour-intensive nature of collecting and generating the Consumer Prices Index means that it cannot be, with its current design, a real-time measure. National statistical offices usually publish it once a month with the understanding that it reflects the reality of a few weeks back.

Computational Social Science allows new opportunities to measure and monitor socio-economic phenomena—from park usage to inflation. Recent research uncovered the value of using social media data and mobile phone app data to measure park visitation (see, e.g., Donahue et al., 2018; Hamstead et al., 2018; Sinclair et al., 2021; Suse et al., 2021). Attempts to create real-time measures of inflation go back more than a decade. In 2010, Google's chief economist, Hal Varian, revealed that the company was working on a Google Price Index—a real-time measure of price changes calculated by monitoring prices online. Although Google never published this measure, it hints at the possibilities of using computational methods and economic expertise to move beyond the inflation measures that we have today.

More generally, Computational Social Science could facilitate a wholescale rethinking of how we measure key socio-economic indicators. As Lazer et al. (2021) reflected in their study of 'Meaningful measures of human society in the twenty-first century':

> Existing measures of key concepts such as gross domestic product and geographical mobility are shaped by the strengths and weaknesses of twentieth century data. If we only evaluate new measures against the old, we simply replicate their shortcomings, mistaking the gold standard of the twentieth century for objective truth.

Traditional social science methods of data analysis tend to perpetuate themselves. Survey researchers, for example, are reluctant to relinquish either long-running surveys or questions within them. This means that over time, surveys become longer and longer and increasingly unsuited to measuring behavioural trends in digital environments (e.g., asking people what they did online is a highly inaccurate way of determining digital behaviour compared with transactional data). Computational Social Science gives us the ability to improve our measurements so that everything—from basic summary stats to the most sophisticated measures—can move away from having to rely on old measurements that are limited by the technologies and data that were available decades ago.

1.4   Prediction

Another tool that Computational Social Science has to offer to policymakers is predictive capability. Machine learning is increasingly used within the private sector to perform prediction and forecasting tasks, as it is well suited to the performance of these tasks. Governments and public sector organisations in general do not have a good record on forecasting and prediction, so this is another area where CSS can add to policymakers' toolkit. Policymakers can use machine learning to spot problematic trends and relationships of concern before they have a detrimental impact and to predict points of failure within a system. One of the most common uses of machine learning by local and central governments is to predict where problems are most likely to arise with the aim of identifying 'objects' (from restaurants and schools to customs forms) for inspection and scrutiny. The largest study on the use of machine learning in US federal government provides the example of the US Food and Drug Administration, which uses machine learning techniques to model relationships between drugs and hepatic liver failure (Engstrom et al., 2020, p. 55), with decision trees and simple neural networks used to predict serious drug-related adverse outcomes. The same agency also uses regularised regression models, random forest, and support vector techniques to construct a rank ordering of reports based on their probability of containing policy-relevant information about safety concerns. This allows the agency to prioritise for attention those that are most likely to reveal problems.

Machine learning can also be used to predict demand, helping policymakers plan for the future. When used in this way, it can be a good way to optimise resources, allowing government agencies to be prescient in terms of service provision and to direct human attention or financial resources where they are most required. For example, some police forces use machine learning to predict where crime hotspots will arise and to anticipate when and where greater police presence will be needed. Recent studies on the use of data science in UK local government (Bright et al., 2019; Vogl et al., 2020) estimate that 15% of UK local authorities were using data science to build some kind of predictive capability in 2018, when the research was carried out.

The use of machine learning for prediction in policymaking is controversial, however. Some have argued that the predictive capacity of Computational Social

Science brings tension to the field, sitting happily with the epistemological aims of computer scientists, but going against the tradition of social science research, which prioritises explanations of individual and collective behaviour, ideally via causal mechanisms (Hofman et al., 2021, p. 181). Kleinberg et al. (2015) argue that some important policy problems do benefit from prediction alone and that machine learning can generate high policy impact as well as theoretical insights (Kleinberg et al., 2015, p. 495). But this use of machine learning generates important ethical questions of fairness and bias (discussed below), as the use of the COMPAS (Correctional Offender Management Profiling for Alternative Sanctions) system for predictive sentencing in the USA has shown (see Hartmann & Wenzelburger, 2021). Furthermore, as Athey (2017) explains, many of the prediction solutions described (e.g., in health care and criminal justice) require some kind of causal inference to achieve payoffs, even where prediction is most commonly cited as beneficial, such as the identification of building sites or other entities for inspection and scrutiny. Overall, she concludes, multidisciplinary approaches are needed that build on the development of machine learning algorithms but also 'bring in the methods and practical learning from decades of multidisciplinary research using empirical evidence to inform policy'. In a similar vein, Hofman et al. (2021) make the case for integrative modelling, developing models that 'explicitly integrate explanatory and predictive thinking', arguing that such an approach is likely to add value over and above what can be achieved with either technique alone and deserves more attention than it has received so far.

1.5 Etiology

The possibilities of detection, measurement, and prediction that CSS methods afford to tackle policy problems do not obviate the need for understanding the underlying causes of observed behaviours, as discussed in the preceding section. Etiology is particularly important when policymakers try to understand human behaviour in digital settings, where they need also to understand how the digital context, including the design of platforms and the algorithms they use, drive behaviour. Wagner et al. (2021) observe that in the 'algorithmically infused society' in which we now live, algorithms shape our behaviour in many contexts: shopping, travelling, socialising, entertainment, and so on. In such a world, the data that we derive from platforms like Twitter gives us useful clues about our interactions, but the social sciences is the only lens through which we can learn to separate what is 'natural' human behaviour and what is algorithm-induced human behaviour. The social sciences are also the domain that gives us the theoretical starting point for re-examining frameworks, models, and theories that were developed when algorithms were not a prevalent part of our lives. We need to understand both how algorithmic amplification (e.g., via recommender systems or other forms of social information) influences relationship formation, while also understanding how social adaptation causes algorithms to change. This understanding is particularly important for regulators, who need to know how digital platforms are influencing consumer preferences and behaviour (e.g., through targeted advertising) and which elements of the behaviours we notice online are attributable to the algorithms themselves. Scientific researchers need to develop this kind of expertise. Although streams of research are being developed within, for example, social media companies, around issues of content moderation and algorithm design, the primary aim of this work is to limit reputational damage. The companies themselves have little motivation to invest in programmes of research that uncover the organisational dynamics of online harms or the impact of such harms on different groups of citizens. They also have limited incentives to share the findings of such research, even if they decide to carry it out.

CSS can also help with etiology via experimental methods. Early social science experiments used survey data or laboratory-based experiments, which were expensive and labour-intensive and quickly resulted in small numbers problems. In contrast, online randomised controlled trials based on large-scale datasets can operate at huge scale and in real time. Such behavioural insights have been used by governments, for example, testing out the effects of redesigning letters and texts urging people to pay tax on time (Hallsworth et al., 2017). Large-scale digital data also offers the possibility of identifying 'natural experiments' (Dunning, 2012) in policy settings, where some disruption of normal activity at a point in time or in a particular location occurs, and the data is analysed after the disruption, as an 'as if random' treatment group. An example is provided by Transport for London's analysis of their Oyster card data to understand the effects of a 2014 industrial dispute which led to a strike of many of the system's train drivers (described in Dunleavy, 2016)). During the strike, millions of passengers switched their journey patterns to avoid their normal lines and stations hit by the strike. Larcom et al. (2017) examined Oyster card data for periods before and after the strike period, linking journeys to cardholders. They found that 1 in 20 passengers changed their journey, and a high proportion of these stayed with their new journey pattern when normal service resumed, suggesting their new route was better for them. The findings suggested that Tube travellers only 'satisfice' and had originally gone with the first acceptable travel solution that they found, later settling on the new route because it saved them time. The analysts also showed that the travel time gains made by the small share of commuters switching routes as a result of the Tube strike more than offset the economic costs to the vast majority (95%), who simply got disrupted on this one occasion. So the strike led to net gains, suggesting that possible side benefits of disruptions might be factored in by policymakers when making future decisions (like whether to close a Tube line wholly in order to accomplish urgent improvements (Dunleavy, 2016)).

Natural experiments like this can be hard to systematise or find. But large-scale observational data can be used to identify causal inference even where there is no identifiable 'as if random' treatment group or no counterfactual control group. Large-scale data analysis offers 'New tricks for Econometrics' (Varian, 2014), for example, where datasets are split into small worlds, creating artificial 'control groups' via a predictive model based on a function of past history and possible predictors of success. CSS methods have developed hugely in this area, especially in economics. Athey and Imbens (2017) discuss a range of such strategies, including regression discontinuity designs, synthetic control and differences-in-differences methods, methods that deal with network effects, and methods that combine experimental and observational data—as well as supplementary analyses (such as sensitivity and robustness analysis)—where the results are intended to convince the reader of the credibility of the primary analysis. They argue that machine learning methods hold great promise for improving the credibility of policy evaluation, particularly through these supplementary strategies.

1.6 Simulation

Another way in which CSS can tackle policy issues is through the development of simulation methods, allowing policymakers to try out interventions before implementing the measures in the real-world and having them give rise to unintended and unanticipated consequences. As noted above, policy choices need to be informed by counterfactuals: if we implemented this measure—or didn't implement it—what would happen?

An increasing range of modelling approaches can now be used for simulation, including complex network analysis and microsimulation, involving highly detailed analysis of, for example, traffic flows, labour mobility, urban industrial agglomeration patterns, or disease spread. One modelling approach that is gaining popularity with the growing availability of large-scale data is agent computing. Agent-based models (ABMs) have been used to study socio-economic phenomena for decades. Thomas Schelling was among the first to use agent-based modelling techniques within the social sciences. In the early 1970s, he published a seminal paper that showed how a simple dynamic model sheds light on how segregation can arise from the interplay of individual choices (see Schelling, 1971). But models like Schelling's—and many others that followed—were 'toy models': formal models without any real-world data to ground them in the socio-economic reality that they were meant to study. In contrast, the agent computing models used now are based on large-scale data, which transforms them into powerful tools for researchers and policymakers alike. Rob Axtell, one of the pioneers of Computational Social Science, recently developed a model of the US private sector, in which 120 million agents self-organise into 6 million firms (Axtell, 2018). Models like Axtell's are extremely powerful tools for studying the dynamics of socio-economic phenomena and carrying out simulations of complex systems, from economies to transport networks. Today's agent computing models can also be used in combination with machine learning methods, where the models provide a practical framework to combine data and theory without constraining oneself with too many unrealistic a priori assumptions about how socio-economic systems behave, such as 'fully rational agents' or 'complete information'.

An agent computing model consists of individual software agents, with states and rules of behaviour and large corpuses of data pertaining to the agents' behaviour and relationships. Running such a model could theoretically amount to instantiating an agent population, letting the agents interact, and monitoring what happens; 'Indeed, in their most extreme form, agent-based computational models will not make any use whatsoever of explicit equations' (Axtell, 2000, p. 3). But models usually involve some combination of data and formulae. Researchers have started to explore the possibilities of 'societal digital twins' (Birks et al., 2020), a combination of spatial computing, agent-based models, and 'digital twins'—virtual data-driven replicas of real-world systems that have become popular for modelling physical systems, in engineering or infrastructure planning, for example. Such 'societal' twins would use agent computing to model the socio-economic world, although the proponents warn that the complexity of socio-economic systems and the slower development of real-time updating means that the societal equivalent of digital twins is 'a long way from being able to simulate real human systems' (Birks et al., 2020, p. 2884).

Agent computing has gained popularity as a tool for transport planning or providing insight for decision-makers in disaster scenarios such as nuclear attacks or pandemics (Waldrop, 2018). UNDP are also trialling the use of an agent computing model to help developing countries work out which policy areas—health, education, transport, and so on—should be prioritised in order to meet the sustainable development goals (Guerrero & Castañeda, 2020). Mainstream economics modelling has struggled to keep pace with the new possibilities brought about by the growing availability of large-scale data, meaning that computational social scientists can and should play a key role in developing collaborations with policymakers and forging a new field of research aimed at enabling governments to design evidence-based policy interventions.

1.7   An Ethics-Driven Computational Social Science

CSS methods are data-driven. Machine learning models in this field are trained on data from human systems. For example, a model to support judicial decision-making will be trained on large datasets generated by earlier judicial decisions. That means that if decision-making in the past or present is biased—clearly the case in some areas, such as policing—then the machine learning algorithms trained on this data will be biased also. The use of the resulting machine learning tools in decision-making processes will reinforce and amplify existing biases. In part for this reason, extensive controversy has accompanied the use of machine learning for decision support, particularly in sensitive areas such as criminal justice (Hartmann & Wenzelburger, 2021; Završnik, 2021) or child welfare (Leslie et al., 2020).

The CSS methods discussed in this chapter raise numerous ethical concerns, from replicating biases to invading people's privacy, limiting individual autonomy, eroding public trust, and introducing unnecessary opaqueness into decision-making processes—to name only a few. To tackle these issues, CSS should take advantage of the work that has been done on the ethical use of AI technologies in government. Guidance on the responsible design, development, and implementation of AI systems in the public sector (Leslie, 2019) and a framework for explaining decisions made with AI (Information Commissioner's Office & The Alan Turing Institute, 2020) are used across UK departments and agencies. These publications focus on how the principles of fairness, sustainability, safety, accountability, and transparency can—and should—guide the responsible design, development, and deployment of AI systems. In contrast, Computational Social Science research has focused far more on the technical details of these data-intensive technologies rather than the ethical concerns, which tend to be underplayed. A recent special issue of *Nature* on CSS,[2] for example, mentioned ethics and responsible innovation only once in the editorial, and none of the articles focused on the topic. So in this case, CSS could have something to learn from recent work on trustworthy and responsible AI innovation for the public sector.

There are significant gains to be had if computational social science makes ethics an integral part of the process of scientific discovery. CSS methods are data-driven, using data generated by existing administrative systems. Rather than replicating biases, CSS can play an important role in shedding light, sometimes for the first time, on the bias endemic in human decision-making. As large-scale data sources become available, CSS could be used to reveal and tackle bias in modern digital public administration and policymaking. Identifying bias and understanding its origins can be a first step towards tackling long-running failings of administration.

1.8 Building Resilience: CSS at the Heart of a Reinvented Policy Toolkit

Nowhere are the possibilities of CSS for public policy—and the importance of realising them—illustrated more starkly than in the coronavirus pandemic of 2020 onwards. Computational Social Science seemed, to these authors at least, to have huge potential for the design of policy interventions and informing decision-making during the pandemic, for example, through undertaking the key tasks of detection, measurement, prediction, etiology, and simulation laid out above. But somehow, the use of CSS in this setting was disappointing. While it was good to see data, modelling, and science in such high relief throughout the pandemic, the use of CSS was limited and many interventions were introduced with no real evidence of their expected payoffs.

The difficulties seemed to be threefold. First, many countries discovered that they did not collect the kind of real-time, fine-grained data that was needed to inform policy design. In the UK, for example, it turned out that there was no availability

---

[2] *Nature* volume 595, issue 7866, 2021

of data on the number of people dying of Covid-19 until weeks after the deaths had taken place, making it impossible to calibrate the use of interventions. Economic policymakers had to design financial support mechanisms such as furlough schemes and stimulus packages without fine grained data about the areas of the economy that would be most affected by social distancing measures and supply chain disruptions. This meant that blanket schemes were applied, helping sectors that benefited from the pandemic (such as delivery companies and many technology firms) along with those that had been devastated (such as travel and hospitality). Policymakers and computational social scientists need to work together to identify the data streams that are likely to be needed in a crisis and 'develop dynamic capabilities' (Mazzucato & Kattel, 2020).

Second, there seemed to be a universal lack of integrated modelling. The focus tended to be on modelling one policy area at a time. There were models that tracked the spread of the virus and separate models that examined the economic effects. These two issues, however, were inextricably intertwined. The absence of integrated models to capture these interdependencies meant that policymakers often pointed to the trade-off between 'public health' and 'economic recovery' but were never able to pinpoint optimal interventions. There is a need for CSS to develop more integrated, generalised models that policymakers could turn to in an emergency. Besides their inability to capture interdependencies between policy areas, many economic models proved to be incapable of dealing with surprises. Models of commodity prices, for example, were based on the assumption that negative oil prices were impossible. During the pandemic, it became clear that not enough attention is given to quantifying uncertainty, which can have a cascading effect in complex multi-level systems. To help policymakers equip themselves for future crises, we need to develop CSS models that are based on robust assumptions and are able to quantify uncertainty. Integrated modelling, data-centric policymaking, causal inference, and uncertainty quantification are all ways in which CSS might build resilience into policymaking processes (MacArthur et al., 2022).

Third, it became clear that the organisational structures involved in policymaking to some extent worked against the kind of computational and modelling expertise that was required during the pandemic. Big departments of state have few incentives to share data, and very little tradition of sharing technical solutions to policy problems. This is unfortunate, because the vertical nature of data-intensive methods means that they lend themselves to being transferred across organisational boundaries. Yet policymakers seeking to meet a generic modelling challenge—such as how to identify vulnerable groups, quantify uncertainty or use machine learning to derive causal explanations as laid out above—are much more likely to seek help in their own department than to turn to departments or agencies in other parts of government. This siloed approach works against building up of expertise.

Overcoming these issues could allow CSS to usher in a new era of policymaking. As we begin to emerge from the pandemic, the word 'resilience' has become widespread in policy circles. Resilience is an organisational value that underpins how a government designs its policymaking systems and processes (Hood, 1991). Governments that value resilience prioritise stability, robustness, and adaptability.

Developing the CSS tools and models we have discussed here, with the focus on detecting and measuring trends and patterns, predicting and understanding human behaviour, and developing integrative modelling techniques that can simulate policy interventions all point in this direction. A resilient approach of this kind could equip policymakers to tackle the aftermath of the pandemic and face future crises (MacArthur et al., 2022).

1.9 Conclusion

This chapter has shown some of the transformational potential of Computational Social Science, bringing analysis of large-scale social and economic data into policymaking. CSS can renew the toolbox of contemporary government, refreshing and sharpening the essential tasks of detection, measurement, prediction, simulation, and etiology. None of these tasks can, alone, transform the policy toolkit. They need to be used in concert and require large-scale, real-time, fine-grained data sources. Measurement, for example, requires detection to be able to observe trends in the variable under scrutiny. Both are needed for prediction, which on its own is of questionable value in policy settings that lack the ability to pinpoint causality. Many researchers are making the case for integrative modelling that incorporates prediction and causal inference. Simulation requires large-scale data and is often used in conjunction with more predictive techniques.

New possibilities for the use of large-scale data about human behaviour bring new responsibilities, in terms of implementing and developing guidelines and frameworks for responsible innovation. Substantial progress has already been made in building ethical frameworks for the growing use of artificial intelligence in government. Guided by these frameworks, CSS researchers have a real opportunity to make explicit long-running biases and entrenched inequalities in public policy and administration. Their scholarship and methodologies have the potential to usher in a new era of policymaking, where interventions and administrative systems are more fair than ever before, as well as more efficient, effective, responsive, and prescient (Margetts & Dorobantu, 2019).

The need to respond to the coronavirus pandemic has raised the profile of data and modelling but has also illustrated missed opportunities in terms of data flows, integrative modelling, and the development of expertise. To face future crises, we need to overcome these challenges, bringing CSS methods to the heart of policymaking and developing models to inform the design of resilient policy interventions.

References

Athey, S. (2017). Beyond prediction: Using big data for policy problems. *Science, 355*(6324), 483–485. https://doi.org/10.1126/science.aal4321

Athey, S., & Imbens, G. W. (2017). The state of applied econometrics: Causality and policy evaluation. *Journal of Economic Perspectives, 31*(2), 3–32. https://doi.org/10.1257/jep.31.2.3

Axtell, R. L. (2000). *Why agents? On the varied motivations for agent computing in the social sciences.* http://www2.econ.iastate.edu/tesfatsi/WhyAgents.RAxtell2000.pdf.

Axtell, R. (2018). Endogenous firm dynamics and labor flows via heterogeneous agents. In C. H. Hommes & B. D. LeBaron (Eds.), *Handbook of computational economics* (Vol. 4, pp. 157–213). Elsevier. https://doi.org/10.1016/bs.hescom.2018.05.001

Bastow, S., Dunleavy, P., & Tinkler, J. (2014). *The impact of the social sciences: How academics and their research make a difference.* SAGE Publications. https://doi.org/10.4135/9781473921511

Birks, D., Heppenstall, A., & Malleson, N. (2020). Towards the development of societal twins. In: *Frontiers in artificial intelligence and applications. 24th European Conference on Artificial Intelligence (ECAI 2020)* (pp. 2883–2884).

Bright, J., Ganesh, B., Vogl, T., & Cathrine, S. (2019). *Data science for local government.* Oxford Internet Institute. https://pure.rug.nl/ws/portalfiles/portal/117364515/SSRN_id3370217.pdf

Cheung, H. K. (1972). A day-use park visitation model. *Journal of Leisure Research, 4*(2), 139–156. https://doi.org/10.1080/00222216.1972.11970070

Donahue, M. L., Keeler, B. L., Wood, S. A., Fisher, D. M., Hamstead, Z. A., & McPhearson, T. (2018). Using social media to understand drivers of urban park visitation in the Twin Cities, MN. *Landscape and Urban Planning, 175*, 1–10. https://doi.org/10.1016/j.landurbplan.2018.02.006

Dunleavy, P. (2016). 'Big data' and policy learning. In G. Stoker & M. Evans (Eds.), *Evidence-based policy making in the social sciences: Methods that matter* (Vol. 143). Policy Press.

Dunning, T. (2012). *Natural experiments in the social sciences: A design-based approach.* Cambridge University Press. https://doi.org/10.1017/CBO9781139084444

Engstrom, D. F., Ho, D. E., Sharkey, C. M., & Cuéllar, M.-F. (2020). Government by algorithm: Artificial intelligence in Federal Administrative Agencies. *SSRN Electronic Journal.* https://doi.org/10.2139/ssrn.3551505

Guerrero, O. A., & Castañeda, G. (2020). Policy priority inference: A computational framework to analyze the allocation of resources for the sustainable development goals. *Data & Policy, 2*, e17. https://doi.org/10.1017/dap.2020.18

Guest, E., Vidgen, B., Mittos, A., Sastry, N., Tyson, G., & Margetts, H. (2021). An expert annotated dataset for the detection of online misogyny. In: *Proceedings of the 16th conference of the European chapter of the Association for Computational Linguistics:* Main Volume (pp. 1336–1350). https://doi.org/10.18653/v1/2021.eacl-main.114

Hallsworth, M., List, J. A., Metcalfe, R. D., & Vlaev, I. (2017). The behavioralist as tax collector: Using natural field experiments to enhance tax compliance. *Journal of Public Economics, 148*, 14–31. https://doi.org/10.1016/j.jpubeco.2017.02.003

Hamstead, Z. A., Fisher, D., Ilieva, R. T., Wood, S. A., McPhearson, T., & Kremer, P. (2018). Geolocated social media as a rapid indicator of park visitation and equitable park access. *Computers, Environment and Urban Systems, 72*, 38–50. https://doi.org/10.1016/j.compenvurbsys.2018.01.007

Hartmann, K., & Wenzelburger, G. (2021). Uncertainty, risk and the use of algorithms in policy decisions: A case study on criminal justice in the USA. *Policy Sciences, 54*(2), 269–287. https://doi.org/10.1007/s11077-020-09414-y

Hofman, J. M., Watts, D. J., Athey, S., Garip, F., Griffiths, T. L., Kleinberg, J., Margetts, H., Mullainathan, S., Salganik, M. J., Vazire, S., Vespignani, A., & Yarkoni, T. (2021). Integrating explanation and prediction in computational social science. *Nature, 595*(7866), 181–188. https://doi.org/10.1038/s41586-021-03659-0

Hood, C. (1991). A public management for all seasons? *Public Administration, 69*(1), 3–19. https://doi.org/10.1111/j.1467-9299.1991.tb00779.x

Hood, C. C., & Margetts, H. Z. (2008). The tools of government in the digital age. *Public Administration, 86*(4), 1137–1138. https://doi.org/10.1111/j.1467-9299.2008.00756_4.x

Information Commissioner's Office, & The Alan Turing Institute. (2020). *Explaining decisions made with AI.* https://ico.org.uk/for-organisations/guide-to-data-protection/key-dp-themes/explaining-decisions-made-with-artificial-intelligence/

Kleinberg, J., Ludwig, J., Mullainathan, S., & Obermeyer, Z. (2015). Prediction policy problems. *American Economic Review, 105*(5), 491–495. https://doi.org/10.1257/aer.p20151023

Larcom, S., Rauch, F., & Willems, T. (2017). The benefits of forced experimentation: Striking evidence from the London underground network. *The Quarterly Journal of Economics, 132*(4), 2019–2055. https://doi.org/10.1093/qje/qjx020

Lazer, D., Hargittai, E., Freelon, D., Gonzalez-Bailon, S., Munger, K., Ognyanova, K., & Radford, J. (2021). Meaningful measures of human society in the twenty-first century. *Nature, 595*(7866), 189–196. https://doi.org/10.1038/s41586-021-03660-7

Lazer, D., Pentland, A., Adamic, L., Aral, S., Barabasi, A. L., Brewer, D., Christakis, N., Contractor, N., Fowler, J., Gutmann, M., Jebara, T., King, G., Macy, M., Roy, D., & Van Alstyne, M. (2009). Life in the network: The coming age of computational social science. *Science (New York, N.Y.), 323*(5915), 721–723. https://doi.org/10.1126/science.1167742

Leslie, D. (2019). *Understanding artificial intelligence ethics and safety: A guide for the responsible design and implementation of AI systems in the public sector.* The Alan Turing Institute. https://doi.org/10.5281/zenodo.3240529

Leslie, D., Holmes, L., Hitrova, C., & Ott, E. (2020). *Ethics review of machine learning in children's social care.* What Works for Children's Social Care. https://whatworks-csc.org.uk/wp%20content/uploads/WWCSC_Ethics_of_Machine_Learning_in_CSC_Jan2020_ Accessible.pdf

MacArthur, B. D., Dorobantu, C. L., & Margetts, H. Z. (2022). Resilient government requires data science reform. *Nature Human Behaviour, 6*(8), 1035–1037. https://doi.org/10.1038/s41562-022-01423-6

Margetts, H. (1999). *Information technology in government: Britain and America.* Routledge.

Margetts, H., & Dorobantu, C. (2019). Rethink government with AI. *Nature, 568*(7751), 163–165. https://doi.org/10.1038/d41586-019-01099-5

Mazzucato, M., & Kattel, R. (2020). COVID-19 and public-sector capacity. *Oxford Review of Economic Policy, 36*(Supplement_1), S256–S269. https://doi.org/10.1093/oxrep/graa031

Muellerleile, C., & Robertson, S. L. (2018). Digital Weberianism: Bureaucracy, information, and the techno-rationality of neoliberal capitalism. *Indiana Journal of Global Legal Studies, 25*(1), 187. https://doi.org/10.2979/indjglolegstu.25.1.0187

Ostmann, F., & Dorobantu, C. (2021). *AI in financial services. The Alan Turing Institute.* https://doi.org/10.5281/zenodo.4916041

Röttger, P., Vidgen, B., Nguyen, D., Waseem, Z., Margetts, H., & Pierrehumbert, J. (2021). HateCheck: Functional tests for hate speech detection models. In: *Proceedings of the 59th Annual Meeting of the Association for Computational Linguistics and the 11th International Joint Conference on Natural Language Processing* (Volume 1: Long Papers, pp. 41–58). https://doi.org/10.5281/zenodo.4916041

Schelling, T. C. (1971). Dynamic models of segregation. *The Journal of Mathematical Sociology, 1*(2), 143–186. https://doi.org/10.1080/0022250X.1971.9989794

Sinclair, M., Zhao, Q., Bailey, N., Maadi, S., & Hong, J. (2021). Understanding the use of greenspace before and during the COVID-19 pandemic by using mobile phone app data. In: *GIScience 2021 Short Paper Proceedings. 11th International Conference on Geographic Information Science. September 27–30, 2021. Poznań*, Poland (Online). https://doi.org/10.25436/E2D59P

Suse, S., Mashhadi, A., & Wood, S. A. (2021). Effects of the COVID-19 Pandemic on Park Visitation Measured by Social Media. In: *Companion Publication of the 2021 Conference on Computer Supported Cooperative Work and Social Computing* (pp. 179–182). https://doi.org/10.1145/3462204.3481754

Taylor, H., Vidgen, B., Anastasiou, Z., Pantazi, M., Inkster, B., & Margetts, H. (2021, June). Investigating vulnerability to online health-related misinformation during COVID-19. In: *MISDOOM 2021: 3rd Multidisciplinary International Symposium on Disinformation in Open Online Media.* https://www.researchgate.net/publication/355436534_Investigating_vulnerability_to_online_health-related_misinformation_during_COVID-19

Varian, H. R. (2014). Big data: New tricks for econometrics. *Journal of Economic Perspectives, 28*(2), 3–28. https://doi.org/10.1257/jep.28.2.3

Vidgen, B., Botelho, A., Broniatowski, D., Guest, E., Hall, M., Margetts, H., Tromble, R., Waseem, Z., & Hale, S. (2020). Detecting East Asian Prejudice on social media. *ArXiv:2005.03909 [Cs].* http://arxiv.org/abs/2005.03909

Vogl, T. M., Seidelin, C., Ganesh, B., & Bright, J. (2020). Smart technology and the emergence of algorithmic bureaucracy: Artificial intelligence in UK local authorities. *Public Administration Review, 80*(6), 946–961. https://doi.org/10.1111/puar.13286

Wagner, C., Strohmaier, M., Olteanu, A., Kıcıman, E., Contractor, N., & Eliassi-Rad, T. (2021). Measuring algorithmically infused societies. *Nature, 595*(7866), 197–204. https://doi.org/10.1038/s41586-021-03666-1

Waldrop, M. M. (2018). Free agents. *Science, 360*(6385), 144–147. https://doi.org/10.1126/science.360.6385.144

Završnik, A. (2021). Algorithmic justice: Algorithms and big data in criminal justice settings. *European Journal of Criminology, 18*(5), 623–642. https://doi.org/10.1177/1477370819876762

Chapter 2
Computational Social Science for the Public Good: Towards a Taxonomy of Governance and Policy Challenges

Stefaan Gerard Verhulst

Abstract Computational Social Science (CSS) has grown exponentially as the process of datafication and computation has increased. This expansion, however, is yet to translate into effective actions to strengthen public good in the form of policy insights and interventions. This chapter presents 20 limiting factors in how data is accessed and analysed in the field of CSS. The challenges are grouped into the following six categories based on their area of direct impact: Data Ecosystem, Data Governance, Research Design, Computational Structures and Processes, the Scientific Ecosystem, and Societal Impact. Through this chapter, we seek to construct a taxonomy of CSS governance and policy challenges. By first identifying the problems, we can then move to effectively address them through research, funding, and governance agendas that drive stronger outcomes.

2.1 Introduction

We live in a digital world, where virtually every realm of our existence has been transformed by a rapid and ongoing process of datafication and computation. Travel, retail, entertainment, finance, and medicine: to these areas of life, all grown virtually unrecognizable in recent years, we must also add the social sciences. In recent years the burgeoning field of Computational Social Science (CSS) has begun changing the way sociologists, anthropologists, economists, political scientists, and others interpret human behaviour and motivations, in the process leading to new insights into human society. Some have gone so far as to herald a "social research revolution" or a "paradigm shift" in the social sciences (Chang et al., 2014; Porter et al., 2020). Recently, the *Economist* magazine proclaimed an era of "third-wave economics",

S. G. Verhulst (✉)
The GovLab, New York University, New York, NY, USA

ISI Foundation, Turin, Italy
e-mail: stefaan@thegovlab.org

© The Author(s) 2023
E. Bertoni et al. (eds.), *Handbook of Computational Social Science for Policy*,
https://doi.org/10.1007/978-3-031-16624-2_2

transformed by the availability of massive amounts of real-time data (Kansas, 2021).

Of course social scientists have always used data to interpret and analyse human beings and the social structures they create. CSS, as a concept, first emerged in the latter half of the twentieth century across the field of social science and STEM (Edelmann et al., 2020). Earlier generations of researchers were well-versed in quantitative methods, as well as in the use of a variety of computational and statistical tools, ranging from SPSS to Excel. What has changed is the sheer quantity of data now available, as well as the easy (and often free) access to sophisticated computational tools to process and analyse that data. To the extent there is indeed a revolution underway in the social sciences, then, it stems in large part from its intersection with the equally heralded Big Data Revolution (McAfee & Brynjolfsson, 2019).

CSS offers some very real opportunities. It enables new forms of research (e.g., large-scale simulations and more accurate predictions), allows social scientists to model and derive findings from a much larger empirical base, and offers the potential for new, cross-disciplinary insights that could lead to innovative and more effective social or economic policy interventions. In recent years, CSS has allowed researchers to better understand, among other phenomena, the roots and patterns of socioeconomic inequalities, how infectious diseases spread, trends in crime and other factors contributing to social malaise, and much more.

As with many technological innovations, however, the rhetoric—and hype—surrounding CSS can sometimes overtake reality (Blosch & Fenn, 2018). For all the undeniable opportunities, there remains a chasm between potential and what CSS is actually doing and revealing. Bridging this chasm could unlock new social insights and also, through more targeted and responsive policy interventions, lead to greater opportunities to enhance public good.

This chapter seeks to take stock of and categorize a variety of governance and policy hurdles that continue to hold back the potential of CSS. In what follows, we outline 20 challenges that limit how data is accessed and analysed in the social sciences. We categorize these into six areas: challenges associated with the *Data Ecosystem*, *Data Governance*, *Research Design*, *Computational Structures and Processes*, the *Scientific Ecosystem*, and those concerned with *Societal Impact* (Fig. 2.1). Albert Einstein once said, "If I had an hour to solve a problem I'd spend 55 minutes thinking about the problem and five minutes thinking about solutions". In the spirit of Einstein's maxim, we do not seek to provide detailed solutions to the identified challenges. Instead, our goal is to design a taxonomy of challenges and issues that require further exploration, in the hope of setting a research, funding, and governance agenda that could advance the field of CSS and help unleash its full potential.

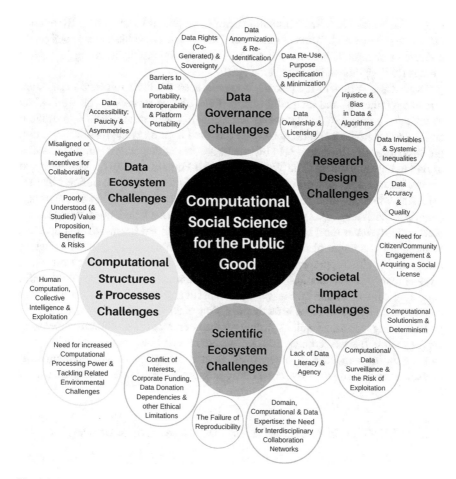

Fig. 2.1 Taxonomy of governance and policy challenges

2.2 Data Ecosystem Challenges

2.2.1 Data Accessibility: Paucity and Asymmetries

Although CSS is enabled by the massive explosion in data availability, in truth access to data remains a serious bottleneck. Accessibility problems can take many forms. In certain cases, accessibility can be limited when certain kinds of data simply don't exist. Such data paucity problems were more common in the early days of CSS but remain a challenge in particular areas of social science research, for example, in the study of certain disaster events (Burger et al., 2019). The challenges posed by data paucity are not limited to an inability to conduct research; the risk of wrong or inappropriate conclusions, built upon shaky empirical foundations, must equally be considered. Such limitations can to an extent be overcome by reliance on new and innovative forms of data—for example, those collected by social media companies or through sensors and other devices on the rapidly growing Internet of Things (IoT) (Hernandez-Suarez et al., 2019).

Even when sufficient data exists, however, accessibility can remain a problem due to asymmetries and inequalities in patterns of data ownership, as well as due to regulatory or policy bottlenecks (OECD, 2019). Recent attention on corporate concentration in the technology industry has shed light on related issues, including the vast stores of siloed data held by private sector entities that remain inaccessible to researchers and others (The World Wide Web Foundation, 2016). The European Union, for example, is working to address this challenge through policies like the Data Act, which attempts to bridge existing inequalities in access to and use of data (Bahrke & Manoury, 2022). While the open data movement and other efforts to spur *data collaboratives (and similar entities)* [1] have made strides in opening up some of these silos, a range of obstacles—reluctance to share data perceived as having competitive value, apprehension about inadvertently violating privacy-protective laws—mean that considerable amounts of private sector data with potential public good applications remain inaccessible (Verhulst et al., 2020b). Access to such large datasets could lead to more effective decision-making in both the corporate and policymaking worlds, as well as stronger transparency and accountability measures across sectors (Russo & Feng, 2021). Concerns around heightened public scrutiny and regulatory exposure as a result of greater transparency and accountability measures are also in part why larger corporations may resist open data policies.

2.2.2   *Misaligned or Negative Incentives for Collaborating*

Misaligned incentives are a common and well-understood problem in the worlds of business and social sciences. Misaligned incentives commonly occur when certain individuals' or groups' incentives are not aligned towards the broader common goal of the collaboration. These incentives can be based on specific parties' interests, as well as on differences between long-term and short-term priorities (Novak, 2011). In a business supply chain, for example, misaligned incentives can cause a number of issues, ranging from operational inefficiency to higher production costs to weak market visibility (Narayanan & Raman, 2004). In order for supply chain relationships to function optimally, incentives must be realigned through trust, transparency, stronger communication, regulation, and clear contracts.

Many of these same concepts apply to the data sharing and data collaboration ecologies and thus to how data is used for CSS. Misaligned incentives can take

---

[1] Data Collaboratives are a new form of collaboration, beyond the public-private partnership model, in which participants from different sectors—in particular companies —exchange their data to create public value. For more information, see https://datacollaboratives.org/.

a number of forms but commonly refer to conflicts or differences between data owners (frequently in the private sector) and those who would potentially benefit or be able to derive insights from access to data (frequently academic researchers, policy analysts, or members of civil society). Data owners may perceive efforts to share data with social scientists as potentially leading to competitive threats, or they may perceive regulatory risk; social scientists, on the other hand, will perceive data collaboration as leading to new insights that can enhance the public good. There are no easy solutions to such misalignments, and alleviating them will rely on a complex interplay of regulation, awareness-raising, and efforts to increase transparency and trust. For the moment, misaligned incentives remain a serious impediment to CSS research.

2.2.3 Poorly Understood (and Studied) Value Proposition, Benefits, and Risks

Misaligned incentives often arise when data owners and social scientists (or others who may benefit from data sharing) have different perceptions about the benefits or risks of sharing. David Lazer et al. note, for instance, that the incidence of data sharing and opening of data may have reduced in the wake of laws designed to protect privacy (e.g., GDPR) (Lazer et al., 2020). This suggests that companies may overestimate the regulatory and other risks of making data available to researchers, while under-valuing the possible benefits. Companies dealing may also face real concerns about data protection and data privacy that are not effectively addressed by laws. Likewise, companies may be reluctant to share data, fearing that doing so will erode a competitive advantage or otherwise harm the bottom line. As our research has shown, this is often a mis-perception (Dahmm, 2020). Data sharing does not operate in a zero-sum ecosystem, and companies willing to open their data to external researchers may ultimately reap the benefits of new insights and new uses for their otherwise siloed datasets.

2.3 Data Governance Challenges

2.3.1 Data Reuse, Purpose Specification, and Minimization

A spate of privacy scandals has led to renewed regulatory oversight of data, data sharing, and data reuse. Such oversight is often justified and very necessary. At the same time, an exclusive focus on privacy risks undermining some of the societal benefits of sharing; we need a more calibrated and nuanced understanding of risk (Verhulst, 2021). Purpose specification and minimization mandates, which seek to narrowly limit the scope of how data may be reused, pose particular challenges to

CSS. Such laws or guidelines do offer greater consumer control over their data and can thus be trust-enhancing. At the same time, serious consideration must be given to the specific circumstances under which it is acceptable to reuse data and the best way to balance potential risk and reward.

Absent such consideration and clear guidelines, a secondary use—for social science research or other purposes—runs the risk of violating regulations, jeopardizing privacy, and de-legitimizing data initiatives by undermining citizen trust. Among the questions that need to be asked are what types of secondary use should be allowed (e.g., only with a clear public benefit), who is permitted to reuse data, are there any types of data that should never be reused (e.g., medical data), and what framework can allow us to weigh the potential benefits of unlocking data against the costs or risks (Verhulst et al., 2020a). The 2019 Finnish Act on the Secondary Use of Health and Social Data is one policy model that effectively addresses these questions (Ministry of Social Affairs and Health, 2019).

To tackle the challenge of purpose specification in data reuse, new processes and notions of stakeholdership must be considered. For example, one emerging vehicle for balancing risk and opportunity is the use of working groups or symposia where thought leaders, public decision-makers, representatives of industry and civil society, and citizens come together to assess and help improve existing approaches and methodologies for data collaboration. [2]

2.3.2   *Data Anonymization and Re-identification*

Data anonymization and/or de-identification refers to the process by which a dataset is sanitized to remove or hide personally identifiable information with the goal of protecting individual privacy (OmniSci, n.d.-a). This process is key to maintaining personal privacy while also empowering actors to expand the ways in which data can be used without violating privacy and data protection laws. As anonymized data becomes more readily accessible and freely available, social scientists are working with large anonymized datasets to answer previously unanswerable questions. In the context of the COVID-19 pandemic, for example, social scientists used mobile phone records and anonymized credit card purchases to understand how people's movement and spending habits shifted in response to the pandemic across numerous sectors of the economy ("The Powers and Perils of Using Digital Data to Understand Human Behaviour", 2021).

In contrast to data anonymization, data re-identification involves matching previously anonymized data with its original owners. The general ease of re-identification means that the promised privacy of data anonymization is a weak

---

[2] See, for example, the World Bank's Open Data Working Group (https://data.worldbank.org/) or the British government's Smart data working group (https://www.gov.uk/government/groups/smart-data-working-group).

commitment and that data privacy laws must also be applied to anonymized data (Ghinita et al., 2009; Ohm, 2010; Rubinstein & Hartzog, 2015). One way to address the risk of re-identification is to prevent the so-called mosaic effect (Czajka et al., 2014). This phenomenon occurs as a result of the re-identification of data by combining multiple datasets containing similar or complementary information. The mosaic effect can pose a threat both to individual and group privacy (e.g., in the case of a small minority demographic group). Groups are frequently established through data analytics and segmentation choices (Mittelstadt, 2017). Under such conditions, individuals are often unaware that their data are being included in the context of a particular group, and decisions made on behalf of a group can limit data holders' control and agency (Radaelli et al., 2018). Children's data and humanitarian data are particularly susceptible to the challenges of group data (Berens et al., 2016; Young, 2020). Mitigation strategies include considering all possible points of intrusion, limiting analysis output details only to what is truly needed, and releasing aggregated information or graphs rather than granular data. In addition, limited access conditions can be established to protect datasets that could potentially be combined (Green et al., 2017).

2.3.3 Data Rights (Co-generated Data) and Sovereignty

CSS research often leads not only to new data but *new forms of data*. In particular, the collaborative process involved in CSS often leads to co-generated or co-created data, processes which raise thorny questions about data rights, data sovereignty, and the very notion of "ownership" (Ducuing, 2020a, 2020b). Without a clear owner, traditional intellectual property laws are difficult and often impossible to apply, which means that CSS may require new models of ownership and governance that promote data sharing and collaborative research while also protecting property rights (Micheli et al., 2020).

In order to tackle the challenge of ownership and governance, stakeholders in the data space have proposed a number of potential models to replace traditional norms of ownership and property. These include adopting a more *collective, rights-based approach to data ownership*, creating public data repositories, and establishing private data cooperatives, data trusts, or data collaboratives. [3] Each of these methods has advantages and certain disadvantages, but they all go beyond the notion of co-ownership towards concepts of co-governance or co-regulation (Richet, 2021; Rubinstein, 2018). Such shared governance models could play a critical role in removing barriers to data and enabling the research potential of CSS.

---

[3] For more information about types of data collaboratives, see "Leveraging Private Data for Public Good: A Descriptive Analysis and Typology of Existing Practices." https://datacollaboratives.org/existing-practices.html.

2.3.4 Barriers to Data Portability, Interoperability, and Platform Portability

Data portability and data interoperability approach the same concept from two different actor perspectives. Data portability refers to the ability of individuals to reuse their personal data by moving it across different service platforms in a secure way (Information Commissioner's Office, n.d.). Data interoperability, on the other hand, allows systems to share and use data across platforms free of any restrictions (OmniSci, n.d.-b). More recently, certain observers have begun to point to the limitations of both these concepts, arguing instead for platform portability, which would, for example, allow consumers to transfer not only their personal data from one social media platform to another but also a broader set of data, including contact lists and other "rich" information (Hesse, 2021).

Such concepts offer great potential for data sharing and more generally for the collaboration and access that are critical to enabling CSS. Yet a series of barriers exist, ranging from technical to regulatory to a general lack of trust among the public (De Hert et al., 2018; Vanberg & Ünver, 2017). Technical barriers are generally surmountable (Kadadi et al., 2014). Regulatory concerns, however, are thornier, with some scholars pointing out that provisions such as Article 20 of the GDPR, the right to data portability, could be interpreted to hamper cross-platform portability and create obstacles in building such partnerships (Hesse, 2021). There are also arguments that applying the new GDPR principles may prove more challenging for small and medium sized enterprises that may lack the resources and technology required to be effective (European Commission, 2020). Such restrictions are linked to a broader set of concerns over privacy and consent. Designed to protect consumer rights, they also have the inadvertent effect of restricting the potential of sharing and collaboration. Once again, they illustrate the difficult challenges involved in balancing a desire to minimize risk while maximizing potential in the data ecosystem.

2.3.5 Data Ownership and Licensing

As noted above, existing notions of data ownership and licensing pose a challenge due to the complex nature of ownership in the data ecosystem (Van Asbroeck, 2019). Traditional notions of ownership (and related concepts of copyright or IP licensing) convey a sense of non-rivalrous control over physical or virtual property. Yet data is more complicated as an entity; data about an individual is often not "owned" or controlled by that individual but rather by an entity—a company, a government organization—that has collected the data and that is now responsible for storing it, ensuring its quality and accuracy, and protecting the data from potential violations. Questions about ownership get even more complicated when we consider the nature of co-creation or co-generation (cf. above) or when we consider the data value chain, by which data is repurposed and mingled with other data to generate new insights and forms of information (Van Asbroeck, 2019). For all these reasons, there have been calls for "more holistic" models and for a recognition of the "intersecting interests" that may define data ownership, particularly of personal information (Kerry & Morris, 2019; Nelson, 2017).

The lack of conceptual and regulatory clarity over data ownership poses serious obstacles to the project of CSS (Balahur et al., 2010). It hinders data collaboration and sharing and prevents the inter-sectoral pooling of data and expertise that are so critical to conducting social science or other forms of research. In the absence of a more robust governance framework, research must often take place on the strength of ad hoc or trust-based relationships between parties—hardly a solid foundation upon which to scale CSS or harness its full potential.

2.4 Research Design Challenges

2.4.1 Injustice and Bias in Data and Algorithms

Datafication—like technology in general—is often accompanied by claims of neutrality. Yet as society becomes increasingly datafied, various forms of bias have emerged more clearly (Baeza-Yates, 2016). Bias can take many forms and present itself at various stages of the data value chain. There can be bias during the process of data collection or processing, as well as in the models or algorithms used to glean insights from datasets. Often, bias replicates existing social or political forms of exclusion. With the rise to prominence of Artificial Intelligence (AI), considerable attention has been paid recently to the issue of algorithmic bias and bias in machine learning models (Krishnamurthy, 2019; Lu et al., 2019; Turner Lee et al., 2019). Bias can also arise from incomplete data that doesn't necessarily replicate societal patterns but that is nonetheless unrepresentative and leads to flawed or discriminatory outcomes. Moreover, biases are not limited to just the data but can also extend into interpretations affecting frames of reference, underlying assumptions and models of analysis to name a few (Jünger et al., 2022).

Bias, in whatever form, poses serious challenges to CSS. One meta-analysis estimates that up to a third of studies using a method known as Qualitative Comparative Analysis (QCA) may be afflicted by bias, one in ten "severely so" (Thiem et al., 2020). Such problems lead to insufficient or incorrect conclusions; when translated into policy, they may result in harmful steps that perpetuate or amplify existing racial, gender, socioeconomic, and other forms of exclusion. Thus the issues posed by bias are deeply tied to questions of power and justice in society and represent some of the more serious challenges to effective, fair, and responsible CSS.

2.4.2 Data Accuracy and Quality

Bias also is one of the main contributors to problems of data quality and accuracy. But these problems are multidimensional—i.e., they are caused by many factors— and inevitably represent a serious challenge to any project involving computational or data-led social studies. Exacerbating matters, the very notions of accuracy and, especially, quality are contested, with definitions and standards varying widely across projects, geographies, and legal jurisdictions.

To an extent, the concept of accuracy can be simplified to a question about whether data is factually correct (facts, of course, are themselves contested). Quality is, however, a more nebulous concept, extending not only to the data itself but to various links in the data chain, including how the data was collected, stored, and processed (Dimitrova, 2021; Herrera & Kapur, 2007). In order to advance the field of CSS, clearer definitions and standards will be required. While doing so, it will be critical to bring data subjects themselves into the conversations, in order to ensure a measure of participatory validation and ensure that any adopted standards have widespread buy-in.

2.4.3 Data Invisibles and Systemic Inequalities

The concept of "data invisibles" refers to individuals who are outside the formal or digital economy and thus systematically excluded from the benefits of that economy (Shuman & Paramita, 2016). Because many of these individuals are located in developing countries, many datasets or algorithmic models trained on such datasets systematically exclude non-Western citizens, gender invisibles, and countless other disadvantaged populations and minority groups and thus pose further challenges to the accuracy of CSS and its findings (D'Ignazio & Klein, 2018; Fisher & Streinz, 2021; Naudts, 2019; Neumayer et al., 2021).

The problem of data invisibility is exacerbated by data governance practices that fail to proactively take into account the need for inclusion (D'Ignazio & Klein, 2018; Fisher & Streinz, 2021; Naudts, 2019; Neumayer et al., 2021). Such practices include insufficient or non-existent guidelines or standards on data quality and representativeness; a lack of robust accountability and auditing mechanisms [4] for algorithms or machine learning models; and the demographic composition of research teams which often lack diversity or representation of those studied. Thus in order to strengthen the practice of CSS, it will be necessary to address the wider ecosystem of data governance.

---

[4] See the Algorithmic Accountability Policy Toolkit, jointly developed by AI Now, the Ada Lovelace Institute and Open Government Partnership (https://ainowinstitute.org/pages/algorithmic-accountability-for-the-public-sector-report.html).

2.5 Computational Structures and Processes Challenges

2.5.1 Human Computation, Collective Intelligence, and Exploitation

Collective intelligence refers to the shared reasoning and insights that arises from our collective participation (both collaborative and competitive) in the data ecosystem (Figueroa & Pérez, 2018; Lévy, 2010). Collective intelligence has emerged as a potentially powerful tool in understanding our societies and in leading to more effective policies and offers tremendous potential for CSS. However, collective intelligence also faces a number of limitations that compromise the quality of its insights. These include bureaucratization that prevents lower-level actors from sharing their insights or expertise; the so-called "common knowledge" effect where participants do not strive to go beyond conventional wisdom and informational pressures which limit independent thoughts and actions.

All of these challenges negatively impact collective intelligence and, indirectly, CSS. A further challenge emerging in this space, especially as collective intelligence intersects with AI, relates to the exploitation of machines, which may be co-participants in the process of collectively generated intelligence (Caverlee, 2013; Melo et al., 2016). Although this challenge remains more hypothetical than actual at the present, it raises complex ethical questions that could ultimately impact how research is conducted and who has the right to take credit (or blame) for its findings.

2.5.2 Need for Increased Computational Processing Power and Tackling Related Environmental Challenges

The massive amounts of data available for social sciences research require equally massive amounts of computational processing power. This raises important questions about equity and inclusion and also poses serious environmental challenges (Lazer et al., 2020). According to a recent study by Harvard's John A. Paulson School of Engineering and Applied Sciences, modern data centres already account for 1% of global energy consumption, a number that is rapidly increasing (Harvard John A. Paulson School of Engineering and Applied Sciences, 2021). The study points out that in addition to energy use, our data economy also contributes indirectly, for example, through e-waste, to pollution. Such problems are only likely to increase with the growing prominence of blockchain and the so-called Web3, which are already making their impact felt in the social sciences (Hurt, 2018). According to the Bitcoin Energy Consumption Index, Bitcoin alone generates as much waste annually as the entire country of Holland. A single Bitcoin transaction uses a similar amount of energy as the consumption of an average US home over 64.61 days ("Bitcoin Energy Consumption Index", n.d.).

Computational processing requirements also pose serious obstacles to participation by less developed countries or marginalized groups within developed countries, both of which may lack the necessary financial and technical resources (Johnson, 2020). Such exclusion may lead, in turn, to unrepresentative or biased social science research and conclusions. One possible solution lies in developing new, less computationally demanding models to analyse data. Solutions of this nature have been developed, for instance, to analyse data from Instagram to monitor social media trends and for natural language processing algorithms that make it easier to process and derive insights from social media data (Pryzant et al., 2018; Riis et al., 2021). Another potential strategy is using volunteer computing, wherein a problem that would ordinarily require the computing power of a super computer is broken down and solved by thousands of volunteers with their personal computers (Toth et al., 2011). As volunteer computing grows in popularity, volunteer numbers must rapidly expand if this solution is to remain viable in the long run. These developments are just a start, but they represent efforts to address current limitations in processing power to help achieve more robust and equitable insights from CSS analyses.

2.6 Scientific Ecosystem Challenges

2.6.1 Domain, Computational, and Data Expertise: The Need for Interdisciplinary Collaboration Networks

As the field of CSS develops, the divide between domain, computational, and data expertise is emerging as a limiting factor. There is a pressing need for interdisciplinary collaboration networks to help bridge this divide and achieve more accurate insights and findings. For example, in order to effectively use large anonymized datasets on credit card purchases to understand shifts in spending patterns, a research team would need the combined expertise of data scientists, economists, sociologists, and anthropologists (relevant skill sets) "bilinguals" from around the world—practitioners across fields who possess both domain knowledge and data science expertise.

One possible way to bridge this gap in CSS applications is by relying on "bilinguals" [5]—scholars and professionals who possess expertise across domains and sectors (Porway, 2019). For example, these individuals can bring the requisite understanding of social sciences alongside strong data know-how required for CSS research. The valuable contribution of bilinguals is evident in the GovLab's 100 Questions initiative, which seeks to identify the most pressing problems facing the world that can be answered by leveraging datasets in a responsible manner ("The 100 Questions Initiative—About", n.d.). Each bilingual brings specific sector

---

[5] "Bilinguals" refer to practitioners from the field who possess domain-specific knowledge, as well as data science expertise. To learn more about bilinguals, visit https://the100questions.org/.

expertise coupled with a strong foundation in data science to draw out not only the most critical questions facing a domain but also to identify questions that can be answered using the current context of data ("The 100 Questions Initiative—About", n.d.). In this way, interdisciplinary collaboration networks and bilinguals can help to bridge the knowledge gap that exists in the field of CSS and to unlock deeper and more insightful outcomes with potentially deeper public impact.

2.6.2 Conflict of Interests, Corporate Funding, Data Donation Dependencies, and Other Ethical Limitations

Conflicts of interest—real or perceived—are a major concern in all social studies research. Such conflicts can skew research results even when they are declared (Friedman & Richter, 2004). Many long-standing ethical concerns are relevant within the field of CSS. These include issues related to funding, conflicts of interest (which may not be limited to financial interests), and scope or type of work. Yet the use of data and emerging computational methods, for which ethical boundaries are often blurred, complicate matters and introduce new concerns. One recent study, for example, points to the difficulties in defence-sector work involving technology, highlighting "the code of ethics of social science organizations and their limits in dealing with ethical problems of new technologies" and "the need to develop an ethical imagination about technological advances and research and develop an appropriately supportive environment for promoting ethical behavior in the scientific community" (Goolsby, 2005). Such recommendations point to the shifting boundaries of ethics in a nascent and rapidly shifting field.

In addition to standard concerns over financial conflicts of interest, CSS practitioners must also consider ethical concerns arising from non-financial contributions, especially shared data. Data donations, for instance, can pose a challenge in terms of quality and transparency creating dependencies and vulnerabilities for the researchers using the data in their work, as was seen in Facebook's Social Science One project (Timberg, 2021). In a collaborative landscape characterized by significant reliance on corporate data, the sources of such data, as well as the motivations involved in sharing it, must be acknowledged, and their potential impact on research thoroughly considered.

2.6.3 The Failure of Reproducibility

Reproducibility is a critical part of the scientific process, as it enables other researchers to verify or challenge the veracity of a study's findings (Coveney et al., 2021). This ensures that high standards of research are maintained and that findings can be corroborated by multiple actors to strengthen their credibility. While the concept has long been used by the scientific community, it only recently began to enter the work of social studies and computational social scientists. The notion of reproducibility has generally been problematic in CSS due to the many difficulties—outlined above—when it comes to data sharing and open software agreements. A lack of transparency in computational research also further aggravates the challenge, making it extremely difficult to implement the practices of reproducibility.

In order to address this challenge, scholars have suggested the use of open trusted repositories as a potential solution (Stodden et al., 2016). Such repositories would enable researchers to share their data, software, and other details of their work in a secure manner to encourage collaboration and reproducibility without compromising the integrity of the original researcher's work. More generally, a stronger culture of collaboration in the ecosystem would also help increase the adoption of reproducibility, which would be to the benefit of computational sciences as a whole (Kedron et al., 2021).

2.7 Societal Impact Challenges

2.7.1 Need for Citizen/Community Engagement and Acquiring a Social License

Trust has emerged as a major issue in the data ecosystem. In order for CSS research to be successful, it requires buy-in from citizens and communities. This is particularly true given the heavy reliance on data sharing, which requires trust and a trust-building culture to sustain the required inter-sectoral collaboration. For instance, a 2012 "Manifesto of Computational Social Science", published in the *European Physical Journal*, emphasizes the importance of involving citizens in gathering data and of "enhancing citizen participation in [the] decision process" (Conte et al., 2012).

In pursuit of such goals, CSS can borrow from the existing methodology of "Citizen Science", which highlights the role of community participation in various stages of social sciences research (Albert et al., 2021). Citizen Science methods can be adapted for—and in some cases strengthened by—the era of big data. New and emerging methods include crowdsourcing through citizen involvement in data gathering (e.g., through the IoT and other sensors); collaborative decision-making processes facilitated by technology that involve a greater range of stakeholders; and technologies to harness the distributed intelligence or expertise of citizens. Recently, some social scientists have also relied on so-called pop-up experiments (or PUEs), defined by one set of Spanish researchers as "physical, light, very flexible, highly adaptable, reproducible, transportable, tuneable, collective, participatory and public experimental set-up for urban contexts" (Sagarra et al., 2016). Indeed, urban settings have proven particularly fertile ground for such methodological innovations, given the density of citizens and data-generating devices.

2.7.2 Lack of Data Literacy and Agency

A lack of public understanding of data and data governance means that the public faces considerable risk associated with mismanagement of their data and exploitative data practices. This is particularly the case given that the current data ecosystem is largely dominated by corporate actors, who control access to large amounts of personal data and may use the data for their gain (Micheli et al., 2020). In order to address the associated inequalities and power asymmetries and to begin democratizing the data ecosystem, data governance methods must improve. Legislation such as the European Union's General Data Protection Regulation (GDPR) is a step in the right direction. In addition to legislative change, the development of data sharing infrastructures and the involvement of government and third sector actors in data collaborations with private actors will help mitigate the challenges of weak data literacy and agency among the public.

A lack of data literacy and agency have both ethical and practical implications for CSS (Chen et al., 2021; Pryzant et al., 2018; Sokol & Flach, 2020). In the context of data, agency refers to the power to make decisions about where and how one's data is used. Without sufficient awareness and agency, it is hard not only for individuals to meaningfully consent to their data being used but also for researchers to effectively and responsibly collect and use data for their studies. Moreover, a lack of data literacy and agency makes it difficult for citizens and others to interpret the results of a study or to implement policy and other concrete steps informed by CSS research. For CSS to achieve its potential, a stronger foundation of data literacy and an understanding of agency will be crucial both among the general public and among key decision-makers.

2.7.3 Computational Solutionism and Determinism

Determinism has a long and problematic history in the social sciences, with concerns historically raised about overly prescriptive or simplistic explanatory frameworks and models for human and social behaviour (Richardson & Bishop, 2002). CSS holds the potential both to improve upon such difficulties and to exacerbate them. The intersection of "technological determinism" and the social sciences is particularly grounds for wariness; any attempt to derive social explanations from technical phenomena must resist the temptation to construct overly deterministic or linear explanations. Models based on unrepresentative or otherwise flawed datasets (as described above) similarly risk flawed solutions and policy interventions.

At the same time, Big Data offers the theoretical potential at least for richer and more complete empirical frameworks. Some have gone so far as to suggest that the interaction of Big Data and the social sciences could spell the "end of theory", offering social scientists a less deterministic and hypothetical framework through which to approach the world (Kitchin, 2014). CSS also offers the potential of more realistic and complex simulations that can help social scientists and policymakers understand phenomena as well as potential outcomes of interventions (Tolk et al., 2018). For such visions to become a reality, however, the challenges posed to collaboration and sharing—many discussed in this paper—need to be mitigated.

2.7.4 Computational/Data Surveillance and the Risk of Exploitation

The final societal impact challenge associated with CSS pertains to the risk of computational and data surveillance (Tufekci, 2014). Considerable concern already exists over the data insights that drive targeted advertising, personalized social media content and disinformation, and more. We live, as Shoshana Zuboff has famously observed, in a "surveillance economy" (Zuboff, 2019).

This economy creates challenges related to misinformation and polarization, and it is a direct result of companies' ability to exploit the wealth of data they hold on their users. While the potential benefits of CSS are manifold, there is also a risk of new forms of exploitation and manipulation, based on new insights and new forms of data (Caled & Silva, 2021). Each case of exploitation has a direct result and also further erodes trust in the broader ecosystem. The only solution is a series of actions—legislative and otherwise—aimed at encouraging responsible data-driven research and CSS. Many potential actions are outlined in this paper. Further research is needed to flesh out some of the proposals and to develop new ones.

To tackle this challenge, new legislation addressing the uses of data and Computational Social Science analyses will be critical.

2.8 Reflections and Conclusion

The intersection of big data, advanced computational tools, and the social sciences is now well established among researchers and policymakers around the world. The potential for dramatic and perhaps even revolutionary insights and impact are clear. But as this paper—and others in this volume—shows, many hurdles remain to achieving that potential. The priority, therefore, is not simply to find ways to leverage data in the pursuit of research but, equally or more importantly, to innovate in how we govern the use of data for the social sciences.

An effective governance framework needs to be multi-tentacled. It would cover the broader ecologies of data, technology, science, and social science. It would address how data is collected and shared and also how research is conducted and transferred into insights and ultimately impact. It would also seek to promote the adoption of more robust data literacy and skills standards and programs. The above touches upon a number of specific suggestions, some of which we hope to expand upon in future research or writing projects. Elements of a responsible governance framework include the need to foster interdisciplinary collaboration; more fairly distribute computational power and technical and financial resources; rethink our notions of ownership and data rights; address misaligned incentives and misunderstood aspects of data reuse and collaboration; and ensure better quality data and representation. Last but not least, a responsible governance framework ought to develop a new research agenda in alignment with emerging concepts and concerns from the data ecosystem.

Perhaps the most urgent priority is the need to gain (or regain) a social license for the reuse of data in the pursuit of social and scientific knowledge. A social license to operate refers to the public acceptance of business practices or operating procedures used by a specific organization or industry (Kenton, 2021). In recent years the tremendous potential of data sharing and collaboration has been somewhat clouded by rising anxiety over misuses of data, with the resulting privacy and surveillance violations. These risks are very real, as are the resulting harms to individual and community rights. They have eroded the trust of the public and policymakers in data and data collaboration and undermined the possibilities offered by data sharing and CSS.

The solution, however, is not to pull away. Rather, we must strengthen the governance framework—and wider norms—within which data reuse and data-driven research take place. This paper represents an initial gesture in that direction. By identifying problems, we hope to take steps towards solutions.

References

Albert, A., Balázs, B., Butkevičienė, E., Mayer, K., & Perelló, J. (2021). Citizen social science: New and established approaches to participation in social research. In K. Vohland, A. Land-Zandstra, L. Ceccaroni, R. Lemmens, J. Perelló, M. Ponti, R. Samson, & K. Wagenknecht (Eds.), *The science of citizen science* (pp. 119–138). Springer International Publishing. https://doi.org/10.1007/978-3-030-58278-4_7

Baeza-Yates, R. (2016). Data and algorithmic bias in the web. *Proceedings of the 8th ACM Conference on Web Science, 1*, 1–1. https://doi.org/10.1145/2908131.2908135

Bahrke, J., & Manoury, C. (2022). *Data act: Commission proposes measures for a fair and innovative data economy.* European Commission - Press Corner. https://ec.europa.eu/commission/presscorner/detail/en/ip_22_1113

Balahur, A., Steinberger, R., Kabadjov, M., Zavarella, V., van der Goot, E., Halkia, M., Pouliquen, B., & Belyaeva, J. (2010). *Sentiment analysis in the news.* European Language Resources Agency (ELRA). http://www.lrec-conf.org/proceedings/lrec2010/index.html

Berens, J., Raymond, N., Shimshon, G., Verhulst, S., & Bernholz, L. (2016). *The humanitarian data ecosystem: The case for collective responsibility.* Stanford Center on Philanthropy and Civil Society. https://pacscenter.stanford.edu/wp-content/uploads/2017/11/humanitarian_data_ecosystem.pdf

Bitcoin Energy Consumption Index. (n.d.). *Digiconomist.* https://digiconomist.net/bitcoin-energy-consumption/

Blosch, M., & Fenn, J. (2018). *Understanding Gartner's hype cycles.* Gartner. https://www.gartner.com/en/documents/3887767

Burger, A., Talha, O. Z., Kennedy, W. G., & Crooks, A. T. (2019). Computational social science of disasters: Opportunities and challenges. *Future Internet, 11*(5), 103. https://doi.org/10.3390/fi11050103

Caled, D., & Silva, M. J. (2021). Digital media and misinformation: An outlook on multidisciplinary strategies against manipulation. *Journal of Computational Social Science, 5*, 123. https://doi.org/10.1007/s42001-021-00118-8

Caverlee, J. (2013). Exploitation in human computation systems. In P. Michelucci (Ed.), *Handbook of human computation* (pp. 837–845). Springer. https://doi.org/10.1007/978-1-4614-8806-4_68

Chang, R. M., Kauffman, R. J., & Kwon, Y. (2014). Understanding the paradigm shift to computational social science in the presence of big data. *Decision Support Systems, 63*, 67–80. https://doi.org/10.1016/j.dss.2013.08.008

Chen, H., Yang, C., Zhang, X., Liu, Z., Sun, M., & Jin, J. (2021). *From symbols to embeddings: A tale of two representations in computational social science.* arXiv.org. https://doi.org/10.48550/ARXIV.2106.14198

Conte, R., Gilbert, N., Bonelli, G., Cioffi-Revilla, C., Deffuant, G., Kertesz, J., Loreto, V., Moat, S., Nadal, J.-P., Sanchez, A., Nowak, A., Flache, A., San Miguel, M., & Helbing, D. (2012). Manifesto of computational social science. *The European Physical Journal Special Topics, 214*(1), 325–346. https://doi.org/10.1140/epjst/e2012-01697-8

Coveney, P. V., Groen, D., & Hoekstra, A. G. (2021). Reliability and reproducibility in computational science: Implementing validation, verification and uncertainty quantification *in silico*. *Philosophical Transactions of the Royal Society A: Mathematical, Physical and Engineering Sciences, 379*(2197) rsta.2020.0409, 20200409. https://doi.org/10.1098/rsta.2020.0409

Czajka, J., Schneider, C., Sukasih, A., & Collins, K. (2014). Minimizing disclosure risk in HHS open data initiatives. *Mathematica Policy Research.* https://aspe.hhs.gov/sites/default/files/private/pdf/77196/rpt_Disclosure.pdf

D'Ignazio, C., & Klein, L. (2018). Chapter one: Bring back the bodies. In *Data feminism.* PubPub. https://mitpressonpubpub.mitpress.mit.edu/pub/zrlj0jqb/release/6

Dahmm, H. (2020). *Laying the Foundation for Effective Partnerships: An examination of data sharing agreements* [Preprint]. *Open Science Framework.* https://doi.org/10.31219/osf.io/t2f36

De Hert, P., Papakonstantinou, V., Malgieri, G., Beslay, L., & Sanchez, I. (2018). The right to data portability in the GDPR: Towards user-centric interoperability of digital services. *Computer Law & Security Review, 34*(2), 193–203. https://doi.org/10.1016/j.clsr.2017.10.003

Dimitrova, D. (2021). The rise of the personal data quality principle. Is it legal and does it have an impact on the right to rectification? *SSRN Electronic Journal.*https://doi.org/10.2139/ssrn.3790602

Ducuing, C. (2020a, May 11). 'Data rights in co-generated data': The ground-breaking proposal under development at ELI and ALI. *KU Leuven Centre for IT and IP Law.* https://www.law.kuleuven.be/citip/blog/data-rights-in-co-generated-data-part-1/

Ducuing, C. (2020b, December 11). 'Data rights in co-generated data': How to legally qualify such a legal 'UFO'? *KU Leuven Centre for IT and IP Law.* https://www.law.kuleuven.be/citip/blog/data-rights-in-co-generated-data-part-2/

Edelmann, A., Wolff, T., Montagne, D., & Bail, C. A. (2020). Computational social science and sociology. *Annual Review of Sociology, 46*(1), 61–81. https://doi.org/10.1146/annurev-soc-121919-054621

European Commission. (2020). *Communication from the Commission to the European Parliament and the Council: Data protection as a pillar of citizens' empowerment and the EU's approach to the digital transition—Two years of application of the general data protection regulation.* https://ec.europa.eu/info/sites/default/files/1_en_act_part1_v6_1.pdf

Figueroa, J. L. P., & Pérez, C. V. (2018). Collective intelligence: A new model of business Management in the big-Data Ecosystem. *European Journal of Economics and Business Studies, 10*(1), 208. https://doi.org/10.26417/ejes.v10i1.p208-219

Fisher, A., & Streinz, T. (2021). Confronting data inequality. *SSRN Electronic Journal.*https://doi.org/10.2139/ssrn.3825724

Friedman, L. S., & Richter, E. D. (2004). Relationship between conflicts of interest and research results. *Journal of General Internal Medicine, 19*(1), 51–56. https://doi.org/10.1111/j.1525-1497.2004.30617.x

Ghinita, G., Karras, P., Kalnis, P., & Mamoulis, N. (2009). A framework for efficient data anonymization under privacy and accuracy constraints. *ACM Transactions on Database Systems, 34*(2), 1–47. https://doi.org/10.1145/1538909.1538911

Goolsby, R. (2005). Ethics and defense agency funding: Some considerations. *Social Networks, 27*(2), 95–106. https://doi.org/10.1016/j.socnet.2005.01.003

Green, B., Cunningham, G., Ekblaw, A., & Kominers, P. (2017). Open data privacy. *SSRN Electronic Journal.*https://doi.org/10.2139/ssrn.2924751

Harvard John A. Paulson School of Engineering and Applied Sciences. (2021, February 3). Environmental impact of computation and the future of green computing. *ScienceDaily.* https://www.sciencedaily.com/releases/2021/03/210302185414.htm

Hernandez-Suarez, A., Sanchez-Perez, G., Toscano-Medina, K., Perez-Meana, H., Portillo-Portillo, J., Sanchez, V., & García Villalba, L. (2019). Using twitter data to monitor natural disaster social dynamics: A recurrent neural network approach with word embeddings and kernel density estimation. *Sensors, 19*(7), 1746. https://doi.org/10.3390/s19071746

Herrera, Y. M., & Kapur, D. (2007). Improving data quality: Actors, incentives, and capabilities. *Political Analysis, 15*(4), 365–386. https://doi.org/10.1093/pan/mpm007

Hesse, M. (2021). *Essays on trust and reputation portability in digital platform ecosystems.* Technische Universität Berlin. https://doi.org/10.14279/DEPOSITONCE-11679

Hurt, M. (2018, May 15). The blockchain and its possible utility for social science investigation. *Deconstructing Korea.* https://medium.com/deconstructing-korea/the-blockchain-and-its-possible-utility-for-social-science-investigation-15d9f2fe6eff

Information Commissioner's Office. (n.d.). *Right to data portability.* https://ico.org.uk/for-organisations/guide-to-data-protection/guide-to-the-general-data-protection-regulation-gdpr/individual-rights/right-to-data-portability/

Johnson, K. (2020, November 11). AI research finds a 'compute divide' concentrates power and accelerates inequality in the era of deep learning. *Venture Beat.* https://venturebeat.com/2020/11/11/ai-research-finds-a-compute-divide-concentrates-power-and-accelerates-inequality-in-the-era-of-deep-learning/

Jünger, J., Geise, S., & Hänelt, M. (2022). Unboxing computational social media research from a Datahermeneutical perspective: How do scholars address the tension between automation and interpretation? *International Journal of Communication, 16,* 1482–1505.

Kadadi, A., Agrawal, R., Nyamful, C., & Atiq, R. (2014). Challenges of data integration and interoperability in big data. *2014 IEEE International Conference on Big Data (Big Data),* 38–40. https://doi.org/10.1109/BigData.2014.7004486

Kansas, S. (2021, October 23). Enter third-wave economics. *The Economist.* https://www.economist.com/briefing/2021/10/23/enter-third-wave-economics

Kedron, P., Li, W., Fotheringham, S., & Goodchild, M. (2021). Reproducibility and replicability: Opportunities and challenges for geospatial research. *International Journal of Geographical Information Science, 35*(3), 427–445. https://doi.org/10.1080/13658816.2020.1802032

Kenton, W. (2021). *Social License to Operate (SLO).* Investopedia. https://www.investopedia.com/terms/s/social-license-slo.asp

Kerry, C. F., & Morris, J. B. J. (2019, June 26). Why data ownership is the wrong approach to protecting privacy. *Brookings.* https://www.brookings.edu/blog/techtank/2019/06/26/why-data-ownership-is-the-wrong-approach-to-protecting-privacy/

Kitchin, R. (2014). Big data, new epistemologies and paradigm shifts. *Big Data & Society, 1*(1), 205395171452848. https://doi.org/10.1177/2053951714528481

Krishnamurthy, P. (2019, December 9). Understanding data bias. *Towards Data Science.* https://towardsdatascience.com/survey-d4f168791e57

Lazer, D. M. J., Pentland, A., Watts, D. J., Aral, S., Athey, S., Contractor, N., Freelon, D., Gonzalez-Bailon, S., King, G., Margetts, H., Nelson, A., Salganik, M. J., Strohmaier, M., Vespignani,

A., & Wagner, C. (2020). Computational social science: Obstacles and opportunities. *Science, 369*(6507), 1060–1062. https://doi.org/10.1126/science.aaz8170

Lévy, P. (2010). From social computing to reflexive collective intelligence: The IEML research program. *Information Sciences, 180*(1), 71–94. https://doi.org/10.1016/j.ins.2009.08.001

Lu, J., Lee, D., & (DK), Kim, T. W., & Danks, D. (2019). Good explanation for algorithmic transparency. *SSRN Electronic Journal.* https://doi.org/10.2139/ssrn.3503603

McAfee, A., & Brynjolfsson, E. (2019, October). Big data: The management revolution. *Harvard Business Review.* https://hbr.org/2012/10/big-data-the-management-revolution

Melo, C. D., Marsella, S., & Gratch, J. (2016). People do not feel guilty about exploiting machines. *ACM Transactions on Computer-Human Interaction, 23*(2), 1–17. https://doi.org/10.1145/2890495

Micheli, M., Ponti, M., Craglia, M., & Berti Suman, A. (2020). Emerging models of data governance in the age of datafication. *Big Data & Society, 7*(2), 2053951720948087. https://doi.org/10.1177/2053951720948087

Ministry of Social Affairs and Health. (2019). *Secondary use of health and social data.* Government of Finland. https://stm.fi/en/secondary-use-of-health-and-social-data

Mittelstadt, B. (2017). From individual to group privacy in big data analytics. *Philosophy & Technology, 30*(4), 475–494. https://doi.org/10.1007/s13347-017-0253-7

Narayanan, V. G., & Raman, A. (2004, November). Aligning incentives in supply chains. *Harvard Business Review.* https://hbr.org/2004/11/aligning-incentives-in-supply-chains

Naudts, L. (2019). How machine learning generates unfair inequalities and how data protection instruments may help in mitigating them. In R. Leenes, V. R. Brakel, S. Gutwirth, & P. de Hert (Eds.), *Data protection and privacy: The internet of bodies* (pp. 71–92). Hart Publishing. https://papers.ssrn.com/sol3/papers.cfm?abstract_id=3468121

Nelson, D. (2017, March 10). The problems of data ownership and data security. *Science Trends.* https://sciencetrends.com/problems-data-ownership-data-security/

Neumayer, C., Rossi, L., & Struthers, D. M. (2021). Invisible data: A framework for understanding visibility processes in social media data. *Social Media + Society, 7*(1), 205630512098447. https://doi.org/10.1177/2056305120984472

Novak, B. (2011, May 16). A series on four overarching themes across acquisition programs: First theme, misaligned incentives. *Carnegie Mellon University: Software Engineering Institute Blog.* https://insights.sei.cmu.edu/blog/a-series-on-four-overarching-themes-across-acquisition-programs-first-theme-misaligned-incentives/

OECD. (2019). *Enhancing access to and sharing of data: Reconciling risks and benefits for data re-use across societies.* OECD. https://doi.org/10.1787/276aaca8-en

Ohm, P. (2010). Broken promises of privacy: Responding to the surprising failure of anonymization. *UCLA Law Review, 57*, 1701.

OmniSci. (n.d.-a). *Data anonymization.* https://www.heavy.ai/technical-glossary/data-anonymization

OmniSci. (n.d.-b). *Interoperability.* https://www.heavy.ai/technical-glossary/interoperability

Porter, N. D., Verdery, A. M., & Gaddis, S. M. (2020). Enhancing big data in the social sciences with crowdsourcing: Data augmentation practices, techniques, and opportunities. *PLoS One, 15*(6), e0233154. https://doi.org/10.1371/journal.pone.0233154

Porway, J. (2019, April 1). DataKind's four hopes for data & AI in 2019. *DataKind.* https://www.datakind.org/blog/datakinds-four-hopes-for-data-ai-in-2019

Pryzant, R., Shen, K., Jurafsky, D., & Wagner, S. (2018). Deconfounded lexicon induction for interpretable social science. *Proceedings of the 2018 Conference of the North American Chapter of the Association for Computational Linguistics: Human Language Technologies, 1.* (Long Papers), 1615–1625. https://doi.org/10.18653/v1/N18-1146

Radaelli, L., Sapiezynski, P., Houssiau, F., Shmueli, E., & de Montjoye, Y.-A. (2018). Quantifying surveillance in the networked age: Node-based intrusions and group privacy. *ArXiv:1803.09007 [Cs].* http://arxiv.org/abs/1803.09007

Richardson, F., & Bishop, R. (2002). Rethinking determinism in social science. In H. Atmanspacher & R. C. Bishop (Eds.), *Between chance and choice: Interdisciplinary perspectives on determinism* (pp. 425–446). Imprint Academic.

Richet, C. (2021, November 1). Towards co-regulation of cyberspace: Between power relationship and sovereignty imperatives. *InCyber*. https://incyber.fr/en/towards-co-regulation-of-cyberspace-between-power-relationship-and-sovereignty-imperatives/

Riis, C., Kowalczyk, D., & Hansen, L. (2021). On the limits to multi-modal popularity prediction on Instagram: A new robust, efficient and explainable baseline. In: *Proceedings of the 13th International Conference on Agents and Artificial Intelligence* (pp. 1200–1209). https://doi.org/10.5220/0010377112001209

Rubinstein, I. S. (2018). The future of self-regulation is co-regulation. In E. Selinger, J. Polonetsky, & O. Tene (Eds.), *The Cambridge handbook of consumer privacy* (1st ed., pp. 503–523). Cambridge University Press. https://doi.org/10.1017/9781316831960.028

Rubinstein, I., & Hartzog, W. (2015). Anonymization and risk. *91 Washington Law Review, 703*. https://papers.ssrn.com/sol3/papers.cfm?abstract_id=2646185

Russo, M., & Feng, T. (2021). *Where is data sharing headed?* BCG. https://www.bcg.com/publications/2021/broad-data-sharing-models

Sagarra, O., Gutiérrez-Roig, M., Bonhoure, I., & Perelló, J. (2016). Citizen science practices for computational social science research: The conceptualization of pop-up experiments. *Frontiers in Physics, 3*. https://doi.org/10.3389/fphy.2015.00093

Shuman, R., & Paramita, F. M. (2016, January 21). Why Your View of the World Is Riddled with Holes. *World Economic Forum*. https://www.weforum.org/agenda/2016/01/data-invisibles-ignore-at-our-peril/

Sokol, K., & Flach, P. (2020). Explainability fact sheets: A framework for systematic assessment of explainable approaches. In: *Proceedings of the 2020 Conference on Fairness, Accountability, and Transparency* (pp. 56–67). https://doi.org/10.1145/3351095.3372870

Stodden, V., McNutt, M., Bailey, D. H., Deelman, E., Gil, Y., Hanson, B., Heroux, M. A., Ioannidis, J. P. A., & Taufer, M. (2016). Enhancing reproducibility for computational methods. *Science, 354*(6317), 1240–1241. https://doi.org/10.1126/science.aah6168

The 100 Questions Initiative—About. (n.d.). *The 100 Questions Initiative*. https://the100questions.org/about.html

The powers and perils of using digital data to understand human behaviour. (2021). *Nature, 595* (7866), 149–150. https://doi.org/10.1038/d41586-021-01736-y

The World Wide Web Foundation. (2016). *Open data barometer global report third edition*. https://opendatabarometer.org/doc/3rdEdition/ODB-3rdEdition-GlobalReport.pdf

Thiem, A., Mkrtchyan, L., Haesebrouck, T., & Sanchez, D. (2020). Algorithmic bias in social research: A meta-analysis. *PLoS One, 15*(6), e0233625. https://doi.org/10.1371/journal.pone.0233625

Timberg, C. (2021). Facebook made big mistake in data it provided to researchers, undermining academic work. *The Washington Post*. https://www.washingtonpost.com/technology/2021/09/10/facebook-error-data-social-scientists/

Tolk, A., Wildman, W. J., Shults, F. L., & Diallo, S. Y. (2018). Human simulation as the lingua Franca for computational social sciences and humanities: Potential and pitfalls. *Journal of Cognition and Culture, 18*(5), 462–482. https://doi.org/10.1163/15685373-12340040

Toth, D., Mayer, R., & Nichols, W. (2011). Increasing participation in volunteer computing. In: *2011 IEEE International Symposium on Parallel and Distributed Processing Workshops and Phd Forum* (pp. 10878–1882). https://doi.org/10.1109/IPDPS.2011.353

Tufekci, Z. (2014). Engineering the public: Big data, surveillance and computational politics. *First Monday*. https://doi.org/10.5210/fm.v19i7.4901

Turner Lee, N., Resnick, P., & Barton, G. (2019, May 22). Algorithmic bias detection and mitigation: Best practices and policies to reduce consumer harms. *Brookings*. https://www.brookings.edu/research/algorithmic-bias-detection-and-mitigation-best-practices-and-policies-to-reduce-consumer-harms/

Van Asbroeck, B. (2019). *Big Data & Issues & opportunities: Data ownership.* Bird&Bird. https://www.twobirds.com/en/insights/2019/global/big-data-and-issues-and-opportunities-data-ownership

Vanberg, A. D., & Ünver, M. B. (2017). The right to data portability in the GDPR and EU competition law: Odd couple or dynamic duo? *European Journal of Law and Technology, 8*(1), 1.

Verhulst, S. (2021). Reimagining data responsibility: 10 new approaches toward a culture of trust in re-using data to address critical public needs. *Data & Policy, 3*, e6. https://doi.org/10.1017/dap.2021.4

Verhulst, S., Safonova, N., Young, A., & Zahuranec, A. (2020a). The data assembly synthesis report and responsible re-use framework. *SSRN Electronic Journal.* https://doi.org/10.2139/ssrn.3937625

Verhulst, S., Young, A., Zahuranec, A., Calderon, A., Gee, M., & Aaronson, S. A. (2020b). The emergence of a third wave of open data: How to accelerate the re-use of data for public interest purposes while ensuring data rights and community flourishing. *SSRN Electronic Journal.* https://doi.org/10.2139/ssrn.3937638

Young, A. (2020). *Responsible group data for children (Good governance of Children's data project) [Issue brief].* UNICEF - Office of Global Insight and Policy. https://www.unicef.org/globalinsight/reports/responsible-group-data-children

Zuboff, S. (2019). *The age of surveillance capitalism: The fight for a human future at the new frontier of power* (1st ed.). PublicAffairs.

Chapter 3
Data Justice, Computational Social Science and Policy

Linnet Taylor

Abstract Big data has increased attention to Computational Social Science (CSS) on the part of policymakers because it has the power to make populations, activities and behaviour visible in ways that were not previously possible. This kind of analysis, however, often has unforeseen implications for those who are the subjects of the research. This chapter asks what a social justice perspective can tell us about the potential, and the risks, of this kind of analysis when it is oriented towards informing policy. Who benefits, and how, when computational methods and new data sources are used to conduct policy-relevant analysis? Should CSS sidestep, through its novelty and its identification with computational and statistical methodologies, sidestep ethical review and the assessments of power asymmetries and methodological justification that are common in social science research? If not, how should these be applied to CSS research, and what kind of assessment is appropriate? The analysis offers two main conclusions: first, that the field of CSS has evolved without an accompanying evolution of debates on ethics and justice and that these debates are long overdue. Second, that CSS is privileged as policy-relevant research precisely because of many of the features which bring up concerns about justice—large-scale datasets, remote data gathering, purely quantitative methods and an orientation towards policy questions rather than the needs of the research subjects.

3.1 Introduction

The rise of Computational Social Science (hereafter CSS) over the last decades has become mingled with the rise of big data, and more recently with that of Artificial Intelligence (AI) and the automated processing of data on an enormous scale. New applications and uses of data arose with the advent of new data sources

L. Taylor (✉)
Global Data Justice Project, TILT, Tilburg University, Tilburg, The Netherlands
e-mail: L.E.M.Taylor@tilburguniversity.edu

© The Author(s) 2023
E. Bertoni et al. (eds.), *Handbook of Computational Social Science for Policy*,
https://doi.org/10.1007/978-3-031-16624-2_3

over the 2000s, and especially the rise of mobile phones and mobile connectivity in most countries of the world. The following decade has borne out many of the predictions that were made when big data was first conceptualised—that it would make populations, activities and behaviour visible in ways that had not previously been possible and that this would have huge impacts on both analysis and intervention across a range of fields, from urban policy to epidemiology and from international development to humanitarian intervention. This chapter examines the use of CSS in relation to national and international policy issues and asks who benefits, and how, when computational methods and new data sources are used to conduct policy-relevant analysis.

What few commentators on big data forecasted was the extent to which big data would represent a private sector revolution (Taylor & Broeders, 2015). The step change in volume, immediacy and power constituted by the new sources of large-scale data did not stem from bureaucratic or academic innovation, but from changes in the commercial world driven by new devices and massive investments in software, hardware and infrastructure. Despite the United Nations' call in 2014 (United Nations, 2014) to use data for the public good, the potential of the new data sources to make people visible and to inform intervention has been led primarily by commercial firms, with policy as a secondary user of what is still largely commercial data. The proprietary nature of much of the data used in CSS is important because it determines what information becomes available and to whom and what kind of analysis and intervention it can inform. It also, as predicted more than a decade ago, creates hierarchies amongst researchers and institutions, since access to data is a privilege to be negotiated (boyd & Crawford, 2012). This has meant that so far the CSS field has been mainly populated by high-status researchers from well-funded institutions in high-income countries, who also tend to be male, white and connected to the well-funded academic disciplines of computer science, quantitative sociology and statistics or to policy interests that tend towards security, population management and economic development.

One example of the increasing hybridisation in the way data is sourced between commercial, international organisations and governments is the case of the 'Premise' app.[1] Developed in Silicon Valley, Premise is a crowdsourcing survey app that pays people small amounts to photograph or report features of their surroundings, from cash machines and construction sites to food prices. Initially marketed as a tool for international development agencies to remotely source information, it then became a tool for businesses to assess market possibilities and competition. It next morphed into a way for intelligence services to collect data covertly, with tasks offered such as photographing the locations of Shiite mosques in Kabul (Tau, 2021).

As the case of Premise suggests, data is not neutral. Unless we understand who it reflects and how it was sourced, there is the potential for harm to what Metcalf and Crawford have termed the 'downstream subjects of research' (Metcalf & Crawford,

---

[1] See https://www.premise.com/contributors/.

2016). Understanding data gathered remotely also poses epistemological problems: without domain and local knowledge to convey ground truth (Dalton et al., 2016), not only its analysis but any interventions it informs are likely to be flawed and unreliable. Digitally informed analysis and intervention also raise issues of power and justice given that powerful interests drive its collection and use. Social data is always attached to people. Analysis often obscures that connection, but it remains throughout the lifecycle of bits and bytes, from information to intervention and evaluation.

Despite its usually explicit aim to have effects on people, Computational Social Science is not, however, subject to the kind of review process that is normal for other research on human subjects. One reason may be that it is designed to inform intervention, but it is not so far classified as constituting intervention itself. This places it in a different category to biomedical research, which is governed through an infrastructure for ethical review at the European, governmental and institutional level (EUREC, 2021). It also tends to escape review within academic institutions because CSS usually does not become registered as social scientific research unless a project lead is employed by a social science faculty. Due to their technical demands, many CSS projects originate in computer science, economics or data science departments and institutes. On the European level, CSS projects undergo an ethics check if the principal investigator flags them at the application stage as using personal data—which may not happen in many cases due to the definitional problem outlined by Metcalf and Crawford (2016). If they do go through review by the European Research Council's ethics committee, they are reviewed for data protection compliance and for classic research ethics issues such as benefit-sharing, but as explored later in this chapter, this may not capture important ethical challenges relating to CSS, particularly where the benefits are defined as relating to policy.

3.2 Background: Computational Social Science and Data Justice

The central aim of using large-scale data and Computational Social Science methods to inform policy is to positively impact society. This aim, however, comes with no definition of which people should benefit and whether those are the same people who are reflected in the data. The unevenness of the new large-scale data sources, their representativeness and their potential for uneven effects when used in policy, therefore, are central concerns for any researcher or policymaker interested in not doing harm.

Over the 2010s first the field of critical data studies and, later, the related field of data justice have taken up these issues methodologically and theoretically. These fields have roots in digital geography, charting how epistemologies of big data (Kitchin, 2014) differ from previous ways of seeing the world through statistics and administrative accounts and how the geography of where and how data is sourced determines whose truth it can tell us (Dalton et al., 2016). They are sceptical of the claims of granularity and representativeness often made about large-scale data, a scepticism also present in the post-colonial strand of critique which has shown how the datafied representation of populations, cities and movement is always filtered through narratives of entrepreneurialism, innovation and modernity, which shape both the starting point and the uses of such analyses (Couldry & Mejias, 2019; Datta & Odendaal, 2019). Similar critiques can be found in sociological research, which take issue with the idea that data can ever be neutral or raw (Gitelman, 2013) and which also expose the underlying ideology of what van Dijck calls 'dataism':

> a widespread *belief* in the objective quantification and potential tracking of all kinds of human behavior and sociality through online media technologies. Besides, dataism also involves *trust* in the (institutional) agents that collect, interpret, and share (meta)data culled from social media, internet platforms, and other communication technologies. (Van Dijck, 2014, p. 198)

Where all these accounts cumulatively tend is towards a statement that not only is big data applied to policy issues not as granular or omniscient as the hype of the early 2010s promised it would be, but that far from being objective, it is fundamentally shaped by the assumptions and standpoint of all the actors (many of them commercial) controlling its trajectory from creation to analysis and use. Not only are the questions asked of data usually oriented towards the needs and perspectives of the most powerful (Taylor & Meissner, 2020), but the data itself is generated, collected and shared in ways that reflect and confirm the status quo in terms of resource distribution, visibility and agency. As AI increasingly becomes an important part of data's potential lifecycle, with data used to train, parameterise and feed models for business and policy, this dynamic where data reflects existing power and its interests becomes magnified. Data is now not only useful for making visible the behaviour and movement of populations, it is useful for optimising them. Correspondingly, any lack of representativeness or understanding of the interests and dynamics the data reflects are translated in this move from modelling to optimising into a direct shaping of subjects' opportunities and possibilities (Kulynych et al., 2020).

Research on these issues of justice has been done in disciplines ranging from computer science (Philip et al., 2012) and information science (Heeks, 2017) to development studies (Taylor, 2017) and media studies (Dencik et al., 2016) and is increasingly affecting how regulators think about the data economy (Slaughter, 2021). How research, and specifically policy-relevant research conducted under the heading of Computational Social Science, intersects with this problem of data justice is the focus of this chapter. The questions that arise from CSS are not confined to data itself or to scientific or policy research methods. Instead, they span issues of democratic decision-making, representative government, the governance of data in general and social justice concerns of recognition, representation and redistribution. As Gangadharan and Niklas have argued (2019), doing justice to the subjects of datafication and datafied policy often implies decentring the technology and being less exceptionalist about data.

3.3 Questions and Challenges

It is possible to group the justice-related issues outlined above around two poles: the effects of CSS on those who use the data and its effects on those whom the data represents. The first deals with how data confers new forms of power on the already powerful through their access not only to data itself but also resources, computing infrastructures and policy attention. The second relates to the way in which making people visible to policy does not automatically benefit them and may instead either amplify existing problems or create entirely new ones, while the remote nature of the research decreases people's agency in relation to policy and decision-making. CSS methods, and the data that fuels them, frequently confer on researchers the power to make social phenomena visible at the aggregate level and continuously— people's behaviour, whereabouts, activities and histories—and on policymakers the power to intervene in new ways.

The optimisation of social systems and its policy predecessors, nudging and governance through statistics, are all ways of intervening that rely on detailed quantitative data. Computational Social Science demonstrates the tendency of this datafied power to be unbalanced in its distribution, favouring those with the resources, infrastructure and power to gather and use data effectively. Like all social science, it involves a power relation between the researcher and the subject, but in the case of CSS, that subject can be an entire population. Large-scale data conveys the power to intervene but also the power to define problems in the first place: what Pentland has termed 'the god's eye view' (Pentland, 2010) brings with it little accountability.

A justice perspective, above all, asks what would shape the power conferred by data towards the public interest. Adding a governance perspective means we should also consider how the negative possibilities of datafied power can be systematically identified and controlled. Computational Social Science, specifically where it has the aim of informing policy, is a relevant field in which to ask these questions for two reasons. First, because the ways in which it accesses data, analyses it and uses it to intervene opaque to the public, taking place in the realm of large producers of data and high-level policymakers. This has meant that CSS has so far been relatively invisible to the kind of ethical or justice-based critiques which have arisen around AI and machine learning over the recent period. Second, we should interrogate it because it increasingly has real and large-scale effects on populations, either local or distant, once translated into policy information.

3.3.1   Who Benefits?

The issue of the distribution of benefits from CSS is both discursive and contested. Discursive because as with all scientific disciplines, there is an argument that fundamental research is justified by the search for knowledge alone, but this is counterbalanced by the responsibility that research on human subjects brings with it. CSS has not so far been categorised as human subject research because, despite its connection to policy and the shaping of social processes, data is collected remotely and the human subjects are not directly connected to the research. This means that CSS research has not so far been subject to the same ethical review process as human subject research, where researchers must explain how any benefits of their research will be distributed. The question is also contested because human subjects of the research, given the chance, will often have very different understandings of what constitute benefits. For example, starting from the assumption that data exists and must therefore be used (Taylor, 2016) is problematic because it addresses data about society as 'terra nullius' (Couldry & Mejias, 2019), a raw resource which exists independently of the people it reflects. In contrast, the subjects of the research (city dwellers, migrants, workers, the subjects of development intervention and others) may disagree that this is true. The 'terra nullius' assumption has also been undermined by work on group privacy (Taylor et al., 2017), which argues that data which facilitates intervention upon people—whether personal data or not—raises the question of when it is justified to shape and optimise behaviour or social conditions. Given that CSS is usually conducted on remote subjects with only the consent of the intermediaries holding the data, this means its legitimacy is usually based on the interests of those intermediaries and the researchers, not the subjects of research themselves (Taylor, 2021).

To offer an example, data stemming from refugees' use of mobile phones was made available by the Turkish national mobile network operator and used remotely by computational social scientists in the Data for Refugees challenge (2019).[2] One group built a model that could identify where people were working informally—something 99% of Syrian refugees in Turkey were doing at the time due to lack of employment permission. The authors explain their logic for conducting the study:

> Refugees don't normally have permission to work and only have access to informal employment. Our results not only provided country-wide statistics of employment but also gave a detailed breakdown of employment characteristics via heatmaps across Turkey. This information is valuable since it would allow GOs and NGOs to refine and target appropriate policy to generate opportunities and economic integration as well as social mobility specific to each area of Turkey. (Reece et al., 2019, p. 13)

It is possible to contest this, however. The fact that Turkey was legally restricting the right of refugees to internal mobility and employment—which the authors note many other countries also do—does not mean that this is in line with international

---

[2] For more on the challenge, see https://datapopalliance.org/publications/data-for-refugees-the-d4r-challenge-on-mobility-of-syrian-refugees-in-turkey/.

human rights law (International Justice Resource Center, 2012). It is doubtful that the Syrians in the dataset would find that creating a way to make visible their mobility and informal employment was in their own interests. The authors' claim that their model allows government and non-governmental organisations to target policy, generate opportunities and economic integration and help refugees become socially mobile rests on the optimistic assumption that these organisations are incentivised to do so. An alternate and more likely result would be that the model would facilitate the authorities' ability to constrain refugees' ability to move and work, an incentive already present in Turkish law.

Whose interests does this analysis serve, then? First, the Turkish government, since the model can help enforce a national law against refugees' moving and working freely. It may be in the interests of NGOs wishing to help refugees, but given the Turkish regime's laws targeting organisations that do so (Deutsche Welle, 2020), it is unlikely. The national telecom provider is a potential beneficiary in terms of positive publicity and potentially governmental approval if the authoritarian government of Turkey sees the researchers' analysis of the data as being useful for its governance of refugee populations. Lastly, the researchers themselves benefit in the form of access to data and ensuing publications. And so we can chart how analysis that claims to be 'Data for Refugees' may in fact be data for government, data for telecom providers and data for academic researchers.

Scholars of data governance have debated the problem of determining interests in, and rights over, data once it enters the public sphere. These include public data commons and data trusts (Micheli et al., 2020), both of which appear at first sight ideal for protecting the rights of data subjects. These approaches are promising under conditions where data is moving within the same jurisdiction (local, national or regional) in which it was created and where there is a fiduciary capable of representing the interests of the people reflected in the data (Delacroix & Lawrence, 2019). In the case of cross-border transfers of data for scientific research, however, this chain is often broken at the starting point. In the case (common in CSS) of mobile data on non-European populations, the data is de-identified and aggregated by the mobile network provider (Taylor, 2016) before it is made available for analysis, placing the network provider in the position of fiduciary. Creating a different fiduciary would in the case explored above mean empowering someone to represent the interests of all Syrian refugees in Turkey.

This hints at several problems: can a fiduciary from a group in a situation of extraordinary vulnerability be expected to have the power to protect that group's interests? What happens when the group in question has, as in this case, a limited set of enforceable rights compared to everyone else with an interest in the data? For example, are the claims of the Western CSS community likely to be effectively contested by a population of refugees primarily engaged with their own survival? It is easy to see how, in cases where people within a population of interest is not able to assert their rights, even fiduciary arrangements quickly come to represent an idea of the public good that may not align with that group's own ideas—if such a diverse group agrees on what is in its interests in the first place.

This case illustrates that, given that the stakes for refugees in being monitored and intervened upon are extraordinarily high, and the CSS in this case actively creates new vulnerabilities, it seems more attention should be given to how far fiduciary-based models can stretch. In situations of radical power asymmetry, it is not clear that the fiduciary model necessarily leads to the legitimate use of data for research. In fact, drawing on discussions of indigenous data sovereignty, it is clear that in the case of people in situations of vulnerability, a model based on the assumption that data will be shared and reused may not be appropriate (Rainie et al., 2019). As indigenous scholars point out (Simpson, 2017), if refusal is not an option on the table for those who have been made vulnerable, further ideas about governance cease to be ethical choices.

3.3.2 Making People Visible: Surveillance as Social Science

Data sourced from platforms, large-scale administrative data from public services or data from monitoring of public space are, in their different ways, all forms of surveillance. They are often quite intimate, drawing a picture of how people use city space or move across borders, how they break rules and create informal ways to support their families in emergency situations and how they catch and pass on infectious diseases, spend their money, interact with each other and use public services. Human activity everywhere is becoming datafied, sometimes with people's knowledge as they engage with platforms and online services, but often without their awareness as they are captured by CCTV, satellites, mobile phone network infrastructures, apps or payment services. Increasingly, these forms of surveillance intersect and feed into each other. Urban space has become securitised through the availability of CCTV and mobile phone data, just as borders have become securitised through satellite surveillance and geospatial sensing. But all these sensing technologies are dual use—either in their potential or in their actual usage by authorities. Urban crowd sensing systems, relying on mobile phone location data and social media analysis, were first created as a way to keep track of crowding during public events and then repurposed to help enforce pandemic public health measures. These functions also, however, support police and security services by showing how public protests evolve, by helping track how people move to and from locations authorities wish to control and by making it possible to identify protesters in real time—something law enforcement used to chilling effect during the Hong Kong protests of 2019–2020 (Zalnieriute, 2021).

Border enforcement activities have also become an important target for Computational Social Science methods. In 2019 the European Asylum Support Office was warned by the European Data Protection Supervisor (EDPS, 2019) that conducting social media analysis of groups assumed to be potential migrants in Africa, with the aim of tracking migration flows towards the EU's borders, was illegal under European data protection law. This was a project the Asylum Support Office had inherited from the United Nations, which had been developing Computational Social Science methods with big data for nearly a decade (Taylor, 2016) using methods developed in collaboration with academic Computational Social Science researchers. Similarly, epidemiological surveillance has a long history of constructing models that show how people move across borders, first in relation to malaria and later dengue and Ebola (Pindolia et al., 2012; Wesolowski et al., 2014). These methods were co-designed and then separately developed by mobility researchers over the 2010s, culminating in the use of mobile phone connectivity for tracking infections (and people's movement in general) during the COVID-19 pandemic (Ferretti et al., 2020). Mobile data in particular can inform many forms of monitoring, from policing borders to political protests, with methods shared between humanitarian technologists, public health specialists, security services and law enforcement.

These interactions between different forms of surveillance suggest two conclusions: first, that an innocuous history and set of uses can always be claimed for any methodology involving surveillance-derived data and, second, that the reverse is also true—all methods and types of data intersect at some point in the data's lifecycle with uses that potentially or actually violate the right to protest anonymously, to move freely, to work, to self-determination and many other rights and entitlements. A justice-based approach illuminates these interactions rather than seeking the innocuous explanations and follows data and methods through their lifecycles to find the points where they generate injustice by rendering people visible in ways that are damaging to their rights and freedoms.

Much of this discussion comes down to the question of who has the right to derive policy-relevant conclusions from data, under what circumstances and on whose behalf. It is not a simple question: should people 'own' data about them (something that is not present in data protection law, or any other, which only confer rights over data to people under some specific circumstances in order to protect from harm), or should the makers and managers of data be free to use it in line with whatever they conceive to be the public benefit? The issue seems mainly to revolve around how the public benefit will be agreed upon, rather than who has the right to data per se. Forced migrants in particular but also those suffering marginalisation or disadvantage of any kind may be generating information that is important not only to them, but also to others—on environmental change, conflicts and humanitarian crises, for example, not to mention living conditions in cities and the adequate provision of public services such as education and transport. What should we say about the shared interests in data that can illuminate problems and inform change?

This is partly a question for democratic discussion—something that has not been well conceptualised so far. It is also, however, a normative question that the EU needs to find a preliminary answer to in order to make possible such a debate. One suggestion from work on data justice is that the normative framing tends to be that of economic growth and technical advancement, whereas an alternative but valid one is that of the good of the groups involved in the data. If the starting point for analysis is the interests of those groups, this demands not only different ways of analysing the ethics of a particular research project or policy advice process but also that democratic processes be set up for determining the interests of the groups in question (Taylor & De Souza, 2021). This becomes a much broader issue of decolonising international relations, reframing the allocation of fundamental rights so that they cover people, for instance, on both sides of the EU's border, and treating people who are in conditions of conflict, forced migration or other precarious situation as if they are the same kind of legal subject as more empowered and vocal research subjects in easier conditions.

3.4 Addressing Justice Concerns: Ethics, Regulation and Governance

The potential and actual justice problems for CSS outlined above are frequently seen as problems of research ethics. If researchers can comply with data protection provisions, the logic goes, they will not violate the rights of those the data reflects. Similarly, if research ethics are followed—again, mainly focusing on the privacy and confidentiality of research data because consent tends to come from the intermediaries offering the data—the subjects of the research will be protected. Both the data protection-compliance and research ethics/privacy approaches, however, are necessary but insufficient to address the justice concerns that arise from CSS methods and the ways in which they inform policy.

As the EDPS' warning to the Asylum Support Office states, the problems caused by remote analysis of data on unaware and often vulnerable populations are not solved by preventing the identification of individual research subjects. In its letter the data protection supervisor's office notes that 'EASO accesses open source info, manually looks at groups and produces reports, which according to them no longer contain personal data' and that 'EASO's monitoring activities subject them to enhanced surveillance due to the mere fact that they are or might become migrants or asylum seekers'. Both these statements accurately describe much of CSS research, hence the relevance of this example. The EDPS names two risks: possible inaccuracy in identifying groups (not individuals) who might attempt to cross borders irregularly—something with potentially serious consequences for the people involved—and the risk of discrimination against those people. The EDPS quotes theory on group privacy, noting 'the risk of group discrimination, i.e. a situation in which inferences drawn from SMM [social media monitoring] can conceivably put a group, as group, at risk, in a way that cannot be covered by ensuring each member's control over their individual data' (EDPS, 2019) (the EDPS also notes, however, that the likelihood of such individuals knowing their data is being used in this way and 'controlling' it is vanishingly small).

The EDPS' analysis of this problem merits serious consideration by CSS researchers, given that it overturns a generation of research ethics based on preserving the individual privacy and confidentiality of research subjects. If we shift the focus from the individual in the dataset—who will often be de-identified anyway—to the consequences of the analysis, a whole different set of concerns opens up, namely, those of rights violations, discrimination and illegitimate intervention on the collective level. In this scenario, it is not enough for researchers to claim that they are merely performing social scientific analysis and that the potential policy uses of their work are not their responsibility. CSS is intimately connected to policy through a history of providing findings on public health, migration dynamics, economic development, urban planning, labour market dynamics and a myriad other areas which connect directly to policy uses.

It is not clear how to govern CSS research so that research ethics is not violated. As experts have pointed out, research ethics practices, and the academic infrastructure of checks and balances that enforce it, urgently require updating for the era of big data research (Metcalf & Crawford, 2016). Given that the field of CSS does not conceptualise itself as 'human subjects research', researchers are not incentivised either to conceptualise the downstream effects on whole populations or to weigh the justification for those effects. Instead they are strongly incentivised to make general statements about how their research will benefit society or institutions, without acknowledging that those benefits come with costs to others, most often the subjects of the research themselves. This lack of alignment between research ethics and much of CSS research does not justify proceeding with business as usual. Instead it sets a challenge to both CSS researchers and the policymakers who use their findings: to place real checks and balances on what research can be done, with processes involving both domain knowledge and rights expertise, and to undertake concentrated work to identify the ways in which projects may create or amplify injustice. Only by doing so can the acceptability and normality of doing unacknowledged dual-purpose research be countered.

This is particularly important given that data's availability will potentially become much greater over the 2020s. New models for data sharing such as those outlined in the EU's data governance act (European Commission, 2020) are designed to contribute to the availability of data for both CSS and AI, both redefining 'public' data as data with possible public uses and setting broader parameters for sharing it between business, government and research. These new models also include new intermediaries to ensure that 'altruistic data sharing' can occur without friction. Once enacted, this vastly greater legal and technical infrastructure will increase the interactions between the public and private sectors, allowing research to more comprehensively inform policy and business. It is likely the line between the two will increasingly blur, as governmental and EU research funding continues to be oriented towards serving business and the EU's economic agenda. It is likely that this blurring of boundaries between the commercial and research worlds will also lead to more policy-relevant research in terms of influencing social behaviour, just as nudging both inherited methods from and contributed to marketing research over the 2000s (Baldwin, 2014). Such a merging of commercial and governmental surveillance and analytical methodologies has already occurred: the Snowden revelations of 2014 (Lyon, 2014) revealed that security surveillance was already based on scanning behavioural and social media data and that it was conducted not by native security technicians but by commercial contractors. More recently the work of the Data Justice Lab in Cardiff, for example, has demonstrated that citizen scoring has transitioned from a commercial to a governmental practice, with the two connected by common methodologies and analytical practices (Dencik et al., 2018).

3.5 The Way Forward

The analysis in this chapter offers two main conclusions: First, that the field of CSS has evolved without an accompanying evolution of debates on ethics and justice and that these debates are long overdue. Second, that CSS is privileged as policy-relevant research precisely because of many of the features which bring up concerns about justice—large-scale datasets, remote data gathering, purely quantitative methods and an orientation towards policy questions rather than the needs of the research subjects.

The hype that has accompanied the discovery of new data sources and new ways of applying statistical methodologies to very large-scale data has frequently eclipsed the question of when doing such analysis is justified and whether the benefits it may create are proportionate to the costs of making people and their activities visible to new (policy) actors. Migration data offers a key lesson here: computational collection and analysis of large-scale data does not aim at identifying individuals and is therefore considered by its practitioners not to be problematic. However, when practised with the aim of providing an 'early warning system' for the approach of irregular migrants to the EU's borders, it has the potential to violate fundamental human rights, both in the form of discrimination and by narrowing the right to claim asylum. Similarly, building models to identify those working irregularly in refugee receiving states may be welcomed by state authorities and by the statistical methods community, but does not represent a contribution to the care and wellbeing of the refugees in question. Once such a model exists, the researcher cannot unpublish it—it is open to the use of anyone with access to the relevant type of data. The responsibility in this case is squarely with the researcher, but accountability is absent.

One step, therefore—if the field of CSS and the policymakers it informs wish to move towards a justice-based approach—is to subject all CSS studies involving data on people and informing any kind of intervention, to the same kind of ethical review that is performed on standard social scientific research projects involving human subjects. This is not enough on its own, however: that ethical review has to also respond to concerns about proportionality, fairness and the appropriateness of the methods to the question, regardless of whether the research is remote or in-person. The examples offered in this chapter suggest that it is time to update research ethics to cover the fields and methods involved in big data and that this is also a concern for policymakers interested in aligning their work with human rights. Demand from CSS researchers and policymakers could provide the necessary stimulus to update academic research review for the 2020s and align checks and balances with contemporary research practices.

A second concern is that CSS is rarely, if ever, performed in circumstances where the individuals implicated by the research either influence the questions asked or have access to the conclusions. A notable exception is 'citizen sensing' methods (Suman, 2021) where people source data about their local environment and use it to create public awareness, policy change or both. There is much room for expanding these methodologies and practices, as well as formalising and standardising them so that they can be a more accessible resource for policymakers (Suman, 2019). Another exception is the informal version of citizen sensing, sousveillance, which has a long history of disrupting the use of digital data for restricting public freedoms. Like citizen sensing, which tends to challenge the business and policy status quo, sousveillance practices are a datafied tool for the marginalised or neglected to assert their rights and claim space in policy debates. Unlike established CSS analysis where people are addressed as passive research subjects generating data which can only meaningfully be analysed at scale, sousveillance analysis tends to be conducted on the micro-level, as, for example, in Akbari's account of Iranian women tracking the moral police through Tehran in order to avoid their scrutiny (Akbari, 2019), van Doorn's account of gig workers in Berlin collecting data to reverse-engineer a platform's fee structures and challenge its labour practices (Doorn, 2020) or AlgorithmWatch's construction of a crowdsourced credit check model in Germany (AlgorithmWatch, 2018).

Although they also employ social science methods and can be rigorous and reliable, the entire point of these sousveillance methods is that they do not scale: they are local and specific, devised in response to particular challenges. They constitute participatory action research, a methodology where the research subject sets the agenda and where the aim is advancing social justice. Such methods constitute a claim to the right to participate, both in research and in society: they are an assertion of the presence and rights of the research subject. It is worth considering the numerous obstacles that this kind of research meets when it claims policy relevance: it has traditionally been rejected as unsystematic, not scalable, and unreliable because it reflects a local, rather than generalised, understanding (Chambers, 2007). These methods can be seen as the antithesis of current CSS in that they present a contradictory set of assumptions about what constitutes reliability, policy-relevance and participation. They also raise the question as to whether CSS in its current policy- and optimisation-oriented form can align with social justice concerns or whether data governance in this sphere should be aiming for legal compliance and harm reduction.

References

Akbari, A. (2019). *Spatial data justice: Mapping and digitised strolling against moral police in Iran* (No. 76; Development Informatics Working Paper). University of Manchester. https://papers.ssrn.com/sol3/papers.cfm?abstract_id=3460224

AlgorithmWatch. (2018). SCHUFA, a black box: OpenSCHUFA results published. *AlgorithmWatch*. https://algorithmwatch.org/en/schufa-a-black-box-openschufa-results-published/

Baldwin, R. (2014). From regulation to behaviour change: Giving nudge the third degree: Giving nudge the third degree. *The Modern Law Review, 77*(6), 831–857. https://doi.org/10.1111/1468-2230.12094

Boyd, D., & Crawford, K. (2012). Critical questions for big data: Provocations for a cultural, technological, and scholarly phenomenon. *Information, Communication & Society, 15*(5), 662–679.

Chambers, R. (2007). *Who counts? The quiet revolution of participation and numbers* (p. 45). Institute of Development Studies.

Couldry, N., & Mejias, U. A. (2019). Data colonialism: Rethinking big data's relation to the contemporary subject. *Television & New Media, 20*(4), 336–349. https://doi.org/10.1177/1527476418796632

Dalton, C. M., Taylor, L., & Thatcher, J. (2016). Critical data studies: A dialog on data and space. *Big Data & Society, 3*(1), 205395171664834. https://doi.org/10.1177/2053951716648346

Datta, A., & Odendaal, N. (2019). *Smart cities and the banality of power*. SAGE Publications Sage UK.

Delacroix, S., & Lawrence, N. D. (2019). Bottom-up data trusts: Disturbing the 'one size fits all' approach to data governance. *International Data Privacy Law, 9*(4), 236–252.

Dencik, L., Hintz, A., & Cable, J. (2016). Towards data justice? The ambiguity of anti-surveillance resistance in political activism. *Big Data & Society, 3*(2), 205395171667967. https://doi.org/10.1177/2053951716679678

Dencik, L., Hintz, A., Redden, J., & Warne, H. (2018). *Data scores as governance: Investigating uses of citizen scoring in public services project report* [Project Report]. Data Justice Lab. http://orca.cf.ac.uk/117517/1/data-scores-as-governance-project-report2.pdf

Deutsche Welle, D. (2020, December 29). *Turkey tightens control over NGOs to 'combat terrorism'|DW|29.12.2020*. DW.COM. https://www.dw.com/en/turkey-tightens-control-over-ngos-to-combat-terrorism/a-56088205

EDPS. (2019). *European Data Protection Supervisor, communication to European Asylum Support Office: Formal consultation on EASO's social media monitoring reports (case 2018–1083)*. European Data Protection Supervisor. https://edps.europa.eu/sites/edp/files/publication/19-11-12_reply_easo_ssm_final_reply_en.pdf

EUREC. (2021). *EUREC - Home*. European Network of Research Ethics Committees - EUREC. http://www.eurecnet.org/index.html

European Commission. (2020). *Proposal for a regulation of the European Parliament and of the Council on European data governance (Data Governance Act) COM(2020) 767 final*. https://eur-lex.europa.eu/legal-content/EN/TXT/PDF/?uri=CELEX:52020PC0767

Ferretti, L., Wymant, C., Kendall, M., Zhao, L., Nurtay, A., Abeler-Dörner, L., Parker, M., Bonsall, D., & Fraser, C. (2020). Quantifying SARS-CoV-2 transmission suggests epidemic control with digital contact tracing. *Science, 368*(6491). https://doi.org/10.1126/science.abb6936

Gangadharan, S. P., & Niklas, J. (2019). Decentering technology in discourse on discrimination. *Information, Communication & Society, 22*(7), 882–899. https://doi.org/10.1080/1369118X.2019.1593484

Gitelman, L. (Ed.). (2013). *'Raw data' is an oxymoron*. The MIT Press.

Heeks, R. (2017). *A structural model and manifesto for data justice for international development*. Development Informatics Working Paper no. 69, Available at SSRN: https://ssrn.com/abstract=3431729 or http://dx.doi.org/10.2139/ssrn.3431729

International Justice Resource Center. (2012, October 10). *Asylum & the rights of refugees*. https://ijrcenter.org/refugee-law/

Kitchin, R. (2014). Big data, new epistemologies and paradigm shifts. *Big Data & Society, 1*(1), 205395171452848. https://doi.org/10.1177/2053951714528481

Kulynych, B., Overdorf, R., Troncoso, C., & Gürses, S. (2020). POTs: Protective Optimization Technologies. In: *ACM FAT* 2019*. Conference on Fairness, Accountability, and Transparency (FAT*), Atlanta, GA, USA.

Lyon, D. (2014). Surveillance, Snowden, and big data: Capacities, consequences, critique. *Big Data & Society, 1*(2), 205395171454186. https://doi.org/10.1177/2053951714541861

Metcalf, J., & Crawford, K. (2016). Where are human subjects in big data research? The emerging ethics divide. *Big Data & Society, 3*(1), 205395171665021. https://doi.org/10.1177/2053951716650211

Micheli, M., Ponti, M., Craglia, M., & Berti Suman, A. (2020). Emerging models of data governance in the age of datafication. *Big Data & Society, 7*(2). https://doi.org/10.1177/2053951720948087

Pentland, A. (2010). Preface: A God's eye view. In A. Pentland (Ed.), *Honest signals: How they shape our world* (pp. 1–15). MIT Press.

Philip, K., Irani, L., & Dourish, P. (2012). Postcolonial computing: A tactical survey. *Science, Technology, & Human Values, 37*(1), 3–29.

Pindolia, D. K., Garcia, A. J., Wesolowski, A., Smith, D. L., Buckee, C. O., Noor, A. M., Snow, R. W., & Tatem, A. J. (2012). Human movement data for malaria control and elimination strategic planning. *Malaria Journal, 11*(1), 205. https://doi.org/10.1186/1475-2875-11-205

Rainie, S. C., Kukutai, T., Walter, M., Figueroa-Rodríguez, O. L., Walker, J., & Axelsson, P. (2019). Indigenous data sovereignty. In T. Davies & B. Walker (Eds.), *The state of open data: Histories and horizons* (pp. 300–319). African Minds and International Development Research Centre.

Reece, S., Duvell, F., Vargas-Silva, C., & Kone, Z. (2019). New approaches to the study of spatial mobility and economic integration of refugees in Turkey. In: *Data for refugees challenge workshop*. D4R, Boğaziçi University.

Simpson, A. (2017). The ruse of consent and the anatomy of 'refusal': Cases from indigenous North America and Australia. *Postcolonial Studies, 20*(1), 18–33.

Slaughter, R. K. (2021). *Algorithms and economic justice: A taxonomy of harms and a path forward for the federal trade commission* (ISP Digital Future Whitepaper & YJoLT Special Publication). Yale University. https://law.yale.edu/sites/default/files/area/center/isp/documents/algorithms_and_economic_justice_master_final.pdf

Suman, A. B. (2019). Between freedom and regulation: Investigating community standards for enhancing scientific robustness of citizen science. In L. Reins (Ed.), *Regulating new technologies in uncertain times* (pp. 31–46). Springer.

Suman, A. B. (2021). Citizen sensing from a legal standpoint: Legitimizing the practice under the Aarhus framework. *Journal for European Environmental & Planning Law, 18*(1–2), 8–38.

Tau, B. (2021, June 24). WSJ News Exclusive|App taps unwitting users abroad to gather open-source intelligence. *Wall Street Journal*. https://www.wsj.com/articles/app-taps-unwitting-users-abroad-to-gather-open-source-intelligence-11624544026

Taylor, L. (2016). The ethics of big data as a public good: Which public? Whose good? *Philosophical Transactions of the Royal Society A: Mathematical, Physical and Engineering Sciences, 374*(2083), 20160126.

Taylor, L. (2017). What is data justice? The case for connecting digital rights and freedoms globally. *Big Data & Society, 4*(2), 205395171773633. https://doi.org/10.1177/2053951717736335

Taylor, L. (2021). Public actors without public values: Legitimacy, domination and the regulation of the technology sector. *Philosophy & Technology, 34*, 897–922. https://doi.org/10.1007/s13347-020-00441-4

Taylor, L., & Broeders, D. (2015). In the name of development: Power, profit and the datafication of the global south. *Geoforum, 64*, 229–237.

Taylor, L., & De Souza, S. (2021). *Should might make right? On data, norms and justice*. UN Data Forum Blog Series. https://unstats.un.org/unsd/undataforum/blog/should-might-make-right-on-data-norms-and-justice/

Taylor, L., Floridi, L., & van der Sloot, B. (2017). *Group privacy: New challenges of data technologies*. Springer International Publishing.

Taylor, L., & Meissner, F. (2020). A crisis of opportunity: Market-making, big data, and the consolidation of migration as risk. *Antipode, 52*(1), 270–290.

United Nations. (2014). *A world that counts*. Report prepared at the request of the United Nations Secretary-General, by the Independent Expert Advisory Group on a Data Revolution for Sustainable Development. https://www.undatarevolution.org/wp-content/uploads/2014/11/A-World-That-Counts.pdf

Van Dijck, J. (2014). Datafication, dataism and dataveillance: Big data between scientific paradigm and ideology. *Surveillance & Society, 12*(2), 197–208.

van Doorn, N. (2020). At what price? Labour politics and calculative power struggles in on-demand food delivery. *Work Organisation, Labour & Globalisation, 14*(1), 136–149.

Wesolowski, A., Buckee, C. O., Bengtsson, L., Wetter, E., Lu, X., & Tatem, A. J. (2014). Commentary: Containing the Ebola outbreak - the potential and challenge of mobile network data. *PLoS Currents.*https://doi.org/10.1371/currents.outbreaks.0177e7fcf52217b8b634376e2f3efc5e

Zalnieriute, M. (2021). *Protests and public space surveillance: From metadata tracking to facial recognition technologies* (p. 9). Human Rights Council. [Submission to the Thematic Report to the 50th Session of the Human Rights Council]. https://papers.ssrn.com/sol3/papers.cfm?abstract_id=3882317

Chapter 4
The Ethics of Computational Social Science

David Leslie

Abstract This chapter is concerned with setting up practical guardrails within the research activities and environments of Computational Social Science (CSS). It aims to provide CSS scholars, as well as policymakers and other stakeholders who apply CSS methods, with the critical and constructive means needed to ensure that their practices are ethical, trustworthy, and responsible. It begins by providing a taxonomy of the ethical challenges faced by researchers in the field of CSS. These are challenges related to (1) the treatment of research subjects, (2) the impacts of CSS research on affected individuals and communities, (3) the quality of CSS research and to its epistemological status, (4) research integrity, and (5) research equity. Taking these challenges as motivation for cultural transformation, it then argues for the incorporation of end-to-end habits of Responsible Research and Innovation (RRI) into CSS practices, focusing on the role that contextual considerations, anticipatory reflection, impact assessment, public engagement, and justifiable and well-documented action should play across the research lifecycle. In proposing the inclusion of habits of RRI in CSS practices, the chapter lays out several practical steps needed for ethical, trustworthy, and responsible CSS research activities. These include stakeholder engagement processes, research impact assessments, data lifecycle documentation, bias self-assessments, and transparent research reporting protocols.

4.1 Introduction

Since its inception, one of the great promises of Computational Social Science (CSS) has been the possibility of leveraging a variety of algorithmic techniques to gain insights and identify patterns in big social data that would have otherwise been unavailable to the researchers and policymakers who had to draw

D. Leslie (✉)
The Alan Turing Institute, London, UK
e-mail: dleslie@turing.ac.uk

E. Bertoni et al. (eds.), *Handbook of Computational Social Science for Policy*,
https://doi.org/10.1007/978-3-031-16624-2_4

on more traditional, non-computational approaches to the study of society. By applying computational methods to the vast amounts of data generated from today's complex, digitised, and datafied society, CSS for policy is well placed to generate empirically grounded inferences, explanations, theories, and predictions about human behaviours, networks, and social systems, which not only effectively manage the volume and high dimensionality of big data but also, in fact, draw epistemic advantage from their unprecedented breadth, quantity, depth, and scale. This chapter is concerned with fleshing out the myriad ethical challenges faced by this endeavour. It aims to provide CSS scholars, as well as policymakers and other stakeholders who apply CSS methods, with the critical and constructive means needed to ensure that their research is ethical, trustworthy, and responsible.

Though some significant attempts to articulate the ethical stakes of CSS have been made by scholars and professional associations over the past two decades,[1] the scarcity of ethics in the mainstream labours of CSS, and across its history,[2] signals a general lack of awareness that is illustrative of several problematic dimensions of current CSS research practices that will motivate the arguments presented in this chapter. It is illustrative insofar as the absence of an active recognition of the ethical issues surrounding the social practice and wider human impacts of CSS may well shed light on a troublesome disconnection that persists between the self-understanding of CSS researchers who implicitly see themselves largely as neutral and disinterested scientists operating within the pure, self-contained confines of the laboratory or lecture hall, on the one hand, and the lived reality of their existence as contextually situated scholars whose framings, subject matters, categories, and methods have been forged in the crucible of history, society, and culture, on the other.

---

[1] See, for instance, the series of Association of Internet Researchers (AoIR) guidelines on internet research ethics published in 2002, 2012, and 2019 as well as the British Sociological Association (BSA) guidance. For scholarly interventions, see (Collmann & Matei, 2016; Dobrick et al., 2018; Ess & Jones, 2004; Eynon et al., 2017; Franzke et al., 2020; Giglietto et al., 2012; Hollingshead et al., 2021; Lomborg, 2013; Markham & Buchanan, 2012; Moreno et al., 2013; Salganik, 2019; Weinhardt, 2020).

[2] For example, Across the four volumes of Nigel Gilbert's magisterial Computational Social Science (2010), none of the 66 contributing chapters are dedicated to ethics. Likewise, no explicit mention or discussion of research ethics appears in Conte et al. (2012). There are only two passing mentions of ethics in the 10 chapters of Cioffi-Revilla's substantial Introduction to Computational Social Science (2014), and the word "ethics" also appears only twice (and only in the final chapter) of Chen's edited volume, Big Data for the Computational Social Sciences and Humanities (2018).

To be sure, when CSS researchers assume a scientistic "view from nowhere"[3] and regard the objects of their study solely through quantitative and computational lenses, they run two significant risks: First, they run the risk of assuming positivistic attitudes that frame the objects of their study through the quantifying and datafying lenses of models, formalisms, behaviours, networks, simulations, and systems thereby setting aside or trivialising ethical considerations in an effort to get to the real science without further ado. When the objects of the study of CSS are treated solely as elements of automated information analysis rather than as human subjects—each of whom possesses a unique dignity and is thus, first and foremost, worthy of moral regard and interpretive care—scientistic subspecies of CSS are liable to run roughshod over fundamental rights and freedoms like privacy, autonomy, meaningful consent, and non-discrimination with a blindered view to furthering computational insight and data quantification (Fuchs, 2018; Hollingshead et al., 2021). Second, they risk seeing themselves as operationally independent or even immune from the conditioning dynamics of the social environments they study and in which their own research activities are embedded (Feenberg, 1999, 2002). This can create conditions of deficient reflexivity—i.e., defective self-awareness of the limitations of one's own standpoint—and ethical precarity (Leslie et al., 2022a). As John Dewey long ago put it, "the notion of the complete separation of science from the social environment is a fallacy which encourages irresponsibility, on the part of scientists, regarding the social consequences of their work" (Dewey, 1938, p. 489).

In the case of CSS for policy, the price of this misperceived independence of researchers from the formative dynamics of their sociohistorical environment has been extremely high. CSS practices have developed and matured in an age of unprecedented sociotechnical sea change—an age of unbounded digitisation, datafication, and mediatisation. The cascading societal effects of these revolutionary transformations have, in fact, directly shaped and implicated CSS in its research trajectories, motivations, objects, methods, and practices. The rise of the veritably limitless digitisation and datafication of social life has brought with it a corresponding impetus—among an expanding circle of digital platforms, private corporations, and governmental bodies—to engage in behavioural capture and manipulation at scale. In this wider societal context, the aggressive extraction and harvesting of data from the digital streams and traces generated by human activities, more often than not, occur without the meaningful consent or active awareness of the people whose

---

[3] As Sorell (2013) has argued, scientism is typified by the privileging of natural or exact scientific language, knowledge, and methods over those of other branches of learning and culture, especially those of the "human sciences" like philosophy, ethics, history, anthropology, and sociology. Such a privileging of exact scientific "ideas, methods, practices, and attitudes" can be especially damaging where these are extended "to matters of human social and political concern" (Olson, 2008, p. 1)—matters that require an understanding of subtle historical, ethical, and sociocultural contexts, contending human values, norms, and purposes, and subjective meaning-complexes of action and interaction (Apel, 1984; Habermas, 1988; Taylor, 2021; von Wright, 2004; Weber, 1978; Wittgenstein, 2009).

digital and digitalised lives[4] are the targets of increasing surveillance, consumer curation, computational herding, and behavioural steering. Such extractive and manipulative uses of computational technologies also often occur neither with adequate reflection on the potential transformative effects that they could have on the identity formation, agency, and autonomy of targeted data subjects nor with appropriate and community-involving assessment of the adverse impacts they could have on civic and social freedoms, human rights, the integrity of interpersonal relationships, and communal and biospheric well-being.

The real threat here, for CSS, is that the prevailing "move fast and break things" attitude possessed by the drivers of the "big data revolution", and by the beneficiaries of its financial and administrative windfalls, will simply be transposed into the key of the data-driven research practices they influence, making a "research fast and break things" posture a predominant disposition. This threat to the integrity of CSS research activity, in fact, derives from the potentially inappropriate dependency relationships which can emerge from power imbalances that exist between the CSS community of practice and those platforms, corporations, and public bodies who control access to the data resources, compute infrastructures, project funding opportunities, and career advancement prospects upon which CSS researchers rely for their professional viability and endurance. Here, the misperceived independence of researchers from their social environments can mask toxic and agenda-setting dependencies.

Taken together, these downstream hazards signal potential deficits in the social responsibility, trustworthiness, and ethical permissibility of its practices. To confront such hazards, this chapter will first provide a taxonomy of ethical challenges faced by CSS researchers. These are (1) challenges related to the treatment of research subjects, (2) challenges related to the impacts of CSS research on affected individuals and communities, (3) challenges related to the quality of CSS research and to its epistemological status, (4) challenges related to research integrity, and (5) challenges related to research equity. Taking these challenges as a motivation for cultural transformation, it will then argue for the incorporation into CSS practices of end-to-end habits of Responsible Research and Innovation (RRI), focusing, in particular, on the role that contextual considerations, anticipatory reflection, public engagement, and justifiable and well-documented action should play across the research lifecycle. The primary goal of this focus on RRI is to centre the understanding of CSS as "science with and for society" and to foster, in turn, critical self-reflection about the consequential role that human values, norms, and purposes play in its discovery and design processes and in considerations of the real-world effects of the insights and tools that these processes yield. In proposing the inclusion of habits of RRI in CSS practices, the chapter lays out several practical steps needed for ethical, trustworthy, and responsible CSS research activities. These include stakeholder engagement processes, research impact assessments, data lifecycle documentation, bias self-assessments, and transparent research reporting protocols.

---

[4] The use of the terms 'digital' and 'digitalised' follows Lazer & Radford (2017).

4.2   Ethical Challenges Faced by CSS

A preliminary step needed to motivate the centring of Responsible Research and Innovation practices in CSS is the identification of the range of ethical challenges faced by its researchers. These challenges can be broken down into five categories:

1. *Challenges related to the treatment of research subjects.* These challenges have to do with the interrelated aspects of confidentiality, data privacy and protection, anonymity, and informed consent.
2. *Challenges related to the impacts of CSS research on affected individuals and communities.* These challenges cover areas such as the potential adverse impacts of CSS research activities on the respect for human dignity and on other fundamental rights and freedoms.
3. *Challenges related to the quality of CSS research and to its epistemological status.* Challenges related to the quality of CSS research include erroneous data linkage, dubious "ideal user assumptions", the infusion of algorithmic influence in observational datasets of digital traces, the "illusion of the veracity of volume", and blind spots vis-à-vis non-human data generation that undermine data quality and integrity. Challenges related to the epistemological status of CSS include the inability of computation-driven techniques to fully capture non-random missingness in datasets and sociocultural conditions of data generation and hence a broader tendency to potentially misrepresent the real social world in the models, simulations, analyses, and predictions it generates.
4. *Challenges related to research integrity.* These challenges are rooted in the asymmetrical dynamics of resourcing and influence that can emerge from power imbalances between the CSS research community and the corporations and public agencies upon whom CSS scholars rely for access to the data resources, compute infrastructures, project funding opportunities, and career advancement prospects they need for their professional subsistence and advancement.
5. *Challenges related to research equity.* These challenges include the potential reinforcement of digital divides and data inequities through biased sampling techniques that render digitally marginalised groups invisible as well as potential aggregation biases in research results that mask meaningful differences between studied subgroups and therefore hide the existence of real-world inequities. Research equity challenges may also derive from long-standing dynamics of regional and global inequality that may undermine reciprocal sharing between research collaborators from more and less resourced geographical areas, universities, or communities of practice.

Let us expand on each of these challenges in turn.

4.2.1 Challenges Related to the Treatment of Research Subjects

When identifying and exploring challenges related to the treatment of research subjects in CSS, it is helpful to make a distinction between participation-based and observation-based research, namely, between CSS research that is gathering data directly from research subjects through their deliberate involvement in digital media (e.g., research that uses online methods to gather data by way of human involvement in surveys, experiments, or participatory activities) and CSS research that is investigating human action and social interaction in observed digital environments, like social media or search platforms, through the recording, measurement, and analysis of digital life, digital traces, and digitalised life (Eynon et al., 2017). Though participation-based and observation-based research raise some overlapping issues related to privacy and data protection, there are notable differences that yield unique challenges.

Several general concerns about privacy preservation, data protection, and the responsible handling and storage of data are common to participation-based and observation-based CSS research. This is because empirical CSS research often explores topics that require the collection, analysis, and management of personal data, i.e., data that can uniquely identify individual human beings. Although CSS research frequently spans different jurisdictions, which may have diverging privacy and data protection laws, responsible research practices that aim to optimally protect the rights and interests of research subjects in light of risks posed to confidentiality, privacy, and anonymity should recur to the highest standards of privacy preservation, data protection, and the responsible handling and storage of data. They should also establish and institute proportionate protocols for attaining informed and meaningful consent that are appropriate to the specific contexts of the data extraction and use and that cohere with the reasonable expectations of the targeted research subjects.

Notwithstanding this common footing for ethics considerations related to data protection and the privacy of research subjects, participation-based and observation-based approaches to CSS research each raise distinctive issues. For researchers who focus on online observation or who use data captured from digital traces or data extracted from connected mobile devices, the Internet of Things, public sensors and recording devices, or networked cyber-physical systems, coming to an appropriate understanding of the reasonable expectations of research subjects regarding their privacy and anonymity is a central challenge. When observed research subjects move through their synchronous digital and connected environments striving to maintain communication flows and coherent social interactions, they must navigate moment-to-moment choices about the disclosure of personal information (Joinson et al., 2007). In physical public spaces and in online settings, the perception of anonymity (i.e., of the ability to speak and act freely without feeling like one is continuously being identified or under constant watch) is an important precondition of frictionless information exchange and, correspondingly, of the exercise of freedoms of movement, expression, speech, assembly, and association (Jiang, 2013; Paganoni, 2019; Selinger & Hartzog, 2020).

On the internet, moreover, an increased sense of anonymity may lead data subjects to more freely disclose personal information, opinions, and beliefs that they may not have shared in offline milieus (Meho, 2006). In all these instances of perceived anonymity, research subjects may act under reasonable expectations of gainful obscurity and "privacy in public" (Nissenbaum, 1998; Reidenberg, 2014). These expectations are responsive to and bounded by the changing contexts of communication, namely, by contextual factors like who one is interacting with, how one is exchanging information, what type of information is being exchanged, how sensitive it is perceived to be, and where and when such exchanges are occurring (Quan-Haase & Ho, 2020). This means, not only, that the protection of privacy must, first and foremost, consider contextual determinants (Collmann & Matei, 2016; Nissenbaum, 2011; Steinmann et al., 2015). It also implies that privacy protection considerations must acknowledge that the privacy preferences of research subjects can change from circumstance to circumstance and are therefore not one-off or one-dimensional decisions that can be made at the entry point to the usage of digital or social media applications through Terms of Service or end-user license agreements—which often go unread—or the initial determination of privacy settings (Henderson et al., 2013). For this reason, the conduct of observation-based research in CSS that pertains to digital and digitalised life should be informed by contextual considerations about the populations and social groups from whom the data are drawn, the character and potential sensitivities of their data, the nature of the research question (as it may be perceived by observed research subjects), research subjects' reasonable expectations of privacy in public, and the data collection practices and protocols of the organisation or company which has extracted the data (Hollingshead et al., 2021). Notably, thorough assessment of these issues by members of a research team may far exceed formal institutional processes for gaining ethics approval, and it is the responsibility of CSS researchers to evaluate the appropriate scale and depth of privacy considerations regardless of minimal legal and institutional requirements (Eynon et al., 2017; Henderson et al., 2013).

Apart from these contextual considerations, the protection of the privacy and anonymity of CSS research subjects also requires that risks of re-identification through triangulation and data linkage are anticipated and addressed. While processes of anonymisation and removal of personally identifiable information from datasets scraped or extracted from digital platforms and digitalised behaviour may seem straightforward when those data are treated in isolation, multiple sources of linkable data points and multiple sites of downstream data collection pose tangible risks of re-identification via the combination and linkage of datasets (de Montjoye et al., 2015; Eynon et al., 2017; Obole & Welsh, 2012). As Narayanan & Shmatikov (2009) and de Montjoye et al. (2015) both demonstrate, the inferential triangulation of social data collected from just a few sources can lead to re-identification even under conditions where datasets have been anonymised in the conventional, single dataset sense. Moreover, when risks of triangulation and re-identification are considered longitudinally, downstream risks of de-anonymisation also arise. In this case, the endurance of the public accessibility of social data on the internet over time means that information that could lead to re-identification is ready-to-hand indefinitely. By the same token, the production and extraction of new data that post-dates the creation and use of anonymised datasets also present downstream opportunities for data linkage and inference creep that can lead to re-identification through unanticipated triangulation (Weinhardt, 2020).

Although many of these privacy and data protection risks also affect participation-based research (especially in cases where observational research is combined or integrated with it), experimental and human-involving CSS projects face additional challenges. Signally, participation-based CSS research must confront several issues surrounding the ascertainment of informed and meaningful consent. The importance of consent has been a familiar part of the "human subjects" paradigm of research ethics from its earliest expressions in the World Medical Association (WMA) Declaration of Helsinki[5] and the Belmont Report.[6] However, the exponentially greater scale and societal penetration of CSS in comparison to more conventional forms of face-to-face, survey-driven, or laboratory-based social scientific research present a new order of hazards and difficulties. First, since CSS researchers, or their collaborators, often control essential digital infrastructure like social media platforms, they have the capability to efficiently target and experiment on previously unimaginable numbers of human subjects, with potential N's approaching magnitudes of hundreds of thousands or even millions of people. Moreover, in the mould of such platforms, these researchers have an unprecedented capacity to manipulate or surreptitiously intervene in the unsuspecting activities and behaviours of such large, targeted groups.

The controversy around the 2014 Facebook emotional contagion experiment demonstrates some of the potential risks generated by this new scale of research capacity (Grimmelmann, 2015; Lorenz, 2014; Puschmann & Bozdag, 2014). In the study, researchers from Facebook, Cornell, and the University of California involved almost 700,000 unknowing Facebook users in what has since been called a "secret mood manipulation experiment" (Meyer, 2014). Users were split into two experimental groups and exposed to negative or positive emotional content to test whether News Feed posts could spread the relevant positive or negative emotion. Critics of the approach soon protested that the failure to obtain consent—or even to inform research subjects about the experiment—violated basic research ethics. Some also highlighted the dehumanising valence of these research tactics: "To Face-book, we are all lab rats", wrote Vindu Goel in the *New York Times* (Goel, 2014). Hyperbole aside, this latter comment makes explicit the internal logic of many of

---

[5] https://www.wma.net/policies-post/wma-declaration-of-helsinki-ethical-principles-for-medical-research-involving-human-subjects/

[6] https://www.hhs.gov/ohrp/regulations-and-policy/belmont-report/index.html

the moral objections to the experiment that were voiced at the time. The Facebook researchers had blurred the relationship between the laboratory and the lifeworld. They had, in effect, unilaterally converted the social world of people connecting and interacting online into a world of experimental objects that subsisted merely as standing reserve for computational intervention and study—a transformation of the interpersonally animated life of the community into the ethically impoverished terrain of an "information laboratory" (Cohen, 2019a). Behind such a degrading conversion was the assertion of the primacy of objectifying and scientistic attitudes over considerations of the equal moral status and due ethical regard of research subjects. The experiment had, on the critical view, *reduced Facebook users to the non-human standing of laboratory rodents*, thereby disregarding their dignity and autonomy and consequently failing to properly consult them so to attain their informed consent to participate.

Even when the consent of research participants is sought by CSS researchers, a few challenges remain. These revolve around the question of how to ensure that participants are fully informed so that they can freely, meaningfully, and knowledgeably consent to their involvement in the research (Franzke et al., 2020). Though diligent documentation protocols for gaining consent are an essential element of ascertaining informed and meaningful consent in any research environment, in the digital or online milieus of CSS, the provision of this kind of text-based information is often inadequate. When consent documentation is provided in online environments through one-way or vertical information flows that do not involve real, horizontal dialogue between researchers and potential research subjects, opportunities to clarify possible misunderstandings of the terms of consent can be lost (Varnhagen et al., 2005). What is more, it becomes difficult under these conditions of incomplete or impeded communication to confirm that research subject actually comprehend what they are agreeing to do as research participants (Eynon et al., 2017). Relatedly, barriers to information exchange in the online environment can prevent researchers from being able to verify the capacity of research subjects to consent freely and knowledgeably (Eynon et al., 2017; Kraut et al., 2004). That is, it is more difficult to detect potential limitations of or impairments in the competence of participants (e.g., from potentially vulnerable subgroups) in giving consent where researchers are at a significant digital remove from research subjects. In all these instances, various non-dialogical techniques for confirming informed consent are available—such as comprehension tests, smart forms that employ branching logic to ensure essential text is completely read, identity verification, etc. Such techniques, however, present varying degrees of uncertainty and drop-out risk (Kraut et al., 2004; Varnhagen et al., 2005), and they do not adequately substitute for interactive mechanisms that could connect researchers directly with participants and their potential questions and concerns.

4.2.2   Challenges Related to the Impacts of CSS Research on Affected Individuals and Communities

While drawing on the formal techniques and methods of mathematics, statistics, and the exact sciences, CSS is a research practice that is policy-oriented, problem-driven, and societally consequential. As an applied science that directly engages with issues of immense social concern like socioeconomic inequality, the spread of infectious disease, and the growth of disinformation and online harm, it impacts individuals and communities with the results, capabilities, and tools it generates. Moreover, CSS is an "instrument-enabled science" (Cioffi-Revilla, 2014, p. 4) that employs computational techniques, which can be applied to large-scale datasets excavated from veritably all societal sectors and spheres of human activity and experience. This makes its researchers the engineers and custodians of a *general purpose research technology* whose potential scope in addressing societal challenges is seemingly unbounded. With this in view, Lazer et al. (2020) call for the commitment of "resources, from public and private sources, that are extraordinary by current standards of social science funding" to underwrite the rapid expansion of CSS research infrastructure, so that its proponents can enlarge their quest to "solve real-world problems" (p. 1062). Beyond the dedication of substantial resources, such an expansion, Lazer et al. (2020) argue, also requires the formulation of "policies that would encourage or mandate the ethical use of private data that preserves public values like privacy, autonomy, security, human dignity, justice, and balance of power to achieve important public goals—whether to predict the spread of disease, shine a light on societal issues of equity and access, or the collapse of the economy" (p. 1061). CSS, along these lines, is not simply an applied social science, a *science for policy*. It is a social impact science *par excellence*.

The mission-driven and impact-oriented perspective conveyed here is, however, a double-edged sword. On the one hand, the drive to improve the human lot and to solve societal problems through the fruits of scientific discovery has constructively guided the impetus of modern scientific research and innovation at least since the seventeenth-century dawning of the Baconian and Newtonian revolutions. In this sense, the practical and problem-solving aspirations for CSS expressed by Lazer et al. (2020) are continuous with a deeper tradition of societally oriented science.

On the other hand, the view that CSS is a mission-driven and impact-oriented science raises a couple of thorny ethical issues that are not necessarily solvable by the application of its own methodological and epistemic resources. First, the assumption of a mission-driven starting point surfaces a difficult set of questions about the relationship of CSS research to the values, interests, and power dynamics that influence the trajectories of its practice: *Whose* missions are driving CSS and w*hose* values and interests are informing the policies that are guiding these missions? To what extent are these values and interests shared by those who are likely to be impacted by the research? To what extent do these values and interests, and the policies they shape, sufficiently reflect the plurality of values and interests that are possessed by members of communities who will potentially be affected by the research (especially those from historically marginalised, discriminated-against, and vulnerable social groups)? Are these missions determined through democratic and community-involving processes or do other parties (e.g., funders, research collaborators, resource providers, principal investigators, etc.) wield asymmetrical agenda-setting power in setting the direction of travel for the research and its outputs? Who are the beneficiaries of these mission-driven research projects and who are at risk of any adverse impacts that they could have? Are these potential risks and benefits equitably distributed or are some stakeholders disparately exposed to harm while others in positions of disproportionate advantage?

Taken together, these questions about the role that values, interests, and power dynamics play in shaping mission-driven research and its potential impacts evoke critical, though often concealed, interdependencies that exist between the CSS community of practice and the social environments in which its research activities, subject matters, and outputs are embedded. They likewise evoke the inadequacy of evasive scientistic tendencies to appeal to neutral or value-free stances when faced with queries about how values, interests, and power dynamics motivate and influence the aims, purposes, and areas of concern that steer vectors of CSS research. Responding appropriately to such questions surrounding the social determinants of research paths and potential impacts demands an inclusive broadening of the conversations that shape, articulate, and determine the missions to be pursued, the problems to be addressed, and the assessment of potential harms and benefits—a broadening both in terms of the types of knowledge and expertise that are integrated into such deliberative processes and in terms of the range of stakeholder groups that should be involved.

Second, the recognition of a mission-driven and impact-oriented starting point elevates the importance of *identifying the potential adverse effects of CSS research* so that these can, as far as possible, be pinpointed at the outset of research projects and averted. Such practices of anticipatory reflection are necessary because the intended and unintended consequences of the societally impactful insights, tools, and capabilities CSS research produces could be negative and injurious rather than positive and mission-supporting. As the short history of the "big data revolution" demonstrates, the rapid and widespread proliferation of algorithmic systems, data-driven technologies, and computation-led analytics has already had numerous deleterious effects on human rights, fundamental freedoms, democratic values, and biospheric sustainability. Such harmful effects have penetrated society at multiple levels including on the planes of individual agency, social interaction, and biospheric integrity. Let us briefly consider these levels in turn.

4.2.2.1 Adverse Impacts at the Individual Level

At the agent level, the predominance "radical behaviourist" attitudes among the academic, industrial, and governmental drivers of data innovation ecosystems have led to the pervasive mobilisation of individual-targeting predictive analytics which have had damaging impacts across a range of human activities (Cardon, 2016;

Cohen, 2019b; Zuboff, 2019). For instance, in the domain of e-commerce and ad-tech, strengthening regimes of consumer surveillance have fuelled the use of "large-scale behavioural technologies" (Ball, 2019) that have enabled incessant practices of hyper-personalised psychographic profiling, consumer curation, and behavioural nudging. As critics have observed, such technologies have tended to exploit the emotive vulnerabilities and psychological weaknesses of targeted people (Helbing et al., 2019), instrumentalising them as monetisable sites of "behavioural surplus" (Zuboff, 2019) and treating them as manipulable objects of prediction and "behavioural certainty" rather than as reflective subjects worthy of decision-making autonomy and moral regard (Ball, 2019; Yeung, 2017). Analogous behaviourist postures have spurred state actors and other public bodies to subject their increasingly datafied citizenries to algorithmic nudging techniques that aim to obtain aggregated patterns of desired behaviour which accord with government generated models and predictions (Fourcade & Gordon, 2020; Hern, 2021). Some scholars have characterised such an administrative ambit as promoting the paternalistic displacement of individual agency and the degradation of the conditions needed for the successful exercise of human judgment, moral reasoning, and practical rationality (Fourcade & Gordon, 2020; Spaulding, 2020).

In like manner, the nearly ubiquitous scramble to capture behavioural shares of user engagement across online search, entertainment, and social media platforms has led to parallel feedback loops of digital surveillance, algorithmic manipulation, and behavioural engineering (Van Otterlo, 2014). The proliferation of the so-called "attention market" business model (Wu, 2019) has prompted digital platforms to measure commercial success in terms of the non-consensual seizure and monopolisation of focused mental activity. This has fostered the deleterious attachment of targeted consumer populations to a growing ecosystem of "distraction technologies" (Syvertsen, 2020; Syvertsen & Enli, 2020) and compulsion-forming social networking sites and reputational platforms, consequently engendering, on some accounts, widespread forms of surveillant anxiety (Crawford, 2014), cognitive impairment (Wu, 2019), mental health issues (Banjanin et al., 2015; Barry et al., 2017; Lin et al., 2016; Méndez-Diaz et al., 2022; Peterka-Bonetta et al., 2019), and diminished adolescent self-esteem and quality of life (Scott & Woods, 2018; Viner et al., 2019; Woods & Scott, 2016).

4.2.2.2 Adverse Impacts at the Social Level

Setting aside the threats to basic individual dignity and human autonomy that these patterns of instrumentalisation, disempowerment, and exploitation present (Aizenberg & van den Hoven, 2020; Halbertal, 2015), the proliferation of data-driven behavioural steering at the collective level has also generated risks to the integrity of social interaction, interpersonal solidarity, and democratic ways of life. In current digital information and communication environments, for example, the predominant steering force of social media and search engine platforms has mobilised opaque computational methods of relevance ranking, popularity sorting, and trend predicting to produce calculated digital publics devoid of any sort of active participatory social or political choice (Beer, 2017; Bogost, 2015; Cardon, 2016; Gillespie, 2014; O'Neil, 2016; Striphas, 2015; Ziewitz, 2016). Rather than being guided by the deliberatively achieved political will of interacting citizens, this vast meshwork of connected digital services shapes these computationally fashioned publics in accordance with the drive to commodify monitored behaviour and to target and capture user attention (Carpentier, 2011; De Cleen & Carpentier, 2008; Dean, 2010; Fuchs, 2021; John, 2013; Zuckerman, 2020). And, as this manufacturing of digital publics is ever more pressed into the service of profit seeking by downstream algorithmic mechanisms of hyper-personalised profiling, engagement-driven filtering, and covert behavioural manipulation, democratic agency and participation-centred social cohesion will be increasingly supplanted by insidious forms of social sorting and digital atomisation (Vaidhyanathan, 2018; van Dijck, 2013; van Dijck et al., 2018). Combined with complimentary dynamics of wealth polarisation and rising inequality (Wright et al., 2021), such an attenuation of social capital, discursive interaction, and interpersonal solidarity is already underwriting the crisis of social and political polarisation, the widespread kindling of societal distrust, and the animus towards rational debate and consensus-based science that have come to typify contemporary post-truth contexts (Cosentino, 2020; D'Ancona, 2017; Harsin, 2018; McIntyre, 2018).

Indeed, as these and similar kinds of computation-based social sorting and management infrastructures continue to multiply, they promise to jeopardise more and more of the formative modes of open interpersonal communication that have enabled the development of crucial relations of mutual trust and responsibility among interacting individuals in modern democratic societies. This is beginning to manifest in the widespread deployment of algorithmic labour and productivity management technologies, where manager-worker and worker-worker relations of reciprocal accountability and interpersonal recognition are being displaced by depersonalising mechanisms of automated assessment, continuous digital surveillance and computation-based behavioural incentivisation, discipline, and control (Ajunwa et al., 2017; Akhtar & Moore, 2016; Kellogg et al., 2020; Moore, 2019). The convergence of the unremitting sensor-based tracking and monitoring of workers' movements, affects, word choices, facial expressions, and other biometric cues, with algorithmic models that purport to detect and correct defective moods, emotions, and levels of psychological engagement and well-being, may not simply violate a worker's sense of bodily, emotional, and mental integrity by rendering their inner life legible and available for managerial intervention as well as productivity optimisation (Ball, 2009). These forms of ubiquitous personnel tracking and labour management can also have so-called panoptic effects (Botan, 1996; Botan & McCreadie, 1990), causing people to alter their behaviour on suspicion it is being constantly observed or analysed and deterring the sorts of open worker-to-worker interactions that enable the development of reciprocal trust, social solidarity, and interpersonal connection. This labour management example merely signals a broader constellation of ethical hazards that are raised by the parallel use of sensor- and location-based surveillance, psychometric and physiognomic profiling (Agüera y Arcas et al., 2017; Barrett et al., 2019; Chen & Whitney, 2019; Gifford, 2020; Hoegen et al., 2019; Stark & Hutson, 2021), and computation-driven technologies of behavioural governance in areas like education (Andrejevic & Selwyn, 2020; Pasquale, 2020), job recruitment (Sánchez-Monedero et al., 2020; Sloane et al., 2022), criminal justice (Brayne, 2020; Pasquale & Cashwell, 2018), and border control (Amoore, 2021; Muller, 2019). The heedless deployment of these kinds of algorithmic systems could have transformative effects on democratic agency, social cohesion, and interpersonal intimacy, preventing people from exercising their freedoms of expression, assembly, and association and violating their right to participate fully and openly in the moral, cultural, and political life of the community.

4.2.2.3 Adverse Impacts at the Biospheric Level

Lastly, at the level of biospheric integrity and sustainability, the exploding computing power—which has played a major part in ushering in the "big data revolution" and the rise of CSS—has also had significant environmental costs that deserve ethical consideration. As Lannelongue et al. (2021) point out, "the contribution of data centers and high-performance computing facilities to climate change is substantial . . . with 100 megatonnes of CO_2 emissions per year, similar to American commercial aviation". At bottom, this increased energy consumption has hinged on the development of large, computationally intensive algorithmic models that ingest abundant amounts of data in their training and tuning, that undergo iterative model selection and hyperparameter experiments, and that require exponential augmentations in model size and complexity to achieve relatively modest gains in accuracy (Schwartz et al., 2020; Strubell et al., 2019). In real terms, this has meant that the amount of compute needed to train complex, deep learning models increased by 300,000 times in 6 years (from 2013 to 2019) with training expenditures of energy doubling every 6 months (Amodei & Hernandez, 2018; Schwartz et al., 2020). Strubell et al. (2019) observe, along these lines, that training Google's large language model, BERT, on GPU, produces substantial carbon emissions "roughly equivalent to a trans-American flight". Though recent improvements in algorithmic techniques, software, and hardware have meant some efficiency gains in the operational energy consumption of computationally hungry, state-of-the-art models, some have stressed that such training costs are increasingly compounded by the carbon emissions generated by hardware manufacturing and infrastructure (e.g., designing and fabricating integrated circuits) (Gupta et al., 2020). Regardless of the sources of emissions, important ethical issues emerge both from the overall contribution of data research and innovation practices to climate change and to the degradation of planetary health and from the differential distribution of the benefits and risks that derive from the design and use of computationally intensive models. As Bender et al. (2021) have emphasised, such allocations of benefits and risks have closely tracked the historical patterns of environmental racism, coloniality, and "slow violence" (Nixon, 2011) that have typified the disproportionate exposure of marginalised communities (especially those who inhabit what has conventionally been referred to as "the Global South") to the pollution and destruction of local ecosystems and to involuntary displacement.

As a whole, these cautionary illustrations of the hazards posed at individual, societal, and environmental levels by ever more ubiquitous computational interventions in the social world should impel CSS researchers to adopt an ethically sober and pre-emptive posture when reflecting on the potential impacts of their projects. The reason for this is not just that many of the methods, tools, capabilities, and epistemic frameworks that they utilise have already operated, in the commercial and political contexts of datafication, as accessories to adverse societal impacts. It is, perhaps more consequentially, that, as Wagner et al. (2021) point out, CSS practices of measurement and corollary theory construction in "algorithmically infused societies . . . indirectly alter behaviours by informing the development of social theories and subsequently influence the algorithms and technologies that draw on those theories" (p. 197). This dimension of the "performativity" of CSS research—i.e., the way that the activities and theories of CSS researchers can function to reformat, reorganise, and shape the phenomena that they purport only to measure and analyse—is crucial (Healy, 2015; Wagner et al., 2021). It enjoins, for instance, an anticipatory awareness that the methodological predominance of measurement-centred and prediction-driven perspectives in CSS can support the noxious proliferation of the scaled computational manipulation and instrumentalisation of large populations of affected people (Eynon et al., 2017; Schroeder, 2014). It also implores cognizance that an unreflective embrace of unbounded sociometrics and the pervasive sensor-based observation and monitoring of research subjects may support wider societal patterns of "surveillance creep" (Lyon, 2003; Marx, 1988) and ultimately have chilling effects on the exercise of fundamental rights and freedoms. The intractable endurance of these kinds of risks of adverse effects and the possibilities for unintended harmful consequences recommends vigilance both in the assessment of the potential impacts of CSS research on affected individuals and communities and in the dynamic monitoring of the effects of the research outputs, and the affordances they create, once these are released into the social world.

4.2.3 Challenges Related to the Quality of CSS Research and to Its Epistemological Status

CSS research that is of dubious quality or that misrepresents the world can produce societal harms by misleading people, misdirecting policies, and misguiding further academic research. Many of the pitfalls that can undermine CSS research quality are precipitated by deficiencies in the accuracy and the integrity of the datasets on which it draws. First off, erroneous data linkage can lead to false theories and conclusions. Researchers face ongoing challenges when they endeavour to connect the data generated by identified research subjects to other datasets that are believed to include additional information about those individuals (Weinhardt, 2020). Mismatches can poison downstream inferences in undetectable ways and lead to model brittleness, hampered explanatory power, and distorted world pictures.

The poisoning of inferences by corrupted, inaccurate, invalid, or unreliable datasets can occur in a few other ways. Where CSS researchers are not sufficiently critical of the "ideal user assumption" (Lazer & Radford, 2017), they can overlook instances in which data subjects intentionally mispresent themselves, subsequently perverting the datasets in which they are included. For example, online actors can multiply their identities as "sock puppets" by creating fake accounts that serve different purposes; they can also engage in "gaslighting" or "catfishing" where intentional methods of deception about personal characteristics and misrepresentation of identities are used to fool other users or to game the system; they can additionally impersonate real internet users to purposefully mislead or exploit others (Bu et al., 2013; Ferrara, 2015; Lazer & Radford, 2017; Wang et al., 2006; Woolley, 2016; Woolley & Howard, 2018; Zheng et al., 2006). Such techniques of deception can be automated or deployed using various kinds of robots (e.g., chat bots, social media bots, robocalls, spam bots, etc.) (Ferrara et al., 2016; Gupta et al., 2015; Lazer & Radford, 2017; Ott et al., 2011). If researchers are not appropriately attentive to the distortions that may arise in datasets as a result of such non-human sources of misleading data, they can end up unintentionally baking the corresponding corruptions of the underlying distribution that are present in the sample into their models and theories, thereby misrepresenting or painting a false picture of the social world (Ruths & Pfeffer, 2014; Shah et al., 2015). Similar blind spots in detecting dataset corruption can arise when sparse attention is paid to how the algorithms, which pervade the curation and delivery of information on online platforms, affect and shape the data that is generated by the users that they influence and steer (Wagner et al., 2021).

Attentiveness to such data quality and integrity issues can be hindered by the illusion of the veracity of volume or, what has been termed, "big data hubris" (Hollingshead et al., 2021; Kitchin, 2014; Lazer et al., 2014; Mahmoodi et al., 2017). This is the misconception that, in virtue of their sheer volume, big data can "solve all problems", including potential deficiencies in data quality, sampling, and research design (Hollingshead et al., 2021; Meng, 2018). When it is believed that "data quantity is a substitute for knowledge-driven methodologies and theories" (Mahmoodi et al., 2017, p. 57), the rigorous and epistemically vetted approaches to social measurement, theory construction, explanation, and understanding that have evolved over decades in the social sciences and statistics can be perilously neglected or even dismissed.

Such a potential impoverishment of epistemic vigour can also result when CSS researchers fall prey to the enticements of the flip side of big data hubris, namely, computational solutionism. Predispositions to computational solutionism have emerged as a result of the coalescence of the rapid growth of computing power and the accelerating development of complex algorithmic modelling techniques that have together complemented the explosion of voluminous data and the big data revolution. This new access to the computational tools availed by potent compute and high-dimensional algorithmic machinery have led to the misconception in some corners of CSS that tools themselves can, by and large, "solve all problems". Rather than confronting the contextual complexities that lie behind the social processes and historical conditions that generate observational data (Shaw, 2015; Törnberg & Uitermark, 2021), and that concomitantly create manifold possibilities for non-random missingness and meaningful noise, the computational solutionist reverts to a toolbox of heuristic algorithms and technical tricks to "clean up" the data, so that computational analysis can forge ahead frictionlessly (Agniel et al., 2018; Leonelli, 2021). At heart, this contextual sightlessness among some CSS researchers originates in scientistic attitudes that tend to naturalise and reify digital trace data (Törnberg & Uitermark, 2021), treating them as primitive and organically given units of measurement that facilitate the analytical capture of "social physics" (Pentland, 2015), "the 'physics of culture'" (Manovich, 2011), or the "physics of society" (Caldarelli et al., 2018). The scientistic aspiration to discover invariant "laws of society" rests on this erroneous naturalisation of social data. Were the confidence of CSS research in such a naturalist purity of data to be breeched and their contextual and sociohistorical origins appropriately acknowledged, then the scientistic metanarratives that underwrite beliefs in "social physics", and in its nomological character, would consequently be subverted. Computational solutionism provides an epistemic strategy for the wholesale avoidance of this problem: it directs researchers to rely solely on the virtuosity of algorithmic tooling and the computational engineering of observational data to address congenital problems of noise, confounders, and non-random missingness rather than employing a genuine methodological pluralism that takes heed of the critical importance of context and of the complicated social and historical conditions surrounding the generation and construction of data. Such a solutionist tack, however, comes at the cost of potentially misapprehending the circumstantial intricacies and the historically contingent evolution of agential entanglements, social structures, and interpersonal relations and of thereby "misrepresenting the real world" in turn (Ruths & Pfeffer, 2014, p. 1063).

In addition to these risks posed to the epistemic integrity of CSS by big data hubris and computational solutionism, CSS researchers face another challenge related to the epistemological status of the claims and conclusion they hold forth. This has to do with the problem of interpretability. As the mathematical models employed in CSS research have come to possess ever greater access both to big data and to increasing computing power, their designers have correspondingly been able to enlarge the feature spaces of these computational systems and to turn to gradually more complex mapping functions in order either to forecast future observations or to explain underlying causal structures or effects. In many cases, this has meant vast improvements in the performance of models that have become more accurate and expressive, but this has also meant the growing prevalence of non-linearity, non-monotonicity, and high-dimensional complexity in an expanding array of so-called "black box" models (Leslie, 2019). Once high-dimensional feature spaces and complex functions are introduced into algorithmic models, the effects of changes in any given input can become so entangled with the values and interactions of other inputs that understanding the rationale behind how individual components are transformed into outputs becomes extremely difficult. The complex and unintuitive curves of many of these models' decision functions preclude linear and monotonic relations between their inputs and outputs. Likewise, the high-dimensionality of their architectures—frequently involving millions of parameters and complex correlations—presents a sweep of compounding statistical associations that range well beyond the limits of human-scale cognition and understanding. Such increasing complexity in input-output mappings creates model opacity and barriers to interpretability. The epistemological problem, here, is that, *as a science that seeks to explain, clarify, and facilitate a better understanding of the human phenomena it investigates*, CSS would seemingly have to avoid or renounce incomprehensible models that obstruct the demonstration of sound scientific reasoning in the conclusions and results attained.

A few epistemic strategies have emerged over the past decade or so to deal with the challenge posed by the problem of interpretability in CSS. First, building on a longstanding distinction originally made by statisticians between the predictive and explanatory functions of computational modelling (Breiman, 2001; Mahmoodi et al., 2017; Shmueli, 2010), some CSS scholars have focused on the importance of predictive accuracy, de-prioritising the goals of discovering and explaining the causal mechanisms and reasons that lie behind the dynamics of human behaviour and social systems (Anderson, 2008; Hindman, 2015; Lin, 2015; Yarkoni & Westfall, 2017). Lin (2015), for instance, makes a distinction between the goal of "better science", i.e., "to reveal insights about the human condition", what Herbert Simon called the "basic science" of explaining phenomena (2002), and the goal of "better engineering", i.e., "to produce computational artifacts that are more effective according to well-defined metrics" (p. 35)—what Simon called the "applied science" of inferring or predicting from known variables to unknown variables (Shmueli, 2010; Simon, 2002). For Lin, if the purpose of CSS, as an applied science, is "better engineering", then "whatever improves those [predictive] metrics should be exploited without prejudice. Sound scientific reasoning, while helpful, is not necessary to improve engineering". Such a positivistic view would, of course, tamp down or even cast aside the desideratum of interpretability.

However, even for scholars that aspire to retain both the explanatory and predictive dimensions of CSS, the necessity of using interpretable models is far from universally embraced. Illustratively, Hofman et al. (2021) argue for "integrating explanation and prediction in CSS" by treating these approaches as complementary (cf. Engel, 2021; James et al., 2013; Mahmoodi et al., 2017). Still, these authors simultaneously claim that explanatory modelling is about "the estimation of causal effects, regardless of whether those effects are explicitly tied to theoretically motivated mechanisms that are interpretable as 'the cogs and wheels of the causal process'" (Hofman et al., 2021, p. 186). To be sure, they maintain that:

interpretability is logically independent of both the causal and predictive properties of a model. That is, in principle a model can accurately predict outcomes under interventions or previously unseen circumstances (out of distribution), thereby demonstrating that it captures the relevant causal relationships, and still be resistant to human intuition (for example, quantum mechanics in the 1920s). Conversely, a theory can create the subjective experience of having made sense of many diverse phenomena without being either predictively accurate or demonstrably causal (for example, conspiracy theories). (pp. 186–187)

These justifications for treating the goal of interpretability as independent from the causal and predictive characteristics of a model raise some concerns. At an epistemic level, the extreme claim that "interpretability is logically independent of both the causal and predictive properties of a model" is unsupported by the observation that people can be deluded into believing false states of affairs. The attempt to cast aside the principal need for the rational acceptability and justification of the assertoric validity claims that explain a model's causal and predictive properties, because it is possible to be misled by "subjective experience", smacks of a curious epistemological relativism which is inconsistent with the basic requisites of scientific reasoning and deliberation. It offends the "no magic doctrine" (Anderson & Lebiere, 1998) of interpretable modelling, namely, that "it needs to be clear how (good) model performance comes about, that the components of the model are understandable and linked to known processes" (Schultheis, 2021). To level off all adjudications of explanatory claims (strong or weak) about a model because humans can be duped by misled feelings of subjective experience amounts to an absurdity: People can be convinced of bad explanations that are not predictively or causally efficacious (look at all those sorry souls who have fallen prey to conspiracy theories), so all explanations of complex models are logically independent of their actual causal and predictive properties. This line of thinking ends up in a ditch of epistemic whataboutism.

Moreover, at an ethical level, the analogy offered by Hofman et al. between the opaqueness of quantum physics and the opaqueness of "black box" predictive models about human behaviours and social dynamics is misguided and unsupportable. Such an erroneous parallelism is based on a scientistic confusion of the properties of natural scientific variables (like the wavelike mechanics of electrons) that function as heuristics for theory generation, testing, and confirmation in the exact physical sciences and the properties of the social variables of CSS whose generation, construction, and correlation are the result of human choices, evolving cultural patterns, and path dependencies created by sociohistorical structures. Unlike the physics data generated, for instance, by firing a spectroscopic light through a perforated cathode and measuring the splitting of the Balmer lines of a radiated hydrogen spectrum, the all-too-human genealogy of social data means that they can harbour discriminatory biases and patterns of sociohistorical inequity and injustice that become buried within the architectures of complex computational models. In this respect, the "relevant causal relationships" that are inaccessible in opaque models might be fraught with objectionable sociohistorical patterns of inequity, prejudice, coloniality, and structural racism, sexism, ablism, etc. (Leslie et al., 2022a). Because "human data encodes human biases by default" (Packer et al., 2018), complex algorithmic models can house and conceal a troubling range of unfair biases and discriminatory associations—from social biases against gender (Bolukbasi et al., 2016; Lucy & Bamman, 2021; Nozza et al., 2021; Sweeney &

Najafian, 2019; Zhao et al., 2017), race (Benjamin, 2019; Noble, 2018; Sweeney, 2013), accented speech (Lawrence, 2021; Najafian et al., 2017), and political views (Cohen & Ruths, 2013 Iyyer et al., 2014; Preoţiuc-Pietro et al., 2017) to structures of encoded prejudice like proxy-based digital redlining (Cottom, 2016; Friedline et al., 2020) and the perpetuation of harmful stereotyping (Abid et al., 2021; Bommasani et al., 2021; Caliskan et al., 2017; Garrido-Muñoz et al., 2021; Nadeem et al., 2020; Weidinger et al., 2021). A lack of interpretability in complex computational models whose performant causal and predictive properties could draw opaquely on secreted discriminatory biases or patterns of inequity is therefore ethically intolerable. As Wallach (2018) observes:

> the use of black box predictive models in social contexts ... [raises] a great deal of concern—and rightly so—that these models will reinforce existing structural biases and marginalize historically disadvantaged populations ... we must [therefore] treat machine learning for social science very differently from the way we treat machine learning for, say, handwriting recognition or playing chess. We cannot just apply machine learning methods in a black-box fashion, as if computational social science were simply computer science plus social data. We need transparency. We need to prioritize interpretability—even in predictive contexts. (p. 44) (cf. Lazer et al., 2020, p. 1062)

4.2.4 Challenges Related to Research Integrity

Challenges related to research integrity are rooted in the asymmetrical dynamics of resourcing and influence that can emerge from power imbalances between the CSS research community and the corporations and government agencies upon whom CSS scholars often rely for access to the data resources, compute infrastructures, project funding opportunities, and career advancement prospects they need for their professional subsistence and advancement. Such challenges can manifest, inter alia, in the exercise of research agenda-setting power by private corporations and governmental institutions, which set the terms of project funding schemes and data sharing agreements, and in the willingness of CSS researchers to produce insights and tools that support scaled behavioural manipulation and surveillance infrastructures.

These threats to the integrity of CSS research activity manifests in a cluster of potentially unseemly alignments and conflicts of interest between its own community of practice and those platforms, corporations, and public bodies who control access to the data resources and compute infrastructures upon which CSS researchers depend (Theocharis & Jungherr, 2021). First, there is the potentially unseemly alignment between the extractive motives of digital platforms, which monetise, monger, and link their vast troves of personal data and marshal inferences derived from these to classify, mould, and behaviourally nudge targeted data subjects, and the professional motivations CSS researchers who desire to gain access to as much of this kind of social big data as possible (Törnberg & Uitermark, 2021). A similar alignment can be seen between the motivations of CSS researchers to accumulate data and the security and control motivations of political bodies, which collect large amounts of personal data from the provision and administration of essential social goods and services often in the service of such motivations (Fourcade & Gordon, 2020). There is also a potentially unseemly alignment between the epistemic leverage and sociotechnical capabilities desired by private corporations and political bodies interested in scaled behavioural control and manipulation and the epistemic leverage and sociotechnical capabilities cultivated, as a vocational *raison d'être*, by some CSS researchers who build predictive tools. This alignment is made all-the-more worrying by the asymmetrical power dynamics that can be exercised by the former organisations over the latter researchers, who not only are increasingly reliant on private companies and governmental bodies for essential data access and computing resources but are also increasingly the obliged beneficiaries of academic-corporate research partnerships and academic-corporate "dual-affiliation" career trajectories that are funded by large tech corporations (Roberge et al., 2019). Finally, there is a broader scale cultural alignment between the way that digital platforms and tech companies pursue their corporate interests through technology practices that privilege considerations of strategic control, market creation, and efficiency and that are thereby functionally liberated from the constraints of social licence, democratic governance, and considerations of the interests of impacted people (Feenberg, 1999, 2002) and the way that CSS scholars can pursue of their professional interests through research practices similarly treated as operationally autonomous and independent from the societal conditions they impact and the governance claims of affected individuals and communities.

4.2.5   Challenges Related to Research Equity

Challenges related to research equity fall under two categories: (1) inequities that arise within the outputs of CSS research in virtue of biases that crop up within its methods and analytical approaches and (2) inequities that arise within the wider field of CSS research that result from material inequalities and capacity imbalances between different research communities. Challenges emerging from the first category include the potential reinforcement of digital divides and data inequities through biased sampling techniques that render digitally marginalised groups invisible as well as potential aggregation biases in research results that mask meaningful differences between studied subgroups and therefore hide the existence of real-world inequities. Challenges emerging from the second category include exploitative data appropriation by well-resourced researchers and the perpetuation of capacity divides between research communities, both of which derive from long-standing dynamics of regional and global inequality that may undermine reciprocal sharing and collaboration between researchers from more and less resourced geographical areas, universities, or communities of practice.

   Issues of sampling or population bias in CSS datasets extracted from social media platforms, internet use, and connected devices arise when the sampled population that is being studied differs from the larger target population in virtue of the non-random selection of certain groups into the sample (Hargittai, 2015, 2020; Hollingshead et al., 2021; Mehrabi et al., 2021; Olteanu et al., 2019; Tufekci, 2014). It has been widely observed that people do not select randomly into social media sites like Twitter (Blank, 2017; Blank & Lutz, 2017), MySpace (boyd, 2011), Facebook (boyd, 2011; Hargittai, 2015), and LinkedIn (Blank & Lutz, 2017; Hargittai, 2015). As Hargittai (2015) shows, in the US context, people with greater educational attainment and higher income were more likely to be users of Twitter, Facebook, and LinkedIn than others of less privilege. Hargittai (2020) claims, more generally, that "big data derived from social media tend to oversample the views of more privileged people" and people who possess greater levels of "internet skill". Earlier studies and surveys have also demonstrated that, at any given time, "different user demographics tend to be drawn to different social platforms" (Olteanu et al., 2019), with men and urban populations significantly over-represented among Twitter users (Mislove et al., 2011) and women over-represented on Pinterest (Ottoni et al., 2013).

The oversampling of self-selecting privileged and dominant groups, and the under-sampling or exclusion of members of other groups who may lack technical proficiency, digital resources, or access to connectivity, for example, large portions of elderly populations (Friemel, 2016; Haight et al., 2014; Quan-Haase et al., 2018), can lead to an inequitable lack of representativity in CSS datasets—rendering those who have been left out of data collection for reason of accessibility, skills, and resource barriers "digitally invisible" (Longo et al., 2017). Such sampling biases can cause deficiencies in the ecological validity of research claims (Olteanu et al., 2019), impaired performance of predictive models for non-majority subpopulations (Johnson et al., 2017), and, more broadly speaking, the failure of CSS models to generalise from sampled behaviours and opinions to the wider population (Blank, 2017; Hargittai & Litt, 2012; Hollingshead et al., 2021). This hampered generalisability can be especially damaging when the insights and results of CSS models, which oversample privileged subpopulations and thus disadvantage those missing from datasets, are applied willy-nilly to society as a whole and used to shape the policymaking approaches to solving real-world problems. As Hollingshead et al. (2021) put it, "the ethical concern here is that, as policymakers and corporate stakeholders continue to draw insights from big data, the world will be recursively fashioned into a space that reflects the material interests of the infinitely connected" (p. 173).[7]

Another research inequity that can crop up within CSS methods and analytical approaches is aggregation bias (Mehrabi et al., 2021; Suresh & Guttag, 2021). This occurs when a model's analysis is applied in a "one-size-fits-all" manner to

---

[7] A similar and compounding form of sampling bias can occur when survey data is linked, through participant consent, to digital trace data from social media networks. Here the dynamic of non-random self-selection manifests in the select group of research subjects (likely those who are privileged and young and more frequently male) who have social media accounts and who consent to having them linked to the survey research (Al Baghal et al., 2020; Stier et al., 2020).

subpopulations that have different conditional distributions, thereby treating the results as "population-level trends" that map inputs to outputs uniformly across groups despite their possession of diverging characteristics (Hollingshead et al., 2021; Suresh & Guttag, 2021). Such aggregation biases can lead models to fit optimally for dominant or privileged subpopulations that are oversampled while underperforming for groups that lack adequate representation. These biases can also conceal patterns of inequity and discrimination that are differentially distributed among subpopulations (Barocas & Selbst, 2016; boyd & Crawford, 2012; Hollingshead et al., 2021; Longo et al., 2017; Olteanu et al., 2019), consequently entrenching or even augmenting structural injustices that are hidden from view on account of the irresponsible statistical homogenisation of target populations.

A different set of research inequities arise within the wider field of CSS research as a consequence of material inequalities and capacity imbalances that exist between different research communities. Long-standing dynamics of global inequality, for instance, may undermine reciprocal sharing between research collaborators from high-income countries (HICs) and those from low-/middle-income countries (LMICs) (Leslie, 2020). Given asymmetries in resources, infrastructure, and research capabilities, data sharing between LMICs and HICs, and transnational research collaboration, can lead to inequity and exploitation (Bezuidenhout et al., 2017; Leonelli, 2013; Shrum, 2005). That is, data originators from LMICs may put immense amounts of effort and time into developing useful datasets (and openly share them) only to have their countries excluded from the benefits derived by researchers from HICs who have capitalised on such data in virtue of greater access to digital resources and compute infrastructure (World Health Organization, 2022). Moreover, data originators from LMICs may generate valuable datasets that they are then unable to independently and expeditiously utilise for needed research, because they lack the aptitudes possessed by researchers from HICs who are the beneficiaries of arbitrary asymmetries in education, training, and research capacitation (Bull et al., 2015; Merson et al., 2015).

This can create a twofold architecture of research inequity wherein the benefits of data production and sharing do not accrue to originating researchers and research subjects and the scientists from LMICs are put in a position of relative disadvantage vis-à-vis those from HICs whose research efficacy and ability to more rapidly convert data into insights function, in fact, to undermine the efforts of their disadvantaged research partners (Bezuidenhout et al., 2017; Crane, 2011). It is important to note, here, that such gaps in research resources and capabilities also exist within HICs where large research universities and technology corporations (as opposed to less well-resourced universities and companies) are well positioned to advance data research given their access to data and compute infrastructures (Ahmed & Wahed, 2020).

In redressing these access barriers, emphasis must be placed on "the social and material conditions under which data can be made useable, and the multiplicity of conversion factors required for researchers to engage with data" (Bezuidenhout et al., 2017, p. 473). Equalising know-how and capability is a vital counterpart to equalising access to resources, and both together are necessary preconditions of just research environments. CSS scholars engaging in international research collaborations should focus on forming substantively reciprocal partnerships where capacity-building and asymmetry-aware practices of cooperative innovation enable participatory parity and thus greater research access and equity.

4.3   Incorporating Habits of Responsible Research and Innovation into CSS Practices

The foregoing taxonomy of the five main ethical challenges faced by CSS is intended to provide CSS researchers with a critical lens that enables them to sharpen their field of vision so that they are equipped to engage in the sort of anticipatory reflection which roots out irresponsible research practices and harmful impacts. However, circumvention of the potential endurance of "research fast and break things" attitudes requires a deeper cultural transformation in the CSS community of practice. It requires the end-to-end incorporation of habits of Responsible Research and Innovation (RRI) into all its research activities. An RRI perspective provides CSS researchers with an awareness that all processes of scientific discovery and problem-solving possess sociotechnical aspects and ethical stakes. Rather than conceiving research as independent from human values, RRI regards these activities as ethically implicated social practices. For this reason, such practices are charged with a responsibility for *critical self-reflection* about the role that these values play both in discovery, engineering, and design processes and in considerations of the real-world effects of the insights and technologies that these processes yield.

Those who have been writing on the ethical dimension of CSS for the past decade have emphasised the importance of precisely these kinds of self-reflective research practices (for instance, British Sociological Association, 2016; Eynon et al., 2017 Franzke et al., 2020; Hollingshead et al., 2021; Lomborg, 2013; Markham & Buchanan, 2012; Moreno et al., 2013; Weinhardt, 2020). Reacting to recent miscarriages of research ethics that have undermined public trust, such as the 2016 mass sharing of sensitive personal information that had been extracted by researchers from the OKCupid dating site (Zimmer, 2016), they have stressed the need for "a bottom-up, case-based approach to research ethics, one that emphasizes that ethical judgment must be based on a sensible examination of the unique object and circumstances of a study, its research questions, the data involved, and the expected analysis and reporting of results, along with the possible ethical dilemmas arising from the case" (Lomborg, 2013, p. 20). What is needed to operationalise such a "bottom-up, case-based approach to research ethics" is the development across the CSS community of habits of RRI. In this section, we will explore how CSS practices can incorporate habits of RRI, focusing, in particular, on the role that contextual considerations, anticipatory reflection, public engagement, and justifiable action should play across the research lifecycle.

Building on research in Science and Technology Studies and Applied Technology Ethics, the RRI view of "science with and for society" has been transformed into helpful general guidance in such interventions as Engineering and Physical Sciences Research Council (EPSRC)'s 2013 AREA framework[8] and the 2014 Rome Declaration[9] (Fisher & Rip, 2013; Owen, 2014; Owen et al., 2012, 2013; Stilgoe et al., 2013; von Schomberg, 2013). More recently, EPSRC's AREA principles (anticipate, reflect, engage, act) have been extended into the fields of data science and AI by the CARE & Act Framework (consider context, anticipate impacts, reflect on purposes, positionality, and power, engage inclusively, act responsibly and transparently) (Leslie, 2020; Leslie et al., 2022b). The application of the CARE & Act principles to CSS aims to provide a handy tool that enables its researchers to continuously sense check the social and ethical implications of their research practices and that helps them to establish and sustain responsible habits of scientific investigation and reporting. Putting the CARE & Act Framework into practice involves taking its several guiding maxims as a launching pad for continuously reflective and deliberate choice-making across the research workflow. Let us explore each of these maxims in turn.

4.3.1 Consider Context

The imperative of considering context enjoins CSS researchers to think diligently about the conditions and circumstances surrounding their research activities and outputs. This involves focusing on the norms, values, and interests that inform the people undertaking the research and that shape and motivate the reasonable expectations of research subject and those who are likely to be impacted by the research and its results: How are these norms, values and interests influencing or steering the project and its outputs? How could they influence research subjects' meaningful consent and expectations of privacy, confidentiality, and anonymity? How could they shape a research project's reception and impacts across impacted communities? Considering context also involves taking into account the specific domain(s), geographical location(s), and jurisdiction(s) in which the research is situated and reflecting on the expectations of affected stakeholders that derive these specific contexts: How do the existing institutional norms and rules in a given domain or jurisdiction shape expectations regarding research goals, practices, and outputs? How do the unique social, cultural, legal, economic, and political environments in which different research projects are embedded influence the conditions of data generation, the intentions and behaviours of the research subjects

---

[8] https://www.ukri.org/about-us/epsrc/our-policies-and-standards/framework-for-responsible-innovation/

[9] https://digital-strategy.ec.europa.eu/en/library/rome-declaration-responsible-research-and-innovation-europe

that are captured by extracted data, and the space of possible inferences that data analytics, modelling, and simulation can yield?

The importance of responsiveness to context has been identified as significant in internet research ethics for nearly two decades (Buchanan, 2011; Markham, 2006) and has especially been emphasised more recently in the *Internet Research: Ethical Guidelines 3.0* of the Association of Internet Researchers (AoIR), where the authors stress that a "basic ethical approach" involves focussing on "on the fine-grained contexts and distinctive details of each specific ethical challenge" (Franzke et al., 2020, p. 4).[10] For Franzke et al., such a

> process- and context-oriented approach... helps counter a common presumption of "ethics" as something of a "one-off" tick-box exercise that is primarily an obstacle to research. On the contrary... taking on board an ongoing attention to ethics as inextricably interwoven with method often leads to better research as this attention entails improvements on both research design and its ethical dimensions throughout the course of a project. (pp. 4–5)

This ongoing attention entails a keen awareness of the need to "respect people's values or expectations in different settings" (Eynon et al., 2017) as well as the need to acknowledge cultural differences, ethical pluralism, and diverging interpretations of moral values and concepts (Capurro, 2005, 2008; C. M. Ess, 2020; Hongladarom & Ess, 2007; Leslie et al., 2022a). Likewise, contextual considerations need to include a recognition of interjurisdictional differences in legal and regulatory requirements (for instance, variations in data protection laws and legal privacy protections across regions and countries whence digital trace data is collected).

All in all, contextual considerations should, at minimum, track three vectors: The first involves considering the contextual determinants of the condition of the production of the research (e.g., thinking about the positionality of the research team, the expectations of the relevant CSS community of practice, and the external influences on the aims and means of research by funders, collaborators, and providers of data and research infrastructure). The second involves considering the context of the subjects of research (e.g., thinking about research subjects' reasonable expectations of gainful obscurity and "privacy in public" and considering the changing contexts of their communications such as with whom they are interacting, where, how, and what kinds of data are being shared). The third involves considering the contexts of the social, cultural, legal, economic, and political environments in which different research projects are embedded as well as the historical, geographic, sectoral, and jurisdictional specificities that configure such environments (e.g., thinking about the ways different social groups—both within and between cultures—understand and define key values, research variables, and studied concepts differently as well as the ways that these divergent understandings place limitations on what computational approaches to prediction, classification, modelling, and simulation can achieve).

---

[10] It is important to note that the importance of contextual considerations has also been present in earlier versions of the AoIR guidelines which date back two decades (Internet Research Ethics—IRE 1.0, 2002; Internet Research Ethics-IRE 2.0, 2012).

4.3.2 Anticipate Impacts

The imperative of anticipating impacts enjoins CSS researchers to reflect on and assess the potential short-term and long-term effects their research may have on impacted individuals (e.g., research participants, data subjects, and the researchers themselves) and on affected communities and social groups, more broadly. The purpose of this kind of anticipatory reflection is *to safeguard the sustainability of CSS projects across the entire research lifecycle*. To ensure that the activities and outputs of CSS research remain socially and environmentally sustainable and support the sustainability of the communities they affect, researchers must proceed with a continuous responsiveness to the real-world impacts that their research could have. This entails concerted and stakeholder-involving exploration of the possible adverse and beneficial effects that could otherwise remain hidden from view if deliberate and structured processes for anticipating downstream impacts were not in place. Attending to sustainability, along these lines, also entails the iterative re-visitation and re-evaluation of impact assessments. To be sure, in its general usage, the word "sustainability" refers to the maintenance of and care for an object or endeavour *over time*. In the CSS context, this implies that building sustainability into a research project is not a "one-off" affair. Rather, carrying out an initial research impact assessment at the inception of a project is only a first, albeit critical, step in a much longer, end-to-end process of responsive re-evaluation and re-assessment. Such an iterative approach enables sustainability-aware researchers to pay continuous attention both to the dynamic and changing character of the research lifecycle and to the shifting conditions of the real-world environments in which studies are embedded.

This demand to anticipate research impacts is not new in the modern academy—especially in the biomedical and social sciences, where Institutional Review Board (IRB) processes for research involving human subjects have been in place for decades (Abbott & Grady, 2011; Grady, 2015). However, the novel human scale, breadth, and reach of CSS research, as well as the new (and often subtler) range of potential harms it poses to impacted individuals, communities, and the biosphere, call into question the adequacy of conventional IRB processes (Metcalf & Crawford, 2016). While the latter have been praised a necessary step forward in protecting the physical, mental, and moral integrity of human research subjects, building public trust in science, and institutionalising needed mechanisms for ethical oversight (Resnik, 2018), critics have also highlighted their unreliability, superficiality, narrowness, and inapplicability to the new set of information hazards posed by the processing of aggregated big data (Prunkl et al., 2021; Raymond, 2019).

A growing awareness of these deficiencies has generated an expanding interest in CSS-adjacent computational disciplines (like machine learning, artificial intelligence, and computational linguistics) to come up with more robust impact assessment regimes and ethics review processes (Hecht et al., 2021; Leins et al., 2020; Nanayakkara et al., 2021). For instance, in 2020, the NeurIPS (Neural

Information Processing Systems) conference introduced a new ethics review protocol that required paper submissions to include an impact statement "discussing the broader impact of their work, including possible societal consequences—both positive and negative" (Neural Information Processing Systems Conference, 2020). Informatively, this protocol was converted into a responsible research practices checklist in 2021 (Neural Information Processing Systems, 2021) after technically oriented researchers protested that they lacked the training and guidance needed to carry out impact assessments effectively (Ashurst et al., 2021; Johnson, 2020; Prunkl et al., 2021). Though there has been recent progress made, in both AI and CSS research communities, to integrate some form of ethics training into professional development (Ashurst et al., 2020; Salganik & The Summer Institutes in Computational Social Science, n.d.) and to articulate guidelines for anticipating ethical impacts (Neural Information Processing Systems, 2022), there remains a lack of institutionalised instruction, codified guidance, and professional stewardship for research impact assessment processes. As an example, conferences such as International AAAI Conference on Web and Social Media, ICWSM (2022); International Conference on Machine Learning, ICML (2022); North American Chapter of the Association for Computational Linguistics, NAACL (2022); and Empirical Methods in Natural Language Processing, EMNLP ((2022) each require some form of research impact evaluation and ethical consideration, but aside from directing researchers to relevant professional guidelines and codes of conduct (e.g., from the Association for Computational Linguistics, ACL; Association for Computing Machinery, ACM; and Association for the Advancement of Artificial Intelligence, AAAI), there is scant direction on how to operationalise impact assessment processes (Prunkl et al., 2021).

What is missing from this patchwork of ethics review requirements and guidance is a set of widely accepted procedural mechanisms that would enable and standardise conscientious research impact assessment practices. To fill this gap, recent research into the governance practices needed to create responsible data research environments has called for a coherent, integrated, and holistic approach to impact assessment that includes several interrelated elements (Leslie, 2019, 2020; Leslie et al., 2021; Leslie et al., 2022c, 2022d, 2022e):

- *Stakeholder analysis*: Diligent research impact assessment practices should include processes that allows researchers to identify and evaluate the salience and contextual characteristics of individuals or groups who may be affected by, or may affect, the research project under consideration (Mitchell et al., 2017; Reed et al., 2009; Schmeer, 1999; Varvasovszky & Brugha, 2000). Stakeholder analysis aims to help researchers understand the relevance of each identified stakeholder to their project and to its use contexts. It does this by providing a structured way to assess the relative interests, rights, vulnerabilities, and advantages of identified stakeholders as these characteristics may be impacted by, or may impact, the research.
- *Establishment of clear normative criteria for impact assessment*: Effective research impact assessment practices should start from a clear set of ethical

values or human rights criteria against which the potential impacts of a project on affected individuals and communities can be evaluated. Such criteria should provide common but non-exclusive point of departure for collective deliberation about the ethical permissibility of the research project under consideration. Adopting common normative criteria from the outset enables reciprocally respectful, sincere, and open discussion about the ethical challenges a research project may face by helping to create a shared vocabulary for informed dialogue and impact assessment. Such a common starting point also facilitates deliberation about how to balance ethical values when they come into tension.

- *Methodical evaluation of potential impacts*: The actual research impact assessment process provides an opportunity for research teams (and engaged stakeholders, where deemed appropriate) to produce detailed evaluations of the potential and actual impacts that the project may have, to contextualise and corroborate potential harms and benefits, to make possible the collaborative assessment of the severity of potential adverse impacts identified, and to facilitate the co-design of an impact mitigation plan.
- *Impact mitigation planning*: Once impacts have been evaluated and the severity of any potential harms assessed, impact prevention and mitigation planning should commence. Diligent impact mitigation planning begins with a scoping and prioritisation stage. Research team members (and engaged stakeholders, where appropriate) should go through all the identified potential adverse impacts and map out the interrelations and interdependencies between them as well as surrounding social factors (such as contextually specific stakeholder vulnerabilities and precariousness) that could make impact mitigation more challenging. Where prioritisation of prevention and mitigation actions is necessary (for instance, where delays in addressing a potential harm could reduce its remediability), decision-making should be steered by the relative severity of the impacts under consideration. As a general rule, while impact prevention and mitigation planning may involve prioritisation of actions, all potential adverse impacts must be addressed. When potential adverse impacts have been mapped out and organised, and mitigation actions have been considered, the research team (and engaged stakeholders, where appropriate) should begin co-designing an impact mitigation plan (IMP). The IMP will become the part of your transparent reporting methodology that specifies the actions and processes needed to address the adverse impacts which have been identified and that assign responsibility for the completions of these tasks and processes. As such, the IMP will serve a crucial documenting function.
- *Establishment of protocols for re-visitation and re-evaluation of the research impact assessment*: Research impact assessments must pay continuous attention both to the dynamic and changing character of the research lifecycles and to the shifting conditions of the real-world environments in which research practices, results, and outputs are embedded. There are two sets of factors that should inform when and how often initial research impact assessments are re-visited to ensure that they remain adequately responsive to factors that could present new potential harms or significantly influence impacts that have been

previously identified: (1) research workflow and production factors: Choices made at any point along the research workflow may affect the veracity of prior impact assessments, leading to a need for re-assessment, reconsideration, and amendment. For instance, research design choices could be made that were not anticipated in the initial impact assessment (such choices might include adjusting the variables that are included in the model, choosing more complex algorithms, or grouping variables in ways that may impact specific groups); (2) environmental factors, changes in project-relevant social, regulatory, policy, or legal environments (occurring during the time in which the research is taking place) may have a bearing on how well the resulting computational model works and on how the research outputs impact affected individuals and groups. Likewise, domain-level reforms, policy changes, or changes in data recording methods may take place in the population of concern in ways that affect whether the data used to train the model accurately portrays phenomena, populations, or related factors in an accurate manner.

4.3.3 Reflect on Purposes, Positionality, and Power

The foregoing elements of research impact assessment presuppose that the CSS researchers who undertake them also engage in reflexive practices that scrutinise the way potential perspectival limitations and power imbalances can exercise influence on the equity and integrity of research projects and on the motivations, interests, and aims that steer them. The imperative of reflecting on purposes, positionality, and power makes explicit the importance of this dimension of inward-facing reflection.

All individual human beings come from unique places, experiences, and life contexts that shape their perspectives, motivations, and purposes. Reflecting on these contextual attributes is important insofar as it can help researchers understand how their viewpoints might differ from those around them and, more importantly, from those who have diverging cultural and socioeconomic backgrounds and life experiences. Identifying and probing these differences enables individual researchers to better understand how their own backgrounds, for better or worse, frame the way they see others, the way they approach and solve problems, and the way they carry out research and engage in innovation. By undertaking such efforts to recognise social position and differential privilege, they may gain a greater awareness of their own personal biases and unconscious assumptions. This then can enable them to better discern the origins of these biases and assumptions and to confront and challenge them in turn.

Social scientists have long referred to this site of self-locating reflection as "positionality" (Bourke, 2014; Kezar, 2002; Merriam et al., 2001). When researchers take their own positionalities into account, and make this explicit, they can better grasp how the influence of their respective social and cultural positions potentially creates research strengths and limitations. On the one hand, one's positionality—with respect to characteristics like ethnicity, race, age, gender, socioeconomic status, education and training levels, values, geographical background, etc.—can have a positive effect on an individual's contributions to a research project; the uniqueness of each person's lived experience and standpoint can play a constructive role in introducing insights and understandings that other team members do not have. On the other hand, one's positionality can assume a harmful role when hidden biases and prejudices that derive from a person's background, and from differential privileges and power imbalances, creep into decision-making processes undetected and subconsciously sway the purposes, trajectories, and approaches of research projects.[11]

4.3.4 Engage Inclusively

While practices of inward-facing reflection on purposes, positionality, and power can strengthen the reflexivity, objectivity, and reasonableness of CSS research activities (D'Ignazio & Klein, 2020; Haraway, 1988; Harding, 1992, 1995, 2008, 2015), practices of outward-facing stakeholder engagement and community involvement can bolster a research project's legitimacy, social license, and democratic governance as well as ensure that its outputs will possess an appropriate degree of public accountability and transparency. A diligent stakeholder engagement process can help research teams to identify stakeholder salience, undertake team positionality reflection, and facilitate proportionate community involvement and input throughout the research project workflow. This process can also safeguard the equity and the contextual accuracy of impact assessments and facilitate appropriate end-to-end processes of transparent project governance by supporting their iterative re-visitation and re-evaluation. Moreover, community-involving engagement processes can empower the public and the CSS community alike by introducing the transformative agency of "citizen science" into research processes (Albert et al., 2021; Sagarra et al., 2016; Tauginienė et al., 2020).

It is important to note, however, that all stakeholder engagement processes can run the risk either of being cosmetic or tokenistic tools employed to legitimate research projects without substantial and meaningful participation or of being insufficiently participatory, i.e., of being one-way information flows or nudging exercises that serve as public relations instruments (Arnstein, 1969; Tritter & McCallum, 2006). To avoid such hazards of superficiality, CSS researchers should shore up a proportionate approach to stakeholder engagement through deliberate

---

[11] When taking positionality into account, researchers should reflect on their own positionality matrix. They should ask: to what extent do my personal characteristics, group identifications, socioeconomic status, educational, training, and work background, team composition, and institutional frame represent sources of power and advantage or sources of marginalisation and disadvantage? How does this positionality influence my (and my research team's) ability to identify and understand affected stakeholders and the potential impacts of my project? For details on this process see Leslie et al. (2022b).

and precise goal setting. Researchers should prioritise the establishment of clear and explicit stakeholder engagement objectives. Relevant questions to pose in establishing these goals include *Why are we engaging with stakeholders? What do we envision the ideal purpose and the expected outcomes of engagement activities to be? How can we best drawn on the insights and lived experience of participants to inform and shape our research?*[12]

4.3.5 Act Transparently and Responsibly

The imperative of acting transparently and responsibly enjoins CSS researchers to marshal the habits of Responsible Research and Innovation cultivated in the CARE processes to produce research that prioritises data stewardship and that is robust, accountable, fair, non-discriminatory, explainable, reproducible, and replicable. While the mechanisms and procedures which are put in place to ensure that these normative goals are achieved will differ from project to project (based on the specific research contexts, research design, and research methods), all CSS researchers should incorporate the following priorities into their governance, self-assessment, and reporting practices:

- *Full documentation of data provenance, lineage, linkage, and sourcing:* This involves keeping track of and documenting both responsible data management practices across the entire research lifecycle, from data extraction or procurement and data analysis, cleaning, and pre-processing to data use, retention, deletion, and updating (Bender & Friedman, 2018; Gebru et al., 2021; Holland et al., 2018). It also involves demonstrating that the data is ethically sourced, responsibly linked, and legally available for research purposes (Weinhardt, 2020) and making explicit measures taken to ensure data quality (source integrity and measurement accuracy, timeliness and recency, relevance, sufficiency of quantity, dataset representativeness), data integrity (attributability, consistency, completeness, contemporaneousness, traceability, and auditability), and FAIR data (findable, accessible, interoperable, and reusable).
- *Full documentation of privacy, confidentiality, consent, and data protection due diligence:* This involves demonstrating that data has been handled securely and responsibly from beginning to end of the research lifecycle so that any potential breaches of confidentiality, privacy, and anonymity have been prevented and any risks of re-identification through triangulation and data linkage mitigated. Regardless of the jurisdictions of data collection and use, researchers should aim to optimally protect the rights and interests of research subjects by adhering to the highest standards of privacy preservation, data protection, and responsible data handling and storage such as those contained in the IRE 3.0 and the National

---

[12] An elaboration on the essential components of a responsible stakeholder engagement process can be found in Leslie et al. (2022b).

Committee for Research Ethics in the Social Sciences and the Humanities (NESH) guidelines (Franzke et al., 2020; National Committee for Research Ethics in the Social Sciences and the Humanities (NESH), 2019). They should also demonstrate that they have sufficiently taken into account contextual factors in meeting the privacy expectations of observed research subjects (like who is involved in observed interactions, how and what type of information is exchanged, how sensitive it is perceived to be, and where and when such exchanges occur). Documentation should additionally include evidence that researchers have instituted proportionate protocols for attaining informed and meaningful consent that are appropriate to the specific contexts of the data extraction and use and that cohere with the reasonable expectations of targeted research subjects.

- *Transparent and accountable reporting of research processes and results and appropriate publicity of datasets:* Research practices and methodological conduct should be carried out deliberately, transparently, and in accordance with recording protocols that enable the interpretability, reproducibility, and replicability of results. For prediction models, the documentation protocols presented in Transparent Reporting of a Multivariable Prediction Model for Individual Prognosis or Diagnosis (TRIPOD) provide a good starting point for best conduct guidelines in research reporting (Collins et al., 2015; Moons et al., 2015).[13] Following TRIPOD, transparent and accountable reporting should demonstrate diligent methodological conduct across all stages and elements of research. For prediction models, this includes clear descriptions of research participants, predictors, outcome variables, sample size, missing data, statistical analysis methods, model specification, model performance, model validation, model updating, and study limitations. While transparent research conduct can facilitate reproducibility and replicability, concerns about the privacy and anonymity of research subjects should also factor into how training data, models, and results are made available to the scientific community. This notwithstanding, CSS researchers should prioritise the publication of well-archived, high-quality, and accessible datasets that enable the replication of results and the advancement of further research (Hollingshead et al., 2021). They should also pursue research design, analysis, and reporting in an interpretability-aware manner that prioritises process transparency, the understandability of models, and the accessibility and explainability of the rationale behind their results.
- *An end-to-end process for bias self-assessment:* This should cover all research stages as well as all sources of biases that could arise in the data; in the data collection; in the data pre-processing; in the organising, categorising, describing, annotating, structuring of data (text-as-data, in particular); and in research design and execution choices. Bias self-assessment processes should cover *social, statistical, and cognitive biases* (Leslie et al., 2022a). An end-to-end process

---

[13] Though the TRIPOD method is intended to be applied in the medical domain, its reporting protocols are largely applicable to CSS studies.

for bias self-assessment should move across the research lifecycle, pinpointing specific forms of social, statistical, and cognitive bias that could arise at each stage (for instance, social biases like representation bias and label bias as well as statistical biases like missing data bias and measurement bias could arise in the data pre-processing stage of a research project).

4.4 Conclusion

This chapter has explored the spectrum of ethical challenges that CSS for policy faces across the myriad possibilities of its application. It has further elaborated on how these challenges can be met head-on only through the adoption of habits of RRI that are instantiated in end-to-end governance mechanisms which set up practical guardrails throughout the research lifecycle. As a quintessential *social impact science*, CSS for policy holds great promise to advance social justice, human flourishing, and biospheric sustainability. However, CSS is also an *all-too-human science*—conceived in particular social, cultural, and historical contexts and pursued amidst intractable power imbalances, structural inequities, and potential conflicts of interest. Its proponents, in both research and policymaking communities, must thus remain continuously self-critical about the role that values, interests, and power dynamics play in shaping mission-driven research. Likewise, they must vigilantly take heed of the complicated social and historical conditions surrounding the generation and construction of data as well as the way that the activities and theories of CSS researchers can function to reformat, reorganise, and shape the phenomena that they purport only to measure and analyse. Such a continuous labour of exposing and redressing the often-concealed interdependencies that exist between CSS and the social environments in which its research activities, subject matters, and outputs are embedded will only *strengthen its objectivity* and ensure that its impacts are equitable, ethical, and responsible. Such a human-centred approach will make CSS for policy a "science with and for society" second-to-none.

References

Abbott, L., & Grady, C. (2011). A systematic review of the empirical literature evaluating IRBs: What we know and what we still need to learn. *Journal of Empirical Research on Human Research Ethics, 6*(1), 3–19. https://doi.org/10.1525/jer.2011.6.1.3

Abid, A., Farooqi, M., & Zou, J. (2021). Persistent Anti-Muslim Bias in Large Language Models. *Proceedings of the 2021 AAAI/ACM Conference on AI, Ethics, and Society*, 298–306. https://doi.org/10.1145/3461702.3462624

Agniel, D., Kohane, I. S., & Weber, G. M. (2018). Biases in electronic health record data due to processes within the healthcare system: Retrospective observational study. *BMJ, 361*, k1479. https://doi.org/10.1136/bmj.k1479

Agüera y Arcas, B., Mitchell, M., & Todorov, A. (2017, May 7). Physiognomy's New Clothes. *Medium.* https://medium.com/@blaisea/physiognomys-new-clothes-f2d4b59fdd6a

Ahmed, N., & Wahed, M. (2020). *The De-democratization of AI: Deep learning and the compute divide in artificial intelligence research.* Cornell University Library, arXiv.org. https://doi.org/10.48550/ARXIV.2010.15581

Aizenberg, E., & van den Hoven, J. (2020). Designing for human rights in AI. *Big Data & Society, 7*(2), 205395172094956. https://doi.org/10.1177/2053951720949566

Ajunwa, I., Crawford, K., & Schultz, J. (2017). Limitless worker surveillance. *California Law Review, 105,* 735. https://doi.org/10.15779/Z38BR8MF94

Akhtar, P., & Moore, P. (2016). The psychosocial impacts of technological change in contemporary workplaces, and trade union responses. *International Journal of Labour Research, 8*(1/2), 101.

Al Baghal, T., Sloan, L., Jessop, C., Williams, M. L., & Burnap, P. (2020). Linking twitter and survey data: The impact of survey mode and demographics on consent rates across three UK studies. *Social Science Computer Review, 38*(5), 517–532. https://doi.org/10.1177/0894439319828011

Albert, A., Balázs, B., Butkevičienė, E., Mayer, K., & Perelló, J. (2021). Citizen social science: New and established approaches to participation in social research. In K. Vohland, A. Land-Zandstra, L. Ceccaroni, R. Lemmens, J. Perelló, M. Ponti, R. Samson, & K. Wagenknecht (Eds.), *The science of citizen science* (pp. 119–138). Springer International Publishing. https://doi.org/10.1007/978-3-030-58278-4_7

Amodei, D., & Hernandez, D. (2018, May 16). AI and Compute. *OpenAI.* https://openai.com/blog/ai-and-compute/

Amoore, L. (2021). The deep border. *Political Geography,* 102547. https://doi.org/10.1016/j.polgeo.2021.102547

Anderson, C. (2008, June 23). The end of theory: The data deluge makes the scientific method obsolete. *Wired Magazine.* https://www.wired.com/2008/06/pb-theory/

Anderson, J. R., & Lebiere, C. (1998). *The atomic components of thought.* Lawrence Erlbaum Associates.

Andrejevic, M., & Selwyn, N. (2020). Facial recognition technology in schools: Critical questions and concerns. *Learning, Media and Technology, 45*(2), 115–128. https://doi.org/10.1080/17439884.2020.1686014

Apel, K.-O. (1984). *Understanding and explanation: A transcendental-pragmatic perspective.* MIT Press.

Arnstein, S. R. (1969). A ladder of citizen participation. *Journal of the American Institute of Planners, 35*(4), 216–224. https://doi.org/10.1080/01944366908977225

Ashurst, C., Barocas, S., Campbell, R., Raji, D., & Russell, S. (2020). *Navigating the broader impacts of AI research.* https://aibroader-impacts-workshop.github.io/

Ashurst, C., Hine, E., Sedille, P., & Carlier, A. (2021). AI ethics statements—Analysis and lessons learnt from NeurIPS broader impact statements. *ArXiv: 2111.01705 [Cs].* http://arxiv.org/abs/2111.01705

Ball, K. (2009). Exposure: Exploring the subject of surveillance. *Information, Communication & Society, 12*(5), 639–657. https://doi.org/10.1080/13691180802270386

Ball, K. (2019). Review of Zuboff's the age of surveillance capitalism. *Surveillance & Society, 17*(1/2), 252–256. 10.24908/ss.v17i1/2.13126.

Banjanin, N., Banjanin, N., Dimitrijevic, I., & Pantic, I. (2015). Relationship between internet use and depression: Focus on physiological mood oscillations, social networking and online addictive behavior. *Computers in Human Behavior, 43,* 308–312. https://doi.org/10.1016/j.chb.2014.11.013

Barocas, S., & Selbst, A. D. (2016). Big data's disparate impact. *SSRN Electronic Journal.* https://doi.org/10.2139/ssrn.2477899

Barrett, L. F., Adolphs, R., Marsella, S., Martinez, A. M., & Pollak, S. D. (2019). Emotional expressions reconsidered: Challenges to inferring emotion from human facial movements. *Psychological Science in the Public Interest, 20*(1), 1–68. https://doi.org/10.1177/1529100619832930

Barry, C. T., Sidoti, C. L., Briggs, S. M., Reiter, S. R., & Lindsey, R. A. (2017). Adolescent social media use and mental health from adolescent and parent perspectives. *Journal of Adolescence, 61*(1), 1–11. https://doi.org/10.1016/j.adolescence.2017.08.005

Beer, D. (2017). The social power of algorithms. *Information, Communication & Society, 20*(1), 1–13. https://doi.org/10.1080/1369118X.2016.1216147

Bender, E. M., & Friedman, B. (2018). Data statements for natural language processing: Toward mitigating system bias and enabling better science. *Transactions of the Association for Computational Linguistics, 6,* 587–604. https://doi.org/10.1162/tacl_a_00041

Bender, E. M., Gebru, T., McMillan-Major, A., & Shmitchell, S. (2021). On the dangers of stochastic parrots: Can language models be too big?*Proceedings of the 2021 ACM Conference on Fairness, Accountability, and Transparency,* 610–623. https://doi.org/10.1145/3442188.3445922

Benjamin, R. (2019). *Race after technology: Abolitionist tools for the new Jim code.* Polity.

Bezuidenhout, L. M., Leonelli, S., Kelly, A. H., & Rappert, B. (2017). Beyond the digital divide: Towards a situated approach to open data. *Science and Public Policy, 44*(4), 464–475. https://doi.org/10.1093/scipol/scw036

Blank, G. (2017). The digital divide among twitter users and its implications for social research. *Social Science Computer Review, 35*(6), 679–697. https://doi.org/10.1177/0894439316671698

Blank, G., & Lutz, C. (2017). Representativeness of social Media in Great Britain: Investigating Facebook, LinkedIn, Twitter, Pinterest, Google+, and Instagram. *American Behavioral Scientist, 61*(7), 741–756. https://doi.org/10.1177/0002764217717559

Bogost, I. (2015, January 15). The Cathedral of computation. *The Atlantic.* https://www.theatlantic.com/technology/archive/2015/01/the-cathedral-of-computation/384300/

Bolukbasi, T., Chang, K.-W., Zou, J., Saligrama, V., & Kalai, A. (2016). Man is to computer programmer as woman is to homemaker? Debiasing word embeddings. *ArXiv:1607.06520 [Cs, Stat].* http://arxiv.org/abs/1607.06520

Bommasani, R., Hudson, D. A., Adeli, E., Altman, R., Arora, S., von Arx, S., Bernstein, M. S., Bohg, J., Bosselut, A., Brunskill, E., Brynjolfsson, E., Buch, S., Card, D., Castellon, R., Chatterji, N., Chen, A., Creel, K., Davis, J. Q., Demszky, D., et al. (2021). *On the opportunities and risks of foundation models.* Cornell University Library, arXiv.org. https://doi.org/10.48550/ARXIV.2108.07258

Botan, C. (1996). Communication work and electronic surveillance: A model for predicting panoptic effects. *Communication Monographs, 63*(4), 293–313. https://doi.org/10.1080/03637759609376396

Botan, & McCreadie. (1990). *Panopticon: Workplace of the information society.* International Communication Association Conference, Dublin, Ireland.

Bourke, B. (2014). Positionality: Reflecting on the research process. *The Qualitative Report, 19,* 1. https://doi.org/10.46743/2160-3715/2014.1026

boyd, danah. (2011). White flight in networked publics? How race and class shaped American teen engagement with MySpace and Facebook. In *Race After the Internet* (pp. 203–222). Routledge.

boyd, d., & Crawford, K. (2012). Critical questions for big data: Provocations for a cultural, technological, and scholarly phenomenon. *Information, Communication & Society, 15*(5), 662–679. https://doi.org/10.1080/1369118X.2012.678878

Brayne, S. (2020). *Predict and Surveil: Data, discretion, and the future of policing* (1st ed.). Oxford University Press. https://doi.org/10.1093/oso/9780190684099.001.0001

Breiman, L. (2001). Statistical Modeling: The two cultures (with comments and a rejoinder by the author). *Statistical Science, 16*(3), 199. https://doi.org/10.1214/ss/1009213726

British Sociological Association. (2016). *Ethics guidelines and collated resources for digital research. Statement of ethical practice annexe.*https://www.britsoc.co.uk/media/24309/bsa_statement_of_ethical_practice_annexe.pdf

Bu, Z., Xia, Z., & Wang, J. (2013). A sock puppet detection algorithm on virtual spaces. *Knowledge-Based Systems, 37,* 366–377. https://doi.org/10.1016/j.knosys.2012.08.016

Buchanan, E. A. (2011). Internet research ethics: Past, present, and future. In M. Consalvo & C. Ess (Eds.), *The handbook of internet studies* (pp. 83–108). Wiley-Blackwell. https://doi.org/ 10.1002/9781444314861.ch5

Bull, S., Cheah, P. Y., Denny, S., Jao, I., Marsh, V., Merson, L., Shah More, N., Nhan, L. N. T., Osrin, D., Tangseefa, D., Wassenaar, D., & Parker, M. (2015). Best practices for ethical sharing of individual-level Health Research data from low- and middle-income settings. *Journal of Empirical Research on Human Research Ethics, 10*(3), 302–313. https://doi.org/10.1177/ 1556264615594606

Caldarelli, G., Wolf, S., & Moreno, Y. (2018). Physics of humans, physics for society. *Nature Physics, 14*(9), 870–870. https://doi.org/10.1038/s41567-018-0266-x

Caliskan, A., Bryson, J. J., & Narayanan, A. (2017). Semantics derived automatically from language corpora contain human-like biases. *Science, 356*(6334), 183–186. https://doi.org/ 10.1126/science.aal4230

Capurro, R. (2005). Privacy. An intercultural perspective. *Ethics and Information Technology, 7*(1), 37–47. https://doi.org/10.1007/s10676-005-4407-4

Capurro, R. (2008). Intercultural information ethics: Foundations and applications. *Journal of Information, Communication and Ethics in Society, 6*(2), 116–126. https://doi.org/10.1108/ 14779960810888347

Cardon, D. (2016). Deconstructing the algorithm: Four types of digital information calculations. In R. Seyfert & J. Roberge (Eds.), *Algorithmic cultures* (pp. 95–110). Routledge. http:// spire.sciencespo.fr/hdl:/2441/19a26i12vl9epootg7j45rfpmk

Carpentier, N. (2011). *Media and participation: A site of ideological-democratic struggle*. Intellect Ltd. https://doi.org/10.26530/OAPEN_606390

Chen, S.-H. (Ed.). (2018). *Big data in computational social science and humanities* (1st ed.). Springer. https://doi.org/10.1007/978-3-319-95465-3

Chen, Z., & Whitney, D. (2019). Tracking the affective state of unseen persons. *Proceedings of the National Academy of Sciences, 116*(15), 7559–7564. https://doi.org/10.1073/pnas.1812250116

Cioffi-Revilla, C. (2014). *Introduction to computational social science*. Springer London. https:// doi.org/10.1007/978-1-4471-5661-1

Cohen, J. E. (2019a). *Between truth and power: The legal constructions of informational capitalism*. Oxford University Press.

Cohen, J. E. (2019b). Review of Zuboff's the age of surveillance capitalism. *Surveillance & Society, 17*(1/2), 240–245. https://doi.org/10.24908/ss.v17i1/2.13144

Cohen, R., & Ruths, D. (2013). Classifying political orientation on twitter: It's not easy! *Proceedings of the International AAAI Conference on Web and Social Media, 7*(1), 91–99.

Collins, G. S., Reitsma, J. B., Altman, D. G., & Moons, K. (2015). Transparent reporting of a multivariable prediction model for individual prognosis or diagnosis (TRIPOD): The TRIPOD statement. *BMC Medicine, 13*(1), 1. https://doi.org/10.1186/s12916-014-0241-z

Collmann, J., & Matei, S. A. (2016). *Ethical reasoning in big data: An exploratory analysis* (1st ed.). Springer. https://doi.org/10.1007/978-3-319-28422-4

Conte, R., Gilbert, N., Bonelli, G., Cioffi-Revilla, C., Deffuant, G., Kertesz, J., Loreto, V., Moat, S., Nadal, J.-P., Sanchez, A., Nowak, A., Flache, A., San Miguel, M., & Helbing, D. (2012). Manifesto of computational social science. *The European Physical Journal Special Topics, 214*(1), 325–346. https://doi.org/10.1140/epjst/e2012-01697-8

Cosentino, G. (2020). *Social media and the post-truth world order: The global dynamics of disinformation*. Springer International Publishing. https://doi.org/10.1007/978-3-030-43005-4

Cottom, T. M. (2016). *Black cyberfeminism: Intersectionality, institutions and digital sociology*. Policy Press.

Crane, J. (2011). Scrambling for Africa? Universities and global health. *The Lancet, 377*(9775), 1388–1390. https://doi.org/10.1016/S0140-6736(10)61920-4

Crawford, K. (2014, May 30). The anxieties of big data. *The New Inquiry*. https:// thenewinquiry.com/the-anxieties-of-big-data/

D'Ancona, M. (2017). *Post truth: The new war on truth and how to fight back*. Ebury Press.

De Cleen, B., & Carpentier, N. (2008). Introduction: Blurring participations and convergences. In N. Carpentier & B. De Cleen (Eds.), *Participation and media production. Critical reflections on content creation* (pp. 1–12). Cambridge Scholars Publishing.

de Montjoye, Y.-A., Radaelli, L., Singh, V. K., & "Sandy" Pentland, A. (2015). Unique in the shopping mall: On the reidentifiability of credit card metadata. *Science, 347*(6221), 536–539. https://doi.org/10.1126/science.1256297

Dean, J. (2010). *Blog theory: Feedback and capture in the circuits of drive*. Polity Press.

Dewey, J. (1938). *Logic: The theory of inquiry*. Holt, Richart and Winston.

D'Ignazio, C., & Klein, L. F. (2020). *Data feminism*. The MIT Press.

Dobrick, F. M., Fischer, J., & Hagen, L. M. (Eds.). (2018). *Research ethics in the digital age*. Springer Fachmedien Wiesbaden. https://doi.org/10.1007/978-3-658-12909-5

Engel, U. (2021). Causal and predictive modeling in computational social science. In I. U. Engel, A. Quan-Haase, S. X. Liu, & L. Lyberg (Eds.), *Handbook of computational social science, volume 1* (1st ed., pp. 131–149). Routledge. https://doi.org/10.4324/9781003024583-10

Ess, C., & Jones, S. (2004). Ethical decision-making and Internet research: Recommendations from the aoir ethics working committee. In *Readings in virtual research ethics: Issues and controversies* (pp. 27–44). IGI Global.

Ess, C. M. (2020). Interpretative pros hen pluralism: From computer-mediated colonization to a pluralistic intercultural digital ethics. *Philosophy & Technology, 33*(4), 551–569. https://doi.org/10.1007/s13347-020-00412-9

Eynon, R., Fry, J., & Schroeder, R. (2017). The ethics of online research. In I. N. Fielding, R. Lee, & G. Blank (Eds.), *The SAGE handbook of online research methods* (pp. 19–37). SAGE Publications, Ltd. https://doi.org/10.4135/9781473957992.n2

Feenberg, A. (1999). *Questioning technology*. Routledge.

Feenberg, A. (2002). *Transforming technology: A critical theory revisited*. Oxford University Press.

Ferrara, E. (2015). *Manipulation and abuse on social media*. Cornell University Library, arXiv.org. https://doi.org/10.48550/ARXIV.1503.03752

Ferrara, E., Varol, O., Davis, C., Menczer, F., & Flammini, A. (2016). The rise of social bots. *Communications of the ACM, 59*(7), 96–104. https://doi.org/10.1145/2818717

Fisher, E., & Rip, A. (2013). Responsible innovation: Multi-level dynamics and soft intervention practices. In R. Owen, J. Bessant, & M. Heintz (Eds.), *Responsible Innovation* (pp. 165–183). Wiley. https://doi.org/10.1002/9781118551424.ch9

Fourcade, M., & Gordon, J. (2020). Learning like a state: Statecraft in the digital age. *Journal of Law and Political Economy, 1*(1). https://doi.org/10.5070/LP61150258

Franzke, Aline Shakti, Bechmann, A., Zimmer, M., Ess, C. M., & the Association of Internet Researchers. (2020). *Internet research: Ethical guidelines 3.0*. https://aoir.org/reports/ethics3.pdf

Friedline, T., Naraharisetti, S., & Weaver, A. (2020). Digital redlining: Poor rural communities' access to fintech and implications for financial inclusion. *Journal of Poverty, 24*(5–6), 517–541. https://doi.org/10.1080/10875549.2019.1695162

Friemel, T. N. (2016). The digital divide has grown old: Determinants of a digital divide among seniors. *New Media & Society, 18*(2), 313–331. https://doi.org/10.1177/1461444814538648

Fuchs, C. (2018). 'Dear Mr. Neo-Nazi, Can You Please Give Me Your Informed Consent So That I Can Quote Your Fascist Tweet?': Questions of social media research ethics in online ideology critique. In G. Meikle (Ed.), *The Routledge companion to media and activism*. Routledge.

Fuchs, C. (2021). *Social media: A critical introduction* (3rd ed.). SAGE.

Garrido-Muñoz, I., Montejo-Ráez, A., Martínez-Santiago, F., & Ureña-López, L. A. (2021). A survey on bias in deep NLP. *Applied Sciences, 11*(7), 3184. https://doi.org/10.3390/app11073184

Gebru, T., Morgenstern, J., Vecchione, B., Vaughan, J. W., Wallach, H., Iii, H. D., & Crawford, K. (2021). Datasheets for datasets. *Communications of the ACM, 64*(12), 86–92. https://doi.org/10.1145/3458723

Gifford, C. (2020, June 15). The problem with emotion-detection technology. *The New Economy*. https://www.theneweconomy.com/technology/the-problem-with-emotion-detection-technology

Giglietto, F., Rossi, L., & Bennato, D. (2012). The open laboratory: Limits and possibilities of using Facebook, twitter, and YouTube as a research data source. *Journal of Technology in Human Services, 30*(3–4), 145–159. https://doi.org/10.1080/15228835.2012.743797

Gilbert, G. N. (Ed.). (2010). *Computational Social Science*. SAGE.

Gillespie, T. (2014). The relevance of algorithms. In T. Gillespie, P. J. Boczkowski, & K. A. Foot (Eds.), *Media technologies* (pp. 167–194). The MIT Press. https://doi.org/10.7551/mitpress/9780262525374.003.0009

Goel, V. (2014, June 29). Facebook tinkers with users' emotions in news feed experiment, stirring outcry. *The New York Times*. https://www.nytimes.com/2014/06/30/technology/facebook-tinkers-with-users-emotions-in-news-feed-experiment-stirring-outcry.html

Grady, C. (2015). Institutional review boards. *Chest, 148*(5), 1148–1155. https://doi.org/10.1378/chest.15-0706

Grimmelmann, J. (2015). The law and ethics of experiments on social media users. *Colorado Technology Law Journal, 13*, 219.

Gupta, P., Srinivasan, B., Balasubramaniyan, V., & Ahamad, M. (2015). Phoneypot: Data-driven understanding of telephony threats. In: *Proceedings 2015 Network and Distributed System Security Symposium*. Network and Distributed System Security Symposium, San Diego, CA. https://doi.org/10.14722/ndss.2015.23176

Gupta, U., Kim, Y. G., Lee, S., Tse, J., Lee, H.-H. S., Wei, G.-Y., Brooks, D., & Wu, C.-J. (2020). *Chasing carbon: The elusive environmental footprint of computing*. Cornell University Library, arXiv.org. https://doi.org/10.48550/ARXIV.2011.02839

Habermas, J. (1988). *On the logic of the social sciences*. MIT Pr.

Haight, M., Quan-Haase, A., & Corbett, B. A. (2014). Revisiting the digital divide in Canada: The impact of demographic factors on access to the internet, level of online activity, and social networking site usage. *Information, Communication & Society, 17*(4), 503–519. https://doi.org/10.1080/1369118X.2014.891633

Halbertal, M. (2015, November 11). *The Dewey lecture: Three concepts of human dignity*. https://www.law.uchicago.edu/news/dewey-lecture-three-concepts-human-dignity

Haraway, D. (1988). Situated knowledges: The science question in feminism and the privilege of partial perspective. *Feminist Studies, 14*(3), 575. https://doi.org/10.2307/3178066

Harding, S. (1992). Rethinking standpoint epistemology: What is 'strong objectivity?'. *The Centennial Review, 36*(3), 437–470. JSTOR.

Harding, S. (1995). 'Strong objectivity': A response to the new objectivity question. *Synthese, 104*(3), 331–349. https://doi.org/10.1007/BF01064504

Harding, S. G. (2008). *Sciences from below: Feminisms, postcolonialities, and modernities*. Duke University Press.

Harding, S. G. (2015). *Objectivity and diversity: Another logic of scientific research*. The University of Chicago Press.

Hargittai, E. (2015). Is bigger always better? Potential biases of big data derived from social network sites. *The Annals of the American Academy of Political and Social Science, 659*(1), 63–76. https://doi.org/10.1177/0002716215570866

Hargittai, E. (2020). Potential biases in big data: Omitted voices on social media. *Social Science Computer Review, 38*(1), 10–24.

Hargittai, E., & Litt, E. (2012). Becoming a tweep: How prior online experiences influence Twitter use. *Information, Communication & Society, 15*(5), 680–702. https://doi.org/10.1080/1369118X.2012.666256

Harsin, J. (2018). Post-truth and critical communication studies. In J. Harsin (Ed.), *Oxford research Encyclopedia of communication*. Oxford University Press. https://doi.org/10.1093/acrefore/9780190228613.013.757

Healy, K. (2015). The performativity of networks. *European Journal of Sociology, 56*(2), 175–205. https://doi.org/10.1017/S0003975615000107

Hecht, B., Wilcox, L., Bigham, J. P., Schöning, J., Hoque, E., Ernst, J., Bisk, Y., De Russis, L., Yarosh, L., Anjum, B., Contractor, D., & Wu, C. (2021). It's time to do something: Mitigating the negative impacts of computing through a change to the peer review process. *ArXiv:2112.09544 [Cs]*. http://arxiv.org/abs/2112.09544

Helbing, D., Frey, B. S., Gigerenzer, G., Hafen, E., Hagner, M., Hofstetter, Y., van den Hoven, J., Zicari, R. V., & Zwitter, A. (2019). Will democracy survive big data and artificial intelligence? In D. Helbing (Ed.), *Towards digital enlightenment* (pp. 73–98). Springer International Publishing. https://doi.org/10.1007/978-3-319-90869-4_7

Henderson, M., Johnson, N. F., & Auld, G. (2013). Silences of ethical practice: Dilemmas for researchers using social media. *Educational Research and Evaluation, 19*(6), 546–560. https://doi.org/10.1080/13803611.2013.805656

Hern, A. (2021, September 8). Study finds growing government use of sensitive data to 'nudge' behaviour. *The Guardian*. https://www.theguardian.com/technology/2021/sep/08/study-finds-growing-government-use-of-sensitive-data-to-nudge-behaviour%23:~:text=Study%20finds%20growing%20government%20use%20of%20sensitive%20data%20to%20'nudge'%20behaviour,-This%20article%20is&text=A%20new%20form%20of%20%E2%80%9Cinfluence,tech%20firms%2C%20researchers%20have%20warned

Hindman, M. (2015). Building better models: Prediction, replication, and machine learning in the social sciences. *The Annals of the American Academy of Political and Social Science, 659*(1), 48–62. https://doi.org/10.1177/0002716215570279

Hoegen, R., Gratch, J., Parkinson, B., & Shore, D. (2019). Signals of emotion regulation in a social dilemma: Detection from face and context. In: *2019 8th International Conference on Affective Computing and Intelligent Interaction (ACII)* (pp. 1–7). https://doi.org/10.1109/ACII.2019.8925478

Hofman, J. M., Watts, D. J., Athey, S., Garip, F., Griffiths, T. L., Kleinberg, J., Margetts, H., Mullainathan, S., Salganik, M. J., Vazire, S., Vespignani, A., & Yarkoni, T. (2021). Integrating explanation and prediction in computational social science. *Nature, 595*(7866), 181–188. https://doi.org/10.1038/s41586-021-03659-0

Holland, S., Hosny, A., Newman, S., Joseph, J., & Chmielinski, K. (2018). The dataset nutrition label: A framework to drive higher data quality standards. *ArXiv:1805.03677 [Cs]*. http://arxiv.org/abs/1805.03677

Hollingshead, W., Quan-Haase, A., & Chen, W. (2021). Ethics and privacy in computational social science. A call for pedagogy. *Handbook of Computational Social Science, 1*, 171–185.

Hongladarom, S., & Ess, C. (2007). *Information technology ethics: Cultural perspectives*. IGI Global. https://doi.org/10.4018/978-1-59904-310-4

Iyyer, M., Enns, P., Boyd-Graber, J., & Resnik, P. (2014). Political ideology detection using recursive neural networks. In: *Proceedings of the 52nd Annual Meeting of the Association for Computational Linguistics (Volume 1: Long Papers)* (pp. 1113–1122). https://doi.org/10.3115/v1/P14-1105

James, G., Witten, D., Hastie, T., & Tibshirani, R. (Eds.). (2013). *An introduction to statistical learning: With applications in R*. Springer.

Jiang, M. (2013). Internet sovereignty: A new paradigm of internet governance. In M. Haerens & M. Zott (Eds.), *Internet censorship (opposing viewpoints series)* (pp. 23–28). Greenhaven Press.

John, N. A. (2013). Sharing and Web 2.0: The emergence of a keyword. *New Media & Society, 15*(2), 167–182. https://doi.org/10.1177/1461444812450684

Johnson, I., McMahon, C., Schöning, J., & Hecht, B. (2017). The effect of population and 'structural' biases on social media-based algorithms: A case study in geolocation inference across the urban-rural spectrum. In: *Proceedings of the 2017 CHI Conference on Human Factors in Computing Systems* (pp. 1167–1178). https://doi.org/10.1145/3025453.3026015

Johnson, K. (2020, February 24). NeurIPS requires AI researchers to account for societal impact and financial conflicts of interest. *Venturebeat*. https://venturebeat.com/ai/neurips-requires-ai-researchers-to-account-for-societal-impact-and-financial-conflicts-of-interest/

Joinson, A. N., Woodley, A., & Reips, U.-D. (2007). Personalization, authentication and self-disclosure in self-administered internet surveys. *Computers in Human Behavior, 23*(1), 275–285. https://doi.org/10.1016/j.chb.2004.10.012

Kellogg, K. C., Valentine, M. A., & Christin, A. (2020). Algorithms at work: The new contested terrain of control. *Academy of Management Annals, 14*(1), 366–410. https://doi.org/10.5465/annals.2018.0174

Kezar, A. (2002). Reconstructing static images of leadership: An application of positionality theory. *Journal of Leadership Studies, 8*(3), 94–109. https://doi.org/10.1177/107179190200800308

Kitchin, R. (2014). Big data, new epistemologies and paradigm shifts. *Big Data & Society, 1*(1), 205395171452848. https://doi.org/10.1177/2053951714528481

Kraut, R., Olson, J., Banaji, M., Bruckman, A., Cohen, J., & Couper, M. (2004). Psychological research online: Report of Board of Scientific Affairs' advisory group on the conduct of research on the internet. *American Psychologist, 59*(2), 105–117. https://doi.org/10.1037/0003-066X.59.2.105

Lannelongue, L., Grealey, J., & Inouye, M. (2021). Green algorithms: Quantifying the carbon footprint of computation. *Advanced Science, 8*(12), 2100707. https://doi.org/10.1002/advs.202100707

Lawrence, H. M. (2021). Siri Disciplines. In T. S. Mullaney, B. Peters, M. Hicks, & K. Philip (Eds.), *Your computer is on fire* (pp. 179–198). The MIT Press. https://doi.org/10.7551/mitpress/10993.003.0013

Lazer, D., Kennedy, R., King, G., & Vespignani, A. (2014). The parable of Google Flu: Traps in big data analysis. *Science, 343*(6176), 1203–1205.

Lazer, D. M. J., Pentland, A., Watts, D. J., Aral, S., Athey, S., Contractor, N., Freelon, D., Gonzalez-Bailon, S., King, G., Margetts, H., Nelson, A., Salganik, M. J., Strohmaier, M., Vespignani, A., & Wagner, C. (2020). Computational social science: Obstacles and opportunities. *Science, 369*(6507), 1060–1062. https://doi.org/10.1126/science.aaz8170

Lazer, D., & Radford, J. (2017). Data ex machina: Introduction to big data. *Annual Review of Sociology, 43*(1), 19–39. https://doi.org/10.1146/annurev-soc-060116-053457

Leins, K., Lau, J. H., & Baldwin, T. (2020). Give me convenience and give her death: Who should decide what uses of NLP are appropriate, and on what basis? *ArXiv:2005.13213 [Cs].* http://arxiv.org/abs/2005.13213

Leonelli, S. (2013). Why the current insistence on open access to scientific data? Big data, knowledge production, and the political economy of contemporary biology. *Bulletin of Science, Technology & Society, 33*(1–2), 6–11. https://doi.org/10.1177/0270467613496768

Leonelli, S. (2021). Data science in times of pan(dem)ic. *Harvard Data Science Review.*https://doi.org/10.1162/99608f92.fbb1bdd6

Leslie, D. (2019). Understanding artificial intelligence ethics and safety. *ArXiv:1906.05684 [Cs, Stat].* doi:https://doi.org/10.5281/zenodo.3240529

Leslie, D. (2020). Tackling COVID-19 through responsible AI innovation: Five steps in the right direction. *Harvard Data Science Review.*https://doi.org/10.1162/99608f92.4bb9d7a7

Leslie, D., Burr, C., Aitken, M., Katell, M., Briggs, M., & Rincón, C. (2021). Human rights, democracy, and the rule of law assurance framework: A proposal. *The Alan Turing Institute.*https://doi.org/10.5281/zenodo.5981676

Leslie, D., Katell, M., Aitken, M., Singh, J., Briggs, M., Powell, R., Rincón, C., Chengeta, T., Birhane, A., Perini, A., Jayadeva, S., & Mazumder, A. (2022a). *Advancing data justice research and practice: An integrated literature review.* Zenodo. https://doi.org/10.5281/ZENODO.6408304

Leslie, D., Katell, M., Aitken, M., Singh, J., Briggs, M., Powell, R., Rincón, C., Perini, A., Jayadeva, S., & Burr, C. (2022c). *Data justice in practice: A guide for developers.* arXiv.org. https://doi.org/10.5281/ZENODO.6428185

Leslie, D., Rincón, C., Burr, C., Aitken, M., Katell, M., & Briggs, M. (2022b). *AI fairness in practice.* The Alan Turing Institute and the UK Office for AI.

Leslie, D., Rincón, C., Burr, C., Aitken, M., Katell, M., & Briggs, M. (2022d). *AI sustainability in practice: Part I.* The Alan Turing Institute and the UK Office for AI.

Leslie, D., Rincón, C., Burr, C., Aitken, M., Katell, M., & Briggs, M. (2022e). *AI sustainability in practice: Part II.* The Alan Turing Institute and the UK Office for AI.

Lin, J. (2015). On building better mousetraps and understanding the human condition: Reflections on big data in the social sciences. *The Annals of the American Academy of Political and Social Science, 659*(1), 33–47. https://doi.org/10.1177/0002716215569174

Lin, L. Y. I., Sidani, J. E., Shensa, A., Radovic, A., Miller, E., Colditz, J. B., Hoffman, B. L., Giles, L. M., & Primack, B. A. (2016). Association between social media use and depression among U.S. young adults. *Depression and Anxiety, 33*(4), 323–331. https://doi.org/10.1002/da.22466

Lomborg, S. (2013). Personal internet archives and ethics. *Research Ethics, 9*(1), 20–31. https://doi.org/10.1177/1747016112459450

Longo, J., Kuras, E., Smith, H., Hondula, D. M., & Johnston, E. (2017). Technology use, exposure to natural hazards, and being digitally invisible: Implications for policy analytics: Policy implications of the digitally invisible. *Policy & Internet, 9*(1), 76–108. https://doi.org/10.1002/poi3.144

Lorenz, T. (2014, March 7). Plugin allows you to recreate Facebook's controversial mood-altering experiment on YOUR News Feed. *The Daily Mail.* https://www.dailymail.co.uk/sciencetech/article-2678561/Facebook-mood-altering-experiment-News-Feed.html

Lucy, L., & Bamman, D. (2021). Gender and representation bias in GPT-3 generated stories. *Proceedings of the Third Workshop on Narrative Understanding* (pp. 48–55). https://doi.org/10.18653/v1/2021.nuse-1.5

Lyon, D. (Ed.). (2003). *Surveillance as social sorting: Privacy, risk, and digital discrimination.* Routledge

Mahmoodi, J., Leckelt, M., van Zalk, M., Geukes, K., & Back, M. (2017). Big data approaches in social and behavioral science: Four key trade-offs and a call for integration. *Current Opinion in Behavioral Sciences, 18*, 57–62. https://doi.org/10.1016/j.cobeha.2017.07.001

Manovich, L. (2011). Trending: The promises and the challenges of big social data. *Debates in the Digital Humanities, 2*(1), 460–475.

Markham, A. (2006). Ethic as method, method as ethic: A case for reflexivity in qualitative ICT research. *Journal of Information Ethics, 15*(2), 37–54. https://doi.org/10.3172/JIE.15.2.37

Markham, A., & Buchanan, E. (2012). *Ethical Decision-Making and Internet Research: Recommendations from the AoIR Ethics Working Committee (Version 2.0).* Association of Internet Researchers. https://aoir.org/reports/ethics2.pdf

Marx, G. T. (1988). Undercover: Police surveillance in America. *University of California Press..* http://site.ebrary.com/id/10676197

McIntyre, L. C. (2018). *Post-truth.* MIT Press.

Meho, L. I. (2006). E-mail interviewing in qualitative research: A methodological discussion. *Journal of the American Society for Information Science and Technology, 57*(10), 1284–1295. https://doi.org/10.1002/asi.20416

Mehrabi, N., Morstatter, F., Saxena, N., Lerman, K., & Galstyan, A. (2021). A survey on bias and fairness in machine learning. *ACM Computing Surveys, 54*(6), 1–35. https://doi.org/10.1145/3457607

Méndez-Diaz, N., Akabr, G., & Parker-Barnes, L. (2022). The evolution of social media and the impact on modern therapeutic relationships. *The Family Journal, 30*(1), 59–66. https://doi.org/10.1177/10664807211052495

Meng, X.-L. (2018). Statistical paradises and paradoxes in big data (I): Law of large populations, big data paradox, and the 2016 US presidential election. *The Annals of Applied Statistics, 12*(2). https://doi.org/10.1214/18-AOAS1161SF

Merriam, S. B., Johnson-Bailey, J., Lee, M.-Y., Kee, Y., Ntseane, G., & Muhamad, M. (2001). Power and positionality: Negotiating insider/outsider status within and across cultures. *International Journal of Lifelong Education, 20*(5), 405–416. https://doi.org/10.1080/02601370120490

Merson, L., Phong, T. V., Nhan, L. N. T., Dung, N. T., Ngan, T. T. D., Kinh, N. V., Parker, M., & Bull, S. (2015). Trust, respect, and reciprocity: Informing culturally appropriate data-sharing practice in Vietnam. *Journal of Empirical Research on Human Research Ethics, 10*(3), 251–263. https://doi.org/10.1177/1556264615592387

Metcalf, J., & Crawford, K. (2016). Where are human subjects in big data research? The emerging ethics divide. *Big Data & Society, 3*(1), 205395171665021. https://doi.org/10.1177/2053951716650211

Meyer, R. (2014, June 28). Everything we know about Facebook's secret mood manipulation experiment. *The Atlantic.* https://www.theatlantic.com/technology/archive/2014/06/everything-we-know-about-facebooks-secret-mood-manipulation-experiment/373648/

Mislove, A., Lehmann, S., Ahn, Y.-Y., Onnela, J.-P., & Rosenquist, J. (2011). Understanding the demographics of Twitter users. *Proceedings of the International AAAI Conference on Web and Social Media, 5*(1), 554–557.

Mitchell, R. K., Lee, J. H., & Agle, B. R. (2017). Stakeholder prioritization work: The role of stakeholder salience in stakeholder research. In D. M. Wasieleski & J. Weber (Eds.), *Business and society 360* (Vol. 1, pp. 123–157). Emerald Publishing Limited. https://doi.org/10.1108/S2514-175920170000006

Moons, K. G. M., Altman, D. G., Reitsma, J. B., Ioannidis, J. P. A., Macaskill, P., Steyerberg, E. W., Vickers, A. J., Ransohoff, D. F., & Collins, G. S. (2015). Transparent reporting of a multivariable prediction model for individual prognosis or diagnosis (TRIPOD): Explanation and elaboration. *Annals of Internal Medicine, 162*(1), W1–W73. https://doi.org/10.7326/M14-0698

Moore, P. V. (2019). E(a)ffective precarity, control and resistance in the digitalised workplace. In D. Chandler & C. Fuchs (Eds.), *Digital objects, digital subjects* (pp. 125–144). University of Westminster Press; JSTOR. http://www.jstor.org/stable/j.ctvckq9qb.12

Moreno, M. A., Goniu, N., Moreno, P. S., & Diekema, D. (2013). Ethics of social media research: Common concerns and practical considerations. *Cyberpsychology, Behavior and Social Networking, 16*(9), 708–713. https://doi.org/10.1089/cyber.2012.0334

Muller, B. J. (2019). Biometric borders. In *Handbook on Critical Geographies of Migration.* Edward Elgar Publishing.

Nadeem, M., Bethke, A., & Reddy, S. (2020). StereoSet: Measuring stereotypical bias in pretrained language models. *ArXiv:2004.09456 [Cs].* http://arxiv.org/abs/2004.09456

Najafian, M., Hsu, W.-N., Ali, A., & Glass, J. (2017). Automatic speech recognition of Arabic multi-genre broadcast media. In: *2017 IEEE Automatic Speech Recognition and Understanding Workshop (ASRU)* (pp. 353–359). https://doi.org/10.1109/ASRU.2017.8268957

Nanayakkara, P., Hullman, J., & Diakopoulos, N. (2021). *Unpacking the expressed consequences of AI research in broader impact statements.* AI, Ethics, and Society. https://doi.org/10.48550/ARXIV.2105.04760

Narayanan, A., & Shmatikov, V. (2009). De-anonymizing Social Networks. In: *2009 30th IEEE Symposium on Security and Privacy* (pp. 173–187). https://doi.org/10.1109/SP.2009.22

National Committee for Research Ethics in the Social Sciences and the Humanities (NESH). (2019). *A guide to internet research ethics.* NESH.

Neural Information Processing Systems. (2021). *NeurIPS 2021 paper checklist guidelines.* https://neurips.cc/Conferences/2021/PaperInformation/PaperChecklist

Neural Information Processing Systems. (2022). *NeurIPS 2022 ethical review guidelines.* https://nips.cc/public/EthicsGuidelines

Neural Information Processing Systems Conference. (2020). *Getting started with NeurIPS 2020.* https://neuripsconf.medium.com/getting-started-with-neurips-2020-e350f9b39c28

Nissenbaum, H. (1998). Protecting privacy in an information age: The problem of privacy in public. *Law and Philosophy, 17*(5), 559–596. https://doi.org/10.1023/A:1006184504201

Nissenbaum, H. (2011). A contextual approach to privacy online. *Daedalus, 140*(4), 32–48. https://doi.org/10.1162/DAED_a_00113

Nixon, R. (2011). *Slow violence and the environmentalism of the poor.* Harvard University Press.

Noble, S. U. (2018). *Algorithms of oppression: How search engines reinforce racism*. New York University Press.

Nozza, D., Bianchi, F., & Hovy, D. (2021). HONEST: Measuring hurtful sentence completion in language models. In: *Proceedings of the 2021 Conference of the North American Chapter of the Association for Computational Linguistics: Human Language Technologies* (pp. 2398–2406). https://doi.org/10.18653/v1/2021.naacl-main.191

Obole, A., & Welsh, K. (2012). The danger of big data: Social media as computational social science. *First Monday, 17*(7). https://doi.org/10.5210/fm.v17i7.3993

Olson, R. (2008). *Science and scientism in nineteenth-century Europe*. University of Illinois Press.

Olteanu, A., Castillo, C., Diaz, F., & Kıcıman, E. (2019). Social data: Biases, methodological pitfalls, and ethical boundaries. *Frontiers in Big Data, 2*, 13.

O'Neil, C. (2016). *Weapons of math destruction: How big data increases inequality and threatens democracy* (1st ed.). Crown.

Ott, M., Choi, Y., Cardie, C., & Hancock, J. T. (2011). Finding deceptive opinion spam by any stretch of the imagination. *ArXiv:1107.4557 [Cs]*. http://arxiv.org/abs/1107.4557

Ottoni, R., Pesce, J. P., Las Casas, D., Franciscani, G., Jr., Meira, W., Jr., Kumaraguru, P., & Almeida, V. (2013). Ladies first: Analyzing gender roles and Behaviors in Pinterest. *Proceedings of the International AAAI Conference on Web and Social Media, 7*(1), 457–465.

Owen, R. (2014). The UK Engineering and Physical Sciences Research Council's commitment to a framework for responsible innovation. *Journal of Responsible Innovation, 1*(1), 113–117. https://doi.org/10.1080/23299460.2014.882065

Owen, R., Macnaghten, P., & Stilgoe, J. (2012). Responsible research and innovation: From science in society to science for society, with society. *Science and Public Policy, 39*(6), 751–760. https://doi.org/10.1093/scipol/scs093

Owen, R., Stilgoe, J., Macnaghten, P., Gorman, M., Fisher, E., & Guston, D. (2013). A framework for responsible innovation. *Responsible Innovation: Managing the Responsible Emergence of Science and Innovation in Society, 31*, 27–50.

Packer, B., Halpern, Y., Guajardo-Céspedes, M., & Mitchell, M. (2018, April 13). Text embeddings contain Bias. Here's why that matters. *Google AI*. https://developers.googleblog.com/2018/04/text-embedding-models-contain-bias.html

Paganoni, M. C. (2019). Ethical concerns over facial recognition technology. *Anglistica AION, 23*(1), 85–94. https://doi.org/10.19231/angl-aion.201915

Pasquale, F. (2020). *New laws of robotics: Defending human expertise in the age of AI*. The Belknap Press of Harvard University Press.

Pasquale, F., & Cashwell, G. (2018). Prediction, persuasion, and the jurisprudence of behaviourism. *University of Toronto Law Journal, 68*(supplement 1), 63–81. https://doi.org/10.3138/utlj.2017-0056

Pentland, A. (2015). *Social physics: How social networks can make us smarter*. Penguin Press.

Peterka-Bonetta, J., Sindermann, C., Elhai, J. D., & Montag, C. (2019). Personality associations with smartphone and internet use disorder: A comparison study including links to impulsivity and social anxiety. *Frontiers in Public Health, 7*, 127. https://doi.org/10.3389/fpubh.2019.00127

Preoţiuc-Pietro, D., Liu, Y., Hopkins, D., & Ungar, L. (2017). Beyond binary labels: Political ideology prediction of Twitter users. In: *Proceedings of the 55th Annual Meeting of the Association for Computational Linguistics* (*Volume 1: Long Papers*, pp. 729–740). https://doi.org/10.18653/v1/P17-1068

Prunkl, C. E. A., Ashurst, C., Anderljung, M., Webb, H., Leike, J., & Dafoe, A. (2021). Institutionalizing ethics in AI through broader impact requirements. *Nature Machine Intelligence, 3*(2), 104–110. https://doi.org/10.1038/s42256-021-00298-y

Puschmann, C., & Bozdag, E. (2014). Staking out the unclear ethical terrain of online social experiments. *Internet Policy Review, 3*(4). https://doi.org/10.14763/2014.4.338

Quan-Haase, A., & Ho, D. (2020). Online privacy concerns and privacy protection strategies among older adults in East York, Canada. *Journal of the Association for Information Science and Technology, 71*(9), 1089–1102. https://doi.org/10.1002/asi.24364

Quan-Haase, A., Williams, C., Kicevski, M., Elueze, I., & Wellman, B. (2018). Dividing the Grey divide: Deconstructing myths about older adults' online activities, skills, and attitudes. *American Behavioral Scientist, 62*(9), 1207–1228. https://doi.org/10.1177/0002764218777572

Raymond, N. (2019). Safeguards for human studies can't cope with big data. *Nature, 568*(7752), 277–277. https://doi.org/10.1038/d41586-019-01164-z

Reed, M. S., Graves, A., Dandy, N., Posthumus, H., Hubacek, K., Morris, J., Prell, C., Quinn, C. H., & Stringer, L. C. (2009). Who's in and why? A typology of stakeholder analysis methods for natural resource management. *Journal of Environmental Management, 90*(5), 1933–1949. https://doi.org/10.1016/j.jenvman.2009.01.001

Reidenberg, J. R. (2014). Privacy in public. *University of Miami Law Review, 69*, 141.

Resnik, D. B. (2018). *The ethics of research with human subjects: Protecting people, advancing science, promoting trust* (1st ed.). Springer. https://doi.org/10.1007/978-3-319-68756-8

Roberge, J., Morin, K., & Senneville, M. (2019). Deep Learning's governmentality: The other black box. In A. Sudmann (Ed.), *The democratization of artificial intelligence* (pp. 123–142). transcript Verlag. https://doi.org/10.1515/9783839447192-008

Ruths, D., & Pfeffer, J. (2014). Social media for large studies of behavior. *Science, 346*(6213), 1063–1064. https://doi.org/10.1126/science.346.6213.1063

Sagarra, O., Gutiérrez-Roig, M., Bonhoure, I., & Perelló, J. (2016). Citizen science practices for computational social science research: The conceptualization of pop-up experiments. *Frontiers in Physics, 3*. https://doi.org/10.3389/fphy.2015.00093

Salganik, M. J. (2019). *Bit by bit: Social research in the digital age*. https://app.kortext.com/Shibboleth.sso/Login?entityID=https%3A%2F%2Felibrary.exeter.ac.uk%2Fidp%2Fshibboleth&target=https://app.kortext.com/borrow/277287

Salganik, M., & The Summer Institutes in Computational Social Science. (n.d.). *Ethics and Computational Social Science*. https://sicss.io/overview/ethics-part-1

Sánchez-Monedero, J., Dencik, L., & Edwards, L. (2020). What does it mean to 'solve' the problem of discrimination in hiring?: Social, technical and legal perspectives from the UK on automated hiring systems. In: *Proceedings of the 2020 Conference on Fairness, Accountability, and Transparency*, (pp. 458–468). doi:https://doi.org/10.1145/3351095.3372849

Schmeer, K. (1999). Stakeholder analysis guidelines. *Policy Toolkit for Strengthening Health Sector Reform, 1*, 1–35.

Schroeder, R. (2014). Big data and the brave new world of social media research. *Big Data & Society, 1*(2), 205395171456319. https://doi.org/10.1177/2053951714563194

Schultheis, H. (2021). Computational cognitive modeling in the social sciences. In U. Engel, A. Quan-Haase, S. X. Liu, & L. Lyberg (Eds.), *Handbook of computational social science, volume 1* (pp. 53–65). Routledge.

Schwartz, R., Dodge, J., Smith, N. A., & Etzioni, O. (2020). Green AI. *Communications of the ACM, 63*(12), 54–63. https://doi.org/10.1145/3381831

Scott, H., & Woods, H. C. (2018). Fear of missing out and sleep: Cognitive behavioural factors in adolescents' nighttime social media use. *Journal of Adolescence, 68*(1), 61–65. https://doi.org/10.1016/j.adolescence.2018.07.009

Selinger, E., & Hartzog, W. (2020). The inconsentability of facial surveillance. *Loyola Law Review, 66*, 33.

Shah, D. V., Cappella, J. N., & Neuman, W. R. (2015). Big data, digital media, and computational social science: Possibilities and perils. *The Annals of the American Academy of Political and Social Science, 659*(1), 6–13. https://doi.org/10.1177/0002716215572084

Shaw, R. (2015). Big data and reality. *Big Data & Society, 2*(2), 205395171560887. https://doi.org/10.1177/2053951715608877

Shmueli, G. (2010). To explain or to predict? *Statistical Science, 25*(3). https://doi.org/10.1214/10-STS330

Shrum, W. (2005). Reagency of the internet, or, how I became a guest for science. *Social Studies of Science, 35*(5), 723–754. https://doi.org/10.1177/0306312705052106

Simon, H. A. (2002). Science seeks parsimony, not simplicity: Searching for pattern in phenomena. In A. Zellner, H. A. Keuzenkamp, & M. McAleer (Eds.), *Simplicity, inference and modelling:*

Keeping it sophisticatedly simple (pp. 32–72). Cambridge University Press. https://doi.org/10.1017/CBO9780511493164.003

Sloane, M., Moss, E., & Chowdhury, R. (2022). A Silicon Valley love triangle: Hiring algorithms, pseudo-science, and the quest for auditability. *Patterns, 3*(2), 100425. https://doi.org/10.1016/j.patter.2021.100425

Sorell, T. (2013). *Scientism: Philosophy and the infatuation with science.* Routledge.

Spaulding, N. W. (2020). Is human judgment necessary?: Artificial intelligence, algorithmic governance, and the law. In M. D. Dubber, F. Pasquale, & S. Das (Eds.), *The Oxford handbook of ethics of AI* (pp. 374–402). Oxford University Press. https://doi.org/10.1093/oxfordhb/9780190067397.013.25

Stark, L., & Hutson, J. (2021). Physiognomic artificial intelligence. *SSRN Electronic Journal.* https://doi.org/10.2139/ssrn.3927300

Steinmann, M., Shuster, J., Collmann, J., Matei, S. A., Tractenberg, R. E., FitzGerald, K., Morgan, G. J., & Richardson, D. (2015). Embedding privacy and ethical values in big data technology. In S. A. Matei, M. G. Russell, & E. Bertino (Eds.), *Transparency in social media* (pp. 277–301). Springer International Publishing. https://doi.org/10.1007/978-3-319-18552-1_15

Stier, S., Breuer, J., Siegers, P., & Thorson, K. (2020). Integrating survey data and digital trace data: Key issues in developing an emerging field. *Social Science Computer Review, 38*(5), 503–516. https://doi.org/10.1177/0894439319843669

Stilgoe, J., Watson, M., & Kuo, K. (2013). Public engagement with biotechnologies offers lessons for the governance of geoengineering research and beyond. *PLoS Biology, 11*(11), e1001707. https://doi.org/10.1371/journal.pbio.1001707

Striphas, T. (2015). Algorithmic culture. *European Journal of Cultural Studies, 18*(4–5), 395–412. https://doi.org/10.1177/1367549415577392

Strubell, E., Ganesh, A., & McCallum, A. (2019). *Energy and policy considerations for deep learning in NLP.* arXiv.org. https://doi.org/10.48550/ARXIV.1906.02243

Suresh, H., & Guttag, J. V. (2021). A framework for understanding sources of harm throughout the machine learning life cycle. *Equity and Access in Algorithms, Mechanisms, and Optimization,* 1–9. https://doi.org/10.1145/3465416.3483305

Sweeney, C., & Najafian, M. (2019). A transparent framework for evaluating unintended demographic bias in word embeddings. In: *Proceedings of the 57th Annual Meeting of the Association for Computational Linguistics* (pp. 1662–1667). https://doi.org/10.18653/v1/P19-1162

Sweeney, L. (2013). Discrimination in online ad delivery. *Communications of the ACM, 56*(5), 44–54. https://doi.org/10.1145/2447976.2447990

Syvertsen, T. (2020). *Digital detox: The politics of disconnecting.* Emerald Publishing.

Syvertsen, T., & Enli, G. (2020). Digital detox: Media resistance and the promise of authenticity. *Convergence: The International Journal of Research into New Media Technologies, 26*(5–6), 1269–1283. https://doi.org/10.1177/1354856519847325

Tauginienė, L., Butkevičienė, E., Vohland, K., Heinisch, B., Daskolia, M., Suškevičs, M., Portela, M., Balázs, B., & Prūse, B. (2020). Citizen science in the social sciences and humanities: The power of interdisciplinarity. *Palgrave Communications, 6*(1), 89. https://doi.org/10.1057/s41599-020-0471-y

Taylor, C. (2021). *The explanation of behaviour.* Routledge.

Theocharis, Y., & Jungherr, A. (2021). Computational social science and the study of political communication. *Political Communication, 38*(1–2), 1–22. https://doi.org/10.1080/10584609.2020.1833121

Törnberg, P., & Uitermark, J. (2021). For a heterodox computational social science. *Big Data & Society, 8*(2), 205395172110477. https://doi.org/10.1177/20539517211047725

Tritter, J. Q., & McCallum, A. (2006). The snakes and ladders of user involvement: Moving beyond Arnstein. *Health Policy, 76*(2), 156–168. https://doi.org/10.1016/j.healthpol.2005.05.008

Tufekci, Z. (2014). *Big questions for social media big data: Representativeness, validity and other methodological pitfalls.* Cornell University Library, arXiv.org. https://doi.org/10.48550/ARXIV.1403.7400

Vaidhyanathan, S. (2018). *Antisocial media: How facebook disconnects US and undermines democracy.* Oxford University Press.

van Dijck, J. (2013). *The culture of connectivity: A critical history of social media.* Oxford University Press.

van Dijck, J., Poell, T., & de Waal, M. (2018). *The platform society.* Oxford University Press.

Van Otterlo, M. (2014). Automated experimentation in Walden 3.0.: The next step in profiling, predicting, control and surveillance. *Surveillance & Society, 12*(2), 255–272. https://doi.org/10.24908/ss.v12i2.4600

Varnhagen, C. K., Gushta, M., Daniels, J., Peters, T. C., Parmar, N., Law, D., Hirsch, R., Sadler Takach, B., & Johnson, T. (2005). How informed is online informed consent? *Ethics & Behavior, 15*(1), 37–48. https://doi.org/10.1207/s15327019eb1501_3

Varvasovszky, Z., & Brugha, R. (2000). A stakeholder analysis. *Health Policy and Planning, 15*(3), 338–345. https://doi.org/10.1093/heapol/15.3.338

Viner, R. M., Gireesh, A., Stiglic, N., Hudson, L. D., Goddings, A.-L., Ward, J. L., & Nicholls, D. E. (2019). Roles of cyberbullying, sleep, and physical activity in mediating the effects of social media use on mental health and wellbeing among young people in England: A secondary analysis of longitudinal data. *The Lancet Child & Adolescent Health, 3*(10), 685–696. https://doi.org/10.1016/S2352-4642(19)30186-5

von Schomberg, R. (2013). A vision of responsible research and innovation. In R. Owen, J. Bessant, & M. Heintz (Eds.), *Responsible Innovation* (pp. 51–74). Wiley. https://doi.org/10.1002/9781118551424.ch3

von Wright, G. H. (2004). *Explanation and understanding.* Cornell University Press.

Wagner, C., Strohmaier, M., Olteanu, A., Kıcıman, E., Contractor, N., & Eliassi-Rad, T. (2021). Measuring algorithmically infused societies. *Nature, 595*(7866), 197–204. https://doi.org/10.1038/s41586-021-03666-1

Wallach, H. (2018). Computational social science ≠ computer science + social data. *Communications of the ACM, 61*(3), 42–44. https://doi.org/10.1145/3132698

Wang, G. A., Chen, H., Xu, J. J., & Atabakhsh, H. (2006). Automatically detecting criminal identity deception: An adaptive detection algorithm. *IEEE Transactions on Systems, Man, and Cybernetics - Part A: Systems and Humans, 36*(5), 988–999. https://doi.org/10.1109/TSMCA.2006.871799

Weber, M. (1978). *Economy and society: An outline of interpretive sociology* (Vol. 2). University of California press.

Weidinger, L., Mellor, J., Rauh, M., Griffin, C., Uesato, J., Huang, P.-S., Cheng, M., Glaese, M., Balle, B., Kasirzadeh, A., Kenton, Z., Brown, S., Hawkins, W., Stepleton, T., Biles, C., Birhane, A., Haas, J., Rimell, L., Hendricks, L. A., … Gabriel, I. (2021). Ethical and social risks of harm from Language Models. *ArXiv:2112.04359 [Cs].* http://arxiv.org/abs/2112.04359

Weinhardt, M. (2020). Ethical issues in the use of big data for social research. *Historical Social Research, 45*(3), 342–368. https://doi.org/10.12759/HSR.45.2020.3.342-368

Wittgenstein, L. (2009). *Philosophical investigations* (P. M. S. Hacker & J. Schulte, Eds.; G. E. M. Anscombe, P. M. S. Hacker, & J. Schulte, Trans.; Rev. 4th ed). Wiley-Blackwell.

Woods, H. C., & Scott, H. (2016). #Sleepyteens: Social media use in adolescence is associated with poor sleep quality, anxiety, depression and low self-esteem. *Journal of Adolescence, 51*(1), 41–49. https://doi.org/10.1016/j.adolescence.2016.05.008

Woolley, S. C. (2016). Automating power: Social bot interference in global politics. *First Monday.* https://doi.org/10.5210/fm.v21i4.6161

Woolley, S., & Howard, P. N. (Eds.). (2018). *Computational propaganda: Political parties, politicians, and political manipulation on social media.* Oxford University Press.

World Health Organization. (2022). *Report of the WHO global technical consultation on public health and social measures during health emergencies: Online meeting, 31 August to 2 September 2021.* World Health Organization. https://apps.who.int/iris/handle/10665/352096

Wright, J., Leslie, D., Raab, C., Ostmann, F., Briggs, M., & Kitagawa, F. (2021). *Privacy, agency and trust in human-AI ecosystems: Interim report (short version).* The Alan Turing Institute. https://www.turing.ac.uk/research/publications/privacy-agency-and-trust-human-ai-ecosystems-interim-report-short-version

Wu, T. (2019). Blind spot: The attention economy and the Law. *Antitrust Law Journal, 82*(3), 771–806.

Yarkoni, T., & Westfall, J. (2017). Choosing prediction over explanation in psychology: Lessons from machine learning. *Perspectives on Psychological Science, 12*(6), 1100–1122. https://doi.org/10.1177/1745691617693393

Yeung, K. (2017). 'Hypernudge': Big data as a mode of regulation by design. *Information, Communication & Society, 20*(1), 118–136. https://doi.org/10.1080/1369118X.2016.1186713

Zhao, J., Wang, T., Yatskar, M., Ordonez, V., & Chang, K.-W. (2017). *Men also like shopping: Reducing gender bias amplification using corpus-level constraints.* arXiv.org. https://doi.org/10.48550/ARXIV.1707.09457

Zheng, R., Li, J., Chen, H., & Huang, Z. (2006). A framework for authorship identification of online messages: Writing-style features and classification techniques. *Journal of the American Society for Information Science and Technology, 57*(3), 378–393. https://doi.org/10.1002/asi.20316

Ziewitz, M. (2016). Governing algorithms: Myth, mess, and methods. *Science, Technology, & Human Values, 41*(1), 3–16. https://doi.org/10.1177/0162243915608948

Zimmer, M. (2016, May 14). OkCupid study reveals the perils of big-data science. *Wired Magazine.* https://www.wired.com/2016/05/okcupid-study-reveals-perils-big-data-science/

Zuboff, S. (2019). *The age of surveillance capitalism: The fight for a human future at the new frontier of power* (1st ed.). Public Affairs.

Zuckerman, E. (2020). *The case for digital public infrastructure.* Springer. https://doi.org/10.7916/D8-CHXD-JW34

Part II
Methodological Aspects

Chapter 5
Modelling Complexity with Unconventional Data: Foundational Issues in Computational Social Science

Magda Fontana and Marco Guerzoni

Abstract The large availability of data, often from unconventional sources, does not call for a data-driven and theory-free approach to social science. On the contrary, (big) data eventually unveil the complexity of socio-economic relations, which has been too often disregarded in traditional approaches. Consequently, this paradigm shift requires to develop new theories and modelling techniques to handle new types of information. In this chapter, we first tackle emerging challenges about the collection, storage, and processing of data, such as their ownership, privacy, and cybersecurity, but also potential biases and lack of quality. Secondly, we review data modelling techniques which can leverage on the new available information and allow us to analyse relationships at the microlevel both in space and in time. Finally, the complexity of the world revealed by the data and the techniques required to deal with such a complexity establishes a new framework for policy analysis. Policy makers can now rely on positive and quantitative instruments, helpful in understanding both the present scenarios and their future complex developments, although profoundly different from the standard experimental and normative framework. In the conclusion, we recall the preceding efforts required by the policy itself to fully realize the promises of computational social sciences.

M. Fontana (✉)
Department of Economics and Statistics, University of Turin, Turin, Italy
e-mail: magda.fontana@unito.it

M. Guerzoni
DEMS, University of Milan-Bicocca, Milano, Italy

BETA, University of Strasbourg, Strasbourg, France
e-mail: marco.guerzoni@unimib.it

© The Author(s) 2023
E. Bertoni et al. (eds.), *Handbook of Computational Social Science for Policy*,
https://doi.org/10.1007/978-3-031-16624-2_5

5.1 Introduction

We define CSS as the development and application of computational methods to complex, typically large-scale, human (sometimes simulated) behavioural data (Lazer et al., 2009). The large availability of data, the development of both algorithms and new modelling techniques, and the improvement of storage and computational power opened up a new scientific paradigm for social scientists willing to take into account the complexity of the social phenomena in their research. In this chapter, we develop the idea that the key transformation in place concerns two specific self-reinforcing events:

- (Big) data unveil the complexity of the world and pull for new modelling techniques
- New modelling techniques based on many data push the development of new tools in data science.

We do not share the view that this vast availability of data can allow science to be purely data-driven as in a word without theory (Anderson, 2008; Prensky, 2009). On the contrary, as other authors suggested, science needs more theory to account for the complexity of reality as revealed by the data (Carota et al., 2014; Gould, 1981; Kitchin, 2014; Nuccio & Guerzoni, 2019) and develop new modelling techniques. Obviously, in this age of abundance of information, data analysis occupies a privileged position and can eventually debate with the theory on a level playing field as it has never happened before.

Social sciences are not yet fully equipped to deal with this paradigm shift towards a quantitative, but positive, analysis. Indeed, economics developed an elegant, but purely normative, approach, while other social sciences, when not colonized by the economics' mainstream positive approach, remained mainly qualitative.

Consequently, this present shift can have a profound impact on the way researchers address research questions and, ultimately, also on policy questions. However, before this scientific paradigm unravels its potential, it needs to wind up any uncertainty about its process, specifically around the following issues:

- **Data as the input of the modelling process.** There are three levels of data-related issues:

  - How to collect the data and from which sources.
  - Data storage which relates with ownership, privacy, and cybersecurity.
  - Data quality and biases in data and data collection.

- **New modelling techniques for new data**, which account for heterogeneous, networked, geo-located, and time-stamped data.
- **Policy as the output of the process**, namely, the type of policy questions that can be addressed, e.g. positive vs. normative and prediction vs. causality.

The chapter is organized as follows: Sect. 5.2 frames the topic in the existing literature; Sect. 5.3 addresses the main issues that revolve around the making of computational social sciences (data sources, modelling techniques, and policy implications); and Sect. 5.4 discusses and concludes.

5.2 Existing Literature

Starting from the late 1980s, sciences have witnessed the increasing influence of complex system analyses. Far from a mechanistic conception of social and economic systems, complex systems pose several challenges to policy making:

- They are comprised of many diverse parts, and, therefore, the use of a representative individual is of no avail (Arthur, 2021).
- They operate on various temporal and spatial scales. It follows that the system behaviour cannot be derived from the mere summation of the behaviour of individual components(Arthur, 2021).
- They operate out of balance, where minor disturbances may lead to events of all dimensions. Thus, most of equilibrium-based theoretical framework used to devise (economic) policy does not apply (Bonabeau, 2002; Fontana, 2012).

In the aftermath of the economic crisis of 2008, the idea that social systems were more complex than what was so far assumed spreads in the policy[1] domain. Meanwhile, the European Union has been placing an increasing focus on complexity-based projects: "In complex systems, even if the local interactions among the various components may be simple, the overall behaviour is difficult and sometimes impossible to predict, and novel properties may emerge. Understanding this kind of complexity is helping to study and understand many different phenomena, from financial crises, global epidemics, propagation of news, connectivity of the internet, animal behaviour, and even the growth and evolution of cities and companies. Mathematical and computer-based models and simulations, often utilizing various techniques from statistical physics are at the heart of this initiative" (Complexity Research Initiative for Systemic InstabilitieS). Furthermore, a growing theoretical literature and the related empirical evidence (Loewenstein & Chater, 2017; Lourenço et al., 2016) spur policy makers to gradually substitute the rational choice framework with the behavioural approach that stresses the limitations in human decision-making. This change in ontology brings about a whole set of new policy features and, subsequently, new modelling challenges. Firstly, since local interaction of heterogeneous agents (consumers, households, states, industries)

---

[1] See also J. Landau, Deputy Governor of the Bank of France "Complex systems exhibit well-known features: non-linearity and discontinuities (a good example being liquidity freezes); path dependency; sensitivity to initial conditions. Together, those characteristics make the system truly unpredictable and uncertain, in the Knightian sense. Hence the spectacular failure of models during the crisis: most, if not all, were constructed on the assumption that stable and predictable (usually normal) distribution probabilities could be used to describe the different states of the financial system and the economy. They collapsed when extreme events occurred with a frequency that no one ever thought would be possible"(Cooper, 2011).

shapes the overall behaviour and performance of systems, ABMs are used to model the heterogeneity of the system's elements and describe their autonomous interaction. ABMs are computer simulations in which a system is modelled in terms of agents and their interactions (Bonabeau, 2002). Agents, which are autonomous, make decision on the basis of a set or rules and, often, adapt their action to the behaviour of other agents. ABM is being used to inform policy or decisions in various contexts. Recent examples include land use and agricultural policy (Dai et al., 2020), ecosystems and natural resource management (González et al., 2018), control of epidemics (Kerr et al., 2021; Truszkowska et al., 2021), economic policy (Chersoni et al., 2022; Dosi et al., 2020), institutional design (Benthall & Strandburg, 2021), and technology diffusion (Beretta et al., 2018). Moreover, ABM rely on the idea that information does not flow freely and homogeneously within systems and they often connect the policy domain to the field of network science (Kenis & Schneider, 2019). The position of an agent (a state, a firm, a decision-maker) within the network determines its ability to affect its neighbours and vice versa, while the overall structure of the network of agent's connection determines both how rapidly a signal travels the network and its resilience to shocks(Sorenson et al., 2006). Although social network analysis was initiated in the early years of the twentieth century, the last two decades have built on the increased availability of data and of computational resources to inaugurate the study of complex networks, i.e. those networks whose structure is irregular and dynamic and whose units are in the order of millions of nodes (Boccaletti et al., 2006). In the policy perspective, spreading and synchronization processes have a pivotal importance. The diffusion of a signal in a network has been used to model processes such as the diffusion of technologies and to explore static and dynamic robustness (Grassberger, 1983) to the removal of central or random nodes. It is worth noting that ABM and networks can be used jointly. Beretta et al. (2018) use ABM and network to show that cultural dissimilarity in Ethiopian Peasant Associations could impair the diffusion of a subsidized efficient technology, while Chersoni et al. (2021) use agent-based modelling and network analysis to simulate the adoption of technologies under different policy scenarios showing that the diffusion is very sensitive to the network topology. Secondly, the abandonment of the rational choice framework renders the mathematical maximization armoury ineffective and calls for new modelling approaches. The wide range of techniques that fall under the big tent of adaptive behaviours have a tight connections with data and algorithms. In addition to the heuristic and statistical models of behaviour, recent developments have perfected machine learning and evolutionary computation. These improve the representation of agents both by identifying patterns of behaviour in data and also by modelling agents' adaptation in simulations (Heppenstall et al., 2021; Runck et al., 2019). By providing agents with the ability to elaborate different data sources to adapt to their environment and to evolve the rules that are the most suitable response to a given set of inputs, machine learning constitutes an interesting tool to overcome the Lucas' critique. That is to say that it allows modelling individual adjustments to policy making, without renouncing the observed heterogeneity of agents and their dispersed interaction.

Thirdly, the information required to populate these models are not (only) the traditional socio-demographic and national account data but is more diverse and multifaceted.

5.3 Addressing Foundational Issues of CSS for Policy

The development and application of CSS implies a rethinking of the approach to policy making. The increased availability of data provides institutions with abundant information that broadens the spectrum of practicable interventions. Yet, the recognition that economic, social, and ecologic systems are complex phenomena imposes a rethinking of the modelling techniques and of the evaluation of their output. A further layer of complexity is that of data management. While traditional data are collected through institutional channels, new data sources require protocols to establish data ownership and privacy protection. In this section, we propose a three-pronged framework to develop an efficient approach to CSS.

5.3.1 Data as the Input of the Process

There are two main sources of data in the modelling process. On the one hand, the wide application of smart technologies in an increasing number of realms of social and economic life made the presence of sensors ubiquitous: For instance, they record information for machines on the shop floor; register pollution, traffic, and weather data in the smart city; and check and store vital parameters of athletes or sick people. On the other hand, the increased amount of activities occurring on the internet allows for detailed registration of the individual's behaviour with fine-grained details. The extraordinary effort by Blazquez and Domenech (2018) to create a taxonomy of all these possible data sources is a vain one since such an enterprise would require a constant update. However, from their work, it clearly emerges how this new world of data presents peculiar characteristics so far unconventional for the social scientist.

- First of all, most of the data collected are at the microlevel, being the unit of analysis about a single person, a firm, or a specific machine.
- Data are almost always geo-located and time-stamped with a very high precision. In other words, each observation with its attributes occupies a small point in a very dense time-space coordinates.
- There is an unprecedented data collection activity with the focus on interactions. We have at disposal for research any information about commercial transactions among both individuals and firms, which could eventually create a map of the economic activity of a system (Einav & Levin, 2014). At the same time, the advent of social platforms allows to register data on social interactions, which represent the networked world of human relationships.

- There are new data on people's behaviour coming from their searches on search engines, online purchasing activities, and reading and entertainment habits (Bello-Orgaz et al., 2016; Renner et al., 2020).

Moreover, the format of data collected is often unconventional. While statistics has been developed to deal with figures, most of the data available today record texts, images, and videos, only eventually transformed into binary figures by the process of digitization. These types of data convey new content of a paramount importance for the social scientist since it allows to analyse information about ideas, opinions, and feelings (Ambrosino et al., 2018; Fontana et al., 2019).

Thus, different from statistics, which evolved over the last century in time of data scarcity, the present state of the art in the use of data leverages precisely on the vast size of datasets in terms of number of observations, of their attributes, and of different data formats (Nuccio et al., 2020). As a consequence, newly collected data is increasingly stored in the same location, and there is a constant effort to link and merge existing datasets in data warehouses or data lakes. The traditional solution in data science for data storage is a data warehouse, in which data is extracted, transformed, and loaded, while more recently many organizations are opting for a data lake solution, which stores heterogeneously structured raw data from various sources (Ravat & Zhao, 2019). A concurrent and partly connected phenomenon is the widespread adoption of big data analytics as a service (BDaaS), that is, when firms and institutions rely on cloud services on online platforms for the storage and analysis of data (Aldinucci et al., 2018). As a result, there is an increasing presence of very large online databases. This present situation raises the following challenges to CSS.

- The collection, storage, and maintenance of vast datasets create competitive advantages for the private sectors, but it is a very costly process. For these reasons, firms are not always willing to share their data with third parties involved in research- or data-driven policy making. Moreover, even in the presence of an open approach to data sharing, privacy laws do not always allow it without the explicit authorization of subjects providing the data, as in the case of the European Union GDPR[2] (Peloquin et al., 2020; Suman & Pierce, 2018). Public organizations are increasingly digitized and becoming an important hub of data collection of fine-grained data. However, even in the presence of large investments, they often lack adequate human resource and organization capabilities for both the deployment of data warehouse and data lakes and their accessibility for research purposes.
- The capacity for investing in data structure and the ability to collect data are very skewed, with few large players owning a tremendous amount of data. Alphabet's yearly investment in production equipment, facilities, and data centres has been around 10 billion dollars for the last 5 years for the maintenance of about 15 exabytes of data (Nuccio & Guerzoni, 2019). This high concentration

---

[2] GDPR: https://gdpr-info.eu/

allows a small handful of players to exploit data at a scale incomparable even with the scientific community. As a result, scientific institutions need to rely on partnership with these private players to effectively conduct research.

- The storage of vast amounts also raises issues in cybersecurity since dataset can become a target for unlawful activities due to the monetary value of detailed and sensitive information. Once again, protection against possible cybersecurity threats requires investment in technologies and human capital which only large firms possess. The cost and the accountability involved might discourage the use of data for scientific purposes (Peloquin et al., 2020).

The availability of data does not free social science from its original curse, that is, employing data created elsewhere for different purposes than research. Data, even if very large, might not be representative of a population due to biases in the selection of the sample or because they are affected by measurement errors. Typically, data collected on the internet over-represent young cohorts—which are more prone to the use of technology—or rich households and their related socio-demographic characteristics, since they are rarely affected by the digital divide. Alternatively, data might lack some variables which represent the true key of a phenomenon under investigation. Important attributes might be missing because they are not measured (say expectation on the future) or not available for privacy concern (say gender or ethnicity) (Demoussis & Giannakopoulos, 2006; Hargittai & Hinnant, 2008). As an exemplary case for the depth of this problem, consider the widespread debate on the alleged racism of artificial intelligence. A prediction model might systematically provide biased estimation for individual in a specific ethnicity class, not because it is racist, but because it might be very efficient in fitting the data provided which (a) describe a racist reality, (b) show (over)under-representation of a specific ethnic group, and (c) lack important features (typically income) which might be the true explanation of a phenomenon and they are highly correlated with ethnicity.

In this case, a model can represent very well the data at disposal, but also its possible distortions. Thus, it will fail in being a correct support for policy making or research. It is thus of a paramount importance to have in place data quality evaluation practices (Corrocher et al., 2021). The next section discusses different methodologies at disposal to deal with this large availability of data.

5.3.2 Modelling Techniques for New Data

This availability of data reveals a no longer deniable complexity of the world and opens up for social scientists a vast array of possibilities under the condition that they go beyond "two-variable problem of simplicity" Weaver (1948). We now discuss some theoretical and empirical data techniques which recently reached their mature stage after decades of incubation. As recalled before, data available today are usually at the microlevel, geo-located, time-stamped, and characterized by attributes that described the interaction of the unit of observations both with other observations and with a non-stationary environment. Take, for instance, the phenomena of localization and diffusion. Geospatial data, initially limited to the study of geographical and environmental issues, are currently increasingly available and accessible. These data are highly complex in that they imply the management of several types of information: physical location of the observation and its attributes and, possibly, temporal information. Complexity further increases since such observations change in accordance to the activities taking place in a given location (e.g. resources depletion, fire diffusion, opinion, and epidemic dynamics) and that the agents undertaking those activities are, in turn, changed by the attributes of the location. The main challenge here is the simultaneous modelling of two independent processes: the interaction of the agents acting in a given location and the adaptation of the attributes of both. Any model attempting to grasp these fined-grained dynamic phenomena should account for these properties.

Agent-Based Modelling

These data are naturally dealt with agent-based modelling and networks. Agent-based modelling describes the system of interest in terms of agents (autonomous individuals with properties, actions, and possibly goals), of their environment (a geometrical, GIS, or network landscape with its own properties and actions), and of agent-agent, agent-environment, and environment-environment interactions that affect the action and internal state of both agents and environment (Wilenski & Rand, 2015). ABMs can be deployed in policy making in several ways. Policy can exploit their ability to cope with complex data, with data and theoretical assumptions (e.g. simulate different diffusion models in an empirical environment), and with interaction and heterogeneity. Literature agrees on two general explanatory mechanisms and three categories of applications. ABMs can be fruitfully applied when there are data or theories on individual behaviour and the overall pattern that emerge from it is unknown, *integrative understanding*, or when there is information on the aggregate pattern and the individual rules of behaviour are not known—*compositional understanding* (Wilenski & Rand, 2015). In both cases, ABM offers insights into policy and interventions in a prospective and/or retrospective framework.[3] Prospective models simulate the design of policies and investigate their potential effects. Since they rely on non-linear out-of-equilibrium theory, they can help in identifying critical thresholds and tipping points, i.e. small interventions that might trigger radical and irreversible changes in the system of interest (Bak et al., 1987). These are hardly treated with more traditional techniques. The identification of early warning signals of impending shifts (Donangelo et al., 2010) relies on the observation of increasing variance and changes in autocorrelation and skewness in time series data; however, traditional data are often too coarse-grained and cover a time window that is too small with respect to the rate of change of the system. Empirically calibrated ABMs instead can simulate long-term dynamics and the related interventions (see, for instance, Gualdi et al. (2015).

---

[3] This classification is proposed and discussed at length by Hammond (2015).

When multiple systems are involved—say, the economy and the environment—ABMs map the trade-offs or synergies of policies across qualitatively different systems. Moreover, they are useful to highlight the unintended or unexpected consequences of the interventions, especially when in vivo or in vitro experiments are expensive, unpractical, or unethical. Retrospective models are useful especially under the compositional understanding framework. Firstly, they can investigate why policy have or have not played out the way they were expected to. This is relevant especially when data do not exist. For instance, Chersoni et al. (2021) study the reasons behind the underinvestment in energy-efficient technologies in Europe in spite the EU-wide range of interventions. They start from data on households and simulate their—unobserved—connections to show that policy should account for behavioural and imitative motives beyond the traditional financial incentives. While retaining the heterogeneity of observations, ABMs can also reveal different effects of policies across sub-samples of the population. Retrospective models can be used in combination with the prospective ones as input in the policy design process.

Network Modelling
Policy can exploit the theoretical and empirical mapping provided by network modelling to improve the knowledge of the structure of connections among the elements of the systems of interest, to reinforce the resulting networks, and to guide the processes that unfold on it. Network modelling elaborate on the mapping by computing metrics (e.g. density, reciprocity, transitivity, centralization, and modularity) to characterize the network and to quantify its dimensions. The features associated with those metrics are key to understand the robustness of network to random or target nodes and to study the speed at which a signal travels on it. Once the structure of the network is known, policy makers can design their intervention in order to foster or prevent the processes that are driven by local interactions. For instance, it has been shown that small world networks maximize diffusion (Schilling & Phelps, 2007) and that policy that encourage the formation of distant connections can sustain the production of scientific knowledge (Chessa et al., 2013). The identification of pivotal nodes, on the other hand, allows the design of policy that target the most central or fragile components of the networks. Network modelling also contributes to the identification of tipping points and to the elaboration of the required preventive policies. If the elements of the systems are connected through a preferential attachment topology (for instance, the world banking system Benazzoli and Di Persio, 2016), then the system could experience radical and irreversible change if the most central nodes are hit, while it is resilient to random node removals (Eckhoff & Morters, 2013).

Explaining, Predicting, and Summarizing
That traditional modelling techniques based on optimization naturally suggest simple closed-form equations apt to be tested with econometric techniques does not come as a surprise. The funding father of econometrics Ragnar Frisch clearly emphasized the ancillary role of data analysis in economics with respect to the neoclassical theorizing by stating that econometrics should achieve "the advance-

ment of economic theory in its relation to statistics and mathematics" (Cowles, 1960) and not vice versa. However, the complexity of the world now revealed and measured by new data and modelled by networks and ABM pushes for an evolution in the analysis of data. There exist three types of approach in data analysis: causal explanation, prediction, and summarization of the data. Guerzoni et al. (2021) explain that the specificity of the three approaches with the most severe consequences is the way they deal with external validity. Standard econometrics techniques rely on inference, and the properties of estimators have been derived under strong assumptions on error distribution and for a small class of simple and usually linear models: the focus is on the creation of reliable sample either via experiments or by employing instrumental variables to account for possible endogeneity of the data. As a consequence, econometrics manages to be robust in terms of identifying specific causal relationships, but at the cost of a reduced model fitness, since simple and mostly linear models are always inappropriate to fit the complexity of the data. Moreover, further issues such as the number of degrees of freedom and multicollinearity reduce the use of a large number of variables. While a scientist might be satisfied with sound evidence on casual relationships, for policy making, this is a truly unfortunate situation. Knowing the causal impact of a policy measure on a target variable is surely important, but useless if this impact accounts for a tiny percentage of the overall variation of the target.

On the contrary, prediction models measure their uncertainty by looking at the accuracy of prediction on out-of-sample data. There are no restrictions in the type or complexity of the models (or combination of models), and the most advanced data processing techniques such as deep learning can fully displace their power. In this way, the prediction of future scenarios became possible at the expense of eliciting specific causal effects. The trade-off, known as bias-variance trade-off, is clear: On the one hand, simple econometric models allow us to identify an unbiased sample average response at the cost of inhibiting any accuracy of fitness. On the other hand, complex prediction models reach remarkable level of accuracy, even on the single future observation, but they are silent on specific causal relations. In this situation, the importance of complex theoretical approaches such as the ABM or network modelling becomes clear. Indeed, predicted result can be used to evaluate the rules of the model and the parameter settings. A complex theoretical model fine-tuned with many data and in line with predictions can be rather safely employed for policy analysis since it incorporates both theoretical insights on causal relationship and a verified prediction power (Beretta et al., 2018).

Lastly, summarizing techniques serve the purpose of classifying and display-ing, often with advanced visualization, properties of the data. Traditionally, the taxonomic approach to epistemology, that is, to create a partition of empirical obser-vations based on their characteristics, has been carried on by a careful qualitative evaluation of data made by the researcher. In the words of most philosophers of science, classification is a mean to "bring related items together" (Wynar et al., 1985, p. 317), "putting together like things" (Richardson, 1935); (Svenonius, 2000, p. 10), and "putting together things that are alike" (Vickery, 1975, p. 1) (see Mai, 2011 for a review). Of course, the antecedent of this approach dates back in the

Aristotelian positive approach to science, which describes and compares vis-á-vis Plato's normative approach (Reale, 1985). More recently, the availability of large dataset made a qualitative approach to the creation of taxonomic possible only at the expense of a sharp a priori reduction of the information in data. However, at the same time, algorithms and computational power allow for an automatic elaboration of the information with the purpose of creating a taxonomy. This approach is known as pattern recognition, unsupervised machine learning, or clustering and has been introduced in science by the anthropologists Driver and Kroeber (1932) and the psychologists Zubin (1938) and Tyron (1939). Typically, unsupervised algorithms are fed by rich datasets in terms of both variables and observations and require as main output the number of groups to be identified from the researcher. On this basis, as output, they provide a classification which minimize within-group variation and maximize between-groups variation, usually captured by some measures of distance in the n-dimensional space of the n variables. Although among these methods in social science the use of K-means algorithm MacQueen et al. (1967) is the most widespread, it has some weaknesses such as a possible dependence by initial condition and the risk of lock-in in local optima. More recently, the *self-organizing maps* (SOM) (Kohonen, 1990) gained attention as a new method in pattern recognition since they improve on K-means and present other advantages such as a clear visualization of the results.[4]

Data Analysis for Unconventional Data Sources

Also, the large share of unconventional data sources such as texts, images, or videos requires new techniques in the scientist's toolbox, and the nature of such information is more informative than figures since it contains ideas, opinions, and judgments. However, as for numerical data, the challenge is to reduce and organize such information in a meaningful way with the purpose of using it for a quantitative analysis, which does not require the time-consuming activity of reading and watching. Concerning text mining, the term "distant reading" attributed to Moretti (2013) could be used as an umbrella definition encompassing the use of automatic information process for books. The large divide for text mining is between the unsupervised and the supervised approach. The former usually deals with corpora of many documents which need to be organized. Techniques such as topic modelling allow to extrapolate the hidden thematic structure of an archive, that is, they highlight topics as specific distribution of words which are likely to occur together (Blei et al., 2003). Moreover, they return also the relative distributions of such topics in each document. Thus, at the same time, it is possible to have a bird's-eye overview on the key concept discussed in an archive, their importance, and when such concepts occur together in the different documents. The exact nature of the topic depends on the exercise at hand and, as in any unsupervised model, is subject to the educated interpretation of the researcher. Ultimately, it is possible

---

[4] For example, in the use of SOM for policy making, see, for instance, Carlei and Nuccio (2014) and Nuccio et al. (2020).

also to automatically assign a topic distribution to a new document, evaluate the emergence of new instances and the disappearing of old ones, and monitor how the relative importance of different instances changes over time (Di Caro et al., 2017). Topic modelling has been employed for the analysis of scientific literature (Fontana et al., 2019), policy evaluation (Wang & Li, 2021), legal documents (Choi et al., 2017), and political writings and speeches (Greene & Cross, 2017).

Documents can also come as an annotated text, that is, a text in which words or sentence of a group of documents is associated with a category. For instance, each word can be assigned to a feeling (bad vs. good), an evaluation (positive vs. negative), or an impact (relevant vs. non-relevant). The annotated text can be used as a training dataset to train a model able to recognize and predict the specific category in new document and analyse them. In this vain, a dataset of annotated tweets returning the feeling of the author can be used to infer the feeling of other users or the average sentiment of a geographical area or a group of people. For instance, Dahal et al. (2019) analyse the sentiment of climate change tweets.

5.3.3 Policy Recommendation as an Output of the Process

Based on the above review, it is possible to discuss which policies can be expected as outcome of a data-driven approach. The main theoretical element brought forward in the previous paragraphs consists in the link between the complexity of the world revealed by the data and the techniques require to deal with such a complexity. Such element constitutes the foundation of CSS and establishes the consequences for its use for policy analysis.

Precisely, since theoretical models such as ABM and network modelling lack the ability to come forward with simple testable equations, the attempt of deriving clear causal links as tool for policy should be abandoned. Nevertheless, it is not necessary to look back in despair since the fine and elegant armoury of causal identification has been developed in a century of scarcity of data and it made the best under such circumstances. However, as discussed above, the use of data was only ancillary to positive theorizing, and such an impoverished use of data science made prediction and quantitative scenario analysis ineffective for the policy maker. Nowadays, the combination of complex modelling with prediction empowers a truly quantitative policy analysis: on the one hand, relation among variables is hypothesized and tested within theoretical, but positive, models taking into account the heterogeneous attributes (also behavioural) of the subjects, the temporal and geographical specificity, and the dense interactions and feedback in the systems. The fine-tuning of their parameters and accuracy of their prediction are evaluated with supervised algorithms. Moreover, the unsupervised approach allows also for a hypothesis-free and easy-to-visualize exploration of data.

Such a positive and quantitative analysis can be helpful in understanding the present scenarios and their future complex development as the result of interactions of complex elements such as in the case of contagion of diseases (Currie et al.,

2020), but also the diffusion of ideas, technologies, and information as the aggregated manifestation of underlying adoption decisions (Beretta et al., 2018). Note that according to data at the disposal, these results can be achieved either by fine-tuned modelling with micro- and behavioural data, which return predictions on aggregate behaviour, or, on the contrary, by theoretical models which infer micro-behaviour when the model can replicate aggregate results in an exercise of compositional understanding. Economic systems can be also depicted as a complex evolving system, and CSS can describe aggregate fluctuation in economics and finances by feeding with data at the microlevel ABM models which can predict aggregate fluctuations with much fine accuracy than present DSGE models (Dosi & Roventini, 2019). Predictions aside, these models can easily incorporate heterogeneity of the agents, such as in income distribution or different behavioural routines such as propensity to save for consumers or to invest for entrepreneurs.

Finally, the discussion holds not only for policy analysis but also for the corresponding process of policy monitoring and evaluation. The current state of the art in scientific literature suggests that it is possible to evaluate the single causal impact of a policy, but this is far to be true: even in a controlled policy field experiment, it is not possible to estimate the external validity of the results when a pilot policy instrument is deployed at the country level, that is, at a different scale of complexity, or repeated in a slightly different situation in which local attributes are different.

5.4 The Way Forward

Data and algorithms applied to CSS can heavily impact upon the way we conceive the process of policy generation. However, the adoption of such tools needs preceding efforts by the policy itself, mainly in the areas of data as an input.

The ability of the public infrastructure to gather and store data for many sources calls for investment in technology, human capital, and a legislation that find a fine balance between citizens' right for privacy and a flexible use of the data.

The storage and the computational power of large amount of data should not rely on foreign service providers since data should be subject to European regulation. Therefore, policies within the European Data Infrastructure such as the European Open Science Cloud are welcome as well as the high prioritization of technological infrastructure in the European Regional Development Fund.[5]

Data collected and stored in Europe are subject to the GDPR which is correctly concerned with citizens' privacy protection. Although art. 89 allows research of certain privileges in data handling, the regulation is silent about the use for research

---

[5] European Data Infrastructure, https://www.eudat.eu/; European Open Science Cloud, https://eosc-portal.eu/; European Regional Development Fund, https://ec.europa.eu/regional_policy/en/funding/erdf/

of privately gathered data, de facto providing a solid reason to a large private platform not to share their data. This is in area in which the policy maker could intervene with the purpose of facilitating public-private data exchange for achieving the purpose of public interest.

The management of data and CSS also requires investments in human capital. The introduction of new professional profiles such as data stewards is required to deal with legislation and technical issue related to data, and the introduction of university curricula in data science should be encouraged. Moreover, due to variegated mix of skills that are required to apply CSS, interdisciplinary research should be supported and promoted.

References

Aldinucci, M., Rabellino, S., Pironti, M., Spiga, F., Viviani, P., Drocco, M., Guerzoni, M., Boella, G., Mellia, M., Margara, P., Drago, I., Marturano, R., Marchetto, G., Piccolo, E., Bagnasco, S., Lusso, S., Vallero, S., Attardi, G., Barchiesi, A., ... Galeazzi, F. (2018). HPC4AI: an ai-on-demand federated platform endeavour. In *Proceedings of the 15th ACM International Conference on Computing Frontiers* (pp. 279–286).

Ambrosino, A., Cedrini, M., Davis, J. B., Fiori, S., Guerzoni, M., & Nuccio, M. (2018). What topic modeling could reveal about the evolution of economics. *Journal of Economic Methodology, 25*(4), 329–348.

Anderson, C. (2008). The end of theory: The data deluge makes the scientific method obsolete. *Wired Magazine, 16*(7), 16–07.

Arthur, W. B. (2021). Foundations of complexity economics. *Nature Reviews Physics, 3*(2), 136–145.

Bak, P., Tang, C., & Wiesenfeld, K. (1987). Self-organized criticality: An explanation of the 1/f noise. *Physical Review Letters, 59*, 381–384. https://doi.org/10.1103/PhysRevLett.59.381. https://link.aps.org/doi/10.1103/PhysRevLett.59.381

Bello-Orgaz, G., Jung, J. J., & Camacho, D. (2016). Social big data: Recent achievements and new challenges. *Information Fusion, 28*, 45–59.

Benazzoli, C., & Di Persio, L. (2016). default contagion in financial networks. *International Journal of Mathematics and Computers in Simulation, 10*, 112–117.

Benthall, S., & Strandburg, K. J. (2021). Agent-based modeling as a legal theory tool. *Frontiers in Physics, 9*, 337. ISSN 2296-424X. https://doi.org/10.3389/fphy.2021.666386. https://www.frontiersin.org/article/10.3389/fphy.2021.666386

Beretta, E., Fontana, M., Guerzoni, M., & A. Jordan. (2018). Cultural dissimilarity: Boon or bane for technology diffusion? *Technological Forecasting and Social Change, 133*, 95–103.

Blazquez, D., & Domenech, J. (2018). Big data sources and methods for social and economic analyses. *Technological Forecasting and Social Change, 130*, 99–113.

Blei, D. M., Ng, A. Y., & Jordan, M. I. (2003). Latent dirichlet allocation. *The Journal of Machine Learning Research, 3*, 993–1022.

Boccaletti, S., Latora, V., Moreno, Y., Chavez, M., & Hwang, D.-U. (2006). Complex networks: Structure and dynamics. *Physics Reports, 424*(4), 175–308. ISSN 0370-1573. https://doi.org/10.1016/j.physrep.2005.10.009. https://www.sciencedirect.com/science/article/pii/S037015730500462X

Bonabeau, E. (2002). Agent-based modeling: Methods and techniques for simulating human systems. *Proceedings of the National Academy of Sciences, 99*(Suppl 3), 7280–7287. ISSN 0027-8424. https://doi.org/10.1073/pnas.082080899. https://www.pnas.org/content/99/suppl_3/7280

Carlei, V., & Nuccio, M. (2014). Mapping industrial patterns in spatial agglomeration: A som approach to italian industrial districts. *Pattern Recognition Letters, 40*, 1–10.

Carota, C., Durio, A., & Guerzoni, M. (2014). An application of graphical models to the innobarometer survey: A map of firms' innovative behaviour. *Italian Journal of Applied Statistics 25*(1), 61–79.

Chersoni, G., Della Valle, N., & Fontana, M. (2021). The role of economic, behavioral, and social factors in technology adoption. In Ahrweiler P. & Neumann M. (Eds.), *Advances in Social Simulation. ESSA 2019. Springer Proceedings in Complexity*. Cham: Springer.. https://doi.org/10.1007/978-3-030-61503-1_44

Chersoni, G., Della Valle, N., & Fontana, M. (2022). Modelling thermal insulation investment choices in the eu via a behaviourally informed agent-based model. *Energy Policy, 163*, 112823.

Chessa, A., Morescalchi, A., Pammolli, F., Pennera, O., Petersen, A. M., & Riccaboni, M. (2013). Is Europe evolving toward an integrated research area? *Scince, 339*, 650–651.

Choi, H. S., Lee, W. S., & Sohn, S. Y. (2017). Analyzing research trends in personal information privacy using topic modeling. *Computers & Security, 67*, 244–253.

Cooper, M. (2011). Complexity theory after the financial crisis: The death of neoliberalism or the triumph of Hayek?. *Journal of Cultural Economy, 4*(4), 371–385.

Corrocher, N., Guerzoni, M., & Nuccio, M. (2021). Innovazione e algoritmi da maneggiare con cura. *Economia & Management: la rivista della Scuola di Direzione Aziendale dell'Università L. Bocconi, 2*, 17–20.

Cowles, A. (1960). Ragnar frisch and the founding of the econometric society. *Econometrica (pre-1986), 28*(2), 173.

Currie, C. S., Fowler, J. W., Kotiadis, K., Monks, T., Onggo, B. S., Robertson, D. A., & Tako, A. A. (2020). How simulation modelling can help reduce the impact of COVID-19. *Journal of Simulation, 14*(2), 83–97.

Dahal, B., Kumar, S. A., & Li, Z. (2019). Topic modeling and sentiment analysis of global climate change tweets. *Social Network Analysis and Mining, 9*(1), 1–20.

Dai, E., Ma, L., Yang, W., Wang, Y., Yin, L., & Tong, M. (2020). Agent-based model of land system: Theory, application and modelling frameworks. *Journal of Geographical Sciences, 30*, 1555–1570.

Demoussis, M., & Giannakopoulos, N. (2006). Facets of the digital divide in europe: Determination and extent of internet use. *Economics of Innovation and New Technology, 15*(03), 235–246.

Di Caro, L., Guerzoni, M., Nuccio, M., & Siragusa, G. (2017). A bimodal network approach to model topic dynamics. *Preprint arXiv:1709.09373*.

Donangelo, R., Fort, H., Dakis, V., Scheffer, M., & Van Nes, E. H. (2010). Early warnings for catastrophic shifts in ecosystems: Comparison between spatial and temporal indicators. *International Journal of Bifurcation and Chaos, 20*(02), 315–321. https://doi.org/10.1142/S0218127410025764

Dosi, G., Pereira, M., Roventini, A., & Virgillito, M. (2020). The labour-augmented k+s model: A laboratory for the analysis of institutional and policy regimes. *Economi A, 21*(2), 160–184. ISSN 1517-7580. https://doi.org/10.1016/j.econ.2019.03.002. https://www.sciencedirect.com/science/article/pii/S151775801830122X

Dosi, G., & Roventini, A. (2019). More is different... and complex! the case for agent-based macroeconomics. *Journal of Evolutionary Economics, 29*(1), 1–37.

Driver, H., & Kroeber, A. (1932). *Quantitative expression of cultural relationships* (Vol. 31, pp. 211–256). University of California publications in American Archaeology and Ethnology.

Ester, M., Kriegel, H. P., Sander, J., & Xu, X. (1996). A density-based algorithm for discovering clusters in large spatial databases with noise. In *Driver 21131 Quantitative Expression of Cultural Relationships 1932*.

Eckhoff, M., & Morters, P. (2013). Vulnerability of robust preferential attachment networks. *Electronic Journal of Probability, 19*, 1–47.

Einav, L., & Levin, J. (2014). Economics in the age of big data. *Science, 346*(6210), 1243089.

Fontana, M. (2012). On policy in non linear economic systems. In Heritier, P. & Silvestri, P. (Eds.), *Good goverment governance and human complexity* (pp. 221–234). Oelscki.

Fontana, M., Montobbio, F., & Racca, P. (2019). Topics and geographical diffusion of knowledge in top economic journals. *Economic Inquiry, 57*(4), 1771–1797. https://doi.org/10.1111/ecin.12815

González, I., D'Souza, G., & Ismailova, Z. (2018). Agent-based modeling: An application to natural resource management. *Journal of Environmental Protection, 9*, 991–1019.

Gould, P. (1981). Letting the data speak for themselves. *Annals of the Association of American Geographers, 71*(2), 166–176.

Grassberger, P. (1983). On the critical behavior of the general epidemic process and dynamical percolation. *Mathematical Biosciences, 63*(2), 157–172. ISSN 0025-5564. https://doi.org/10.1016/0025-5564(82)90036-0. https://www.sciencedirect.com/science/article/pii/0025556482900360

Greene, D., & Cross, J. P. (2017). Exploring the political agenda of the european parliament using a dynamic topic modeling approach. *Political Analysis, 25*(1), 77–94.

Gualdi, S., Tarzia, M., Zamponi, F., & Bouchaud, J.-P. (2015). Tipping points in macroeconomic agent-based models. *Journal of Economic Dynamics and Control, 50*, 29–61. ISSN 0165-1889. https://doi.org/10.1016/j.jedc.2014.08.003. https://www.sciencedirect.com/science/article/pii/S0165188914001924. Crises and Complexity.

Guerzoni, M., Nava, C. R., & Nuccio, M. (2021). Start-ups survival through a crisis. combining machine learning with econometrics to measure innovation. *Economics of Innovation and New Technology, 30*(5), 468–493.

Hammond, R. (2015). *Considerations and best practices in agent-based modeling to inform policy*. Wahsington, DC, USA: National Academies Press.

Hargittai, E., & Hinnant, A. (2008). Digital inequality: Differences in young adults' use of the internet. *Communication Research, 35*(5), 602–621.

Heppenstall, A., Crooks, A., Malleson, N., Manley, E., Ge, J., & Batty, M. (2021). Future developments in geographical agent-based models: Challenges and opportunities. *Geographical Analysis, 53*(1), 76–91. https://doi.org/10.1111/gean.12267. https://onlinelibrary.wiley.com/doi/abs/10.1111/gean.12267

Kenis, P., & Schneider, V. (2019). *Analyzing policy-making II: Policy network analysis* (pp. 471–491). Springer. ISBN 9783030160647. https://doi.org/10.1007/978-3-030-16065-4_27.

Kerr, C. C., Stuart, R. M., Mistry, D., Abeysuriya, R. G., Rosenfeld, K., & Hart, G. R. (2021). Covasim: An agent-based model of COVID-19 dynamics and interventions. *PLoS Computational Biology, 17*(7), e1009149.

Kitchin, R. (2014). Big data, new epistemologies and paradigm shifts. *Big Data & Society, 1*(1), 2053951714528481.

Kohonen, T. (1990). The self-organizing map. *Proceedings of the IEEE, 78*(9), 1464–1480.

Lazer, D., Pentland, A., Adamic, L., Aral, S., Barabasi, A.-L., Brewer, D., Christakis, N., Contractor, N., Fowler, J., Gutmann, M., Jebara, T., King, G., Macy, M., Roy, D., & Van Alstyne, M. (2009). Social science. computational social science. *Science (New York, NY), 323*(5915), 721–723.

Loewenstein, G., & Chater, N. (2017). Putting nudges in perspective. *Behavioural Public Policy, 1*(1), 26–53. https://doi.org/10.1017/bpp.2016.7

Lourenço, J. S., Ciriolo, E., Rafael Almeida, S., & Troussard, X. (2016). *Behavioural insights applied to policy, european report 2016*. EUR 27726.

MacQueen, J. (1967). Some methods for classification and analysis of multivariate observations. In *Proceedings of the Fifth Berkeley Symposium on Mathematical Statistics and Probability* (Vol. 1, pp. 281–297). Oakland, CA, USA.

Mai, J.-E. (2011). The modernity of classification. *Journal of Documentation, 67*(4), 710–730.

Moretti, F. (2013). *Distant reading*. Verso Books.

Nuccio, M., & Guerzoni, M. (2019). Big data: Hell or heaven? Digital platforms and market power in the data-driven economy. *Competition & Change, 23*(3), 312–328.

Nuccio, M., Guerzoni, M., Cappelli, R., & Geuna, A. (2020). Industrial pattern and robot adoption in European regions. *Department of Management, Università Ca'Foscari Venezia Working Paper, 1*(3), 33.

Peloquin, D., DiMaio, M., Bierer, B., & Barnes, M. (2020). Disruptive and avoidable: GDPR challenges to secondary research uses of data. *European Journal of Human Genetics, 28*(6), 697–705.

Prensky, M. (2009). H. sapiens digital: From digital immigrants and digital natives to digital wisdom. *Innovate: Journal of Online Education, 5*(3).

Ravat, F., & Zhao, Y. (2019). Data lakes: Trends and perspectives. In *International Conference on Database and Expert Systems Applications* (pp. 304–313). Springer.

Reale, G. (1985). *A History of Ancient philosophy II: Plato and Aristotle* (Vol. 2). Suny Press.

Renner, K.-H., Klee, S., & von Oertzen, T. (2020). Bringing back the person into behavioural personality science using big data. *European Journal of Personality, 34*(5), 670–686.

Richardson, E. C. (1935). *Classification*. New York: H. W. Wilson.

Runck, B., Manson, S., Shook, E., Gini, M., & Jordan, N. (2019). Using word embeddings to generate data-driven human agent decision-making from natural language. *GeoInformatica, 23*, 221–242.

Schilling, M. A., & Phelps, C. C. (2007). Interfirm collaboration networks: The impact of large-scale network structure on firm innovation. *Management Science, 53*(7), 1113–1126. https://doi.org/10.1287/mnsc.1060.0624.

Sorenson, O., Rivkin, J. W., & Fleming, L. (2006). Complexity, networks and knowledge flow. *Research Policy, 35*(7), 994–1017.

Suman, A. B., & Pierce, R. (2018). Challenges for citizen science and the eu open science agenda under the gdpr. *European Data Protection Law Review, 4*, 284.

Svenonius, E. (2000). *The intellectual foundation of information organization*. MIT Press.

Truszkowska, A., Behring, B., Hasanyan, J., Zino, L., Butail, S., Caroppo, E., Jiang, Z.-P., Rizzo, A., & Porfiri, M. (2021). High-resolution agent-based modeling of COVID-19 spreading in a small town. *Advanced Theory and Simulations, 4*(3), 2000277. https://doi.org/10.1002/adts.202000277

Tyron, R. C. (1939). *Cluster analysis*. Ann Arbor, MI: Edwards Brothers.

Vickery, B. C. (1975). *Classification and indexing in science* (3rd ed.).

Wang, Q., & Li, C. (2021). An evolutionary analysis of new energy and industry policy tools in china based on large-scale policy topic modeling. *Plos one, 16*(5), e0252502.

Weaver, W. (1948). There is a large literature on the subject of complexity, for example. *Science and Complexity, 36pp*, 536–544.

Wilenski, U., & Rand, W. (2015). *An introduction to agent-based modeling modeling natural, social, and engineered complex systems with NetLogo*. Massachusetts London, England,: The MIT Press Cambridge.

Wynar, B. S., Taylor, A. G., & Osborn, J. (1985). *Introduction to cataloging and classification* (Vol. 8). Libraries Unlimited Littleton.

Zubin, J. (1938). A technique for measuring like-mindedness. *The Journal of Abnormal and Social Psychology, 33*(4), 508.

Chapter 6
From Lack of Data to Data Unlocking

Computational and Statistical Issues in an Era of Unforeseeable Big Data Evolution

Nuno Crato

Abstract Reliable cross-section and longitudinal data at national and regional level are crucial for monitoring the evolution of a society. However, data now available have many new features that allow for much more than to just monitor large aggregates' evolution. Administrative data now collected has a degree of granularity that allows for causal analysis of policy measures. As a result, administrative data can support research, political decisions, and an increased public awareness of public spending. Unstructured big data, such as digital traces, provide even more information that could be put to good use. These new data is fraught with risks and challenges, but many of them are solvable. New statistical computational methods may be needed, but we already have many tools that can overcome most of the challenges and difficulties. We need political will and cooperation among the various agents. In this vein, this chapter discusses challenges and progress in the use of new data sources for policy causal research in social sciences, with a focus on economics. Its underlying concerns are the challenges and benefits of causal analysis for the effectiveness of policies. A first section lists some characteristics of the new available data and considers basic ethical perspectives. A second section discusses a few computational statistical issues on the light of recent experiences. A third section discusses the unforeseeable evolution of big data and raises a note of hope. A final section briefly concludes.

The author gratefully acknowledges the guiding questions and many helpful suggestions of Paolo Paruolo, as well as the constructive criticisms of Michele Vespe and the editors. The usual disclaimer applies.

N. Crato (✉)
ISEG, Cemapre, University of Lisbon, Lisbon, Portugal
e-mail: ncrato@iseg.ulisboa.pt

6.1   Introduction: Data for Causal Policy Analysis

A few decades ago, researchers and policymakers would struggle to get access to information. A student in time series would frequently have difficulty in getting data with 100 data points. A statistician willing to experiment with novel methods would frequently need to type data by hand, after collecting tables from dozens of print publications. An economist willing to compare the evolution of macroeconomic variables in different countries would need to search for days and would usually get series built with different criteria and with different length.

In the mid-1990s, things changed dramatically. Internet started working as an open means for communication and information access, although too many data sets were proprietary, as too many still are, and too often researchers would need to beg statistical officers or other researchers for getting appropriate data sets.

In parallel to an increasing data availability, a culture of openness spread slowly across countries and fields of activity. Driven by some governmental and institutional examples, by researcher pressure, and by public political tension, data that would previously be safely hidden in institution's departments become progressively available to researchers and the public.

Scientific journals could start avoiding systematically one of the obstacles to scientific reproducibility. Many journals adopted the policy of requiring authors to make data available upon request or by posting the data files at journals' websites.

In official statistics things started also changing. During the first years of the twenty-first century, the idea of using confidential microdata for research gained momentum (Jackson, 2019). This recent interest in original highly granular data officially collected, in brief, in administrative data, prompted the promise of a revolution in econometrics and social statistics studies.[1]

Microdata is usually defined as data 'collected at the individual level of units considered in the database. For instance, a national unemployment database is likely to contain microdata providing information about each unemployed (or employed) person'.[2] Modern administrative data provides access to microdata at an unprecedented level.

This revolution in studies using administrative data was backed by a scientific "credibility revolution" in social statistics. Economists Angrist and Pischke (2015) described this "revolution" in empirical economics as the current "rise of a design-based approach that emphasizes the identification of causal effects". In fact, methods such as regression discontinuity, differences in differences, and others, which have been maturing in areas of statistical analysis as different as psychometrics or biometrics, registered a renewed interest as they become recognized as tools for assessing and isolating social variables influences and for looking for causal factors in overly complex environments. As already expressed in Crato and Paruolo (2019),

---

[1] For additional insights, please refer to the chapter by Signorelli et al. (2023) in the present Handbook.

[2] Glossary in Crato and Paruolo (2019, pp. 10–12).

this means that "Public policy can derive benefit from two modern realities: the increasing availability and quality of data and the existence of modern econometric methods that allow for a causal impact evaluation of policies. These two fairly new factors mean that policymaking can and should be increasingly supported by evidence".

By the end of the twentieth century, collected data volumes increased in such a way that researchers started using the phrase "big data". This phrase usually encompasses data sets with sizes beyond the ability of commonly used hardware and software tools to collect, manage, and process them within a reasonable time (Snijders et al., 2012). The expression encompasses unstructured, semi-structured, and structured data; however, the usual focus is on unstructured data (Dedić & Stanier, 2017).

Administrative data can be considered big data in volume, although usually it is highly structured and so it departs form this common characteristics of the big data classification.

This distinction is important as unstructured big data is evolving at an incredible speed, and it is by essence varied and difficult to characterize. What may be applicable to a big data set may not be applicable to a different big data set, and things are evolving at such a pace that new applications for big data are appearing every day. Very recently, the Covid-19 pandemic demonstrated the usefulness of new sources of data, such as students' logins to sites or the search for specific medical information. It will help our discussion to characterize the types of data we are discussing.

6.1.1 The Variety of Data

In this volume, the chapter by Manzan (2023) provides a valuable discussion of various sources of data and how they have been instrumental for advancements of knowledge in several fields of economics. Our purpose here is more schematic. In Table 6.1, we summarize the characteristics of different data types.

For our purposes, it is also interesting to characterize data according to their level of structuring. An attempt appears on Table 6.2.

For social research, policy design, and democratic public scrutiny, it is important to have access to as much data as possible, both in volume and variety. This is particularly important for data produced and kept by the public sector.

6.1.2 Underlying Statistical Issue: The Culture of Open Access

The idea that information should be available to the public is a democratic and an old one. The following well-known excerpt from James Madison, the father

Table 6.1 Types of data according to their origin, partially based on Connelly et al. (2016)

Origin → Characteristics ↓	General survey	Experimental survey	Administrative data	Other big data types
Research questions	Data addresses multiple questions	Data addresses specific questions	Data collected for non-research purposes	Data collected for non-research purposes
Structure	Highly systematic	Highly systematic	Systematicity varies	Very unsystematic
Dimensions	Large and complex	Reduced size and scope	Large and complex, but messy and fragmented	Very large and very complex
Sampling	Known sample and/or population	Known sample and/or population	Known sample and population	Unknown relationship sample population
Linkage	Difficult linkage	Linkage possible	Unique identifiers simplify linkage	Difficult linkage

Table 6.2 Types of data according to their structure (definitions and examples), loosely inspired by National Academies of Sciences, Engineering, and Medicine (2017)

Structured data from census and surveys	Structured public and private data	Semi-structured data	Unstructured data
Data from a population or a designed probability survey	Data collected by public administrations or from private companies	Data that have flexible structures that made it hard to relate them and need hard scrubbing and transformation for comparability	Data such as images, videos, and texts without any structure requiring value to be extracted and organized for processing and analysis
Examples			
Official censuses, academic and market research surveys, and other well-designed data collections	Tax records, school enrolments, unemployment, salaries, and other public records; commercial transactions, medical records, stock prices, and other private records	GPS and utility company sensors, tide and atmospheric sensors records, mobile texting volumes, web logs, web searches, and others	Internet searches, webcam traffic, security videos, medical data from personal sensors, social network interactions, and other data from IoT records (IoT: Internet of Things refers to devices that can communicate among themselves using the internet as the common transmission protocol)

of the American Constitution, has been recurrently quoted as an indictment of the withholding of government information (Doyle, 2022).

> A popular Government, without popular information, or the means of acquiring it, is but a Prologue to a Farce or a Tragedy; or, perhaps both. Knowledge will forever govern ignorance: And a people who mean to be their own Governors, must arm themselves with the power which knowledge gives.

More than two centuries later, similar concerns were clearly expressed in a report by President Obama's executive office (The White House, 2014), which considers "data as a public resource" and ultimately recommends that government data should be "securely stored, and to the maximum extent possible, open and accessible" (p. 67).

In the European Union, there have been analogous concerns and recommendations. Among other statements, the European Commission has also pledged that, where appropriate, "information will be made more easily accessible" (2016, p. 5).

In addition to the issue of public access to nonconfidential data, there is the issue of data access for research purposes. This latter issue is an old one, but it took a completely different development in the twenty-first century with the rise of two factors: firstly, the availability of very rich, longitudinal, historically ordered, and granular administrative data; secondly, the development of the so-called counterfactual methods for detecting casual relations among complex social data.

In the United States, researcher's call to access to administrative data reached the National Science Foundation (Card et al., 2010; US Congress, 2016 ; The White House, 2014), which established a Commission on Evidence-Based Policymaking, with a composition involving (a) academic researchers and (b) experts on the protection of personally identifiable information and on data minimization.

Similar developments happened in Europe regarding the use of admin data for policy research purposes, albeit with heterogeneity across states. A few countries, namely, the UK and The Netherlands, already make considerable use of admin data for policy research. The European Commission (2016) issued a directive establishing that data, information, and knowledge should be shared as widely as possible within the Commission and promoting cross-cutting cooperation between the Commission and Member States for the exchange of data, aiming at better policymaking.

This research access has been discussed in general terms but has been dominated by policy concerns.[3] We are still far from regularly having the disclosure of administrative data and independent systematic analysis of policies. Too often, policy design is based on ideology, group interests, and particular policy matters,

---

[3] In science in general, the disclosure of scientific data and ideas has also benefited from the digitalization and the internet. The existence of scientific electronic archives that are nonrefereed and with open access, such as arxiv.org, and a variety of preprint archives is an open culture answer to the scientific priority concerns, making available data, experimental data, and ideas, is a way to establish priority (Watt, 2022).

without regard to its efficiency in terms of the intended goals. The possibility of measuring the impact of policies and correcting their course is certainly a very valid one and deserves all efforts for opening the access to data.

Although it is not clear whether this push for evidence-based policy impact evaluation is changing the panorama of policy design, it certainly is increasingly visible.

All these recent developments raise many questions and pose many opportunities and issues. In what follows, I will discuss three particular issues, trying to contribute to specific relevant policy questions raised by JRC scientists and collected by the editors of the volume in Bertoni et al. (2022). A first issue is how to take advantage of the different types of data by adding or consolidating the information available from each type of data set, ideally by linking them. A second issue is the scientific replicability of studies that access propriety data or data that evolves and are no longer retrievable. A third issue is confidentiality. With access to huge volumes of microdata, sensible personal or organizational information may be spread in a nonethical and undesired way. How can we navigate in this changing sea of opportunities without threatening legitimate privacy rights? These three main issues are tightly linked, as we can see in the following discussion.

6.2 Computational Statistical Issues

6.2.1 Statistical Issues with Merging Big Data

In contrast to organized administrative data, nonstructured or loosely structured big data are difficult to link with common probability linkage methods, namely, with those that are used to fix occasional misaligned units (Shlomo, 2019). There are, however, a few promising experiences.

A relatively old problem that can benefit from big data corrections is the so-called problem of the "missing rich", i.e. the paradoxical fact that too often data underestimates the size and wealth of people and families in the upper tail of the income distributions (Lustig, 2020). This has been a well-known problem in household surveys and other type of data collection in various countries.

The "missing rich" problem affects many types of data, not only in income distribution.[4] The expression now stands for issues that affect upper tails of social statistics, namely, underreporting, under covering and non-responses. For proceeding with estimates corrections, social statisticians have used methods that rely on within survey methods, looking for inconsistencies. More recently, there have been renewed interest in methods that rely on external sources, such as media lists and tax records. Researchers have used both parametric and nonparametric methods for these corrections. Corrections can be made by simple reweighing or

---

[4] See, e. g. Lustig (2020) and the references therein.

by adding items. In the first case, we are facing a trend to the use of model-based statistics, which have been common in areas as diverse as national statistics and student's standardized tests. In the second case, we are using selected administrative data linkage, as it has been done for a certain time in France for the EU-SILC survey.

Adamiak and Szyda (2021) work provide another example of merging official statistics with unstructured big data. They studied the distribution of worldwide tourism destinations by complementing the World Tourism Organization (UNWTO) data with two big data sources: a gridded population database and geo-referenced data on Airbnb accommodation offers. Their results emphasize the predominance of domestic tourism in the global tourism movements, an often-hidden phenomenon, which is revealed by a finer granular analysis of locations and types of tourism preferences. Global statistics with movements across borders cannot reveal the true scale of domestic movements.

Other researchers have explored similar big data sources for tracking dynamic changes in almost real time. For monitoring passenger fluxes, hotel stays, and car rentals, various researchers have successfully used booking data, Google searches, mobile device data, remote account logins, card payments, and other similar data. See, e.g. Napierała et al. (2020) and Gallego and Font (2021) as well as the work by Romanillos Arroyo and Moya-Gómez (2023) in the present Handbook.

Alsunaidi et al. (2021) provide a good synthesis of studies for tracking COVID-19 infections by using big data analysis. The pandemic prompted the surge of big data studies which were useful for estimation or prediction of risk score, healthcare decision-making, and pharmaceutical research and use estimation. Data sources for these studies have been incredibly varied, ranging from body sensors and wearable technology to location data for estimating the spread risks of COVID-19.

Additional data sources have been developed and should be most important in a foreseeable future. Among those, activity tracking and health monitoring through smart watches is proving to become an important tool. By using collected disperse data, researchers can now develop real-time diagnosis tools that could be used in the future. In his chapter in this volume, Manzan (2023) provides some other examples of microdata uses.

6.2.2 The Statistical Issue of Replicability and Data Security

The pandemic brought startling scientific advances in medicine and related areas but also in social statistics and in statistics in general.

A surprising reality that hit everybody was the uncertainty regarding many factors and variables in the pandemic. In early October 2020, the comparison of various estimates for the rate of Covid-19 spread in the United Kingdom revealed a degree of uncertainty masked by each individual estimate. Figure 6.1 shows the nine estimates considered at the time by the UK Scientific Pandemic Influenza Group on Modelling. The point estimates ranged from 1.2 to 1.5, i.e. widely different rates of

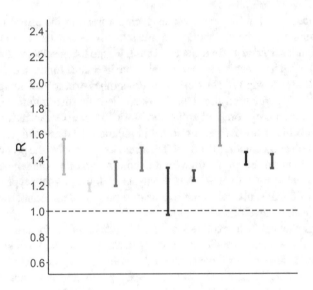

Fig. 6.1 Confidence bands at 90% for estimates for the reproduction rate R of Covid-19 in the UK in October 2022. Graph adapted from Scientific Pandemic Influenza Group on Modelling. (2020)

growth. Even more startling is the fact that different 90% confidence intervals do not overlap. The estimate represented as the fifth from the left on the graph admitted in the corresponding confidence interval the possibility that the pandemic is receding, while the highest estimate, the seventh on the graph, suggested that 100 people infect 166 others.

This example is not unique and similar results have recently been reported in other areas. A recent project in finance that collectively involved 164 teams tested six hypotheses widely discussed in financial economics (Menkveld et al., 2021). The hypotheses were on the existence of trends in the market efficiency, the realized bid-ask spread, the gross trading revenue of clients, and other measurable and testable characteristics of the markets. Additionally, used data were the same *Deutsche Boerse* sample.

Reporting the results from different teams, the authors note a sizeable dispersion in results. For the first hypothesis, for instance, which was that "market efficiency has not changed over time", the global standard error for the estimate was 20.6%, while the variability across researchers' estimates was 13.6%. This is certainly non-negligible.

The authors of this study propose to make a distinction between the traditional standard errors from parameters estimates, computed by using well-established statistical methods, and what they call "the non-standard errors", due to variability in methods used by researchers.

Along the same lines, a recent article in Nature (Wagenmakers et al., 2022) provides startling examples of different conclusions drawn from the same data with different statistical tools. Consequently, they argue persuasively on the need to contrast different research conclusions obtained through different statistical methods.

This would obviously be a particular form of triangulation, a concept worth revisiting.

Following the Oxford Bibliography by Drisko (2017), "triangulation in social science refers to efforts to corroborate or support the understanding of an experience, a meaning, or a process by using multiple sources or types of data, multiple methods of data collection, and/or multiple analytic or interpretive approaches". The concept was arguably first introduced by Campbell and Fiske (1959) and usually comprises four types of triangulation identified by Denzin in the 1970s: (1) data triangulation; (2) investigator triangulation; (3) theory triangulation; and (4) methodological or method triangulation.

As a way to apply triangulation and reaching more robust statistical conclusions in social sciences, Aczel et al. (2021) present a "consensus-based guidance" method and argue that a broad adoption of such "multi-analyst approach" can strengthen the robustness of results and conclusions in basic and applied research.

Wagenmakers et al. (2021) also argue that limitations of single analysis call for contrasting analyses and recommend seven concrete statistical procedures: (1) visualizing data; (2) quantifying inferential uncertainty; (3) assessing data pre-processing choices; (4) reporting multiple models; (5) involving multiple analysts; (6) interpreting results modestly; and (7) sharing data and code. For our purposes, this seventh recommendation is of paramount importance and consequences.

Let us highlight it again: for robustness of statistical inferences in social sciences, it is essential to share data and to share code. These have been practiced for decades in physical sciences. In particular, high-energy physics and astronomy have a long tradition of sharing data and procedures, so that other teams can replicate and corroborate, or contradict the analyses. A similar practice exists in climate research. Why is this such a novelty and odd thing to request in the social sciences?

A serious issue, though, is the security of sensitive data. Should data be completely free, easily available upon request, maybe entailing only a responsibility of a sworn statement, or should it be more rigorously restricted? There is no simple answer to this concern. But there are multiple practical solutions.

One practical solution is the availability to researchers of verified scripts only, with which studies could be done. This way, researchers do not deal directly with data and only get the statistical results. There are some inconveniences to this solution, namely, the difficulty in accessing data in this step-by-step way, while research usually needs to be done in an interactive way.

Another practical solution is the creation of safe environments in which only accredited researchers may have access and in which all interactions with data are recorded. With ethical and peer pressure from the scientific and technical community, this solution is feasible, although not without risks.

As a great provider of reliable data, public authorities should face in a very serious way the issue of safely organizing their data. A governmental example worth following is the X-Road, a centrally created and managed systematic data exchanger between information systems. It is extensively used in Estonia[5] and followed by Finland in 2017, when the exchange systems from both countries were interconnected.

6.2.3   Statistical Issues Risen by Anonymity Concerns and Related Challenges

Privacy is often quoted as the main concern for restricting the use of big data in various settings. This is obviously an important issue, but often shown through biased perspectives.

Firstly, it should be highlighted that tax collection, lack of respect for democratic rules in some countries, and the involuntary or unconscious supply of sensitive data to internet-based companies provide a much higher anonymity threat than big data studies operated by researchers following ethical protocols.

Secondly, the anonymity issue is often a convenient political pretext for not collecting data, not revealing data, nor assessing the impact analyses of public policies.

Thirdly, and most importantly, there are now methods of anonymizing data and realizing studies that do not reveal any personal sensitive data but provide the public with important knowledge about social issues.

Other issues are worth noting, namely, information correctness and replicability. Missing data and incorrect data can lead to biased findings (Richardson et al., 2020). And these incorrect findings can be replicated and induce larger mistakes. Additionally, data collected by businesses often change the sampling and processing methods and do not report it adequately (Vespe et al., 2021). All these issues are even more serious as they mean that replicability is often difficult and so the scientific debate can be hindered.

As we discuss big data availability and issues, it is obligatory to note that a wealth of administrative data of great use and of technically easy access exists and should be available to researchers and interested citizen groups. In this regard, if there are difficulties, they could easily be removed with sufficient governance will.

Rossiter (2020) has noted that access to education data is essential for institutions accountability. This could hardly be overstated as education arguably is one of the most important public policies issues and education budgets are among the most important in any country. What is a stake is highly important for a country's future and for the taxpayer, and what is at stake is the use of substantial public resources.

Read and Atinc (2017) listed the availability of education administrative data in 133 low- and middle-income countries and noted that 61 of these have no available data and 43 have only data at the national level. Of the 29 countries that have desegregate data, they were most in non-machine reading format, and only 16 of

---

[5] https://e-estonia.com/solutions/interoperability-services/x-road/

these provide data from student assessment. The consequent limitation can hardly be overstated: student results are the most—some can even say the only—important data regarding any education system.

This "underutilization of administrative data" has serious consequences form educational development. As Rossiter (2020) again points out, for many educational decisions findings cannot be imported. When there are conflicting evidence results, in particular, then "non-experimental results from the right context are very often a better guide to policy than experimental results from elsewhere".

We should thus look for solutions.

How can we replicate results if data are confidential and restricted to particular groups of researchers? We can address this issue by fostering communities of practice. This way, access to confidential data is guaranteed to trusted researchers under appropriate conditions. This would allow and nudge researchers to independently study the same data set and contrast conclusions.

Public and statistical authorities are among those more reticent to this type of data sharing. However, this is the best way to reach robust conclusions that can illuminate policy evaluation and public policy decisions.

In case a team of researchers claims that policy X had effect Y, one could ask a team of "research team of verifiers" to replicate or reanalyse the data to validate findings, similarly with what happens in physical sciences.

The "research team of verifiers" could even be reimbursed, as they provide a public service. But this could be done in exchange of similar work done by others (reciprocity), or as normal peer review work, which is often done for free.

In an ideal future, access to non-public administrative data could be regimented in a way that forces varied teams access and varied methods. This happens in public tenders. Why should not data access be granted mandatorily to more than a single research team? This prerequisite for data use would foster social sciences, public policy evaluation, and, ultimately, democracy. Publicly collected data is a public good.

A good example to this practice is what has been put in place by some scientific societies and scientific journals[6]: Data sharing is a requirement for paper publication.

A simple proposal is as follows. Similarly to what happens in scientific journals, official analysis of policies impact could have as a normal prerequisite the verification by independent researchers. In these cases, the analyses could involve much more computational and teamwork than normal paper refereeing. It would be of public interest that the promoter of the study includes in the initial budget a provision for paying teams of verifiers that could constitute an accredited pool.

---

[6] See, e. g. Committee on Professional Ethics of the American Statistical Association (2018).

6.3 The Way Forward

As discussed in Callegaro and Yang (2018) inter alia, variability is an important characteristic of big data. This means that gathering, analysing, and interpreting big data requires technical expertise that is always evolving. This also means that methods are evolving, and it is difficult or even impossible to have a fixed set of tools that will allow the use and merging of data, when we deal with this particular type of data.

Researchers have used relatively old or, at least, well-established techniques such as propensity score analysis, regression discontinuity, and differences-in-differences methods.

Another research worth noting is Chen et al. (2020). The authors note that the "challenge of low participation rates and the ever-increasing costs for conducting surveys using probability sampling methods, coupled with technology advances, has resulted in a shift of paradigm". At this moment, even government statistical agencies need to pay attention to non-probability survey samples, i.e. samples that are not random or that do not derive from a known probabilistic rule. One example is the so-called opt-in panels, for which volunteers are recruited. These authors propose a general framework for statistical inferences with this type of samples, by coupling them with auxiliary information available from a reference probability sample survey. In this setting, they propose a novel procedure for the estimation of propensity scores. All their procedure supposes the availability of high-quality probability sample surveys to allow for the pairing.

At this moment, data sources are evolving at such a speedy pace that it is difficult or even impossible to establish general rules. Each data collection method is providing new types of data with different characteristics, different insufficiencies, different challenges, and different possibilities. The general rules we may offer are (1) to apply established scientific rules and methods to the analysis of data and (2) to cross validate conclusions through open science, namely, through data and code sharing.

Is this a pessimistic or an optimistic view? I think it is an optimistic one.

6.4 Conclusion

This chapter discussed the recent evolution of data existence and use. It contrasted the previous lack of data with the current big data moment, in which we are facing a new issue, the issue of unlocking the power of existing data.

There are many types of data that fall under the classification of big data. This distinction is important, as methods to access, analyse, and use these types of data are different according to data structure. However, more than a practical issue, the wide use of data by the society is an ethical imperative. As such, this chapter argues that it is our duty as researchers to contribute to find ways of overcoming the many existing obstacles to full use of data.

There are many technical issues with data use, from anonymity issues to inference issues. This chapter lists some recent experiences and argues that some well-established scientific practices can be extended to data use and analysis, particularly when data are used for causal inference on policy measures. This can be done without increasing risks to data use and adding benefits to the scientific quality of the analyses. Scientific social studies and society will be the great beneficiaries.

References

Aczel, B., Szaszi, B., Nilsonne, G., van den Akker, O. R., Albers, C. J., van Assen, M. A., Bastiaansen, J. A., Benjamin, D., Boehm, U., Botvinik-Nezer, R., Bringmann, L. F., Busch, N. A., Caruyer, E., Cataldo, A. M., Cowan, N., Delios, A., van Dongen, N. N., Donkin, C., van Doorn, J. B., et al. (2021). Consensus-based guidance for conducting and reporting multi-analyst studies. *eLife, 10*, e72185. https://doi.org/10.7554/eLife.72185

Adamiak, C., & Szyda, B. (2021). Combining conventional statistics and big data to map global tourism destinations before Covid-19. *Journal of Travel Research, 004728752110514*. https://doi.org/10.1177/00472875211051418

Alsunaidi, S. J., Almuhaideb, A. M., Ibrahim, N. M., Shaikh, F. S., Alqudaihi, K. S., Alhaidari, F. A., Khan, I. U., Aslam, N., & Alshahrani, M. S. (2021). Applications of big data analytics to control COVID-19 pandemic. *Sensors, 21*(7), 2282. https://doi.org/10.3390/s21072282

American Statistical Association. (2018). Ethical guidelines for statistical practice prepared by the Committee on Professional Ethics of the American Statistical Association approved by the ASA Board in April 2016. http://www.amstat.org/ASA/Your-Career/Ethical-Guidelines-for-Statistical-Practice.aspx

Angrist, J. D., & Pischke, J.-S. (2015). *Mastering metrics: The path from cause to effect*. Princeton University Press.

Bertoni, E., Fontana, M., Gabrielli, L., Signorelli, S., & Vespe, M. (Eds). (2022). *Mapping the demand side of computational social science for policy*. EUR 31017 EN, Luxembourg, Publication Office of the European Union. ISBN 978-92-76-49358-7, https://doi.org/10.2760/901622

Callegaro, M., & Yang, Y. (2018). The role of surveys in the era of "big data". In D. L. Vannette & J. A. Krosnick (Eds.), *The Palgrave handbook of survey research* (pp. 175–192). Springer International Publishing. https://doi.org/10.1007/978-3-319-54395-6_23

Campbell, D. T., & Fiske, D. W. (1959). Convergent and discriminant validation by the multitrait-multimethod matrix. *Psychological Bulletin, 56*(2), 81–105. https://doi.org/10.1037/h0046016

Card, D. E., Chetty, R., Feldstein, M. S., & Saez, E. (2010). Expanding access to administrative data for research in the United States. *SSRN Electronic Journal*. https://doi.org/10.2139/ssrn.1888586

Chen, Y., Li, P., & Wu, C. (2020). Doubly robust inference with nonprobability survey samples. *Journal of the American Statistical Association, 115*(532), 2011–2021. https://doi.org/10.1080/01621459.2019.1677241

Connelly, R., Playford, C. J., Gayle, V., & Dibben, C. (2016). The role of administrative data in the big data revolution in social science research. *Social Science Research, 59*, 1–12. https://doi.org/10.1016/j.ssresearch.2016.04.015

Crato, N., & Paruolo, P. (2019). The power of microdata: An introduction. In N. Crato & P. Paruolo (Eds.), *Data-driven policy impact evaluation* (pp. 1–14). Springer International Publishing. https://doi.org/10.1007/978-3-319-78461-8_1

Dedić, N., & Stanier, C. (2017). Towards differentiating business intelligence, big data, data analytics and knowledge discovery. In F. Piazolo, V. Geist, L. Brehm, & R. Schmidt (Eds.), *Innovations in enterprise information systems management and Engineering* (Vol. 285, pp. 114–122). Springer International Publishing. https://doi.org/10.1007/978-3-319-58801-8_10

Doyle, M. (2022). Misquoting Madison. *Legal Affairs*, July/August. https://www.legalaffairs.org/issues/July-August-2002/scene_doyle_julaug2002.msp

Drisko, J. (2017). *Triangulation [Data set]*. Oxford University Press. https://doi.org/10.1093/obo/9780195389678-0045

European Commission. (2016). *Communication to the Commission 'data, information and knowledge management at the European Commission.*https://ec.europa.eu/info/publications/communication-data-information-and-knowledge-management-european-commission_en

Gallego, I., & Font, X. (2021). Changes in air passenger demand as a result of the COVID-19 crisis: Using big data to inform tourism policy. *Journal of Sustainable Tourism, 29*(9), 1470–1489. https://doi.org/10.1080/09669582.2020.1773476

Jackson, P. (2019). From 'intruders' to 'partners': The evolution of the relationship between the research community and sources of official administrative data. In N. Crato, & P. Paruolo (Eds), Data-driven policy impact evaluation. Springer. https://doi.org/10.1007/978-3-319-78461-8_2

Lustig, N. (2020). *The "Missing Rich" in household surveys: Causes and correction approaches* [Preprint]. SocArXiv. https://doi.org/10.31235/osf.io/j23pn.

Manzan, S. (2023). Big data and computational social science for economic analysis and policy. In *Handbook of computational social science for policy*. Springer International publishing.

Menkveld, A. J., Dreber, A., Holzmeister, F., Huber, J., Johanneson, M., Kirchler, M., Razen, M., Weitzel, U., Abad, D., Abudy, M., Adrian, T., Ait-Sahalia, Y., Akmansoy, O., Alcock, J., Alexeev, V., Aloosh, A., Amato, L., Amaya, D., Angel, J. J., et al. (2021). Non-Standard Errors. *SSRN Electronic Journal*. https://doi.org/10.2139/ssrn.3961574

Napierała, T., Leśniewska-Napierała, K., & Burski, R. (2020). Impact of geographic distribution of COVID-19 cases on hotels' performances: Case of Polish cities. *Sustainability, 12*(11), 4697. https://doi.org/10.3390/su12114697

National Academies of Sciences, Engineering, and Medicine. (2017). *Innovations in Federal statistics: Combining data sources while protecting privacy* (p. 24652). National Academies Press. https://doi.org/10.17226/24652

Read, L., & Atinc, T. M. (2017). Information for accountability: Transparency and citizen engagement for improved service delivery in education systems. *Brookings Working Paper, 99*. https://www.brookings.edu/wp-content/uploads/2017/01/global_20170125_information_for_accountability.pdf

Richardson, S., Hirsch, J. S., Narasimhan, M., Crawford, J. M., McGinn, T., Davidson, K. W., the Northwell COVID-19 Research Consortium, Barnaby, D. P., Becker, L. B., Chelico, J. D., Cohen, S. L., Cookingham, J., Coppa, K., Diefenbach, M. A., Dominello, A. J., Duer-Hefele, J., Falzon, L., Gitlin, J., Hajizadeh, N., et al. (2020). Presenting characteristics, comorbidities, and outcomes among 5700 patients hospitalized with Covid-19 in the New York City area. *JAMA, 323*(20), 2052. https://doi.org/10.1001/jama.2020.6775

Romanillos Arroyo, G., & Moya-Gómez, B. (2023). New data and computational methods opportunities to enhance the knowledge base of tourism. In *Handbook of computational social science for policy*. Springer International Publishing.

Rossiter, J. (2020). *Link it, open it, use it CDG note*. https://www.cgdev.org/publication/link-it-open-it-use-it-changing-how-education-data-are-used-generate-ideas

Shlomo, N. (2019). Overview of data linkage methods for policy design and evaluation. In N. Crato & P. Paruolo (Eds.), *Data-driven policy impact evaluation* (pp. 47–65). Springer International Publishing. https://doi.org/10.1007/978-3-319-78461-8_4

Signorelli, S., Fontana, M., Gabrielli, L., & Vespe, M. (2023). Challenges for official statistics in the digital age. In *Handbook of computational social science for policy*. Springer.

Snijders, C., Matzat, U., & Reips, U.-D. (2012). 'Big data': Big gaps of knowledge in the field of internet science. *International Journal of Internet Science, 7*(1), 1–5.

The White House. (2014). Big data: Seizing opportunities, preserving values. *Executive Office of the President*.

US Congress. (2016). *Evidence-based policymaking commission act of 2016, H.R. 1831, 114th Congress*.

Vespe, M., Iacus, S. M., Santamaria, C., Sermi, F., & Spyratos, S. (2021). On the use of data from multiple mobile network operators in Europe to fight Covid-19. *Data & Policy, 3*, e8. https://doi.org/10.1017/dap.2021.9

Wagenmakers, E.-J., Sarafoglou, A., Aarts, S., Albers, C., Algermissen, J., Bahník, Š., van Dongen, N., Hoekstra, R., Moreau, D., van Ravenzwaaij, D., Sluga, A., Stanke, F., Tendeiro, J., & Aczel, B. (2021). Seven steps toward more transparency in statistical practice. *Nature Human Behaviour, 5*(11), 1473–1480. https://doi.org/10.1038/s41562-021-01211-8

Wagenmakers, E.-J., Sarafoglou, A., & Aczel, B. (2022). One statistical analysis must not rule them all. *Nature, 605*(7910), 423–425. https://doi.org/10.1038/d41586-022-01332-8

Watt, F. (2022, April 22). If you want science to move forward, you have to share it. *EMBL*. https://www.embl.org/news/lab-matters/if-you-want-science-to-move-forward-you-have-to-share-it/#:~:text=In%20December%202021%2C%20EMBL%20announced,research%20across%20the%20life%20sciences

Chapter 7
Natural Language Processing for Policymaking

Zhijing Jin and Rada Mihalcea

Abstract Language is the medium for many political activities, from campaigns to news reports. Natural language processing (NLP) uses computational tools to parse text into key information that is needed for policymaking. In this chapter, we introduce common methods of NLP, including text classification, topic modelling, event extraction, and text scaling. We then overview how these methods can be used for policymaking through four major applications including data collection for evidence-based policymaking, interpretation of political decisions, policy communication, and investigation of policy effects. Finally, we highlight some potential limitations and ethical concerns when using NLP for policymaking.

7.1 Introduction

Language is an important form of data in politics. Constituents express their stances and needs in text such as social media and survey responses. Politicians conduct campaigns through debates, statements of policy positions, and social media. Government staff needs to compile information from various documents to assist in decision-making. Textual data is also prevalent through the documents and debates in the legislation process, negotiations and treaties to resolve international conflicts, and media such as news reports, social media, party platforms, and manifestos.

Natural language processing (NLP) is the study of computational methods to automatically analyse text and extract meaningful information for subsequent analysis. The importance of NLP for policymaking has been highlighted since the

Z. Jin
Max Planck Institute for Intelligent Systems, Tübingen, Germany

ETH Zürich, Zürich, Switzerland
e-mail: zhij.jin@gmail.com

R. Mihalcea (✉)
University of Michigan, Ann Arbor, MI, USA
e-mail: mihalcea@umich.edu

E. Bertoni et al. (eds.), *Handbook of Computational Social Science for Policy*,
https://doi.org/10.1007/978-3-031-16624-2_7

last century (Gigley, 1993). With the recent success of NLP and its versatility over tasks such as classification, information extraction, summarization, and translation (Brown et al., 2020; Devlin et al., 2019), there is a rising trend to integrate NLP into the policy decisions and public administrations (Engstrom et al., 2020; Misuraca et al., 2020; Van Roy et al., 2021). Main applications include extracting useful, condensed information from free-form text (Engstrom et al., 2020), and analysing sentiment and citizen feedback by NLP Biran et al. (2022) as in many projects funded by EU Horizon projects (European Commission, 2017). Driven by the broad applications of NLP (Jin et al., 2021a), the research community also starts to connect NLP with various social applications in the fields of computational social science (Engel et al., 2021; Lazer et al., 2009; Luz, 2022; Shah et al., 2015) and political science in particular (Glavaš et al., 2019; Grimmer & Stewart, 2013).

We show an overview of NLP for policymaking in Fig. 7.1. According to this overview, the chapter will consist of three parts. First, we introduce in Sect. 7.2 NLP methods that are applicable to political science, including text classification, topic modelling, event extraction, and score prediction. Next, we cover a variety of cases where NLP can be applied to policymaking in Sect. 7.3. Specifically, we cover four stages: analysing data for evidence-based policymaking, improving policy communication with the public, investigating policy effects, and interpreting political phenomena to the public. Finally, we will discuss limitations and ethical considerations when using NLP for policymaking in Sect. 7.4.

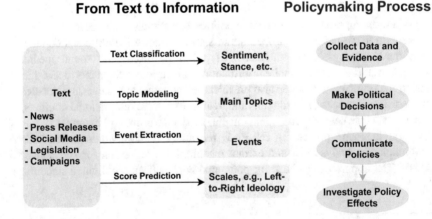

Fig. 7.1 Overview of NLP for policymaking

7.2 NLP for Text Analysis

NLP brings powerful computational tools to analyse textual data (Jurafsky & Martin, 2000). According to the type of information that we want to extract from the text, we introduce four different NLP tools to analyse text data: text classification (by which the extracted information is the *category* of the text), topic modelling (by which the extracted information is the *key topics* in the text), event extraction (by which the extracted information is the list of *events* mentioned in the text), and score prediction (where the extracted information is a *score* of the text). Table 7.1 lists each method with the type of information it can extract and some example application scenarios, which we will detail in the following subsections.

7.2.1 Text Classification

As one of the most common types of text analysis methods, text classification reads in a piece of text and predicts its category using an NLP text classification model, as in Fig. 7.2.

Table 7.1 Four common NLP methods, the type of information extracted by each of them, and example applications

NLP method	Information to extract	Example applications
Text classification	Category of text	Identify the sentiment, stance, etc.
Topic modelling	Key topics in text	Summarize topics in political agenda
Event extraction	List of events	Extract news events, international conflicts
Score prediction	Score	Text scaling

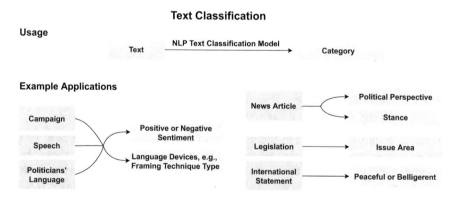

Fig. 7.2 The usage and example applications of text classification on political text

There are many off-the-shelf existing tools for text classification (Brown et al., 2020; Loria, 2018; Yin et al., 2019) such as the implementation[1] using the Python package `transformers` (Wolf et al., 2020). A well-known subtask of text classification is sentiment classification (also known as sentiment analysis or opinion mining), which aims to distinguish the subjective information in the text, such as positive or negative sentiment (Pang & Lee, 2007). However, the existing tools only do well in categories that are easy to predict. If the categorization is customized and very specific to a study context, then there are two common solutions. One is to use dictionary-based methods, by a list of frequent keywords that correspond to a certain category (Albaugh et al., 2013) or using general linguistic dictionaries such as the Linguistic Inquiry and Word Count (LIWC) dictionary (Pennebaker et al., 2001). The second way is to adopt the data-driven pipeline, which requires human hand coding of documents into a predetermined set of categories, then train an NLP model to learn the text classification task (Sun et al., 2019), and verify the performance of the NLP model on a held-out subset of the data, as introduced in Grimmer and Stewart (2013). An example of adapting the state-of-the-art NLP models on a customized dataset is demonstrated in this guide.[2]

Using the text classification method, we can automate many types of analyses in political science. As listed in the examples in Fig. 7.2, researchers can detect political perspective of news articles (Huguet Cabot et al., 2020), the stance in media on a certain topic (Luo et al., 2020), whether campaigns use positive or negative sentiment (Ansolabehere & Iyengar, 1995), which issue area is the legislation about (Adler & Wilkerson, 2011), topics in parliament speech (Albaugh et al., 2013; Osnabrügge et al., 2021), congressional bills (Collingwood & Wilkerson, 2012; Hillard et al., 2008) and political agenda (Karan et al., 2016), whether the international statement is peaceful or belligerent (Schrodt, 2000), whether a speech contains positive or negative sentiment (Schumacher et al., 2016), and whether a US Circuit Courts case decision is conservative or liberal (Hausladen et al., 2020). Moreover, text classification can also be used to categorize the type of language devices that politicians use, such as what type of framing the text uses (Huguet Cabot et al., 2020), and whether a tweet uses political parody (Maronikolakis et al., 2020).

7.2.2 Topic Modelling

Topic modelling is a method to uncover a list of frequent topics in a corpus of text. For example, news articles that are against vaccination might frequently mention the topic "autism", whereas news articles supporting vaccination will be more likely to mention "immune" and "protective". One of the most widely used models is the

---

[1] https://discuss.huggingface.co/t/new-pipeline-for-zero-shot-text-classification/681.

[2] https://skimai.com/fine-tuning-bert-for-sentiment-analysis/.

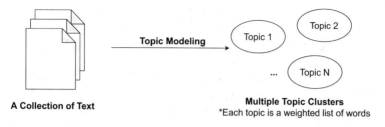

Fig. 7.3 Given a collection of text documents, topic modelling generates a list of topic clusters

Latent Dirichlet Allocation (LDA) (Blei et al., 2001) which is available in the Python packages NLTK and Gensim, as in this guide.[3]

Specifically, LDA is a probabilistic model that models each topic as a mixture of words, and each textual document can be represented as a mixture of topics. As in Fig. 7.3, given a collection of textual documents, LDA topic modelling generates a list of topic clusters, for which the number N of topics can be customized by the analyst. In addition, if needed, LDA can also produce a representation of each document as a weighted list of topics. While often the number of topics is predetermined by the analyst, this number can also be dynamically determined by measuring the perplexity of the resulting topics. In addition to LDA, other topic modelling algorithms have been used extensively, such as those based on principal component analysis (PCA) (Chung & Pennebaker, 2008).

Topic modelling, as described in this section, can facilitate various studies on political text. Previous studies analysed the topics of legislative speech (Quinn et al., 2006, 2010), Senate press releases (Grimmer, 2010a), and electoral manifestos (Menini et al., 2017).

7.2.3   Event Extraction

Event extraction is the task of extracting a list of events from a given text. It is a subtask of a larger domain of NLP called information extraction (Manning et al., 2008). For example, the sentence "Israel bombs Hamas sites in Gaza" expresses an event "*Israel* $\xrightarrow{bombs}$ *Hamas sites*" with the location "*Gaza*". Event extraction usually incorporates both entity extraction (e.g. Israel, Hamas sites, and Gaza in the previous example) and relation extraction (e.g. "bombs" in the previous example).

Event extraction is a handy tool to monitor events automatically, such as detecting news events (Mitamura et al., 2017; Walker et al., 2006) and detecting international conflicts (Azar, 1980; Trappl, 2006). To foster research on event extraction, there are tremendous efforts into textual data collection (McClelland,

---

[3] https://skimai.com/fine-tuning-bert-for-sentiment-analysis/.

1976; Merritt et al., 1993; Raleigh et al., 2010; Schrodt & Hall, 2006; Sundberg & Melander, 2013), event coding schemes to accommodate different political events (Bond et al., 1997; Gerner et al., 2002; Goldstein, 1992), and dataset validity assessment (Schrodt & Gerner, 1994).

As for event extraction models, similar to text classification models, there are off-the-shelf tools such as the Python packages `stanza` (Qi et al., 2020) and `spaCy` (Honnibal et al., 2020). In case of customized sets of event types, researchers can also train NLP models on a collection of textual documents with event annotations (Hogenboom et al., 2011; Liu et al., 2020, inter alia).

7.2.4 Score Prediction

NLP can also be used to predict a score given input text. A useful application is political text scaling, which aims to predict a score (e.g. left-to-right ideology, emotionality, and different attitudes towards the European integration process) for a given piece of text (e.g. political speeches, party manifestos, and social media posts) (Gennaro & Ash, 2021; Laver et al., 2003; Lowe et al., 2011; Slapin & Proksch, 2008, inter alia).

Traditional models for text scaling include Wordscores (Laver et al., 2003) and WordFish (Lowe et al., 2011; Slapin & Proksch, 2008). Recent NLP models represent the text by high-dimensional vectors learned by neural networks to predict the scores (Glavaš et al., 2017b; Nanni et al., 2019). One way to use the NLP models is to apply off-the-shelf general-purpose models such as InstructGPT (Ouyang et al., 2022) and design a prompt to specify the type of the scaling to the API,[4] or borrow existing, trained NLP models if the same type of scaling has been studied by previous researchers. Another way is to collect a dataset of text with hand-coded scales, and train NLP models to learn to predict the scale, similar to the practice in Gennaro and Ash (2021); Slapin and Proksch (2008), inter alia.

7.3 Using NLP for Policymaking

In the political domain, there are large amounts of textual data to analyse (NEUEN-DORF & KUMAR, 2015), such as parliament debates (Van Aggelen et al., 2017), speeches (Schumacher et al., 2016), legislative text (Baumgartner et al., 2006; Bevan, 2017), database of political parties worldwide (Döring & Regel, 2019), and expert survey data (Bakker et al., 2015). Since it is tedious to hand-code all textual data, NLP provides a low-cost tool to automatically analyse such massive text.

---

[4] https://beta.openai.com/docs/introduction.

In this section, we will introduce how NLP can facilitate four major areas to help policymaking: before policies are made, researchers can use NLP to analyse data and extract key information for evidence-based policymaking (Sect. 7.3.1); after policies are made, researchers can interpret the priorities among and reasons behind political decisions (Sect. 7.3.2); researchers can also analyse features in the language of politicians when communicating the policies to the public (Sect. 7.3.3); and finally, after the policies have taken effect, researchers can investigate the effectiveness of the policies (Sect. 7.3.4).

7.3.1   Analysing Data for Evidence-Based Policymaking

A major use of NLP is to extract information from large collections of text. This function can be very useful for analysing the views and needs of constituents, so that policymakers can make decisions accordingly.

As in Fig. 7.4, we will explain how NLP can be used to analyse data for evidence-based policymaking from three aspects: data, information to extract, and political usage.

Data Data is the basis of such analyses. Large amounts of textual data can reveal information about constituents, media outlets, and influential figures. The data can come from a variety of sources, including social media such as Twitter and Facebook, survey responses, and news articles.

Information to Extract Based on the large textual corpora, NLP models can be used to extract information that are useful for political decision-making, ranging from information about people, such as sentiment (Rosenthal et al., 2015; Thelwall et al., 2011), stance (Gottipati et al., 2013; Luo et al., 2020; Stefanov et al., 2020; Thomas et al., 2006), ideology (Hirst et al., 2010; Iyyer et al., 2014; Preoţiuc-Pietro et al., 2017), and reasoning on certain topics (Camp et al., 2021; Demszky et al.,

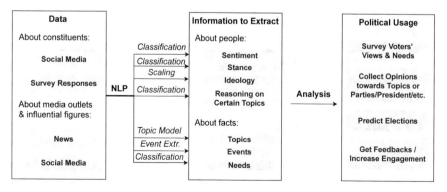

Fig. 7.4 NLP to analyse data for evidence-based policymaking

2019; Egami et al., 2018), to factual information, such as main topics (Gottipati et al., 2013), events (Ding & Riloff, 2018; Ding et al., 2019; Mitamura et al., 2017; Trappl, 2006), and needs (Crayton et al., 2020; Paul & Frank, 2019; Sarol et al., 2020) expressed in the data. The extracted information cannot only be about people but also about political entities, such as the left-right political scales of parties and political actors (Glavaš et al., 2017b; Slapin & Proksch, 2008), which claims are raised by which politicians (Blessing et al., 2019; Padó et al., 2019), and the legislative body's vote breakdown for state bills by backgrounds such as gender, rural-urban, and ideological splits (Davoodi et al., 2020).

To extract such information from text, we can often utilize the main NLP tools introduced in Sect. 7.2, including text classification, topic modelling, event extraction, and score prediction (especially text scaling to predict left-to-right ideology). In NLP literature, social media, such as Twitter, is a popular source of textual data to collect public opinions (Arunachalam & Sarkar, 2013; Pak & Paroubek, 2010; Paltoglou & Thelwall, 2012; Rosenthal et al., 2015; Thelwall et al., 2011).

Political Usage Such information extracted from data is highly valuable for political usage. For example, voters' sentiment, stance, and ideology are important supplementary for traditional polls and surveys to gather information about the constituents' political leaning. Identifying the needs expressed by people is another important survey target, which helps politicians understand what needs they should take care of and match the needs and availabilities of resources (Hiware et al., 2020).

Among more specific political uses is to understand the public opinion on parties/president, as well as on certain topics. The public sentiment towards parties (Pla & Hurtado, 2014) and president (Marchetti-Bowick & Chambers, 2012) can serve as a supplementary for the traditional approval rating survey, and stances towards certain topics (Gottipati et al., 2013; Luo et al., 2020; Stefanov et al., 2020) can be important information for legislators to make decisions on debatable issues such as abortion, taxes, and legalization of same-sex marriage. Many existing studies use NLP on social media text to predict election results (Beverungen & Kalita, 2011; Mohammad et al., 2015; O'Connor et al., 2010; Tjong Kim Sang & Bos, 2012; Unankard et al., 2014). In general, big data-driven analyses can facilitate decision-makers to collect more feedback from people and society, enabling policymakers to be closer to citizens, and increase transparency and engagement in political issues (Arunachalam & Sarkar, 2013).

7.3.2 Interpreting Political Decisions

After policies are made, political scientists and social scientists can use textual data to interpret political decisions. As in Fig. 7.5, there are two major use cases: mining political agendas and discovering policy responsiveness.

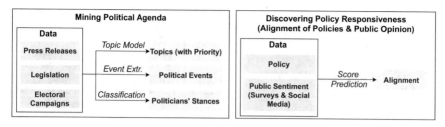

Fig. 7.5 NLP to interpret political decisions

Mining Political Agendas Researchers can use textual data to infer a political agenda, including the topics that politicians prioritize, political events, and different political actors' stances on certain topics. Such data can come from press releases, legislation, and electoral campaigns. Examples of previous studies to analyse the topics and prioritization of political bodies include the research on the prioritization each senator assigns to topics using press releases (Grimmer, 2010b), topics in different parties' electoral manifestos (Glavaš et al., 2017a), topics in EU parliament speeches (Lauscher et al., 2016) and other various types of text (Grimmer, 2010a; Hopkins & King, 2010; King & Lowe, 2003; Roberts et al., 2014), as well as political event detection from congressional text and news (Nanni et al., 2017).

Research on politicians' stances include identifying policy positions of politicians (Laver et al., 2003; Lowe et al., 2011; Slapin & Proksch, 2008; Winter & Stewart, 1977, inter alia), how different politicians agree or disagree on certain topics in electoral campaigns (Menini & Tonelli, 2016), and assessment of political personalities (Immelman, 1993).

Further studies look into how political interests affect legislative behaviour. Legislators tend to show strong personal interest in the issues that come before their committees (Fenno, 1973), and Mayhew (2004) identifies that senators replying on appropriations secured for their state have a strong incentive to support legislations that allow them to secure particularistic goods.

Discovering Policy Responsiveness Policy responsiveness is the study of how policies respond to different factors, such as how changes in public opinion lead to responses in public policy (Stimson et al., 1995). One major direction is that politicians tend to make policies that align with the expectations of their constituents, in order to run for successful re-election in the next term (Canes-Wrone et al., 2002). Studies show that policy preferences of the state public can be a predictor of future state policies (Caughey & Warshaw, 2018). For example, Lax and Phillips (2009) show that more LGBT tolerance leads to more pro-gay legislation in response.

A recent study by Jin et al. (2021b) uses NLP to analyse over 10 million COVID-19-related tweets targeted at US governors; using classification models to obtain the public sentiment, they study how public sentiment leads to political decisions of COVID-19 policies made by US governors. Such use of NLP on massive textual

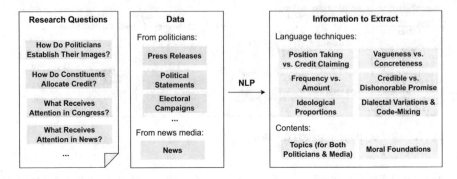

Fig. 7.6 NLP to analyse policy communication

data contrasts with the traditional studies of policy responsiveness which span over several decades and use manually collected survey results (Caughey & Warshaw, 2018; Lax & Phillips, 2009, 2012).

7.3.3 Improving Policy Communication with the Public

Policy communication is the study to understand how politicians present the policies to their constituents. As in Fig. 7.6, common research questions in policy communication include how politicians establish their images (Fenno, 1978) such as campaign strategies (Petrocik, 1996; Sigelman & Buell Jr, 2004; Simon, 2002), how constituents allocate credit, what receives attention in Congress (Sulkin, 2005), and what receives attention in news articles (Armstrong et al., 2006; McCombs & Valenzuela, 2004; Semetko & Valkenburg, 2000).

Based on data from press releases, political statements, electoral campaigns, and news articles,[5] researchers usually analyse two types of information: the language techniques politicians use and the contents such as topics and underlying moral foundations in these textual documents.

Language Techniques Policy communication largely focuses on the types of languages that politicians use. Researchers are interested in first analysing the language techniques in political texts, and then, based on these techniques, researchers can dive into the questions of why politicians use them and what are the effects of such usage.

For example, previous studies analyse what portions of political texts are position-taking versus credit-claiming (Grimmer, 2013; Grimmer et al., 2012),

---

[5] Other data sources used in policy communication research include surveys of senate staffers (Cook, 1988), newsletters that legislators send to constituents (Lipinski, 2009), and so on.

whether the claims are vague or concrete (Baerg et al., 2018; Eichorst & Lin, 2019), the frequency of credit-claiming messages versus the actual amount of contributions (Grimmer et al., 2012), and whether politicians tend to make credible or dishonourable promises (Grimmer, 2010b). Within the political statements, it is also interesting to check the ideological proportions (Sim et al., 2013) and how politicians make use of dialectal variations and code-mixing (Sravani et al., 2021).

The representation styles usually affect the effectiveness of policy communication, such as the role of language ambiguity in framing the political agenda (Campbell, 1983; Page, 1976) and the effect of credit-claiming messages on constituents' allocation of credit (Grimmer et al., 2012).

Contents The contents of policy communication include the topics in the political statements, such as what senators discuss in floor statements (Hill & Hurley, 2002) and what presidents address in daily speeches (Lee, 2008), and also the moral foundations used by politicians underlying their political tweets (Johnson & Goldwasser, 2018).

Using the extracted content information, researchers can explore further questions such as whether competing politicians or political elites emphasize the same issues (Gabel & Scheve, 2007; Petrocik, 1996) and how the priorities politicians articulate co-vary with the issues discussed in the media (Bartels, 1996). Another open research direction is to analyse the interaction between newspapers and politicians' messages, such as how often newspapers cover a certain politician's message and in what way and how such coverage affects incumbency advantage.

Meaningful Future Work Apart from analysing the language of existing political texts that aims to maximize political interests, an advanced question that is more meaningful to society is how to improve policy communication to steer towards a more beneficial future for society as a whole. There is relatively little research on this, and we welcome future work on this meaningful topic.

7.3.4 Investigating Policy Effects

After policies are taken into effect, it is important to collect feedback or evaluate the effectiveness of policies. Existing studies evaluate the effects of policies along different dimensions: one dimension is the change in public sentiment, which can be analysed by comparing the sentiment classification results before and after policies, following a similar paradigm in Sect. 7.3.1. There are also studies on how policies affect the crowd's perception of the democratic process (Miller et al., 1990).

Another dimension is how policies result in economic changes. Calvo-González et al. (2018) investigate the negative consequences of policy volatility that harm long-term economic growth. Specifically, to measure policy volatility, they first obtain main topics by topic modelling on presidential speeches and then analyse how the significance of topics changes over time.

7.4 Limitations and Ethical Considerations

There are several limitations that researchers and policymakers need to take into consideration when using NLP for policymaking, due to the data-driven and black-box nature of modern NLP. First, the effectiveness of the computational models relies on the quality and comprehensiveness of the data. Although many political discourses are public, including data sources such as news, press releases, legislation, and campaigns, when it comes to surveying public opinions, social media might be a biased representation of the whole population. Therefore, when making important policy decisions, the traditional polls and surveys can provide more comprehensive coverage. Note that in the case of traditional polls, NLP can still be helpful in expediting the processing of survey answers.

The second concern is the black-box nature of modern NLP models. We do not encourage decision-making systems to depend fully on NLP, but suggest that NLP can assist human decision-makers. Hence, all the applications introduced in this chapter use NLP to compile information that is necessary for policymaking instead of directly suggesting a policy. Nonetheless, some of the models are hard to interpret or explain, such as text classification using deep learning models (Brown et al., 2020; Yin et al., 2019), which could be vulnerable to adversarial attacks by small paraphrasing of the text input (Jin et al., 2020). In practical applications, it is important to ensure the trustworthiness of the usage of AI. There could be a preference for transparent machine learning models if they can do the work well (e.g. LDA topic models and traditional classification methods using dictionaries or linguistic rules) or tasks with well-controlled outputs such as event extraction to select spans of the given text that mention events. In cases where only the deep learning models can provide good performance, there should be more detailed performance analysis (e.g. a study to check the correlation of the model decisions and human judgments), error analysis (e.g. different types of errors, failure modes, and potential bias towards certain groups), and studies about the interpretability of the model (e.g. feature attribution of the model, visualization of the internal states of the model).

Apart from the limitations of the technical methodology, there are also ethical considerations arising from the use of NLP. Among the use cases introduced in this chapter, some applications of NLP are relatively safe as they mainly involve analysing public political documents and fact-based evidence or effects of policies. However, others could be concerning and vulnerable to misuse. For example, although effective, truthful policy communication is beneficial for society, it might be tempting to overdo policy communication and by all means optimize the votes. As it is highly important for government and politicians to gain positive public perception, overly optimizing policy communication might lead to propaganda, intrusion of data privacy to collect more user preferences, and, in more severe cases, surveillance and violation of human rights. Hence, there is a strong need for policies to regulate the use of technologies that influence public opinions and pose a challenge to democracy.

7.5 Conclusions

This chapter provided a brief overview of current research directions in NLP that provide support for policymaking. We first introduced four main NLP tasks that are commonly used in text analysis: text classification, topic modelling, event extraction, and text scaling. We then showed how these methods can be used in policymaking for applications such as data collection for evidence-based policymaking, interpretation of political decisions, policy communication, and investigation of policy effects. We also discussed potential limitations and ethical considerations of which researchers and policymakers should be aware.

NLP holds significant promise for enabling data-driven policymaking. In addition to the tasks overviewed in this chapter, we foresee that other NLP applications, such as text summarization (e.g. to condense information from large documents), question answering (e.g. for reasoning about policies), and culturally adjusted machine translation (e.g. to facilitate international communications), will soon find use in policymaking. The field of NLP is quickly advancing, and close collaborations between NLP experts and public policy experts will be key to the successful use and deployment of NLP tools in public policy.

References

Adler, E. Scott, & Wilkerson, J. (2011). Congressional bills project. *NSF 00880066 and 00880061*. http://www.congressionalbills.org/

Albaugh, Q., Sevenans, J., Soroka, S., & Loewen, P. J. (2013). The automated coding of policy agendas: A dictionary-based approach. In *The 6th Annual Comparative Agendas Conference, Antwerp, Belgium*.

Ansolabehere, S., & Iyengar, S. (1995). *Going negative: How political advertisements shrink and polarize the electorate* (Vol. 95). New York: Simon & Schuster.

Armstrong, E. M., Carpenter, D. P., & Hojnacki, M. (2006). Whose deaths matter? Mortality, advocacy, and attention to disease in the mass media. *Journal of Health Politics, Policy and Law, 31*(4), 729–772.

Arunachalam, R., & Sarkar, S. (2013). The new eye of government: citizen sentiment analysis in social media. In *Proceedings of the IJCNLP 2013 Workshop on Natural Language Processing for Social Media (SocialNLP)* (pp. 23–28). Nagoya, Japan: Asian Federation of Natural Language Processing. https://www.aclweb.org/anthology/W13-4204

Azar, E. E. (1980). The conflict and peace data bank (COPDAB) project. *Journal of Conflict Resolution, 24*(1), 143–152.

Baerg, N., Duell, D., & Lowe, W. (2018). Central bank communication as public opinion: experimental evidence. *Work in Progress*.

Bakker, R., De Vries, C., Edwards, E., Hooghe, L., Jolly, S., Marks, G., Polk, J., Rovny, J., Steenbergen, M., & Vachudova, M. A. (2015). Measuring party positions in Europe: The Chapel Hill expert survey trend file, 1999–2010. *Party Politics, 21*(1), 143–152.

Bartels, L. M. (1996). Politicians and the press: Who leads, who follows. In *Annual Meeting of the American Political Science Association* (pp. 1–60).

Baumgartner, F. R., Green-Pedersen, C., & Jones, B. D. (2006). Comparative studies of policy agendas. *Journal of European Public Policy, 13*(7), 959– 974.

Bevan, S. (2017). Gone fishing: The creation of the comparative agendas project master code-book. *Comparative Policy Agendas: Theory, Tools, Data*. http://sbevan.%20com/cap-master-codebook.html

Beverungen, G., & Kalita, J. (2011). Evaluating methods for summarizing Twitter posts. In *Proceedings of the 5th AAAI ICWSM*.

Biran, O., Feder, O., Moatti, Y., Kiourtis, A., Kyriazis, D., Manias, G., Mavrogiorgou, A., Sgouros, N. M., Barata, M. T., Oldani, I., Sanguino, M. A. & Kranas, P. (2022). PolicyCLOUD: A prototype of a cloud serverless ecosystem for policy analytics. *CoRR, abs/2201.06077*. https://arxiv.org/abs/2201.06077

Blei, D. M., Ng, A. Y., & Jordan, M. I. (2001). Latent dirichlet allocation. In T. G. Dietterich, S. Becker, & Z. Ghahramani (eds.), *Advances in Neural Information Processing Systems 14 [Neural Information Processing Systems: Natural and Synthetic, NIPS 2001, December 3–8, 2001, Vancouver, British Columbia, Canada]* (pp. 601–608). MIT Press. https://proceedings.neurips.cc/paper/2001/hash/296472c9542ad4d4788d543508116cbc-Abstract.html

Blessing, A., Blokker, N., Haunss, S., Kuhn, J., Lapesa, G., & Padó, S. (2019). An environment for relational annotation of political debates. In *Proceedings of the 57th Annual Meeting of the Association for Computational Linguistics: System Demonstrations* (pp. 105–110). Florence, Italy: Association for Computational Linguistics. https://doi.org/10.18653/v1/P19-3018. https://aclanthology.org/P19-3018

Bond, D., Jenkins, J. C., Taylor, C. L., & Schock, K. (1997). Mapping mass political conflict and civil society: issues and prospects for the automated development of event data. *Journal of Conflict Resolution, 41*(4), 553–579.

Brown, T. B., Mann, B., Ryder, N., Subbiah, M., Kaplan, J., Dhariwal, P., Neelakantan, A., Shyam, P., Sastry, G., Askell, A., Agarwal, S., Herbert-Voss, A., Krueger, G., Henighan, T., Child, R., Ramesh, A., Ziegler, D. M., Wu, J., Winter, C., . . . , Amodei, D. (2020). Language models are fewshot learners. In H. Larochelle, M. Ranzato, R. Hadsell, M. Balcan, & H.-T. Lin (Eds.), *Advances in Neural Information Processing Systems 33: Annual Conference on Neural Information Processing Systems 2020, Neurips 2020, December 6–12, 2020, Virtual*. https://proceedings.neurips.cc/paper/2020/hash/1457c0d6bfcb4967418bfb8ac142f64a-Abstract.html

Calvo-González, O., Eizmendi, A., & Reyes, G. J. (2018). Winners never quit, quitters never grow: Using text mining to measure policy volatility and its link with long-term growth in latin America. *World Bank Policy Research Working Paper* (8310).

Camp, N. P., Voigt, R., Jurafsky, D., & Eberhardt, J. L. (2021). The thin blue waveform: racial disparities in officer prosody undermine institutional trust in the police. *Journal of Personality and Social Psychology, 121*, 1157–1171.

Campbell, J. E. (1983). Ambiguity in the issue positions of presidential candidates: A causal analysis. *American Journal of Political Science, 27*, 284–293.

Canes-Wrone, B., Brady, D. W., & Cogan, J. F. (2002). Out of step, out of office: Electoral accountability and house members' voting. *American Political Science Review, 96*, 127–140.

Caughey, D., & Warshaw, C. (2018). Policy preferences and policy change: Dynamic responsiveness in the American states, 1936–2014. *American Political Science Review, 112*, 249–266.

Chung, C. K., & Pennebaker, J. W. (2008). Revealing dimensions of thinking in OpenEnded self-descriptions: An automated meaning extraction method for natural language. *Journal of Research in Personality, 42*(1), 96–132.

Collingwood, L., & Wilkerson, J. (2012). Tradeoffs in accuracy and efficiency in supervised learning methods. *Journal of Information Technology & Politics, 9*(3), 298–318.

Cook, T. E. (1988). Press secretaries and media strategies in the house of representatives: Deciding whom to pursue. *American Journal of Political Science, 32*, 1047–1069.

Crayton, A., Fonseca, J., Mehra, K., Ng, M., Ross, J., Sandoval-Castañeda, M., & von Gnechten, R. (2020). Narratives and needs: Analyzing experiences of cyclone amphan using Twitter discourse. *CoRR, abs/2009.05560*. https://arxiv.org/abs/2009.05560

Davoodi, M., Waltenburg, E., & Goldwasser, D. (2020). Understanding the language of political agreement and disagreement in legislative texts. In *Proceedings of the 58th Annual*

Meeting of the Association for Computational Linguistics (pp. 5358–5368). Online: Association for Computational Linguistics. https://doi.org/10.18653/v1/2020.acl-main.476. https://aclanthology.org/2020.acl-main.476

Demszky, D., Garg, N., Voigt, R., Zou, J., Shapiro, J., Gentzkow, M., & Jurafsky, D. (2019). Analyzing polarization in social media: method and application to tweets on 21 mass shootings. In *Proceedings of the 2019 Conference of the North American Chapter of the Association for Computational Linguistics: Human Language Technologies, volume 1 (Long and Short Papers)* (pp. 2970–3005). Minneapolis, Minnesota: Association for Computational Linguistics. https://doi.org/10.18653/v1/N19-1304. https://aclanthology.org/N19-1304

Devlin, J., Chang, M.-W., Lee, K., & Toutanova, K. (2019). BERT: Pre-training of deep bidirectional transformers for language understanding. In J. Burstein, C. Doran, & T. Solorio (Eds.), *Proceedings of the 2019 Conference of the North American Chapter of the Association for Computational Linguistics: Human Language Technologies, NAACL-HLT 2019, Minneapolis, MN, USA, June 2–7, 2019, Volume 1 (Long and Short Papers)* (pp. 4171–4186). Association for Computational Linguistics. https://doi.org/10.18653/v1/n19-1423

Ding, H., & Riloff, E. (2018). Human needs categorization of affective events using labeled and unlabeled data. In *Proceedings of the 2018 Conference of the North American Chapter of the Association for Computational Linguistics: Human Language Technologies, Volume 1 (Long papers)* (pp. 1919–1929). New Orleans, Louisiana: Association for Computational Linguistics. https://doi.org/10.18653/v1/N18-1174. https://aclanthology.org/N18-1174

Ding, H., Riloff, E., & Feng, Z. (2019). Improving human needs categorization of events with semantic classification. In *Proceedings of the Eighth Joint Conference on Lexical and Computational Semantics (*SEM 2019)* (pp. 198–204). Minneapolis, Minnesota: Association for Computational Linguistics. https://doi.org/10.18653/v1/S19-1022. https://aclanthology.org/S19-1022

Döring, H., & Regel, S. (2019). Party facts: A database of political parties worldwide. *Party Politics, 25*(2), 97–109.

Egami, N., Fong, C. J., Grimmer, J., Roberts, M. E., & Stewart, B. M. (2018). How to make causal inferences using texts. *CoRR, abs/1802.02163*. http://arxiv.org/abs/1802.02163

Eichorst, J., & Lin, N. C. N. (2019). Resist to commit: Concrete campaign statements and the need to clarify a partisan reputation. *The Journal of Politics, 81*(1), 15–32.

Engel, U., Quan-Haase, A., Liu, S. X., & Lyberg, L. (2021). *Handbook of computational social science* (Vol. 2). Taylor & Francis.

Engstrom, D. F., Ho, D. E., Sharkey, C. M., & Cuéllar, M. (2020). Government by algorithm: Artificial intelligence in federal administrative agencies. *NYU School of Law, Public Law Research Paper* (20–54).

European Commission (2017). COM(2011) 808 Final: Horizon 2020 — the framework programme for research and innovation. In https://eur-lex.europa.eu/legal-content/EN/ALL/?uri$=$CELEX%5C%3A52011PC0809 (15 May, 2022).

Fenno, R. F. (1973). *Congressmen in committees*. In Boston: Little Brown & Company.

Fenno, R. F. (1978). *Home style: House members in their districts*. Boston: Addison Wesley.

Gabel, M., & Scheve, K. (2007). Estimating the effect of elite communications on public opinion using instrumental variables. *American Journal of Political Science, 51*(4), 1013–1028.

Gennaro, G., & Ash, E. (2021). Emotion and reason in political language. *The Economic Journal, 132*(643), 1037–1059. https://doi.org/10.1093/ej/ueab104

Gerner, D. J., Schrodt, P. A., Yilmaz, O., & Abu-Jabr, R. (2002). Conflict and mediation event observations (cameo): A new event data framework for the analysis of foreign policy interactions. In *International Studies Association, New Orleans*.

Gigley, H. M. (1993). Projected government needs in human language technology and the role of researchers in meeting them. In *Human Language Technology: Proceedings of a Workshop Held at Plainsboro, New Jersey, March 21–24, 1993*. https://aclanthology.org/H93-1056

Glavaš, G., Nanni, F., & Ponzetto, S. P. (2017a). Cross-lingual classification of topics in political texts. In *Proceedings of the Second Workshop on NLP and Computational Social Science* (pp. 42–46). Vancouver, Canada: Association for Computational Linguistics. https://doi.org/10.18653/v1/W17-2906. https://aclanthology.org/W17-2906

Glavaš, G., Nanni, F., & Ponzetto, S. P. (2017b). Unsupervised cross-lingual scaling of political texts. In *Proceedings of the 15th Conference of the European Chapter of the Association for Computational Linguistics: Volume 2, Short Papers* (pp. 688–693). Valencia, Spain: Association for Computational Linguistics. https://aclanthology.org/E17-2109.

Glavaš, G., Nanni, F., & Ponzetto, S. P. (2019). Computational analysis of political texts: bridging research efforts across communities. In *Proceedings of the 57th Annual Meeting of the Association for Computational Linguistics: Tutorial Abstracts* (pp. 18–23). Florence, Italy: Association for Computational Linguistics. https://doi.org/10.18653/v1/P19-4004. https://aclanthology.org/P19-4004

Goldstein, J. S. (1992). A conflict-cooperation scale for weis events data. *Journal of Conflict Resolution, 36*(2), 369–385.

Gottipati, S., Qiu, M., Sim, Y., Jiang, J., & Smith, N. A. (2013). Learning topics and positions from Debatepedia. In *Proceedings of the 2013 Conference on Empirical Methods in Natural Language Processing* (pp. 1858–1868). Seattle, Washington, USA: Association for Computational Linguistics. https://aclanthology.org/D13-1191

Grimmer, J. (2010a). A Bayesian hierarchical topic model for political texts: Measuring expressed agendas in Senate press releases. *Political Analysis, 18*(1), 1–35.

Grimmer, J. (2013). Appropriators not position takers: The distorting effects of electoral incentives on congressional representation. *American Journal of Political Science, 57*(3), 624–642.

Grimmer, J., Messing, S., & Westwood, S. J. (2012). How words and money cultivate a personal vote: The effect of legislator credit claiming on constituent credit allocation. *American Political Science Review, 106*(4), 703–719.

Grimmer, J., & Stewart, B. M. (2013). Text as data: The promise and pitfalls of automatic content analysis methods for political texts. *Political Analysis, 21*(3), 267–297.

Grimmer, J. R. (2010b). *Representational style: The central role of communication in representation*. Harvard University.

Hausladen, C. I., Schubert, M. H., & Ash, E. (2020). Text classification of ideological direction in judicial opinions. *International Review of Law and Economics, 62*, 105903. https://doi.org/10.1016/j.irle.2020.105903. https://www.sciencedirect.com/science/article/pii/S0144818819303667

Hill, K. Q., & Hurley, P. A. (2002). Symbolic speeches in the us senate and their representational implications. *Journal of Politics, 64*(1), 219–231.

Hillard, D., Purpura, S., & Wilkerson, J. (2008). Computer-assisted topic classification for mixed-methods social science research. *Journal of Information Technology & Politics, 4*(4), 31–46.

Hirst, G., Riabinin, Y., & Graham, J. (2010). Party status as a confound in the automatic classification of political speech by ideology. In *Proceedings of the 10th International Conference on Statistical Analysis of Textual Data (JADT 2010)* (pp. 731–742)

Hiware, K., Dutt, R., Sinha, S., Patro, S., Ghosh, K., & Ghosh, S. (2020). NARMADA: Need and available resource managing assistant for disasters and adversities. In *Proceedings of the Eighth International Workshop on Natural Language Processing for Social Media* (pp. 15–24). Online: Association for Computational Linguistics. https://doi.org/10.18653/v1/2020.socialnlp-1.3. https://aclanthology.org/2020.socialnlp-1.3

Hogenboom, F., Frasincar, F., Kaymak, U., & de Jong, F. (2011). An overview of event extraction from text. In M. van Erp, W. R. van Hage, L. Hollink, A. Jameson, & R. Troncy (Eds.), *Proceedings of the Workhop on Detection, Representation, and Exploitation of Events in the Semantic Web (Derive 2011), Bonn, Germany, October 23, 2011* (CEUR Workshop Proceedings) (Vol. 77, pp. 948–57). CEUR-WS.org. http://ceur-ws.org/Vol-779/derive2011%5C_submission%5C_1.pdf

Honnibal, M., Montani, I., Van Landeghem, S., & Boyd, A. (2020). spaCy: Industrial-strength Natural Language Processing in Python. https://doi.org/10.5281/zenodo.1212303

Hopkins, D. J., & King, G. (2010). A method of automated nonparametric content analysis for social science. *American Journal of Political Science, 54*(1), 229–247.

Huguet Cabot, P.-L., Dankers, V., Abadi, D., Fischer, A., & Shutova, E. (2020). The pragmatics behind politics: Modelling metaphor, framing and emotion in political discourse. In *Findings of the association for computational linguistics: emnlp 2020* (pp. 4479–4488). Online: Association for Computational Linguistics. https://doi.org/10.18653/v1/2020.findings-emnlp. 402. https://aclanthology.org/2020.findings-emnlp.402

Immelman, A. (1993). The assessment of political personality: A psychodiagnostically relevant conceptualization and methodology. *Political Psychology, 14*, 725–741.

Iyyer, M., Enns, P., Boyd-Graber, J., & Resnik, P. (2014). Political ideology detection using recursive neural networks. In *Proceedings of the 52nd Annual Meeting of the Association for Computational Linguistics (Volume 1: Long Papers)* (pp. 1113–1122). Baltimore, Maryland: Association for Computational Linguistics. https://doi.org/10.3115/v1/P14-1105. https://aclanthology.org/P14-1105

Jin, D., Jin, Z., Zhou, J. T., & Szolovits, P. (2020). Is BERT really robust? A strong baseline for natural language attack on text classification and entailment. In *The Thirty-Fourth AAAI Conference on Artificial Intelligence, AAAI 2020, the Thirty-Second Innovative Applications of Artificial Intelligence Conference, IAAI 2020, the Tenth AAAI Symposium on Educational Advances in Artificial Intelligence, EAAI 2020, New York, NY, USA, February 7–12, 2020* (pp. 8018–8025). AAAI Press. https://aaai.org/ojs/index.php/AAAI/article/view/6311

Jin, Z., Chauhan, G., Tse, B., Sachan, M. & Mihalcea, R. (2021a). How good is NLP? A sober look at NLP tasks through the lens of social impact. In *Findings of the association for computational linguistics: ACL-IJCNLP* (pp. 3099–3113). Online: Association for Computational Linguistics. https://doi.org/10.18653/v1/2021.findings-acl.273

Jin, Z., Peng, Z., Vaidhya, T., Schoelkopf, B., & Mihalcea, R. (2021b). Mining the cause of political decision-making from social media: A case study of COVID-19 policies across the US states. In *Findings of the 2021 Conference on Empirical Methods in Natural Language Processing, EMNLP 2021*. Association for Computational Linguistics.

Johnson, K., & Goldwasser, D. (2018). Classification of moral foundations in microblog political discourse. In *Proceedings of the 56th Annual Meeting of the Association for Computational Linguistics (Volume 1: Long Papers)* (pp. 720–730). Melbourne, Australia: Association for Computational Linguistics. https://doi.org/10.18653/v1/P18-1067. https://aclanthology.org/P18-1067

Jurafsky, D., & Martin, J. H. (2000). *Speech and language processing An introduction to natural language processing, computational linguistics, and speech recognition*. Prentice Hall Series in Artificial Intelligence. Prentice Hall.

Karan, M., Šnajder, J., Širinić, D., & Glavaš, G. (2016). Analysis of policy agendas: Lessons learned from automatic topic classification of Croatian political texts. In *Proceedings of the 10th SIGHUM Workshop on Language Technology for Cultural Heritage, Social Sciences, and Humanities* (pp. 12–21). Berlin, Germany: Association for Computational Linguistics. https://doi.org/10.18653/v1/W16-2102. https://aclanthology.org/W16-2102

King, G., & Lowe, W. (2003). An automated information extraction tool for international conflict data with performance as good as human coders: A rare events evaluation design. *International Organization, 57*(3), 617–642.

Lauscher, A., Fabo, P. R., Nanni, F., & Ponzetto, S. P. (2016). Entities as topic labels: Combining entity linking and labeled lda to improve topic interpretability and evaluability. *IJCoL. Italian Journal of Computational Linguistics, 2*(2–2), 67–87.

Laver, M., Benoit, K., & Garry, J. (2003). Extracting policy positions from political texts using words as data. *American Political Science Review, 97*(2), 311–331.

Lax, J. R., & Phillips, J. H. (2009). Gay rights in the states: Public opinion and policy responsiveness. *American Political Science Review, 103*(3), 367–386.

Lax, J. R., & Phillips, J. H. (2012). The democratic deficit in the states. *American Journal of Political Science, 56*(1), 148–166.

Lazer, D., Pentland, A., Adamic, L., Aral, S., Barabási, A.-L., Brewer, D., Christakis, N., Contractor, N., Fowler, J., Gutmann, M., & Jebara, T. (2009). Computational social science. *Science, 323*(5915), 721–723.

Lee, F. E. (2008). Dividers, not uniters: Presidential leadership and senate partisanship, 1981-2004. *The Journal of Politics, 70*(4), 914–928.

Lipinski, D. (2009). *Congressional communication: Content and consequences.* University of Michigan Press.

Liu, K., Chen, Y., Liu, J., Zuo, X., & Zhao, J. (2020). Extracting events and their relations from texts: A survey on recent research progress and challenges. *AI Open, 1,* 22–39. https://doi.org/10.1016/j.aiopen.2021.02.004. https://www.sciencedirect.com/science/article/pii/S266665102100005X

Loria, S. (2018). TextBlob documentation. *Release 0.15* 2.

Lowe, W., Benoit, K., Mikhaylov, S., & Laver, M. (2011). Scaling policy preferences from coded political texts. *Legislative Studies Quarterly, 36*(1), 123–155.

Luo, Y., Card, D., & Jurafsky, D. (2020). Detecting stance in media on global warming. In *Findings of the association for computational linguistics: EMNLP 2020* (pp. 3296–3315). Online: Association for Computational Linguistics. https://doi.org/10.18653/v1/2020.findings-emnlp.296. https://aclanthology.org/2020.findings-emnlp.296

Luz, S. (2022). Computational linguistics and natural language processing. English. In F. Zanettin & C. Rundle (Eds.), *The Routledge handbook of translation and methodology*. United States: Routledge.

Manning, C. D., Raghavan, P., & Schütze, H. (2008). *Introduction to information retrieval.* Cambridge University Press. https://doi.org/10.1017/CBO9780511809071. https://nlp.stanford.edu/IR-book/pdf/irbookprint.pdf

Marchetti-Bowick, M., & Chambers, N. (2012). Learning for microblogs with distant supervision: political forecasting with Twitter. In *Proceedings of the 13th Conference of the European Chapter of the Association for Computational Linguistics* (pp. 603–612). Avignon, France: Association for Computational Linguistics. https://www.aclweb.org/anthology/E12-1062

Maronikolakis, A., Villegas, D. S., Preotiuc-Pietro, D., & Aletras, N. (2020). Analyzing political parody in social media. In *Proceedings of the 58th Annual Meeting of the Association for Computational Linguistics* (pp. 4373–4384). Online: Association for Computational Linguistics. https://doi.org/10.18653/v1/2020.acl-main.403. https://aclanthology.org/2020.acl-main.403

Mayhew, D. R. (2004). *Congress: The electoral connection.* Yale University Press.

McClelland, C. A. (1976). *World event/interaction survey codebook.*

McCombs, M., & Valenzuela, S. (2004). *Setting the agenda: Mass media and public opinion.* Wiley.

Menini, S., Nanni, F., Ponzetto, S. P., & Tonelli, S. (2017). Topic-based agreement and disagreement in US electoral manifestos. In *Proceedings of the 2017 Conference on Empirical Methods in Natural Language Processing* (pp. 2938–2944). Copenhagen, Denmark: Association for Computational Linguistics. https://doi.org/10.18653/v1/D17-1318. https://aclanthology.org/D17-1318

Menini, S., & Tonelli, S. (2016). Agreement and disagreement: Comparison of points of view in the political domain. In *Proceedings of COLING 2016, the 26th International Conference on Computational Linguistics: Technical Papers* (pp. 2461–2470). Osaka, Japan: The COLING 2016 Organizing Committee. https://aclanthology.org/C16-1232

Merritt, R. L., Muncaster, R. G., & Zinnes, D. A. (1993). *International event-data developments: DDIR phase II.* University of Michigan Press.

Miller, W. L., Clarke, H. D., Harrop, M., LeDuc, L., & Whiteley, P. F. (1990). *How voters change: The 1987 british election campaign in perspective.* Oxford University Press.

Misuraca, G., van Noordt, C., & Boukli, A. (2020). The use of AI in public services: Results from a preliminary mapping across the EU. In Y. Charalabidis, M. A. Cunha, & D. Sarantis (Eds.), *ICEGOV 2020: 13th International Conference on Theory and Practice of Electronic Governance, Athens, Greece, 23–25 September, 2020* (pp. 90–99). ACM. https://doi.org/10.1145/3428502.3428513

Mitamura, T., Liu, Z., & Hovy, E. H. (2017). Events detection, coreference and sequencing: what's next? Overview of the TAC KBP 2017 event track. In *Proceedings of* the 2017 Text Analysis Conference, TAC 2017, Gaithersburg, Maryland, USA, November 13–14, 2017. NIST. https://tac.nist.gov/publications/2017/additional.papers/TAC2017.KBP %5C_Event%5C_Nugget%5C_overview.proceedings.pdf

Mohammad, S. M., Zhu, X., Kiritchenko, S., & Martin, J. D. (2015). Sentiment, emotion, purpose, and style in electoral tweets. *Information Processing & Management, 51*(4), 480–499. https://doi.org/10.1016/j.ipm.2014.09.003

Nanni, F., Glavas, G., Ponzetto, S. P., & Stuckenschmidt, H. (2019). Political text scaling meets computational semantics. *CoRR, abs/1904.06217*. http://arxiv.org/abs/1904.06217

Nanni, F., Ponzetto, S. P., & Dietz, L. (2017). Building entitycentric event collections. In *2017 ACM/IEEE Joint Conference on Digital Libraries, JCDL 2017, Toronto, ON, Canada, June 19–23, 2017* (pp. 199–208). IEEE Computer Society. https://doi.org/10.1109/JCDL.2017.7991574

Neuendorf, K. A., & Kumar, A. (2015). Content analysis. *The International Encyclopedia of Political Communication, 8*, 1–10.

O'Connor, B., Balasubramanyan, R., Routledge, B. R., & Smith, N. A. (2010). From tweets to polls: Linking text sentiment to public opinion time series. In W. W. Cohen & S. Gosling (Eds.), *Proceedings of the Fourth International Conference on Weblogs and Social Media, ICWSM 2010, Washington, DC, USA, May 23–26, 2010*. The AAAI Press. http://www.aaai.org/ocs/index.php/ICWSM/ICWSM10/paper/view/1536

Osnabrügge, M., Ash, E., & Morelli, M. (2021). Cross-domain topic classification for political texts. *Political Analysis*, 1–22. https://doi.org/10.1017/pan.2021.37

Ouyang, L., Wu, J., Jiang, X., Almeida, D., Wainwright, C. L., Mishkin, P., Zhang, C., Agarwal, S., Slama, K., Ray, A., Schulman, J., Hilton, J., Kelton, F., Miller, L., Simens, M., Askell, A., Welinder, P., Christiano, P. F., Leike, J., & Lowe, R. (2022). Training language models to follow instructions with human feedback. *CoRR, abs/2203.02155*. https://doi.org/10.48550/arXiv.2203.02155

Padó, S., Blessing, A., Blokker, N., Dayanik, E., Haunss, S., & Kuhn, J. (2019). Who sides with whom? Towards computational construction of discourse networks for political debates. In *Proceedings of the 57th Annual Meeting of the Association for Computational Linguistics* (pp. 2841–2847). Florence, Italy: Association for Computational Linguistics. https://doi.org/10.18653/v1/P19-1273. https://aclanthology.org/P19-1273

Page, B. I. (1976). The theory of political ambiguity. *American Political Science Review, 70*(3), 742–752.

Pak, A., & Paroubek, P. (2010). Twitter as a corpus for sentiment analysis and opinion mining. In *Proceedings of the Seventh International Conference on Language Resources and Evaluation (LREC'10)*. Valletta, Malta: European Language Resources Association (ELRA). http://www.lrec-conf.org/proceedings/lrec2010/pdf/385_Paper.pdf

Paltoglou, G., & Thelwall, M. (2012). Twitter, myspace, digg: Unsupervised sentiment analysis in social media. *ACM Transactions on Intelligent Systems and Technology (TIST), 3*(4), 66:1–66:19. https://doi.org/10.1145/2337542.2337551

Pang, B., & Lee, L. (2007). Opinion mining and sentiment analysis. *Foundations and Trends® in Information Retrieval, 2*(1–2), 1–135. https://doi.org/10.1561/1500000011

Paul, D., & Frank, A. (2019). Ranking and selecting multi-hop knowledge paths to better predict human needs. In *Proceedings of the 2019 Conference of the North American Chapter of the Association for Computational Linguistics: Human Language Technologies, Volume 1 (Long and Short Papers)* (pp. 3671–3681). Minneapolis, Minnesota: Association for Computational Linguistics. https://doi.org/10.18653/v1/N19-1368. https://aclanthology.org/N19-1368

Pennebaker, J. W., Francis, M. E., & Booth, R. J. (2001). Linguistic inquiry and word count: Liwc 2001. *Mahway: Lawrence Erlbaum Associates, 71*(2001). 2001.

Petrocik, J. R. (1996). Issue ownership in presidential elections, with a 1980 case study. *American Journal of Political Science, 40*, 825–850.

Pla, F., & Hurtado, L.-F. (2014). Political tendency identification in Twitter using sentiment analysis techniques. In *Proceedings of COLING 2014, the 25th International Conference on Computational Linguistics: Technical Papers* (pp. 183–192). Dublin, Ireland: Dublin City

University & Association for Computational Linguistics. https://www.aclweb.org/anthology/C14-1019

Preoţiuc-Pietro, D., Liu, Y., Hopkins, D., & Ungar, L. (2017). Beyond binary labels: Political ideology prediction of Twitter users. In *Proceedings of the 55th Annual Meeting of the Association for Computational Linguistics (Volume 1: Long Papers)* (pp. 729–740). Vancouver, Canada: Association for Computational Linguistics. https://doi.org/10.18653/v1/P17-1068. https://aclanthology.org/P17-1068

Qi, P., Zhang, Y., Zhang, Y., Bolton, J., & Manning, C. D. (2020). Stanza: a Python natural language processing toolkit for many human languages. In *Proceedings of the 58th Annual Meeting of the Association for Computational Linguistics: System Demonstrations*. https://nlp.stanford.edu/pubs/qi2020stanza.pdf

Quinn, K. M., Monroe, B. L., Colaresi, M., Crespin, M. H. & Radev, D. R. (2006). An automated method of topic-coding legislative speech over time with application to the 105th-108th US Senate. In *Midwest political science association meeting* (pp. 1–61).

Quinn, K. M., Monroe, B. L., Colaresi, M., Crespin, M. H., & Radev, D. R. (2010). How to analyze political attention with minimal assumptions and costs. *American Journal of Political Science, 54*(1), 209–228.

Raleigh, C., Linke, A., Hegre, H., & Karlsen, J. (2010). Introducing ACLED-Armed conflict location and event data. *Journal of Peace Research, 47*(5), 651–660. https://journals.sagepub.com/doi/10.1177/0022343310378914

Roberts, M. E., Stewart, B. M., Tingley, D., Lucas, C., Leder-Luis, J., Gadarian, S. K., Albertson, B., & Rand, D. G. (2014). Structural topic models for open-ended survey responses. *American Journal of Political Science, 58*(4), 1064–1082.

Rosenthal, S., Nakov, P., Kiritchenko, S., Mohammad, S., Ritter, A., & Stoyanov, V. (2015). Semeval-2015 task 10: Sentiment analysis in Twitter. In D. M. Cer, D. Jurgens, P. Nakov, & T. Zesch (Eds.), *Proceedings of the 9th International Workshop on Semantic Evaluation, semeval@naacl-hlt 2015, Denver, Colorado, USA, June 4–5, 2015* (pp. 451–463). The Association for Computer Linguistics. https://doi.org/10.18653/v1/s15-2078

Sarol, M. J., Dinh, L., Rezapour, R., Chin, C.-L., Yang, P., & Diesner, J. (2020). An empirical methodology for detecting and prioritizing needs during crisis events. In *Findings of the association for computational linguistics: EMNLP 2020*, 4102–4107. Online: Association for Computational Linguistics. https://doi.org/10.18653/v1/2020.findings-emnlp.366. https://aclanthology.org/2020.findings-emnlp.366

Schrodt, P. A. (2000). Pattern recognition of international crises using Hidden Markov Models. In *Political Complexity: Nonlinear Models of Politics*, 296–328, University of Michigan Press.

Schrodt, P. A., & Gerner, D. J. (1994). Validity assessment of a machinecoded event data set for the middle east, 1982-92. *American Journal of Political Science, 38*, 825–854.

Schrodt, P. A., & Hall, B. (2006). Twenty years of the kansas event data system project. *The Political Methodologist, 14*(1), 2–8.

Schumacher, G., Schoonvelde, M., Traber, D., Dahiya, T., & Vries, E. D. (2016). EUSpeech: A new dataset of EU elite speeches. In *Proceedings of the International Conference on the Advances in Computational Analysis of Political Text (Poltext 2016)* (pp. 75–80).

Semetko, H. A., & Valkenburg, P. M. (2000). Framing European politics: A content analysis of press and television news. *Journal of Communication, 50*(2), 93–109.

Shah, D. V., Cappella, J. N., & Neuman, W. R. (2015). Big data, digital media, and computational social science: possibilities and perils. *The ANNALS of the American Academy of Political and Social Science, 659*(1), 6–13.

Sigelman, L., & Buell Jr., E. H. (2004). Avoidance or engagement? Issue convergence in us presidential campaigns, 1960–2000. *American Journal of Political Science, 48*(4), 650–661.

Sim, Y., Acree, B. D. L., Gross, J. H., & Smith, N. A. (2013). Measuring ideological proportions in political speeches. In *Proceedings of the 2013 Conference on Empirical Methods in Natural Language Processing* (pp. 91–101). Seattle, Washington, USA: Association for Computational Linguistics. https://aclanthology.org/D13-1010

Simon, A. F. (2002). *The winning message: Candidate behavior, campaign discourse, and democracy*. Cambridge University Press.

Slapin, J. B., & Proksch, S.-O. (2008). A scaling model for estimating time-series party positions from texts. *American Journal of Political Science, 52*(3), 705–722.

Sravani, D., Kameswari, L., & Mamidi, R. (2021). Political discourse analysis: A case study of code mixing and code switching in political speeches. In *Proceedings of the Fifth Workshop on Computational Approaches to Linguistic Code-Switching* (pp. 1–5). Online: Association for Computational Linguistics. https://doi.org/10.18653/v1/2021.calcs-1.1. https://aclanthology.org/2021.calcs-1.1

Stefanov, P., Darwish, K., Atanasov, A., & Nakov, P. (2020). Predicting the topical stance and political leaning of media using tweets. In *Proceedings of the 58th Annual Meeting of the Association for Computational Linguistics* (pp. 527–537). Online: Association for Computational Linguistics. https://doi.org/10.18653/v1/2020.acl-main.50. https://aclanthology.org/2020.acl-main.50

Stimson, J. A., MacKuen, M. B., & Erikson, R. S. (1995). Dynamic representation. *American Political Science Review, 89*, 543–565.

Sulkin, T. (2005). *Issue politics in congress*. Cambridge University Press.

Sun, C., Qiu, X., Xu, Y., & Huang, X. (2019). How to fine-tune BERT for text classification? In M. Sun, X. Huang, H. Ji, Z. Liu & Y. Liu (Eds.), *Chinese Computational Linguistics - 18th China National Conference, CCL 2019, Kunming, China, October 18–20, 2019, Proceedings*, Lecture Notes in Computer Science (Vol. 11856, pp. 194–206). Springer. https://doi.org/10.1007/978-3-030-32381-3%5C_16

Sundberg, R., & Melander, E. (2013). Introducing the ucdp georeferenced event dataset. *Journal of Peace Research, 50*(4), 523–532.

Thelwall, M., Buckley, K., & Paltoglou, G. (2011). Sentiment in Twitter events. *Journal of the American Society for Information Science and Technology, 62*(2), 406–418. https://doi.org/10.1002/asi.21462

Thomas, M., Pang, B., & Lee, L. (2006). Get out the vote: Determining support or opposition from congressional floor-debate transcripts. In *Proceedings of the 2006 Conference on Empirical Methods in Natural Language Processing* (327–335). Sydney, Australia: Association for Computational Linguistics. https://aclanthology.org/W06-1639

Tjong Kim Sang, E., & Bos, J. (2012). Predicting the 2011 Dutch senate election results with Twitter. In *Proceedings of the Workshop on Semantic Analysis in Social Media* (pp. 53–60). Avignon, France: Association for Computational Linguistics. https://www.aclweb.org/anthology/W12-0607

Trappl, R. (2006). *Programming for Peace: Computer-Aided Methods for International Conflict Resolution and Prevention* (Vol. 2). Springer Science & Business Media.

Unankard, S., Li, X., Sharaf, M. A., Zhong, J., & Li, X. (2014). Predicting elections from social networks based on sub-event detection and sentiment analysis. In B. Benatallah, A. Bestavros, Y. Manolopoulos, A. Vakali & Y. Zhang (Eds.), *Web Information Systems Engineering - WISE 2014 - 15th International Conference, Thessaloniki, Greece, October 12–14, 2014, Proceedings, Part II*. Lecture Notes in Computer Science (Vol. 8787, pp. 1–16). Springer. https://doi.org/10.1007/978-3-319-11746-1%5C_1.

Van Aggelen, A., Hollink, L., Kemman, M., Kleppe, M. & Beunders, H. (2017). The debates of the European Parliament as linked open data. *Semantic Web, 8*(2), 271–281.

Van Roy, V., Rossetti, F., Perset, K., & Galindo-Romero, L. (2021). *AI watch - national strategies on artificial intelligence: A European perspective, 2021 edition*. Scientific Analysis or Review, Policy Assessment, Country report KJ-NA-30745-EN-N (online). Luxembourg (Luxembourg). https://doi.org/10.2760/069178(online)

Walker, C., Strassel, S., Medero, J., & Maeda, K. (2006). Ace 2005 multilingual training corpus. *Linguistic Data Consortium, Philadelphia, 57*.

Winter, D. G., & Stewart, A. J. (1977). Content analysis as a technique for assessing political leaders. In *A psychological examination of political leaders* (pp. 27–61).

Wolf, T., Debut, L., Sanh, V., Chaumond, J., Delangue, C., Moi, A., Cistac, P., Rault, T., Louf, R., Funtowicz, M., Davison, J., Shleifer, S., von Platen, P., Ma, C., Jernite, Y., Plu, J., Xu, C., Scao, T. L., Gugger, S., . . . , Rush, A. (2020). Transformers: state-of-the-art natural language processing. In *Proceedings of the 2020 Conference on Empirical Methods in Natural Language Processing: System Demonstrations* (pp. 38–45). Online: Association for Computational Linguistics. https://doi.org/10.18653/v1/2020.emnlp-demos.6. https://aclanthology.org/2020.emnlp-demos.6

Yin, W., Hay, J., & Roth, D. (2019). Benchmarking zero-shot text classification: Datasets, evaluation and entailment approach. In *Proceedings of the 2019 Conference on Empirical Methods in Natural Language Processing and the 9th International Joint Conference on Natural Language Processing (EMNLPIJCNLP)* (pp. 3914–3923). Hong Kong, China: Association for Computational Linguistics. https://doi.org/10.18653/v1/D19-1404. https://aclanthology.org/D19-1404

Chapter 8
Describing Human Behaviour Through Computational Social Science

Giuseppe A. Veltri

Abstract The possibilities offered by digital and Computational Social Science can improve our understanding of human behaviour as never before. The availability of behavioural data in a society where the digital has been widely adopted is because of two reasons: first, the vast amount of digital traces produced by people in their daily lives and related behaviours and, second, the possibility of running online experiments that can cover a large segment of a target population (we have seen online experiments with hundreds of thousands of participants). This chapter will discuss the opportunity offered by online large behavioural experiments. The implications for policymakers of this shift are the possibility of having behavioural insights both across different societies and better understanding and capturing within a country heterogeneity. In other words, large-scale online experiments combined with computational methods allow for unprecedented cognitive and behavioural based segmentation.

8.1 Introduction

This chapter describes the role of Computational Social Science in enhancing our understanding of human behaviour. We will highlight the importance of behavioural data compared to the more common self-reported ones and how the increased availability of digital traces of human behaviour is crucial in the new potential analysis.

The digital revolution that has affected the social sciences in the past decade or so (e.g. Salganik, 2018; Veltri, 2020) created the context for three possible forms of studies about human behaviour:

1. The use of large online behavioural experiments, which will be the focus of this chapter

G. A. Veltri (✉)
Department of Sociology and Social Research, University of Trento, Trento, Italy
e-mail: giuseppe.veltri@unitn.it

© The Author(s) 2023
E. Bertoni et al. (eds.), *Handbook of Computational Social Science for Policy*,
https://doi.org/10.1007/978-3-031-16624-2_8

163

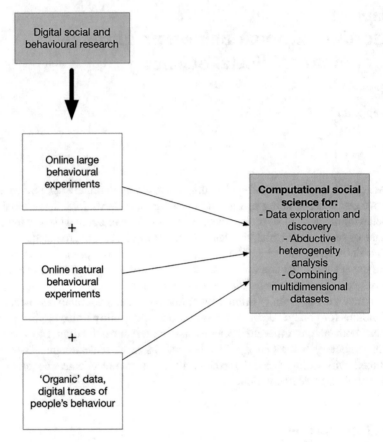

Fig. 8.1 Different forms of behavioural data available due to the 'digital revolution' in the social sciences and the role of CSS in it

2. Similarly to point one, the online has become a large field in which natural experiments occur
3. Many digital platforms collect behavioural data of their users and these can be repurposed for social scientific purposes

The amount, type, and complexity of data generated from the three approaches above require innovation from the analytical point of view. This is where Computational Social Science methods are being applied (Fig. 8.1). Finally, we will discuss one promising form of modelling that we believe is particularly relevant for studies of human behaviour and decision-making.

Large-scale online experiments can test the behavioural response to intervention and explore the heterogeneity of treatment effects across different social strata helping to tailor differentiated options. The use of computational models to identify subgroups in large datasets is a growing area of interest. Of particular interest is the form of online experiments that combine the lessons learned from online surveys. The so-called population-based survey experiments or PBSEs aim to address this problem through research design rather than analyses, combining the best aspects of both approaches, capitalizing on their strengths, and eliminating many weaknesses. We will discuss the potential of unstructured data such as text and the use of text mining techniques that, combined with other types of data, can further enrich our understanding of social behaviour. The contribution of Computational Social Science to our understanding of social behaviour is not limited to data availability. However, it includes the opening to analytical approaches developed in computer science, particularly in machine learning, which brings a new 'culture' of statistical modelling that bears considerable potential for the social scientist and informs policy harnessing heterogeneity. Segmentation is widely used in decision-making. It is usually based on sociodemographic factors (e.g. age, gender, income, geographical location). However, cognitive and behavioural differences within the population are an important source of variability that needs to be considered in policy design and implementation because how one reacts to a public policy is conditioned by what cognitive and cultural patterns are used by those targeted. The target population is behaviourally and cognitively plural: people vary in how they feel, think, and act. Moreover, each citizen interprets reality according to different cognitive and cultural schemas specific to the individual and active in the cultural environment. Therefore, public policies need to be designed in ways that allow for the flexibility to take into account differences in the target population and other dimensions (e.g. sociodemographic, linguistic, and so on).

8.2 Data in the Digital World

To appreciate the transformative nature of digital data and computational methods in the social sciences, we need to draw some fundamental distinctions between the types of data that social scientists have been dealing with. A great deal of social science research has been produced based on self-reported data. Self-reported data stands for the accounts and reporting people do about their views, psychological states, and behaviours. However, the biggest challenge to self-reported data has come from a shift in the model of human behaviour in the wider social sciences except for psychology. Since the late 1990s, psychologists have distinguished between two systems of thought with different capacities and processes (Kahneman, 2012; Kahneman & Frederick, 2002; Lichtenstein & Slovic, 2006; Metcalfe & Mischel, 1999; Sloman, 1996; Smith & DeCoster, 2000), which have been referred to as System 1 and System 2 (Evans & Stanovich, 2013). System 1 (S1) is made up of intuitive thoughts of great capacity, is based on associations acquired through experience and quickly and automatically calculates information.

On the other hand, System 2 (S2) involves low-capacity reflective thinking based on rules acquired through culture or formal learning and calculates information in a relatively slow and controlled manner. The processes associated with these systems have been defined as Type 1 (fast, automatic, unconscious) and Type 2 (slow, conscious, controlled), respectively. The so-called dual model of the mind is now the most supported way of understanding human behaviour at the individual level and in continuous evolution (De Neys, 2018). The model has also been applied outside psychology, for example, in sociology (Lizardo et al., 2016; Moore, 2017) and political science (Achen & Bartels, 2017). The implications of Kahneman and Tversky's work have led to the research programme labelled behavioural economics, which has dramatically impacted traditional microeconomics theory.

A more precise human behaviour and decision-making model has implications for social science research methodology, particularly for the distinction between self-reported and observational/behavioural data. The dual mode of thinking brings back the importance of unconscious thought processes and contextual and environmental influences on the latter in the broader context of the social sciences, which is highly problematic in studies only using self-reported measures and instruments. Traditionally, collecting behavioural data has been very difficult and expensive for social scientists. Keeping track of people's actual behaviour could be done only for small groups of people and for a minimal amount of time. However, the availability of digital data has brought us a significant increase in behavioural data; we now have digital traces of people's actual behaviour that were never available before.

The combined effect of a relatively new and powerful foundational model of human behaviour and decision-making offered by the dual model and the availability of behavioural data thanks to the digital traces recorded by a multitude of services and tools is very promising for social scientists. Before continuing this line of argument, let's clarify one point that might be the object of criticism—considering human behaviour as the outcome of mutual influences of conscious acting and unconscious heuristics, biases, and environmental influences are not a return to reductionism. People's opinions count for nothing. Self-reported data will remain an essential source of information for social scientists, but, at the same time, the availability of behavioural data will function as complementary data to understand complex social phenomena.

8.3   Behavioural Digital Data

The distinction between self-reported and behavioural data is no longer mainly theoretical because the new opportunities for collecting the latter are unprecedented. Such an option opens new research opportunities and the possibility of reviewing current theories and existing models. However, the increased availability of collecting data about people's behaviour does not free us from biases generated by the design and aims of digital platforms. People's behaviour is constrained by the platform they use; for example, it is impossible to write an essay on Twitter unless we decide to write it using many individual tweets. There are, therefore, several potential sources of confounding factors, as we will further elaborate in the section below on construct validity.

Another distinction is relevant about the different levels of analysis: the one between static and dynamic data. The large majority of data collected in the social sciences have been 'static'—that is, data collection has been carried out at a given time. This is because longitudinal data collection, data collected over some time, was challenging and expensive. Digital data introduce a much-increased capacity for recording and using longitudinal data for social scientific purposes. Digital data have not been historically around for many decades, but future researchers might have at their disposal longitudinal datasets that were absent in the past.

Behavioural digital data are the object of attention of a new generation of social scientists who believe in their potential to regenerate the current theories and framework that were developed in a condition of data scarcity, with different models of human behaviour and using only self-reported data. It is too early to say what changes will bring increased data availability, but this is the most exciting aspect of the use of digital data for social scientific research. However, the nature of the data collected from the digital world is not without problems, and it poses specific challenges to researchers.

The distinction between self-reported and behavioural data touched briefly on one feature of digital data: their nonreactive nature in terms of data collection. We can distinguish digital data as the outcome of unobtrusive or obtrusive data collection methods (Webb et al., 1966). The distinction between these two data collection modalities is essential in the social sciences because people 'react' to researchers' measurements and can figure out what a researcher's goals are. Two of the most common problems are people's reactions to measurements, the Hawthorne effect and the social desirability effect. The Hawthorne effect, as mentioned before, refers to the fact that individuals modify their behaviour in response to their awareness of being observed. Recent scandals related to social media and privacy, in which users' data have been harvested for commercial or political campaigning purposes, have made people more conscious that their online behaviour is observed and recorded. Social desirability is the tendency of some respondents to report an answer in a way they deem to be more socially acceptable if they believe they are under observation than would be their 'true' answer, where true means aligned to current dominant social norms. They do this to project a favourable image of themselves and to avoid receiving negative evaluations. The outcome of the strategy results in the over-reporting of socially desirable behaviours or attitudes and the under-reporting of socially undesirable behaviours or attitudes (Nederhof, 1985). Social media are particularly affected by social desirability bias because people manage their presence online to generate a positive self-image. This process leads to a positivity bias in the content present on social media (Spottswood & Hancock, 2016).

The availability of behavioural data in a society where the digital has been widely adopted is because of two reasons: first, the vast amount of digital traces produced by people in their daily lives and related behaviours and, second, the possibility of running online experiments that can cover a large segment of a target population (we have seen online experiments with hundreds of thousands of participants). Next, we will discuss the opportunity offered by the online large behavioural experiments, particularly in the form of population-based survey experiments or PBSE.

8.4 Online Population-Based Survey Experiments

The limitations of laboratory experiments and the opportunities of the digital as a field in which to conduct research have prompted researchers to develop online experiments both in academia and in the private research world. Of particular interest is the form of online experiments that combine lessons learned from online surveys. The aim of so-called population-based survey experiments, or PBSEs (Mutz, 2011), is to address this problem through research design rather than analysis, combining the best aspects of both approaches, capitalizing on their strengths, and eliminating many of their weaknesses.

Defined in the most rudimentary terms, a population-based survey experiment is an experiment that is administered to a representative sample of the population. Another common term for this approach is simply 'survey experiment', but this abbreviated form can be misleading because it is not always clear what the term 'survey' means. The use of survey methods does not distinguish this approach from other combinations of survey and experimental methods. After all, many experiments already involve survey methods in the administration of pre-test and post-test questionnaires, but this is not what is meant here. Population-based survey experiments are not defined by their use of interview techniques, whether written or oral nor by their location in a setting other than a laboratory. Instead, a population-based experiment uses sampling methods to produce a set of experimental subjects that is representative of the target population of interest for a particular theory, whether that population is a country, a state, an ethnic group, or some other subgroup. The population represented by the sample should be representative of the population to which the researcher intends to extend his results. In population-based survey experiments, experimental subjects are randomly assigned to conditions by the researcher, and treatments are administered as in any other experiment. Nevertheless, participants are generally not required to show up in a laboratory to participate. Theoretically, they could, but population-based experiments are infinitely more practical when representative samples are not required to appear in one place.

Strictly speaking, population-based survey experiments are more experiments than surveys. By design, population-based experiments are experimental studies that draw on the power of random assignment to establish unbiased causal inferences. They are also administered to randomly selected representative samples of the target population of interest, just as a survey would be. However, population-based experiments do not need (and often have not relied on) nationally representative population samples. The population of interest could be members of a particular ethnic group, parents of children under 18, people who watch television news, or others. Still, the key is that convenience samples are abandoned in favour of samples representing the target population of interest.

The advantage of population-based survey experiments is that theories can be tested on samples that are representative of the populations to which they are said to apply. The downside of this trade-off is that most researchers have little experience administering experimental treatments outside of a laboratory setting, so new techniques and considerations come into play, as (Veltri, 2020) described extensively. In a sense, population-based survey experiments are by no means new; simplified versions of them have existed since at least the early years of research. However, technological developments in survey research, combined with the development of innovative techniques in experimental design, have made highly complex and methodologically sophisticated population-based experiments increasingly accessible to social scientists from many disciplines.

The development of the digital has made implementable the possibilities of population-based experiments. With the diffusion of pre-recruited online panels that are built according to the golden standards of sampling, the ability to exploit such dynamic data collection tools has expanded social scientists' methodological repertoire and inferential range in many fields (e.g. Veltri et al., 2020). The many advances in interview technology offer social science researchers the potential to introduce some of its most important hypotheses into virtual laboratories scattered across countries. Whether evaluating theoretical hypotheses, examining the robustness of laboratory results, or testing empirical hypotheses of other varieties, the ability of scientists to experiment on large and diverse groups of subjects allows them to address critical social and behavioural phenomena more effectively and efficiently.

Population-based experiments can be used by social scientists in sociology, political science, psychology, economics, cognitive science, law, public health, communication, and public policy, to name just a few of the main fields that find this approach appealing. Although most social scientists recognize the enormous benefits of experimentation, the traditional laboratory setting is unsuitable for all important research questions. Experiments have always been more prevalent in some social science fields than in others. To a large extent, the emphasis on experimental versus investigative methods reflects a field's emphasis on internal versus external validity, with fields such as psychology more oriented towards the former and fields such as political science and sociology more oriented towards the latter. For researchers, population-based experiments provide a means of establishing causality that is unmatched by any large-scale data collection effort, no matter how extensive.

Conducting online population-based survey experiments can benefit from the latest development of survey design and, in particular, adaptive survey design or ASD. ASD (Wagner, 2010) is based on the premise that samples are heterogeneous, and the optimal survey protocol may not be the same for each individual. For example, a particular survey design feature such as incentives may appeal to some individuals but not to others (Groves et al., 2000; Groves & Heeringa, 2006), leading to design-specific response propensity for each individual. Similarly, relative to interviewer-administration, a self-administered mode of data collection may elicit less measurement error bias for some individuals but more measurement error bias for others. The general objective in ASD is to tailor the protocol to sample members to improve targeted survey outcomes. The basic premise of adaptive interventions is shared by ASDs—tailoring methods to individuals based on interim outcomes. We label these *dynamic* adaptive designs to reflect the dynamic nature of the optimization and *static* adaptive designs when they are based solely on information available prior to the start of data collection. A tailoring variable is used to inform the decision to change treatments, such as the type of concerns the sample member may have raised at the contact moment. Decision rules would include the matching of information from the tailoring variables (concerns about time, not worth their effort) to interventions (a shorter version of the task, a larger incentive). Finally, the decision points need to be defined, such as whether to apply the rules and intervene at the time of the interaction or at a given point in the data collection period.

What is noteworthy is that either of these approaches and much more complex experimental designs are easily implemented in the context of use of online platforms. The ability to make strong causal inferences has little to do with the laboratory environment itself and much to do with the ability to control the random assignment of people to different experimental treatments. By moving the possibilities of experimentation out of the laboratory in this way, population-based experiments strengthen the internal validity of social science research and provide the potential to interest a much wider group of social scientists in the possibilities of experimentation. Of course, the fact that it can be done outside the laboratory is not itself a good reason to do so. Therefore, we will review some of the key advantages of online population-based experiments, starting with four advantages over traditional laboratory experiments and then ending with some of their more general benefits for accumulating valuable social scientific knowledge.

The main strategic advantage of an online experiment over a laboratory experiment is the greater possibility of generalization (external validity), the greater statistical power and possibly the quality of the data produced. Web-based studies, having larger samples, usually have greater statistical power than laboratory studies. Data quality can be defined by variable error, constant error, reliability, or validity. Comparisons of power and some quality measures have found cases where web data are of higher quality for one or other of these definitions than comparable laboratory data, although not always (Birnbaum, 2004). Many web researchers are convinced that data obtained via the web can be 'better' than data obtained from students (Reips, 2002), despite the laboratory's obvious advantage for control. The main disadvantage of an online experiment compared to a laboratory experiment is the lack of complete environmental control. Participants in online experiments may answer questions and perform behavioural tasks in very different environments (a room with light and silence, versus their own desk at work with less light and surrounded by much noise) and with different equipment (a participant may use a browser that does not display visual stimuli correctly or may have a slow connection, thus delaying task completion and increasing fatigue, frustration and 'noisy' responses). Most importantly, as lab assistants do not monitor participants, there is more chance that they will engage in automatic responses and task completion, which introduces noise into the data. This can be controlled with control questions and is less of a problem for between-subjects design with randomization of treatments and control conditions.

Other technical/tactical issues can be controlled for in the online experiment (multiple submissions, drop-outs, self-selection). Still, the main trade-off between online experiments and laboratory is to trade off greater generalizability and power of data for less experimental control. Therefore, it is not surprising that experiments are often repeated with the same outcome measures both online and in the laboratory to check the quality and validity of the data.

8.5 Heterogeneity Analysis and Computational Methods

Extending experiments to large samples, both national and international, increases the potential heterogeneity present in response to our treatments. Therefore, identifying and studying such heterogeneity is a crucial step in the world of online behavioural experiments. New analytical techniques have emerged in computational and computer sciences that are very promising to achieve this goal. One of the best examples of how social science can benefit from analytical approaches developed in computational methods is the development of model-based recursive partitioning. This approach improves the use of classification and regression trees. The latter also is a method from the 'algorithmic culture' of modelling that has valuable applications in the social sciences but is essentially data-driven (Berk, 2006; Hand & Vinciotti, 2003).

In summary, classification and regression trees are based on a purely data-driven paradigm. Without using a predefined statistical model, such algorithmic methods recursively search for groups of observations with similar response variable values by constructing a tree structure. Thus, they are instrumental in data exploration and express their best utility in the context of very complex and large datasets. However, such techniques make no use of theory in describing a pattern of how the data was generated and are purely descriptive, although far superior to the 'traditional' descriptive statistics used in the social sciences when dealing with large datasets.

Model-based recursive partitioning (Zeileis et al., 2008) represents a synthesis of a theoretical approach and a set of data-driven constraints for theory validation and further development. In summary, this approach works through the following steps. Firstly, a parametric model is defined to express a set of theoretical assumptions (e.g. through a linear regression). Second, this model is evaluated according to the recursive partitioning algorithm, which checks whether other important covariates that would alter the parameters of the initial model have been omitted. Third, the same regression or classification tree structure is produced. This time, instead of partitioning by different patterns of the response variable, model-based recursive partitioning finds different patterns of associations between the response variable and other covariates that have been pre-specified in the parametric model. In other words, it creates different versions of the parametric model in terms of beta (β) estimation, depending on the different important values of the covariates (for the technical aspects of how this is done, see Zeileis & Hornik, 2007). In other words, the presence of splits indicates that the parameters of the initial theory-driven definition are unstable and that the data are too heterogeneous to be explained by a single global model. The model does not describe the entire dataset.

Classification trees look for different patterns in the response variable based on the available covariates. Since the sample is divided into rectangular partitions defined by the values of the covariates and since the same covariate can be selected for several partitions, classification trees can also evaluate complex interactions, non-linear and non-monotonic patterns. Furthermore, the structure of the underlying data generation process is not specified in advance but is determined in an entirely data-driven way. These are the key distinctions between classification and regression trees and classical regression models.

Model-based recursive partitioning was developed as an advancement of classification and regression trees. Both methods originate from machine learning, which is influenced by both statistics and computer science. Classification and regression trees are purely data-driven and exploratory—and thus mark the complete opposite of the model specification theory approach prevalent in the empirical social sciences. However, the advanced model-based recursive partitioning method combines the advantages of both approaches: at first, a parametric model is formulated to represent a theory-driven research hypothesis. Then this parametric model is handed over to the model-based recursive partitioning algorithm, which checks whether other relevant covariates have been omitted that would alter the model parameters of interest.

Technically, the tree structure obtained from the classification and regression trees remains the same for model-based recursive partitioning. However, the application of model-based recursive partitioning offers new impulses for research in the social, educational, and behavioural sciences. For the interpretation of model-based recursive partitioning, we would like to emphasize the connection to the principle of parsimony: following the fundamental research paradigm that theories developed in the social sciences must produce falsifiable hypotheses, these are translated into statistical models. The aim of model building is thus to simplify complex reality. What is the advantage of having such information? The answer to this question relates to the initial distinction that was introduced about the two modelling cultures. In the predominant (in the social sciences) data modelling culture, comparing different models has always been complex and problematic. The hybrid approach of model-based recursive partitioning modelling can help review models that work for the whole dataset and do not neglect such information that imposes on the models as 'global' straitjackets. Furthermore, suppose the researcher in question values the 'Ockham's razor' rule (that a model should not be more complex than necessary but must be complex enough to describe the empirical data). In that case, model-based recursive partitioning can be used to evaluate different models.

Another valuable piece of information generated by this approach is that the recursive model-based method allows for identifying particular segments of the sample under investigation that might merit further investigation. That is, the possibility of identifying segments of our sample (and, therefore, presumably segments of the population if our sample is representative) that have a different version of the general theoretical model we have employed, in the form of statistical regression, to explain a given phenomenon Y. This possibility of identifying 'local' models of the population is not just a matter of chance. When applied to independent variables involving the measurement of attitudes and preferences, this possibility of identifying' local' models as defined above allows us to identify subgroups characterized by a particular cognitive pattern shared by that group. Such a group could very well be transversal to traditional sociodemographic categories (the young, the old, the middle class, etc.). Applied to experiments, it represents an advanced form of heterogeneity of treatment effects analysis that, with sufficient cases, can be very informative about the presence of general and local effects of a treatment.

This approach is very promising but has a 'cost' in methodological terms. To work well, it needs large samples and, even better, samples collected in several countries. Only with a sufficient number of cases, we can identify noteworthy subgroups. In contrast, if we have a few hundred cases, we cannot be sure of the statistical validity of the partitioning, besides the fact that we are talking about subgroups consisting of a few tens of cases are uninteresting as results.

This brief overview of model-based recursive partitioning illustrates the general point discussed in the previous sections: the complexity, quantity, and availability of digital data have highlighted the need to use analytical approaches other than those considered conventional in the social sciences. Therefore, Computational Social Science is, among other things, an attempt to adapt these new computational techniques and their associated 'modelling culture' to the research goals and questions of social scientists (Veltri, 2017). In other words, it is not only a matter of having more data of different types, which is important but also of innovating modelling techniques that can bring about transformative changes in the social sciences. Of course, there will also be methodological problems. Still, the ability to answer old questions with alternative approaches and ask new questions is the most attractive feature of Computational Social Science.

8.6 Conclusions

The possibilities offered by the new turn of digital and Computational Social Science can improve our understanding of human behaviour as never before. We move from data scarcity and local studies to potential large-scale, complex, and international ones. The implications for policymakers of this shift are the possibility of having behavioural insights both across different societies and better understanding and capturing within a country heterogeneity. In other words, large-scale online experiments combined with computational methods like the one discussed do allow for unprecedented cognitive and behavioural based segmentation (see recent example Steinert et al., 2022).

Consequently, such differences can be used to differentiate the population to identify subsets of people, each characterized by a particular cognitive style. Segmentation is usually associated with profiling—the description of the relevant characteristics of the identified segment—sociodemographic characteristics, occupational status, geographical and spatial location, health status, attitudes towards essential aspects of the public policy in question. It is clear that cognitive and cultural segmentation also interacts with classical forms of classification resulting from affiliations such as occupations, generations, social classes, and status groups. Still, it cannot be taken for granted that they coincide. An example of such cultural segmentation is the analysis of the Brexit vote in the UK and how different cognitive-cultural styles are predictive of that vote (Veltri et al., 2019).

Behavioural segmentation is a potential tool for policy development. It is particularly suitable for the ex ante phase because it refers to a segmentation strategy of the target population and during the monitoring of the intervention *in progress* because it allows identifying the mismatch between the policy objectives and the citizen's interpretation of the policy. Similarly, cognitive-behavioural segmentation helps both the effectiveness and efficiency of policy interventions. In the first case, it helps to tailor instruments to the cognitive and cultural variability within the target population. An analogy here is precision medicine, an emerging approach for the treatment and prevention of diseases that considers individual variability in genes, environment, and lifestyle. In the context of public policy, the unit cannot be the individual but subgroups of the target population that will respond differently to the same public policy intervention. Thus, cognitive and behavioural segmentation plays an important role in improving efficiency. It can warn against implementing policy interventions that are likely to be ineffective with specific subgroups and thus help develop solutions that take cognitive and behavioural specificities into account. The other great opportunity comes from the use of digital traces and unstructured data. The sheer amount of this type of data provides insights into people's behaviour. However, because we are repurposing existing data collected for other purposes, some challenges are present. The first is entirely methodological: the criterion validity of these data types is still unclear (McDonald, 2005). The second concerns the ethical and privacy dimension of covert research, meaning that people are not often fully aware of the extend of their digital traces and how third-parties use them. Computational Social Science is no longer a complementary addition to or an embellishment in the social scientific study of society. Instead, it is changing the nature of social research because the digital has changed our societies. This is the starting point, we believe, that should accompany social scientists from now on.

References

Achen, C. H., & Bartels, L. M. (2017). *Democracy for realists: Why elections do not produce responsive government*. Princeton University Press. https://doi.org/10.2307/j.ctvc7770q

Berk, R. A. (2006). An introduction to ensemble methods for data analysis. *Sociological Methods & Research, 34*(3), 263–295.

Birnbaum, M. H. (2004). Human research and data collection via the internet. *Annual Review of Psychology, 55*(1), 803–832.

De Neys, W. (2018). *Dual process theory 2.0* (1st ed.). Routledge.

Evans, J. S. B., & Stanovich, K. E. (2013). Dual-process theories of higher cognition: Advancing the debate. *Perspectives on Psychological Science, 8*(3), 223–241. https://doi.org/10.1177/1745691612460685

Groves, R. M., & Heeringa, S. G. (2006). Responsive design for household surveys: Tools for actively controlling survey errors and costs. *Journal of the Royal Statistical Society: Series A (Statistics in Society), 169*(3), 439–457. https://doi.org/10.1111/j.1467-985X.2006.00423.x

Groves, R. M., Singer, E., & Corning, A. (2000). Leverage-saliency theory of survey participation: Description and an illustration. *The Public Opinion Quarterly, 64*(3), 299–308.

Hand, D. J., & Vinciotti, V. (2003). Local versus global models for classification problems. *The American Statistician, 57*(2), 124–131.

Kahneman, D. (2012). *Thinking, fast and slow*. Penguin Books.

Kahneman, D., & Frederick, S. (2002). Representativeness revisited: Attribute substitution in intuitive judgment. *Heuristics and Biases: The Psychology of Intuitive Judgment, 49*, 81.

Lichtenstein, S., & Slovic, P. (Eds.). (2006). *The construction of preference*. Cambridge University Press. https://doi.org/10.1017/CBO9780511618031

Lizardo, O., Mowry, R., Sepulvado, B., Stoltz, D. S., Taylor, M. A., Van Ness, J., & Wood, M. (2016). What are dual process models? Implications for cultural analysis in sociology. *Sociological Theory, 34*(4), 287–310.

McDonald, M. P. (2005). Validity, data sources. In K. Kempf-Leonard (Ed.), *Encyclopedia of social measurement* (pp. 939–948). Elsevier. https://doi.org/10.1016/B0-12-369398-5/00046-3

Metcalfe, J., & Mischel, W. (1999). A hot/cool-system analysis of delay of gratification: Dynamics of willpower. *Psychological Review, 106*(1), 3–19. https://doi.org/10.1037/0033-295X.106.1.3

Moore, R. (2017). Fast or slow: Sociological implications of measuring dual-process cognition. *Sociological Science, 4*, 196–223. https://doi.org/10.15195/v4.a9

Mutz, D. C. (2011). *Population-based survey experiments*. Princeton University Press.

Nederhof, A. J. (1985). Methods of coping with social desirability bias: A review. *European Journal of Social Psychology, 15*(3), 263–280. https://doi.org/10.1002/ejsp.2420150303

Reips, U. D. (2002). Standards for internet-based experimenting. *Experimental Psychology (Formerly Zeitschrift Für Experimentelle Psychologie), 49*(4), 243–256.

Salganik, M. J. (2018). *Bit by bit: Social research in the digital age*. Princeton University Press.

Sloman, S. A. (1996). The empirical case for two systems of reasoning. *Psychological Bulletin, 119*(1), 3–22. https://doi.org/10.1037/0033-2909.119.1.3

Smith, E. R., & DeCoster, J. (2000). Dual-process models in social and cognitive psychology: Conceptual integration and links to underlying memory systems. *Personality and Social Psychology Review, 4*(2), 108–131. https://doi.org/10.1207/S15327957PSPR0402_01

Spottswood, E. L., & Hancock, J. T. (2016). The positivity bias and prosocial deception on facebook. *Computers in Human Behavior, 65*, 252–259. https://doi.org/10.1016/j.chb.2016.08.019

Steinert, J. I., Sternberg, H., Prince, H., Fasolo, B., Galizzi, M. M., Büthe, T., & Veltri, G. A. (2022). COVID-19 vaccine hesitancy in eight European countries: Prevalence, determinants, and heterogeneity. *Science Advances, 8*(17), eabm 9825. https://doi.org/10.1126/sciadv.abm9825

Veltri, G. A. (2017). Big data is not only about data: The two cultures of modelling. *Big Data & Society, 4*(1), 205395171770399. https://doi.org/10.1177/2053951717703997

Veltri, G. A. (2020). *Digital social research*. Polity Press.

Veltri, G. A., Lupiáñez-Villanueva, F., Folkvord, F., Theben, A., & Gaskell, G. (2020). The impact of online platform transparency of information on consumer's choices. *Behavioural Public Policy*, 1–28. https://doi.org/10.1017/bpp.2020.11

Veltri, G. A., Redd, R., Mannarini, T., & Salvatore, S. (2019). The identity of Brexit: A cultural psychology analysis. *Journal of Community & Applied Social Psychology, 29*(1), 18–31. https://doi.org/10.1002/casp.2378

Wagner, J. (2010). The fraction of missing information as a tool for monitoring the quality of survey data. *Public Opinion Quarterly, 74*(2), 223–243. https://doi.org/10.1093/poq/nfq007

Webb, E. J., Campbell, D. T., Schwartz, R. D., & Sechrest, L. (1966). *Unobtrusive measures: Nonreactive research in the social sciences.* (pp. xii, 225). Rand Mcnally.

Zeileis, A., & Hornik, K. (2007). Generalized M-fluctuation tests for parameter instability. *Statistica Neerlandica, 61*(4), 488–508. https://doi.org/10.1111/j.1467-9574.2007.00371.x

Zeileis, A., Hothorn, T., & Hornik, K. (2008). Model-based recursive partitioning. *Journal of Computational and Graphical Statistics, 17*(2), 492–514.

Chapter 9
Data and Modelling for the Territorial Impact Assessment (TIA) of Policies

Eduardo Medeiros

Abstract Territorial Impact Assessment (TIA) is still a 'new kid on the block' on the panorama of policy evaluation methodologies. In synthesis, TIA methodologies are thematically holistic and multi-dimensional and require the analysis of a wide pool of data, not only of economic character but also related with social, environmental, governance and planning processes, in all territorial scales. For that, TIA requires a wealth of comparable and updated territorialised data. Here, data availability is often scarce in many of the selected analytic dimensions and respective components, to assess territorial impacts in a given territory, in particular in the domains of governance, planning and environment. In this context, this chapter presents a list of non-traditional potential indicators which can be used in existing TIA methodologies. Moreover, the analysis was able to show how important can be the use of non-traditional data, to complement mainstream statistical indicators associated with socioeconomic development trends. However, for the interested scientist, the dispersal of existing non-traditional data per a multitude of sources can pose a huge challenge. Hence the need of an online platform which centralises and updates non-traditional data for the use of all interested in implementing TIA methodologies.

9.1 Introduction

Academia and public and private entities are being flooded with 'tsunamis' of traditional and non-traditional data for their research. This data is collected via multichannel business environments (Baesens, 2014) and via, for instance, 'sensors, smartphones, internet, social media, and administrative systems'.[1] The central

---

[1] https://ec.europa.eu/jrc/en/research/centre-advanced-studies/css4p

E. Medeiros (✉)
Instituto Universitário de Lisboa (ISCTE-IUL), DINÂMIA'CET—IUL, Lisbon, Portugal
e-mail: Eduardo.Medeiros@iscte-iul.pt

© The Author(s) 2023
E. Bertoni et al. (eds.), *Handbook of Computational Social Science for Policy*,
https://doi.org/10.1007/978-3-031-16624-2_9

question for this chapter is how and which available non-traditional data can increase the effectiveness of the Territorial Impact Assessment (TIA) (see Medeiros, 2020b, 2020d) of projects, programmes and policies, via existing or novel TIA methodologies. This chapter is written in a necessary condensed and focused way and is guided by the following three policy questions raised by European Commission (EC) Joint Research Centre (JRC) (Bertoni et al., 2022):

1. How can CSS help evaluate the measure and its territorial impacts, either in an ex ante way (using simulation methods) or in an ex post one (using data science ones)?
2. Which sources of data could be used to better consider EU territorial heterogeneity?
3. What are the challenges of this approach, for example, how can we establish the correct level of spatial granularity, trading off the optimality of the targeted policy measure with the costs/timeliness of the decision?

All these questions are, in our view, relevant and reflect emerging axioms on the importance of considering the territorial dimension in analysing and assessing the implementation of policies, at different policy phases (ex ante, mid-term, and ex post), that have begun to permeate the policy evaluation discourse over the past decades. The first question provides an insightful emphasis to debate the potential positive and complementary contribution of non-traditional data to analyse all policy phases of TIA, in order to improve its effectiveness. The second aims to identify concrete sources of non-traditional data which can complement mainstream traditional data when implementing a TIA methodology. This is particularly relevant for TIAs since they should consider a broad and comprehensive set of indicators covering all dimensions and components of territorial development (Medeiros, 2014a). Finally, the third question touches a critical foundation of the implementation of TIAs: how to identify the appropriate territorial level for the TIA analysis and the dimensions and components for the analysed policy, in order to increase the efficiency and effectiveness of TIA evaluations. All these questions will be further scrutinised in Sect. 9.3. In this regard, and based on past TIA evaluations (Medeiros, 2014b, 2016a, 2017b), non-traditional data can provide crucial inputs on components related to the territorial governance and spatial planning dimensions of territorial development, which are difficult to obtain via traditional data sources.

9.2 TIA: A Literature Review

What is and why TIAs? These fundamental questions are answered in existing literature in various manners, from the first known report which unveiled the first TIA model (TEQUILA—see ESPON 3.2, 2006), through to a recent book which explains each one of the existing TIA methodologies (Medeiros, 2020d). From the first to the last, no more than 15 years have passed. This formally makes TIA methodologies 'new kids on the block' of policy evaluation (Medeiros, 2020c).

Table 9.1 TIA methodologies and main pros and cons

Name	Pros	Cons
TEQUILA	First TIA. Robust from a methodological standpoint	Does not apply counterfactual evaluation analysis
STEMA	Very complete set of indicators	Relatively simplistic and difficult to spatialise
EATIA	Goes beyond the use of negative/positive impacts	Weak from a methodological rationale–pain free
TARGET_TIA	Flexible, robust, sound and applies counterfactual analysis	Is not to be used quickly if robust impact scores are needed
QUICK_TIA	Presents online attractive cartography at several spatial scales	Does not really produce reliable and sound impact scores
TERRITORIAL FORESIGHT	Based on a wide participatory engagement—future trends	Largely dependent on the knowledge of participants
LUISA TERRITORIAL MODELING	Produce various scenarios of regional development	Largely dependent on the statistical data

Source: Own elaboration

Mostly driven by the ESPON programme, the TIAs are now entering a more mature phase, which is testified by several methodological upgrades from some of the ESPON TIA methodologies (TEQUILA, STEMA, etc.—see Tables 9.1 and 9.2). Even so, current ESPON TIAs are profoundly preconditioned by their erroneous rationale which means it is possible to obtain a valid and sound TIA score in a quick manner (Medeiros, 2016c).

Inevitably, any state-of-the-art literature review of TIA methodologies must start with the first one: the pioneering quantitative TIA model known as TEQUILA. This multi-criteria model is supported by a quantitative database on EU NUTS 3, to assess ex ante impacts of EU directives. According to the authors of this methodology, the criteria to select the TEQULA data refers to the main dimensions of territorial cohesion, territorial efficiency, territorial quality and territorial identity, and their sub-dimensions, measurable by multiple indicators (Camagni, 2020), particularly economic-related ones (Table 9.2). Also devised within the first ESPON programme, the STEMA TIA model is based on an original qualitative-quantitative methodological approach, returning ex ante and ex post impact scores. Just like the TEQUILA model, the STEMA uses traditional sources of data, mostly related to the economic dimension of development (Prezioso, 2020). The same goes for the ESPON EATIA (Marot et al., 2020) and the simplified QUICK_TIA (Ferreira & Verschelde, 2020). Crucially, all these four ESPON TIA models are supported by existing sources of quantitative databases at the EU level (mostly NUTS 2 and 3), collected from several sources and organised in the ESPON database, which has data related to agriculture and fisheries, economy, education, environment and energy, governance, health and safety, information society, labour market, population

Table 9.2 TIA methodologies. Evaluation phases, type of data, sources of data, computational methods and dimensions of territorial development

Name	Evaluation phase		Type of data		Sources of data			Computational method			Dimensions of territorial development				
	Ex Ante	Ex Post	Qualitative	Quantitative	Classical	Mix	Novel	Webgis	Excel	Other	Eco	Soc	Env	Gov	Pla
TEQUILA	X			X	X			X	XX	XX	XX	XX	XX	X	X
STEMA	X	X	X	X	X			X	XXX	X	XXX	XX	XX		
EATIA	X		X	X	X			X	XX	XX	XXX	XX	X	X	
TARGET_TIA	X	X	X	X	X			X	XXX	X	XX	XX	XX	XX	XX
QUICK_TIA	X		X	X	X			XXX	X	X	XXX	XX	X	X	
T. FORESIGHT	X		X		X			X	X	XXX	XXX	XX	XX	X	X
LUISA T. M.	X			X	X	X		XXX	X	X	XXX	XX	XX		X

Source: Own elaboration

Note: *ECO* economic competitiveness, *SOC* social cohesion, *ENV* environmental sustainability, *GOV* territorial governance, *PLA* spatial planning, XXX strong, XX average, X weak, *WEBGIS* use of online geographical information system platform to present the impact scores, *EXCEL* use of Microsoft excel or alike, to obtain the impact scores and living conditions, science and technology, territorial structure, transport and accessibility.

Soon after the creation of the first ESPON TIAs, other TIAs or similar policy assessment methodologies were designed to assess territorial impacts. The first was the TARGET_TIA which, unlike the ESPON TIAs, was specifically designed to assess the ex post impacts of EU Cohesion Policy. Just like the TEQUILA, however, the TARGET_TIA selected the quantitative indicators based on the concept of Territorial Cohesion, but with different analytic dimensions (socioeconomic cohesion, territorial governance and cooperation, polycentricity and environmental sustainability). It uses mainly traditional sources of data for socioeconomic and environmental dimensions. This data is complemented with non-traditional statistical sources of data collected for the dimensions of cooperation/governance and polycentrism, in databases like the INTERACT KEEP database and other sources available in different national and EU entities. In the meantime, the TARGET_TIA was already tested in specific EU programmes like the EU INTERREG-A (Medeiros, 2017a). For this case, the selected quantitative data referred to the two main dimensions of cross-border cooperation (territorial development and reduction of border barriers) and respective components. In this regard, it goes without saying that the collection of data on persisting border barriers required the access to non-mainstream sources of data, which are available in distinct regional and national entities.

Finally, the two remaining types of TIA methodologies mentioned in this chapter are designed for specific policy evaluation contexts. The first, known as Territorial Foresight, is used when the analysis of long-term developments is required. For this, qualitative data is collected via questionnaires, comprising three elements: content, geography and time (Böhme et al., 2020). Conversely the LUISA model 'is based on the concept of 'land function' for cross-sector integration and for the representation of complex system dynamics' (Lavalle et al., 2020). It is fundamentally supported by territorial indicators collected from several external models and presented via an online tool: the Urban Data Platform. This means that it explores higher spatial granularities than other TIA tools, since it provides information at the urban level.

9.3 Computational Guidelines on TIA

9.3.1 The Main Contribution of Computational Social Science for Territorial Impact Assessment

As seen in the previous section, existing TIA methodologies are supported by traditional sources of quantitative data. These are retrieved from EU national and sometimes regional statistical entities such as the Eurostat, ESPON and JRC databases. In some cases, specific data is obtained directly from non-mainstream data sources, especially for measuring components associated with governance and spatial planning dimensions of territorial development. In this context, there is a wide scope for incorporating non-traditional sources of data (see McQueen, 2017) in the implementation of TIA methodologies, in the following domains.

9.3.1.1 Complementarity

Territorial impact assessment analysis is generally related to analysing policy impacts on territorial development or territorial cohesion trends. It can, of course, also tackle other policy arenas, such as territorial cooperation or territorial integration (on territoriality, see Medeiros, 2020a). The problem here is that, as a holistic concept, territory encompasses basically all aspects related to the concept of development (Potter, 2008). This scenario implies a constant struggle to find, in traditional sources of data, a balanced set of indicators for all the analytical dimensions of, for instance, territorial development (Medeiros, 2019), hence, the potential benefits of usage of non-traditional data (e.g. digital footprint, digital tracking data, etc.) to complement largely incomplete traditional sources of data in implementing a TIA methodology. Here, besides the economic-related pool of statistics, which are normally relatively abundant at several territorial levels, the remaining policy dimensions of development can be enriched by non-traditional sources of data. These include social statistics, like 'quality of life' indicators, which often depend on an individual perception, which can be acquired via enquiries made with mobile phones. Furthermore, environmental-related data, such as the potential 'carbon footprint' of each individual in a given territory, can be acquired by means of online questionnaires via mobile phones or even by data on road congestion and public transport data. In the latter case, online applications such as the Flightradar24 (flightradar24.com) or the UCL Energy Institute portal to visualise the world's shipping routes can be used to estimate a carbon footprint impact score for each intended territorial scale. These are just a few examples that can also be applied in other dimensions of territorial development, such as territorial governance (e.g. to identify social engagement and participation in a given domain via the analysis of social network geo-tagged information) and spatial planning (e.g. to determine the compacity of urban areas via the visualisation of Google Maps).

9.3.1.2 Real-Time Information

One of the main advantages of non-traditional sources of data is the possibility to analyse territorial flows of data in real time. One aforementioned example is Flightradar24, which presents the current location of all commercial airplanes at any given time. The same goes for data which can be collected from some public transport operators and mobile phone companies tracking the exact location of individuals in a real-time context. This data, once aggregated and anonymised, can be particularly useful, for instance, to assess cross-border flows, which are a crucial element to understanding the territorial impacts of cross-border cooperation (Medeiros, 2018), or urban mobility processes (Pucci et al., 2015).

9.3.1.3 Spatial Accuracy

Another advantage related to digital tracking is the collection of highly accurate spatial data (Christl & Spiekermann, 2016) which is normally absent in traditional sources of data. However, this data collection should comply with the right of citizens to minimise their digital footprint (Bronskill & McKie, 2016). One domain in which spatial accuracy for TIA is particularly relevant is the analysis of all sorts of flows, especially in urban areas. As Cao (2018) puts it: 'data science can also fundamentally change the way political policies are made, evaluated, reviewed and adjusted by providing evidence-based informed policy making, evaluation, and governance'.

9.3.2 Sources of Data Towards an Analysis of EU Territorial Heterogeneity

I still remember the wise words of a former university professor on research methodologies stating that 'before you think you will not find the data you need, try hard and you will be amazed on what data is out there'. Indeed, data of all kinds and sources is waiting to be found in a myriad of places, to be treated and used in various studies. In the case of TIAs, it would appear reasonable to surmise how important it is to have access to a wide pool of updated and georeferenced data at several territorial levels and at several policy domains. In this regard, the writing of this particular chapter confirmed the premise that it is possible to access a wider pool of data to be used in TIA methodologies, to complement the ones commonly available in traditional data sources (regional, national and EU statistical entities and databases).

What is more striking, as seen in Table 9.3, is that it was possible to find alternative non-traditional sources of data that have already been explored and presented in scientific literature. These data covers basically all dimensions and respective components of a central concept for elaborating TIA analysis: territorial cohesion. Here, the economic-related indicators were basically the exception as regards the availability of relevant non-traditional data which can be used to assess territorial cohesion trends in a given territory. Also, it goes without saying that what this research found does not necessarily equate precisely to all potential non-traditional indicators which can eventually be found and applied in assessing each of the territorial cohesion analytic components. Moreover, many other non-traditional data sources can be found and used to analyse other topics which can be assessed via TIA methodologies, such as cross-border cooperation programmes, and urban, rural or regional development policies, among several other policy domains.

The selection of the territorial cohesion concept (Medeiros, 2016b) serves as a concrete and optimal example to explain the potential selection of sources of non-traditional data towards an analysis of EU territorial heterogeneity. Firstly, territorial cohesion is a multi-dimensional concept which encompasses a wide array of policy arenas, which can, in its own right, be also subject to a stand-alone TIA analysis, as is the case of environmental sustainability-related policies. Secondly, territorial cohesion can be analysed at different territorial levels, and some of them, especially at the urban and local levels, can greatly benefit from the new spatial granularity provided by some of the already available non-traditional sources of data.

In detail, Table 9.3 provides at least one example of a potential indicator and respective data source which can be used to assess most of the identified territorial cohesion components. This is particularly valid for analysing social cohesion, environmental sustainability, territorial governance and cooperation and trends in morphological polycentricity. A large part of these novel and non-traditional data, which can be used as complementary to existing traditional data, is linked to mobile technologies (i.e. phones). Due to the large amount of presented examples, a more detailed explanation of each one of these sources of alternative data can be found on the presented literature references. One can, however, highlight the tremendous possibilities provided by mobile technologies to study commuter flows using public transport in a given territory, which can deliver a very precise location at different times of the day, and even real-time information. Another example is the collection of data from certain operators on the production and use of renewable sources of energy at any given time, in different locations. This data can be particularly useful since traditional sources of statistics do not yet provide detailed information, per territorial sub-national unit, on the production and use of renewable energy. Most instructive in the polycentricity analytic domain of territorial cohesion is the possibility to use geospatial data sources to assess the degree of urban compactness, which is otherwise difficult to analyse by means of traditional sources of data. Finally, it is interesting to see the number of digital sources of information which can be used to analyse and measure governance and cooperation-related analytic components such as social participation and interaction. How far and how this data is spatially detailed and how it can be updated is, however, a discussion topic for subsequent analysis.

9.3.3 Main Challenges on Using Non-traditional Sources of Data on Implementing TIA Methodologies

The previous topic unveiled a wealth of non-traditional sources of information to implement TIA methodologies, mostly based on the use of territorial cohesion as a central concept for the TIA analysis, as would be the case in assessing the

Table 9.3 Non-traditional data for analysing territorial cohesion

Dimension	Component	Traditional indicators	Potential non-traditional indicator	Data and source
Socioeconomic cohesion	Productivity	Work productivity	–	–
	Income	GDP per capita	–	–
	Employment	Employment rate	–	–
	Innovation	Patents granted	Patents in ITC-related technology classes	EU ESSLait project Micro Moments Database (OECD, 2014)
	Infrastructure	Industrial parks	–	–
	Entrepreneurship	Startups	Number of Startups per university	University internet sites
	Education	Tertiary education (%)	Mobile phones for educational assessment	Mobile phone data (Şahin & Mentor, 2016)
	Health	Physicians per capita	Mobile phone data for informing public health actions	Mobile phone data (Oliver et al., 2020)
	Culture	Culture expenses (%)	Digital dimension of cultural capital	e.g. Reading IRT score (Paino & Renzulli, 2013)
	Exclusion/inclusion	Poverty rate	Digital poverty	e.g. digital interaction (Barrantes, 2007)
	Basic infrastructure	Public transports	Digital data for public transport	Mobile phone data and Smart card data (Zannat & Choudhury, 2019)

(continued)

Table 9.3 (continued)

Dimension	Component	Traditional indicators	Potential non-traditional indicator	Data and source
Environmental sustainability	Security	Criminality rate	Urban crime	Mobile Phone Data (Traunmueller et al., 2014)
	Climate change	CO_2 emissions	Environmental extremes and population	Mobile network data from MDEEP project (Lu et al., 2016)
	Energy	Renewable energy production	Renewable energy usage	Mobile Data via network operators using renewable energy (Syed et al., 2021)
	Nature and biodiversity	Protected areas (%)	Visitation in parks and protected areas	Mobile phone data (Monz et al., 2019)
	Environment and economy	Share of expenses with environment	Environmental quality	Client-server architecture of the WISE-MUSE mobile application for analysing environment (Akhmetov & Aitimov, 2015)
	Natural resources/garbage	Share of urban waste collected	Sustainability	Large-scale social data and machine-learning based on 12,720 electric vehicle (EV) charging stations (Asensio et al., 2020)
	Environment and health	Waste collection and treatment	Environmental health decision support	Sensors integrated into mobile devices (Bae et al., 2012)
Territorial cooperation and governance	Horizontal cooperation	Interreg-C programmes	Territorial ties	International call traffic (Blumenstock, 2011)
	Vertical cooperation	Territorial cooperation entities	–	–
	Participation	Elections participation rate	Social participation	Internet social participation (Anderberg et al., 2021)
	Involvement	NGOs + territorial partnerships	Social interactions.	Social network geo-tagged information (Keusch et al., 2019)

		Public consultation processes	Measuring information skills	Navigation, evaluation and management (ITU, 2018)
Morphologic Polycentricity	Hierarchy/ranking	Urban dwellers (%) /City ranking	Urban commuting flows	Data from mobile phones (Yu et al., 2020)
	Density	Population density	Population density distribution	Network Topology data + MS Counters (Joint Research Centre et al., 2015)
	Connectivity	Apartments with cable per capita	Network measurement	Cellular data network measurement for mobile applications e.g. mobile commerce (Wittie et al., 2007)
	Distribution/shape	City compactness index	Urban compactness	Geospatial data sources (Lan et al., 2021)

Information

Source: Own elaboration

territorial impacts of EU Cohesion Policy in a given territory. In almost every way, however, the use of these 'novel and digital' sources of data comes with known challenges, mainly related to the goal of establishing the necessary correct level of spatial granularity provided by spatial analysis, as is the case of a TIA. Alike and complementary challenges can be exposed when trying to find and use such sources of data.

9.3.3.1   The Relevance of the Sample

Collected data for TIA studies must be sound, reliable, comparable and georeferenced. As such, it is crucial that non-traditional data selected for TIA methodologies represent a relevant number (or sample) of the population (individuals, entities) on several territorial levels (from local to national if possible). Furthermore, existing data should be regularly updated, at least each year. For that, individuals and entities which are asked to provide their positions via mobile or non-mobile technological platforms should be convinced of the common benefits to change policies from transmitting the requested information on a regular basis.

9.3.3.2   Precise Location and Low Cost of Collected Data

Entities which use digital technological means to gather data should provide the produced data at distinct spatial granularities preferably via a free or low-cost online framework. This is, of course, challenging, particularly in establishing the correct level of spatial granularity and optimality of the targeted policy measure and costs/timeliness of the decision. These challenges depend on what policy is being assessed via a TIA methodology. In the case of assessing EU cross-border cooperation programmes, for instance, the level of spatial granularity would require the use of EU NUTS 3-related data. In this case, the cost and time associated with the acquisition of non-traditional data on cross-border commuting for each border NUTS 3, for instance, could be financially and timely viable in view of the analytic added value it would provide to the overall TIA analysis, based on our experience (Medeiros, 2017a). Indeed, one of the potential advantages of using data collected via the activation of the GPS location of mobile devices, or via digital questionnaires requesting the exact location of the individual, is the possibility to produce precise spatial analysis, which is vital for analysing certain territorial processes, such as metropolitan and cross-border commuting patterns.

9.3.3.3   Easy Access and Real-Time Data

One of the tantalising challenges associated with accessing non-traditional data sources is its dispersion by a myriad of different sources. In this regard, already existing statistical entities such as Eurostat and national statistic entities could centralise non-traditional data sources in their existing online platforms for data consulting. This would facilitate the access to data to all interested. Another possibility is to have an internet platform with links available to non-traditional sources of data divided by policy domains. Some of these sources are already provided on internet sites and a few demonstrate quite interesting real-time spatialised data (e.g. Flightradar24). To have a platform with the collection of all available real-time spatialised data sites would significantly reduce the time and, inevitably, costs associated with the search for non-traditional data to elaborate a TIA.

9.4 The Way Forward

In the context of policy evaluation, TIAs are relatively new. A cursory glance across existing TIAs also confirms their continuous modification and perfection process towards improved effectiveness in assessing the main ex ante (mostly) and ex post territorial impacts of projects, programmes and policies. In this evolving methodological context, the scientific relevance of using non-traditional data is particularly important for TIA, for several reasons. Firstly, by covering all dimensions of territorial development, TIAs require a wide set of comparable territorialised data which are often difficult to get via traditional data sources (regional, national and European statistic entities). In this regard, it is routinely contended that some dimensions and respective components of territorial development, such as territorial governance and spatial planning, have limited comparable and spatialised data, which can complement abundant data from socioeconomic development-related components. Secondly, non-traditional sets of data do not embrace real-time and spatial accuracy qualities, which can be of great value when assessing territorial impacts of certain policy areas, such as cross-border cooperation processes.

When contemplating the potential advantages of using non-traditional data in TIAs, which include their complementarity with traditional sources of data and the possibility of using real-time information and more detailed spatial accuracy, it is easy to demonstrate the potential advantages for existing TIA methods to not only provide more comprehensive and coherent TIA impact policy scores but to also improve overall policy forecast accuracy, both at ex ante and ex post evaluation phases. There are several open avenues for research on how to conciliate the use of traditional and non-traditional data to be used in TIA methodologies, which is still very much absent in current TIA related literature. There is a wealth of academic literature on the potential use of non-traditional data in many aspects of territorial development.

Amid this ever-growing body of literature discussing the potential use of non-traditional data for policy evaluation in specific policy areas, this chapter compiled, for the first time, a collection of potential non-traditional indicators, proposed in academic literature, which can be used in all existing TIA methodologies. There are, for sure, far more such indicators of this kind which can complement and complete the use of traditional datasets to be used in TIA analysis. What is striking are the tremendous possibilities to obtain non-traditional indicators for analysing the dimensions and components of territorial development as normally there are fewer options available with traditional data. It was indeed, a great surprise that it was possible to find a myriad of potential non-traditional indicators in components related to the analysis of, for instance, territorial governance, which imply wider possibilities to better understand social participation and involvement related processes. The same goes for increasing possibilities to better understand spatial planning trends via the analysis of specific components such as commuting flows, detailed analysis of demographic density and urban compactness. Likewise, the analysis of environmental sustainability trends on related components can be greatly improved using novel non-traditional data in areas such as renewable energy, environmental quality and sustainability. But even domains which are normally relatively robust in terms of data availability, such as the economic and social indicators, can be complemented by existing non-traditional sources of data in certain domains such as innovation, entrepreneurship, education, health, culture and security.

I have to admit that, prior to writing this chapter, I was not fully aware of the sea of possibilities offered by the potential use of non-traditional data indicators which can be used by TIA methodologies. Hence, what this chapter offers to the interested readers is a necessarily short and simplified introduction to the potential advantages of using non-traditional data when implementing TIA methodologies, as well as a wide number of potential non-traditional indicators and respective literature. Future analysis can detail even more the availability of such types of data to be used in assessing the territorial impacts of policies. Given the speed in which science evolves nowadays, I would not be surprised if 10 years from now, the number of non-traditional indicators that could potentially be used for TIA analysis has grown exponentially. But more importantly, in our opinion, existing and future sources of non-traditional data should be compiled on a regular basis and formatted in a sound, reliable, comparable and georeferenced manner, to be used in TIAs. By implication, these novel data should be easily accessible in online platforms and preferably free of charge, so they can be easily collected and used by all interested. In this regard, the EC can play a vital role in defining norms and regulations similar to the ones used for traditional data and use entities such as Eurostat and the JRC, as platforms to make it available to the general public in an organised manner, not only in datasets but also via Web Geographical Information Systems presenting real-time information.

To some extent, data science and technology are at the heart of an ongoing scientific and technological revolution and globalisation transformation. Even more starkly, the past decades saw a drastic change in data availability for policy evaluation. Indeed, around 30 years ago, the implementation of a TIA would be almost impossible since comparable spatialised data only existed for certain social and economic indicators. This means that it was only possible to assess socioeconomic impacts of a given policy. Instead, territorial impact analysis implies a balanced collection of not only socioeconomic but also environmental, governance and spatial planning related indicators. This context explains why TIA analyses are relatively recent. They gravely depend on data availability in several policy domains. For all involved in territorial analysis and specifically in implementing TIA methodologies, data availability is still a major challenge. This is particularly evident for ex post TIA analysis which require a crucial use of comparable quantitative data to verify territorial trends of the analysed territory using a wide set of indicators.

By proposing at least one potential non-traditional data indicator for almost all the components of territorial cohesion, to be used on TIA analysis, this chapter underlines a rosier foresight for TIA evaluations, no matter which methodology and selected time framework (ex ante, mid-term or ex post). This crucial positive implication of using non-traditional sources of data to implement TIAs in a more effective manner remains, however, to be seen in a practical manner, since there are still several challenges ahead to make them usable in scientific research, as previously mentioned. These challenges are also rooted in pre-conceptions related to the potential unreliability and incomparability of certain non-traditional data sources. Even so, the potential gains from using them for territorial analysis are evident. The idea, for instance, of using data from mobile phones and related mobile sources, to analyse metropolitan and cross-border commuting patterns is widely appealing for policy makers and evaluators. Similarly, data obtained from satellites can provide a very detailed spatial granularity, often absent from traditional sources of data. Hence, the use of programmes or software to automate the analysis of territorial impacts (programmatic scope), with a complementary use of non-traditional sources, heralds a battery of choices which are widely promising, but that are yet to be fully understood and tested. This is an appealing testing ground for future research for all involved in TIA implementation.

References

Akhmetov, B., & Aitimov, M. (2015). Data collection and analysis using the mobile application for environmental monitoring. *Procedia Computer Science, 56*, 532–537. https://doi.org/10.1016/j.procs.2015.07.247

Anderberg, P., Abrahamsson, L., & Berglund, J. S. (2021). An instrument for measuring social participation to examine older adults' use of the internet as a social platform: Development and validation study. *JMIR Aging, 4*(2), e23591. https://doi.org/10.2196/23591

Asensio, O. I., Alvarez, K., Dror, A., Wenzel, E., Hollauer, C., & Ha, S. (2020). Real-time data from mobile platforms to evaluate sustainable transportation infrastructure. *Nature Sustainability, 3*(6), 463–471. https://doi.org/10.1038/s41893-020-0533-6

Bae, W. D., Alkobaisi, S., Narayanappa, S., & Liu, C. C. (2012). A mobile data analysis framework for environmental health decision support. In *2012 Ninth International Conference on Information Technology–New Generations* (pp. 155–161). https://doi.org/10.1109/ITNG.2012.31.

Baesens, B. (2014). *Analytics in a big data world: The essential guide to data science and its applications*. Wiley.

Barrantes, R. (2007). Analysis of ICT demand: What is digital poverty and how to measure it? In *Digital poverty: Latin American and Caribbean perspectives* (29–53).

Bertoni, E., Fontana, M., Gabrielli, L., Signorelli, S., & Vespe, M. (Eds). (2022). *Mapping the demand side of computational social science for policy*. EUR 31017 EN, Luxembourg,

Publication Office of the European Union. ISBN 978-92-76-49358-7, https://doi.org/10.2760/901622

Blumenstock, J. E. (2011). Using mobile phone data to measure the ties between nations. *Proceedings of the 2011 IConference* (pp. 195–202). https://doi.org/10.1145/1940761.1940788.

Böhme, K., Lüer, C., & Holstein, F. (2020). From territorial impact assessment to territorial foresight. In E. Medeiros (Ed.), *Territorial impact assessment* (pp. 157–176). Springer International Publishing. https://doi.org/10.1007/978-3-030-54502-4_9

Bronskill, J., & McKie, D. (2016). *Your right to privacy. Minimize your digital footprint.* Self-Counsel Press.

Camagni, R. (2020). The pioneering quantitative model for TIA: TEQUILA. In E. Medeiros (Ed.), *Territorial impact assessment* (pp. 27–54). Springer International Publishing. https://doi.org/10.1007/978-3-030-54502-4_3

Cao, L. (2018). Data science thinking. In I. L. Cao (Ed.), *Data science thinking* (pp. 59–90). Springer International Publishing. https://doi.org/10.1007/978-3-319-95092-1_3

Christl, W., & Spiekermann, S. (2016). Networks of control. *A report on corporate surveillance, digital tracking, big data & privacy Facultas.*

ESPON 3.2. (2006). *Spatial scenarios and orientations in relation to the ESDP and cohesion policy* (Vol. 5). ESPON Luxemburg.

Ferreira, R. C. B., & Verschelde, N. (2020). Enhancing cross-border cooperation through TIA implementation. In E. Medeiros (Ed.), *Territorial impact assessment* (pp. 143–154). Springer International Publishing. https://doi.org/10.1007/978-3-030-54502-4_8

ITU. (2018). *Measuring the information society report.* International Telecommunication Union Publications.

Joint Research Centre, Institute for Environment and Sustainability, Widhalm, P., Pantisano, F., Craglia, M., & Ricciato, F. (2015). *Estimating population density distribution from network-based mobile phone data.* EU Publications. https://data.europa.eu/doi/10.2788/162414

Keusch, F., Struminskaya, B., Antoun, C., Couper, M. P., & Kreuter, F. (2019). Willingness to participate in passive mobile data collection. *Public Opinion Quarterly, 83*(S1), 210–235. https://doi.org/10.1093/poq/nfz007

Lan, T., Shao, G., Xu, Z., Tang, L., & Sun, L. (2021). Measuring urban compactness based on functional characterization and human activity intensity by integrating multiple geospatial data sources. *Ecological Indicators, 121*, 107177. https://doi.org/10.1016/j.ecolind.2020.107177

Lavalle, C., Silva, F. B. E., Baranzelli, C., Jacobs-Crisioni, C., Kompil, M., Perpiña Castillo, C., Vizcaino, P., Ribeiro Barranco, R., Vandecasteele, I., Kavalov, B., Aurambout, J.-P., Kucas, A., Siragusa, A., & Auteri, D. (2020). The LUISA territorial modelling platform and urban data platform: An EU-wide holistic approach. In E. Medeiros (Ed.), *Territorial impact assessment* (pp. 177–194). Springer International Publishing. https://doi.org/10.1007/978-3-030-54502-4_10

Lu, X., Wrathall, D. J., Sundsøy, P. R., Nadiruzzaman, M., Wetter, E., Iqbal, A., Qureshi, T., Tatem, A. J., Canright, G. S., Engø-Monsen, K., & Bengtsson, L. (2016). Detecting climate adaptation with mobile network data in Bangladesh: Anomalies in communication, mobility and consumption patterns during cyclone Mahasen. *Climatic Change, 138*(3–4), 505–519. https://doi.org/10.1007/s10584-016-1753-7

Marot, N., Golobič, M., & Fischer, T. B. (2020). The ESPON EATIA: A qualitative approach to territorial impact assessment. In E. Medeiros (Ed.), *Territorial impact assessment* (pp. 77–99). Springer International Publishing. https://doi.org/10.1007/978-3-030-54502-4_5

McQueen, B. (2017). *Big data analytics for connected vehicles and smart cities.* Artech House.

Medeiros, E. (2014a). Territorial impact assessment (TIA). *The process, methods and techniques.*

Medeiros, E. (2014b). Assessing territorial impacts of the EU cohesion policy: The Portuguese case. *European Planning Studies, 22*(9), 1960–1988. https://doi.org/10.1080/09654313.2013.813910

Medeiros, E. (2016a). EU cohesion policy in Sweden (1995-2013)–A territorial impact assessment. *European Structural and Investment Funds Journal, 3*(4), 254–275.

Medeiros, E. (2016b). Territorial cohesion: An EU concept. *European Journal of Spatial Development, 60*, 1–30.

Medeiros, E. (2016c). Territorial impact assessment and public policies: The case of Portugal and the EU. *Public Policy Portuguese Journal*, 51–61.

Medeiros, E. (2017a). Cross-border cooperation in inner Scandinavia: A territorial impact assessment. *Environmental Impact Assessment Review, 62*, 147–157. https://doi.org/10.1016/j.eiar.2016.09.003

Medeiros, E. (2017b). European Union cohesion policy and Spain: A territorial impact assessment. *Regional Studies, 51*(8), 1259–1269. https://doi.org/10.1080/00343404.2016.1187719

Medeiros, E. (2018). Focusing on cross-border territorial impacts. In E. Medeiros (Ed.), *European territorial cooperation* (pp. 245–265). Springer International Publishing. https://doi.org/10.1007/978-3-319-74887-0_13

Medeiros, E. (2019). Spatial planning, territorial development, and territorial impact assessment. *Journal of Planning Literature, 34*(2), 171–182. https://doi.org/10.1177/0885412219831375

Medeiros, E. (2020a). Fake or real EU territorialicy? Debating the territorial universe of EU policies. *Europa XXI*, 38. https://doi.org/10.7163/Eu21.2020.38.4.

Medeiros, E. (2020b). Introduction: A handbook on territorial impact assessment (TIA). In E. Medeiros (Ed.), *Territorial impact assessment* (pp. 1–6). Springer International Publishing. https://doi.org/10.1007/978-3-030-54502-4_1

Medeiros, E. (2020c). TARGET_TIA: A complete, flexible and sound territorial impact assessment tool. In E. Medeiros (Ed.), *Territorial impact assessment* (pp. 9–25). Springer International Publishing. https://doi.org/10.1007/978-3-030-54502-4_2

Medeiros, E. (2020d). *Territorial Impact Assessment, Advances in Spatial Science*. Springer Cham. https://doi.org/10.1007/978-3-030-54502-4

Monz, C., Mitrovich, M., D'Antonio, A., & Sisneros-Kidd, A. (2019). Using mobile device data to estimate visitation in parks and protected areas: An example from the nature reserve of Orange County, California. *Journal of Park and Recreation Administration, 37*(4), 92–109. https://doi.org/10.18666/JPRA-2019-9899

OECD. (2014). *Measuring the digital economy: A new perspective*. OECD Publishing. https://doi.org/10.1787/9789264221796-en

Oliver, N., Lepri, B., Sterly, H., Lambiotte, R., Deletaille, S., De Nadai, M., Letouzé, E., Salah, A. A., Benjamins, R., Cattuto, C., Colizza, V., de Cordes, N., Fraiberger, S. P., Koebe, T., Lehmann, S., Murillo, J., Pentland, A., Pham, P. N., Pivetta, F., et al. (2020). Mobile phone data for informing public health actions across the COVID-19 pandemic life cycle. *Science Advances, 6*(23), eabc 0764. https://doi.org/10.1126/sciadv.abc0764

Paino, M., & Renzulli, L. A. (2013). Digital dimension of cultural capital: The (in)visible advantages for students who exhibit computer skills. *Sociology of Education, 86*(2), 124–138. https://doi.org/10.1177/0038040712456556

Potter, R. B. (2008). *Geographies of development: An introduction to development studies* (3rd ed.). Pearson Education.

Prezioso, M. (2020). STeMA: A sustainable territorial economic/environmental management approach. In E. Medeiros (Ed.), *Territorial impact assessment (pp. 55–76)*. Springer International Publishing. https://doi.org/10.1007/978-3-030-54502-4_4

Pucci, P., Manfredini, F., & Tagliolato, P. (2015). *Mapping urban practices through mobile phone data*. Springer International Publishing. https://doi.org/10.1007/978-3-319-14833-5

Şahin, F., & Mentor, D. (2016). Using mobile phones for assessment in contemporary classrooms. In *Handbook of research on mobile learning in contemporary classrooms* (pp. 116–138). IGI Global.

Syed, S., Arfeen, A., Uddin, R., & Haider, U. (2021). An analysis of renewable energy usage by mobile data network operators. *Sustainability, 13*(4), 1886. https://doi.org/10.3390/su13041886

Traunmueller, M., Quattrone, G., & Capra, L. (2014). Mining mobile phone data to investigate urban crime theories at scale. In L. M. Aiello & D. McFarland (Eds.), *Social Informatics* (Vol. 8851, pp. 396–411). Springer International Publishing. https://doi.org/10.1007/978-3-319-13734-6_29

Wittie, M. P., Stone-Gross, B., Almeroth, K. C., & Belding, E. M. (2007). MIST: Cellular data network measurement for mobile applications. In *2007 Fourth International Conference on Broadband Communications, Networks and Systems (BROADNETS'07)* (pp. 743–751). https://doi.org/10.1109/BROADNETS.2007.4550508.

Yu, Q., Li, W., Yang, D., & Zhang, H. (2020). Mobile phone data in urban commuting: A network community detection-based framework to unveil the spatial structure of commuting demand. *Journal of Advanced Transportation, 2020*, 1–15. https://doi.org/10.1155/2020/8835981

Zannat, K. E., & Choudhury, C. F. (2019). Emerging big data sources for public transport planning: A systematic review on current state of art and future research directions. *Journal of the Indian Institute of Science, 99*(4), 601–619. https://doi.org/10.1007/s41745-019-00125-9

Chapter 10
Challenges and Opportunities of Computational Social Science for Official Statistics

Serena Signorelli, Matteo Fontana, Lorenzo Gabrielli, and Michele Vespe

Abstract The vast amount of data produced everyday (so-called digital traces) and available nowadays represent a gold mine for the social sciences, especially in a computational context, that allows to fully extract their informational and knowledge value. In the latest years, statistical offices have made efforts to profit from harnessing the potential offered by these new sources of data, with promising results. But how difficult is this integration process? What are the challenges that statistical offices would likely face to profit from new data sources and analytical methods? This chapter will start by setting the scene of the current official statistics system, with a focus on its fundamental principles and dimensions relevant to the use of non-traditional data. It will then present some experiments and proofs of concept in the context of data innovation for official statistics, followed by a discussion on prospective challenges related to sustainable data access, new technical and methodological approaches and effective use of new sources of data.

10.1 Introduction

Official statistics can be defined as the ensemble of all indicators, statistics and indices that are produced and disseminated by national statistical authorities (OECD

The views expressed are purely those of the authors and may not in any circumstances be regarded as stating an official position of the European Commission.

S. Signorelli (✉) · M. Fontana · L. Gabrielli
Scientific Development Unit, Centre for Advanced Studies, Science and Art, European Commission - Joint Research Centre, Ispra, Italy
e-mail: Serena.SIGNORELLI@ec.europa.eu; Matteo.FONTANA@ec.europa.eu; Lorenzo.GABRIELLI@ec.europa.eu

M. Vespe
Digital Economy Unit, European Commission - Joint Research Centre, Ispra, Italy
e-mail: Michele.VESPE@ec.europa.eu

et al., 2002). Right now, in their operations, official statistics tend to rely on so-called traditional data sources, namely, *census data*, *surveys* and *administrative data*.[1]

Yet, in an era characterised by increasing amounts of time spent living with connected devices, large amounts of new data are generated and collected every day. The places that we live in or that we visit can be inferred by analysing the position marked by our smartphones, our passions and relationship networks inferred from what we write on social media, and our health status from physiological data gathered through smart watches.

By living in a world that is a hybrid between its real and virtual instances, every day we leave traces and footprints of our life that are digital and can thus be collected, stored and processed.

Is it possible for statistical offices to draw on these "digital trace" data for creating new statistical indicators or for improving speed, quality and resolution of old ones in the field of social sciences? Such questions are very timely and high policy relevance, as shown by the collective exercise carried out at the Joint Research Centre of the European Commission (Bertoni et al., 2022) with the aim of mapping the demand side of Computational Social Science for Policy and its specific chapter on data innovation for official statistics. In this chapter, we will address main challenges and needs that statistical authorities will have to face in order to harness the full potential of these new data sources and illustrate some successful examples, with a focus on Computational Social Science.

10.2 Current Official Statistics Systems

In a recent report (2019, Chap. 7), the United Nations define three different types of data sources that are or could be used in official statistics:

1. Statistical data sources, composed by data collections created primarily for statistical purposes. This category includes surveys and census data.
2. Administrative data sources, which, differently from the former sources, are primarily set up for administrative purposes by public sector bodies.
3. Other data sources represented by all other sources created for commercial, market research or other private purposes.

The third source of data is the one that usually is referred to as the term "big data".

In the following section, we introduce the official statistics principles and how these new data sources relate to them. The section provides an overview of the steps that statistical agencies have undertaken so far to discover their potential and leverage their value and represent the foundations for Computational Social Science to provide input to policy through Official Statistics.

---

[1] Examples include birth and death registers in demographic statistics or the registries of real estate transactions in housing market statistics.

10.2.1 Statistical Principles

To fulfil their mission of providing timely and reliable data, the National Statistical Systems must comply with a set of principles that were formalised and adopted for the first time in 1991 by the Conference of European Statisticians (1991), revised afterwards and adopted globally by the UN Statistical Commission (1994),[2] with the name of Fundamental Principles of Official Statistics. Subsequently these principles have been updated periodically: the most recent version dates to 2013 (UN Economic and Social Council, 2013).

Together with the principles above, the concept of "quality" of official statistics needs to be taken into account. Brackstone (1999) defined quality in statistical agencies as "embracing those aspects of the statistical outputs of a NSO [National Statistical Office] that reflect their fitness for use by clients" and, as this concept is not capable of giving an operational definition, defined six dimensions of the broader concept to quality (see Table 10.1).

These six dimensions have been adapted by the main International Statistical Organizations to their own needs, as detailed in the table published by UNECE (Vale, 2010) (see Table 10.2).

When dealing with new data sources, one of the key elements to be considered is **timeliness**. Surveys and censuses usually require a substantial timeframe between the collection phase and the publication of results, while different sources like, for example, mobile phone data, could be available, at least theoretically, in near-real time. Together with timeliness, this highlights an additional feature offered by new data sources that is the potential to improve **frequency** or periodicity of the data collected. The time between observation can be reduced almost arbitrarily below the yearly or monthly that are typical in current official statistics.

As Brackstone (1999) points out in Table 10.1, "Timeliness is typically involved in a trade-off against accuracy". In fact, this is specifically true for traditional data sources such as survey or census data. With reference to new sources of data, they usually do not constitute a representative sample of the population marking an intrinsic limitation to accuracy. In the case of new data sources, accuracy is less linked to timeliness given the availability of information that occurs in almost real time. When dealing with innovative data sources, other kind of trade-offs may emerge; as an example, when dealing with mobility data gathered via mobile networks (as done in Iacus et al. (2020)), accuracy could be in trade-off with resolution, since the increase in granularity may further reduce the representativeness of the information.

This representation issue constitutes one of the differences between data coming from research institutions and from commercial companies highlighted by Liu et al. (2016). Private companies do not necessarily follow scientific data collections procedures or statistical sampling schemes, as their main objective is to streamline

---

[2] The United Nations Statistical Commission represents he highest body of the global statistical system and brings together the Chief Statisticians from member states from around the world.

Table 10.1 The six dimensions of data quality, from Brackstone (1999)

Relevance	"The *relevance* of statistical information reflects the degree to which it meets the real needs of clients. It is concerned with whether the available information sheds light on the issues of most importance to users". Assessing relevance is a subjective matter dependent upon the varying needs of users. The NSO's challenge is to weigh and balance the conflicting needs of different users to produce a program that goes as far as possible in satisfying the most important needs and users within given resource constraints".
Accuracy	"The *accuracy* of statistical information is the degree to which the information correctly describes the phenomena it was designed to measure. It is usually characterized in terms of error in statistical estimates and is traditionally decomposed into bias (systematic error) and variance (random error) components. It may also be described in terms of the major sources of error that potentially cause inaccuracy (e.g., coverage, sampling, nonresponse, response)".
Timeliness	"The *timeliness* of statistical information refers to the delay between the reference point (or the end of the reference period) to which the information pertains, and the date on which the information becomes available. It is typically involved in a trade-off against accuracy. The timeliness of information will influence its relevance".
Accessibility	"The *accessibility* of statistical information refers to the ease with which it can be obtained from the NSO. This includes the ease with which the existence of information can be ascertained, as well as the suitability of the form or medium through which the information can be accessed. The cost of the information may also be an aspect of accessibility for some users".
Interpretability	"The *interpretability* of statistical information reflects the availability of the supplementary information and metadata necessary to interpret and utilize it appropriately. This information normally covers the underlying concepts, variables and classifications used, the methodology of collection, and indications of the accuracy of the statistical information".
Coherence	"The *coherence* of statistical information reflects the degree to which it can be successfully brought together with other statistical information within a broad analytic framework and over time. The use of standard concepts, classifications and target populations promotes coherence, as does the use of common methodology across surveys. Coherence does not necessarily imply full numerical consistency".

processes such as billing (e.g., call detail records—CDR—from mobile network operators) or optimise services as product recommendations and advertising (e.g., social media advertising platform data), ultimately maximising their profit. Another accuracy aspect that Liu et al. (2016) highlight is the fact that private companies could "change the sampling methods and processing algorithms at any time and without any notice", adding uncertainty and risk to accuracy. Examples of this were reported when accessing mobility data from multiple mobile network operators in Europe to help fight COVID-19 (Vespe et al., 2021). Finally, Liu et al. (2016) emphasise how the validity of data itself could be at risk, as "commercial platforms have no obligation or motivation to ensure the authenticity and validity of the data they collected".

Table 10.2 Mapping quality components used by International Statistical Organisations, from Vale (2010)

UNECE	OECD	EUROSTAT	IMF
Relevance	Relevance	Relevance	Prerequisites of quality (part)
			Methodological soundness
Accuracy	Accuracy	Accuracy	Accuracy and reliability
Timeliness	Timeliness	Timeliness and punctuality	Serviceability (part)
Punctuality			
Accessibility	Accessibility	Accessibility and clarity	Accessibility
Clarity	Interpretability		Assurances of integrity (part)
Comparability	Coherence	Comparability	Serviceability (part)
		Coherence	
(Considered more relevant at the level of the organisation)	Credibility		
			Prerequisites of quality (part)
			Assurances of integrity (part)

The dimensions described above are not the only ones affected by the uptake of innovative data sources in official statistics; we need to consider **accessibility** issues, as new data sources—often privately held—may be difficult or expensive to procure. At the same time, it is also true that the data sources currently used in official statistics (surveys, generally) already present an increase in nonresponse rates, that leads to a reduction in the quality of data and consequently to an increase in the associated costs (Luiten et al., 2020). In order to improve accessibility, a switch to new data sources could be framed as a possible way to address rising costs associated with traditional data collections. Costs would probably not be reduced, but financial resources could be invested into new sources that could complement (or even replace) existing ones. Nevertheless, in many other cases, data may not be yet available on the market for several reasons (e.g., non-clear reputational or monetisation advantages over risks of non-compliance after sharing the data), requiring additional efforts to improve such data flows, including regulatory ones (e.g., the EU Data Governance Act[3] or the EU Data Act[4]).

As mentioned, the use of such new data sources for official statistics would be a *secondary* one with respect to the reasons for which they were conceived and collected. For example, CDR data could be employed for mobility analysis (Blondel et al., 2015), while social media advertising platform data could be used to estimate population flows (Spyratos et al., 2019). This requires a certain amount of additional processing and interpretation in order to lead to meaningful indicators.

---

[3] https://eur-lex.europa.eu/legal-content/EN/TXT/?uri=CELEX%3A52020PC0767

[4] https://digital-strategy.ec.europa.eu/en/library/data-act-proposal-regulation-harmonised-rules-fair-access-and-use-data

Interpretability[5] will therefore play a significant role in the future of official statistics with new data, as it will not be straightforward as currently is with surveys and census data, designed and set up to describe the phenomenon they are supposed to measure.

Also, **coherence** will be affected, as it will be important for these data sources to be sustainable over time, making them available continuously and with constant underlying methodology (this links again to accuracy), or at least with full knowledge of it to be constantly updated as part of the production process.

Going back to the Fundamental Principles of Official Statistics (UN Economic and Social Council, 2013), which deal with issues like accountability, relevance, impartiality and transparency, among others, it can be observed that a process of adaptation of these guidelines to a new paradigm will be needed. For example, principle no. 2 states that "to retain trust in official statistics, the statistical agencies need to decide according to strictly professional considerations, including scientific principles and professional ethics, on the methods and procedures for the collection, processing, storage and presentation of statistical data". This principle refers to transparency in official statistics, which is assessed and guaranteed by a set of guidelines that must be fulfilled by professionals handling data in statistical offices.

Nevertheless, the concept of transparency applied to new sources and digital trace data should not only be seen from a "data handler" perspective, but it must be complemented by a set of rules that refer to procedures and codes used to produce insights, calling for open-source practices and FAIR (findable, accessible, interpretable and reusable) data principles (Wilkinson et al., 2016) to ensure interpretability and reproducibility.

The structure of the statistical system may need to adapt when using digital trace data: the "survey design" part would become less relevant in this context, possibly superseded by a "data ingestion" and "data processing" sections, while processing becomes central.

The nature and composition of the tasks a NSO needs to perform to deliver reliable official statistics starting from big data may call for an adaptation of the organisational structure as well as of the competences needed by NSOs.

Many statistical offices have begun this transformation with exploratory exercises with the exception of sporadic cases.[6] This is a challenge that statistical offices may need to face. As an example, in terms of computer code, with a one-off analysis (as done with scientific research), it is sufficient to publish the code as open source, while in regular production settings of official statistics, the code itself needs to be maintained, implementing regular edits and versioning. This translation of statistical

---

[5] Interpretability here has to be intended in the broader sense used by Brackstone (1999) and not as in the machine learning context [see, e.g., Murdoch et al. (2019)], where it is more related to algorithmic transparency.

[6] International travel statistics in Estonia: https://statistika.eestipank.ee/failid/mbo/valisreisid_eng.html and foreign visitor statistics in Indonesia: https://www.bps.go.id/subject/16/pariwisata.html#subjekViewTab1, both using mobile positioning data.

methodology into software code has been introduced by Ricciato (2022) with the name of *softwarisation of statistical methodologies*.

10.2.2 Recognition of the Value of New Data Sources

In Europe, the European Statistical System[7] (ESS) has been involved in recognising the existence of digital trace data and its value since nearly a decade.

Two documents have paved the way to the use of innovative data sources in official statistics. The *Scheveningen Memorandum on Big Data and Official Statistics* (DGINS, 2013) represents the first statement through which the ESS recognised the importance of these new data sources and highlighted the main issues related to their use.

The *Bucharest Memorandum on Official Statistics in a Datafied Society (Trusted Smart Statistics)* (DGINS, 2018) represents an updated version of the former document, where the ESS underlines the need for "amendments to the statistical business architecture, processes, production models, IT infrastructures, methodological and quality frameworks, and the corresponding governance structures".

Moreover, in 2021 Eurostat[8] started a revision process of Regulation 223/2009[9] (the EU legal framework for European statistics) considering the new needs of official statistics. The updated version of the Regulation is expected to be finalised by the end of 2022. One of the explicit goals of the revision process is to set the legal framework for the reuse of privately held data for the development, production and dissemination of official statistics in Europe (Baldacci et al., 2021).

On a more international perspective, the Organisation for Economic Co-operation and Development (OECD) collected a series of examples of statistical applications (OECD, 2015) that made use of new data sources, as well as a list of limitations of this type of data. More importantly, the report introduces the implications for statistical offices when using these new data sources. Specifically, they envision three different possible roles for statistical offices, which may:

1. Act as certificatory institutions (giving the so-called trust mark to datasets)
2. Act as dissemination institutions (in a way that all statistics produced with non-traditional data are stored and disseminated in a central agency—the National Statistical Office)
3. Become active users of non-traditional data sources, for complementing the traditional ones as well as to create standalone statistical series

---

[7] The partnership between the European Community statistical authority, composed by Eurostat, the national statistical offices (NSOs) and other national authorities in each EU Member State that are responsible for the development, production and dissemination of European statistics.

[8] The statistical office of the European Union.

[9] https://eur-lex.europa.eu/legal-content/EN/ALL/?uri=CELEX%3A32009R0223

The OECD has already underlined some sensitive issues that will need to be taken into consideration for a successful adoption of non-traditional data in the workflow of national statistical offices. The main challenges are represented by the acquisition of skills needed to work with non-traditional data, the relevant data governance principles as well as privacy concerns (OECD, 2015). They also see space for partnerships of National Statistical Offices with universities and research organisations to best exploit the new opportunities brought by data innovation and to become collectors and disseminators of best practices.

10.2.3   Some Proof of Concepts and Experiences

In 2014, the United Nations established a Global Working Group (GWG) on Big Data for Official Statistics[10] with the aim of promoting the practical use of big data sources as well as building trust in the use of these sources for official statistics.

One of the outputs of the group was a handbook on the use of mobile phone data for official statistics (UN Global Working Group on Big Data for Official Statistics, 2019), which put forward a series of practical examples of the use of this data source in different statistical domains (tourism, population, migration, commuting, traffic flow and employment). Many countries (Estonia, Japan, Sri Lanka, among others) launched pilots and projects that have some potential for statistics in the mentioned statistical domains.

Most practical examples of applications have been carried out by European countries, where a partnership between Eurostat, NSOs and other National authorities that are responsible for the development, production and dissemination of European statistics was implemented with the name of European Statistical System (ESS).

One of the first attempts identified is ESSnet Big Data I,[11] composed by 22 NSOs. The objective of this initiative is to integrate big data into the regular production of official statistics. This is achieved via the development of projects that could explore the potential of these data sources, carried out from February 2016 to May 2018.

One of these projects was carried out with the help of six national statistical institutes (and afterwards other four joined) and investigated the feasibility of using job advertisement data scraped from the Web to improve official estimates of job vacancy statistics.[12] The activity consisted in the comparison between online job advertisement and job vacancy surveys. Some cases demonstrated a high correlation, while others showed only a loose relationship between the two. Nevertheless, this appears to be a promising area where innovative data can complement traditional survey data by potentially producing flash estimates or

---

[10] https://unstats.un.org/bigdata/

[11] https://ec.europa.eu/eurostat/cros/essnet-big-data-1_en

[12] https://ec.europa.eu/eurostat/cros/content/wp1-reports-milestones-and-deliverables1_en

increasing the frequency survey-based statistics but also to produce additional insights about occupations, required skills and labour demand in local areas.

Another ESSnet example aimed at inferring enterprise characteristics by accessing their websites through Web scraping techniques.[13] Six NSOs were involved, and their activity focused on six different use cases (URLs retrieval, e-commerce/web sales, social media detection, job advertisement detection, NACE[14] detection, SDGs detection) using both deterministic and machine learning methods. The predicted values can be used at unit level, to enrich the information contained in the register of the population of interest, and at population level, to produce estimates. The activity resulted in a series of output indicators, published as experimental statistics).[15]

A third example in the framework of the ESS network (European Statistical System, 2017) concerned the use of scanner data or web-scraping for Consumer Price Index (made by NSOs in France, Italy, the Netherlands, Poland and Portugal), the use of mobile phone data to study population and the study of tourist accommodations offered by individuals (French NSO), an analysis on the identification of inhabited addresses through electricity providers data to reduce survey costs (NSOs in Poland and Estonia) and the use of credit and debit cards data in the National Accounts (Portuguese NSO).

A deeper analysis was carried out specifically on tourism statistics. Eurostat has made an extended analysis of data sources having potential relevance for measuring tourism. In a recent report (2017), Demunter develops a taxonomy of big data sources relevant to tourism, including communication systems (e.g., MNO data, social media posts), web (e.g., web activity data), business process generated data (e.g., flight bookings, financial transactions), sensors (e.g., earth observation, vessel tracking systems, smart energy meters) and crowdsourcing (e.g., Wikipedia, OpenStreetMap).

An attempt to develop a hybrid between one-off analyses and regular production statistics has been undertaken by some statistical offices in the form of experimental statistics. Among the examples that can be identified, a very notable one was carried out by Eurostat.[16] These statistics cover 14 topics,[17] ranging from collaborative economy platforms to skills mismatch. All these experiments are listed and can be further explored.[18] They are deemed *experimental* as they "have not reached full maturity in terms of harmonisation, coverage or methodology". Nevertheless, the potential in terms of provided insights and knowledge of such solutions is clearly disruptive. Moreover, in a spirit of experimentation and co-creation, Eurostat and the single NSOs invite users to submit feedback and suggestions to improve them.

---

[13] https://ec.europa.eu/eurostat/cros/content/wp2-reports-milestones-and-deliverables1_en

[14] NACE stands for the statistical classification of economic activities in the European Community.

[15] https://ec.europa.eu/eurostat/web/experimental-statistics/

[16] https://ec.europa.eu/eurostat/web/experimental-statistics

[17] At the moment of publishing.

[18] https://ec.europa.eu/eurostat/web/experimental-statistics/overview/ess

The UK Office for National Statistics published on its website a guide on experimental statistics,[19] defining the features of this kind of statistics, namely:

- new methods, which are being tested and still subject to modification;
- partial coverage (for example, of industries) at that stage of the development programme;
- potential modification following user feedback about their usefulness and credibility compared with other available statistical sources.

10.3 The Need for Change

The above considerations and examples show the significant attention posed by statistical offices on the use of novel data sources since almost a decade, as well as the readiness and will to innovate. But what does this shift mean in practice for them?

With the availability of new data sources, the statistical system may need to adapt, as it was traditionally designed to work with data of a different nature (surveys and administrative data). This comes from the fact that data from new sources (that we will call *non-traditional data* for convenience) are quite different from traditional ones:

- Firstly, surveys are designed to produce specific statistics, whereas non-traditional data are collected for other purposes (see Section *Current Official Statistics systems*).
- Secondly, while non-traditional data tend to capture human behaviour, they are not directly generated by humans, but by automated systems and machines with which humans interact. These peculiarities require an additional effort in terms of translating machine logs into information about human actions and then connecting such actions to human behaviour.
- Thirdly, the process of translation of machine logs into information requires many choices to be done by researchers that will in some way influence the result (Ricciato, 2022; Ricciato et al., 2021). Having understood the context to properly address these issues, we can observe how the analysis and use of new data for official statistics requires a dual set of competences: both in terms of modelling and inference of human behaviour (in its many possible dimensions) as well as the technical capabilities needed to manage and analyse such big and complex data sources. This specific skillset is the one required by the emerging field of Computational Social Science. Statistical authorities may need to develop and strengthen these skills to benefit from the information included into non-traditional data.

---

[19] https://www.ons.gov.uk/methodology/methodologytopicsandstatisticalconcepts/guidetoexperimentalstatistics

These new data sources represent a huge opportunity for statistical offices to innovate while increasing openness. Nonetheless, challenges relevant to data access, adaptation of processes and effective uses of the data will have to be addressed.

10.3.1 Data Access

The great majority of the data sources that could be harnessed for official statistics purposes resides with the private sector. The debate on the access to such data is broad and vivid, with different opinions arising, in favour and against the mandatory obligation for private companies of giving access to the data.

The European Commission is addressing this issue in its legislative process and has recently proposed a regulation in the framework of the European Data Strategy, the Data Act[20] that, among other provisions, aims at fostering business-to-government data sharing for the public interest, supporting business-to-business data sharing and evaluating the Intellectual Property Rights (IPR) framework with a view to further enhance data access and use. The legislative process started in May 2021 and included a public consultation carried out during summer of 2021 that led many affected parties to the publication of a number of position papers. From the perspective of statistical offices, the ESS called on the need for the Data Act to ensure that European Statistical Offices and Eurostat can be granted access to privately held data for the development, production and dissemination of official statistics (European Statistical System, 2021). On the other hand, private sector data holders stressed on a lack of incentives to share data and an unclear impact that this sharing would have in practice (Bitkom, 2021) but also on voluntary sharing of data (and not an obligation) (AmCham EU, 2021; ETNO, 2021; Orgalim, 2021) as well as legitimate business interest around data to be protected. The Data Act was proposed by the Commission on 23 February 2022 (European Commission, 2022), providing means for public sector bodies, EU institutions, agencies or bodies to access and use privately held data in exceptional circumstances such as in emergencies. Such data may be shared to carry out scientific research activities compatible with the purpose for which the data was requested by the public sector body or with national statistical institutes for the compilation of official statistics.

Guidelines and best practices are also being published in the literature, such as by researchers from the Bank of Italy, highlighting the three main challenges that characterise the access and use of new data sources: trust, usability and sustainability (Biancotti et al., 2021). Moreover, the authors developed a set of principles that should guide data partnerships and that concern general aspects, principles specifically directed to statistical agencies and to private sectors' data collectors (Biancotti et al., 2021). The principles directly related to statistical offices build

---

[20] https://digital-strategy.ec.europa.eu/en/library/data-act-proposal-regulation-harmonised-rules-fair-access-and-use-data

around three main notions: responsibility and accountability (on process, output and methodology), safeguard (of individual and business interests), coordination and standardisation (the "collect only once" principle, to avoid the same request to the same data provider).

10.3.2 Adapting the Official Statistics System

In a recent paper (2020), Ricciato and co-authors highlight a set of important challenges that statistical offices may need to address when confronted with the possibility of using non-traditional data, which imply a series of changes "[...] in almost every aspect of the statistical system: processing methodologies, computation paradigms, data access models, regulations, organizational aspects, communication and disseminations approaches, and so forth".

Going more into practical details and on specific issues, one of the most critical is **privacy** that must be protected via, e.g., privacy-enhancing technologies (PETs).

Borrowing greatly from the work of Ricciato and co-authors (2019a), the UN Big Data Working Group defined in 2019 the three goals that need to be taken as guidelines when dealing with privacy concerns: input privacy, output privacy and policy enforcement (Big Data UN Global Working Group, 2019). In particular, "one or more Input Parties provide sensitive data to one or more Computing Parties who statistically analyse it, producing results for one or more Result Parties" (Big Data UN Global Working Group, 2019). The first goal, *input privacy*, must ensure that Computing parties are not able to access (or to indirectly derive with specific techniques and mechanism) any input value provided by Input Parties. At the other end of the process, *output privacy* has to ensure that published results do not contain identifiable input data. The third goal, introduced by the Big Data UN Global Working Group (2019), *policy enforcement*, represents the meeting point of the first two, as it is able to assure that they are automatically assured in a privacy-preserving statistical analysis system. Without entering into many details, this goal is concretised if there exists a mechanism that allows input parties to exercise positive control over computations that can be performed on sensitive inputs and over the publication of results; the just mentioned positive control is "[...] expressed in a formal language that identifies participants and the rules by which they participate" and carried out through a series of rules and decision points.

The report then presents five different PETs for statistics: Secure Multiparty Computation, (Fully) Homomorphic Encryption, Trusted Execution Environments, Differential Privacy and Zero Knowledge Proofs. In light of the abovementioned system, for each PET they describe which of the three goals it supports and in which way.

Another important issue is the **transparency** of National Statistical Offices. Luhmann et al. (2019) propose a new paradigm called STATPRO (shared, transparent, auditable, trusted, participative, reproducible and open). The authors make an open call to all National Statistical Offices about the need of implementing these seven principles, in order to achieve the goal of having a transparent and defensible evidence-based data-informed policymaking. In particular, some best practices from the open-source software (OSS) community are needed for the development and deployment of statistical processes. As an example, they suggest that algorithms and methods should be available and accessible to anyone, with adequate level of documentation, and versioning should be introduced for environments.

After looking at specific issues, we need to focus on how to practically adapt the production system with the new requirements brought by the use of non-traditional data. One possible approach is proposed by Grazzini et al. (2018) through the so-called *plug and play* design. This approach was thought to handle the changes needed in production systems, and it is based on software components, which are modular and customisable, that are subsequently assembled together. This design has the advantage of allowing the integration of existing systems, operations and components with the new ones needed to embrace new data and/or models. Being modular, it also allows to overcome the constraint usually present on the choice of platform used for the implementation.

One practical proposal that has been theorised and discussed in Europe in recent years is the introduction of Trusted Smart Statistics. One of the main principles behind this proposal relies on the idea of "pushing computation out instead of pulling data in" (Ricciato et al., 2019b). The concept is often referred to as "in situ data processing" (Martens et al., 2021). This implies that the new data sources that statistical offices wish to analyse and integrate with traditional ones do not necessarily need to leave the premises of the data holders. Instead, the algorithm will reach the latter in order to perform computations, and afterwards only aggregated and processed data will be led to statistical offices to produce official statistics. On the one hand, this new paradigm will allow to preserve the privacy principle (as the data are not leaving their premises), but on the other hand, more attention must be paid to transparency and accountability. One way to address these issues is the way already paved by the OPen ALgorithm project (OPAL), which declared algorithmic transparency as its foundational principle.[21] The proposal consists in making open by default all the software code along the whole data processing chain and allowing everybody to see it and, eventually, audit it (Ricciato et al., 2019b).

10.3.3 Effective Use of the New Sources

Once the first two issues are addressed (access to the data and changes in the statistical system), an important one (if not the most important) remains: what are the new statistical products that could only be developed using these new data sources? And why would be the responsibility of statistical offices to take care of this (and not, e.g., a local authority)?

---

[21] http://www.opalproject.org/

This implies that the focus must now go to the demand side and to the identification of the questions that statistical offices could tackle with these data sources, in line with what proposed by Bertoni et al. (2022) in the Computational Social Science for policy mapping exercise. This represents a challenging task, as statistical offices need to take some time and reflect on what to highlight, but also why this relies in their mandate, and not among some other institution's activity.

Some examples of these new "needs" are clearly shown for instance in Romanillos Arroyo & Moya-Gómez (2023), Napierala & Kvetan (2023), Manzan (2023) and Crato (2023). Concerning tourism, for example, after an introduction about new data sources and new computational methods for the tourism sector, the authors propose a series of potential applications (in the form of KPIs) on environmental impact and socio-economic resilience of tourism. By looking at the KPIs proposed to monitor land use related to the tourism activities, for instance, one of the indicators put forward aims at quantifying the presence of short-term rentals platforms (like Airbnb) through the analysis of accommodation platform data or similar. This indicator would allow to get more insights about a phenomenon that is increasing and that is not captured through traditional data sources in the tourism sector (viz. surveys) (Romanillos Arroyo & Moya-Gómez, 2023).[22]

Another example concerns direct and indirect water consumption at tourism destinations, a KPI that could be useful for the management of resources consumption related to leisure places. In this case different datasets could be used: from smart meters to food consumption data that in turn can be inferred from credit card data (Romanillos Arroyo & Moya-Gómez, 2023). As can be seen, these new proposed indicators require prior agreement to accessing the data, and therefore the three issues we presented in this chapter again show their very close connectedness.

10.4 The Way Forward

Summarising the issues highlighted in this chapter on the use of Computational Social Science for official statistics, the focus goes to the three main enablers:

- The access to the data
- The adaptation needed by official statistical system
- New statistical products that could be developed using these new data sources

Concerning this last point, a proposal to facilitate the implementation of this could be the institution of specific committees or steering groups with the aim to discuss possible solutions to the issues presented. Something that needs to be

---

[22] The phenomenon of short-term rental accommodation in tourism is already under the lens of the European Commission, that will shortly propose a regulation about it (https://ec.europa.eu/info/law/better-regulation/have-your-say/initiatives/13108-Tourist-services-short-term-rental-initiative/public-consultation_en).

underlined is the fact that even if data access could come for free (following specific partnerships or law provisions), the processing of these new data sources has a cost.

As a concluding remark, these new data sources have enormous potential for the official statistics world in terms of improved timeliness and granularity, but they can only be considered as a complementary source and not pure substitutes of the traditional ones. As it is thoroughly explained in this chapter, due to the strict statistical requirements in terms of quality of the data used in official statistics we think that these new sources of data could improve and complement the existing ones.

References

AmCham EU. (2021). *Data Act—Feedback to the European Commission's Inception Impact Assessment.* https://www.amchameu.eu/system/files/position_papers/iia_data_act.pdf

Baldacci, E., Ricciato, F., & Withmann, A. (2021). A reflection on the re(use) of new data sources for official statistics. *Revista de Estadística y Sociedad, 83*, 8–11.

Bertoni, E., Fontana, M., Gabrielli, L., Signorelli, S., & Vespe, M. (Eds.). (2022). *Mapping the demand side of computational social science for policy.* EUR 31017 EN, Luxembourg, Publication Office of the European Union. ISBN 978-92-76-49358-7, https://doi.org/10.2760/901622

Biancotti, C., Borgogno, O., & Veronese, G. (2021, November 1). *Principled data access: Building public-private data partnerships for better official statistics.* https://blogs.worldbank.org/opendata/principled-data-access-building-public-private-data-partnerships-better-official

Big Data UN Global Working Group. (2019). *UN handbook on privacy-preserving computation techniques.* https://unstats.un.org/bigdata/task-teams/privacy/UN%20Handbook%20for%20Privacy-Preserving%20Techniques.pdf

Bitkom. (2021). *Public consultation on the Data Act.* https://www.bitkom.org/sites/default/files/2021-10/2021011-bitkom-data-act-public-consultation-1.pdf

Blondel, V. D., Decuyper, A., & Krings, G. (2015). A survey of results on mobile phone datasets analysis. *EPJ Data Science, 4*(1), 10. https://doi.org/10.1140/epjds/s13688-015-0046-0

Brackstone, G. (1999). Managing data quality in a statistical agency. *Survey Methodology, 25*(2), 139–149.

Conference of European Statisticians. (1991). *Draft of a resolution on the fundamental principles of official statistics in the region of the Economic Commission for Europe: Document prepared by an Expert Group of Conference of European Statisticians*

Crato, N. (2023). From lack of data to data unlocking. In Bertoni, E., Fontana, M., Gabrielli, L., Signorelli, S., & Vespe, M. (Eds.), *Handbook of computational social science for policy.* Springer.

Demunter, C. (2017). *Tourism statistics: Early adopters of big data?* Eurostat, European Union. https://ec.europa.eu/eurostat/documents/3888793/8234206/KS-TC-17-004-EN-N.pdf

DGINS. (2013). *Scheveningen memorandum big data and official statistics.* https://ec.europa.eu/eurostat/documents/13019146/13237859/Scheveningen-memorandum-27-09-13.pdf/2e730cdc-862f-4f27-bb43-2486c30298b6?t=1401195050000

DGINS. (2018). *Bucharest memorandum on official statistics in a datafied society (Trusted smart statistics).* https://ec.europa.eu/eurostat/documents/13019146/13237859/The+Bucharest+Memorandum+on+Trusted+Smart+Statistics+FINAL.pdf/7a8f6a8f-9805-e77c-a409-eb55a2b36bce?t=1634144384767

ETNO. (2021). *ETNO comments to the public consultation on the data act.* https://www.etno.eu//downloads/positionpapers/etno%20paper%20-%20data%20act%20consultation.pdf

European Commission (2022). *Proposal for a Regulation of the European Parliament and of the Council, on harmonised rules on fair access to and use of data (Data Act), COM(2022) 68 final*. https://eur-lex.europa.eu/legal-content/EN/TXT/PDF/?uri=CELEX:52022PC0068

European Statistical System. (2017). *Position paper on access to privately held data which are of public interest*. https://ec.europa.eu/eurostat/documents/13019146/13346094/ESS+Position+Paper+on+Access+to+privately+held+data+final+-+Nov+2017.pdf/6ef6398f-6580-4731-86ab-9d9d015d15ae?t=1511447619000

European Statistical System. (2021). *European Statistical System (ESS) position paper on the future Data Act proposal*. https://ec.europa.eu/eurostat/documents/13019146/13405116/main+ESS+position+paper+on+future+Data+Act+proposal.pdf/37f3b5c7-abfd-5a05-6be2-fdc4b87ee7d2?t=1631695372906

Grazzini, J., Lamarche, P., Gaffuri, J., & Museux, J.-M. (2018). *"Show me your code, and then I will trust your figures": Towards software-agnostic open algorithms in statistical production*. doi:https://doi.org/10.5281/zenodo.3240282

Iacus, S., Santamaria, C., Sermi, F., Spyratos, S., Tarchi, D., & Vespe, M. (2020). *Mapping Mobility Functional Areas (MFA) using mobile positioning data to inform COVID-19 policies: A European regional analysis*. Publications Office. https://data.europa.eu/doi/10.2760/076318

Liu, J., Li, J., Li, W., & Wu, J. (2016). Rethinking big data: A review on the data quality and usage issues. *ISPRS Journal of Photogrammetry and Remote Sensing, 115*, 134–142. https://doi.org/10.1016/j.isprsjprs.2015.11.006

Luhmann, S., Grazzini, J., Ricciato, F., Meszaros, M., Giannakouris, K., Museux, J.-M., & Hahn, M. (2019). *Promoting reproducibility-by-design in statistical offices*.

Luiten, A., Hox, J., & de Leeuw, E. (2020). Survey nonresponse trends and fieldwork effort in the 21st century: Results of an international study across countries and surveys. *Journal of Official Statistics, 36*(3), 469–487. https://doi.org/10.2478/jos-2020-0025

Manzan, S. (2023). Big data and computational social science for economic analysis and policy. In Bertoni, E., Fontana, M., Gabrielli, L., Signorelli, S., & Vespe, M. (Eds.), *Handbook of computational social science for policy*. Springer.

Martens, B., Parker, G., Petropoulos, G., & Van Alstyne, M. W. (2021, November 3). Towards efficient information sharing in network markets. TILEC discussion paper no. DP2021-014. Available at SSRN: https://ssrn.com/abstract=3956256 or http://dx.doi.org/10.2139/ssrn.3956256

Murdoch, W. J., Singh, C., Kumbier, K., Abbasi-Asl, R., & Yu, B. (2019). Definitions, methods, and applications in interpretable machine learning. *Proceedings of the National Academy of Sciences, 116*(44), 22071–22080. https://doi.org/10.1073/pnas.1900654116

Napierala, J., & Kvetan, V. (2023). Changing job skills in a changing world. In Bertoni, E., Fontana, M., Gabrielli, L., Signorelli, S., & Vespe, M. (Eds.), *Handbook of computational social science for policy*. Springer.

OECD. (2015). *The proliferation of "big data" and implications for official statistics and statistical agencies: A preliminary analysis* (OECD Digital Economy Papers No. 245). doi:https://doi.org/10.1787/5js7t9wqzvg8-en

OECD, International Labour Organization, International Monetary Fund, & International Statistical Committee of the Commonwealth of Independent States. (2002). *Measuring the non-observed economy: A handbook*. OECD. https://doi.org/10.1787/9789264175358-en

Orgalim. (2021). *Orgalim input to the European Commission consultation on the Data Act*. https://orgalim.eu/position-papers/digital-transformation-orgalim-input-european-commission-consultation-data-act

Ricciato, F. (2022). *A reflection on methodological sensitivity,quality and transparency in the processingof new "big" data sources*. European Conference on Quality in Official Statistics (Q2022), Vilnius. https://www.researchgate.net/publication/361284108_A_reflection_on_methodological_sensitivity_quality_and_transparency_in_the_processing_of_new_big_data_sources

Ricciato, F., Bujnowska, A., Wirthmann, A., Hahn, M., & Barredo Capelot, E. (2019a, August). A reflection on privacy and data confidentiality in Official Statistics. *ISI*

World Statistics Conference, Kuala Lumpur. https://ec.europa.eu/eurostat/cros/system/files/isi_paper_ricciato_bujnowska_final.pdf

Ricciato, F., Wirthmann, A., Giannakouris, K., Reis, F., & Skaliotis, M. (2019b). Trusted smart statistics: Motivations and principles. *Statistical Journal of the IAOS, 35*(4), 589–603. https://doi.org/10.3233/SJI-190584

Ricciato, F., Wirthmann, A., & Hahn, M. (2020). Trusted Smart Statistics: How new data will change official statistics. *Data & Policy, 2*, e7. https://doi.org/10.1017/dap.2020.7

Ricciato, F., Grazzini, J., & Museux, J.-M. (2021, September 3). *Public manuals and open-source code: Rethinking methodological documentation for new data sources*. New Techniques and Technologies for Statistics (NTTS) 2021. https://coms.events/NTTS2021/data/x_abstracts/x_abstract_51.pdf

Romanillos Arroyo, G., & Moya-Gómez, B. (2023). New data and computational methods opportunities to enhance the knowledge base of tourism. In Bertoni, E., Fontana, M., Gabrielli, L., Signorelli, S., & Vespe, M. (Eds.), *Handbook of computational social science for policy*. Springer.

Spyratos, S., Vespe, M., Natale, F., Weber, I., Zagheni, E., & Rango, M. (2019). Quantifying international human mobility patterns using Facebook Network data. *PLoS One, 14*(10), e0224134. https://doi.org/10.1371/journal.pone.0224134

UN Economic and Social Council. (2013). *2013/21. Fundamental principles of official statistics*. https://unstats.un.org/unsd/dnss/gp/FP-Rev2013-E.pdf

UN Global Working Group on Big Data for Official Statistics. (2019). *Handbook on the use of mobile phone data for official statistics*. https://unstats.un.org/bigdata/task-teams/mobile-phone/MPD%20Handbook%2020191004.pdf

UN Statistical Commission. (1994). *Statistical Commission: Report on the special session, 11–15 April 1994*. vi, 35 p.

United Nations. (2019). *United Nations national quality assurance frameworks manual for official statistics: Including recommendations, the framework and implementation guidance*. UN. https://doi.org/10.18356/1695ffd8-en

Vale, S. (2010). *Statistical Data Quality in the UNECE*.

Vespe, M., Iacus, S. M., Santamaria, C., Sermi, F., & Spyratos, S. (2021). On the use of data from multiple mobile network operators in Europe to fight COVID-19. *Data & Policy, 3*, e8. https://doi.org/10.1017/dap.2021.9

Wilkinson, M. D., Dumontier, M., Aalbersberg, I. J., Appleton, G., Axton, M., Baak, A., Blomberg, N., Boiten, J.-W., da Silva Santos, L. B., Bourne, P. E., Bouwman, J., Brookes, A. J., Clark, T., Crosas, M., Dillo, I., Dumon, O., Edmunds, S., Evelo, C. T., Finkers, R., et al. (2016). The FAIR Guiding Principles for scientific data management and stewardship. *Scientific Data, 3*(1), 160018. https://doi.org/10.1038/sdata.2016.18

Part III
Applications

Chapter 11
Agriculture, Food and Nutrition Security: Concept, Datasets and Opportunities for Computational Social Science Applications

T. S. Amjath-Babu, Santiago Lopez Riadura, and Timothy J. Krupnik

Abstract Ensuring food and nutritional security requires effective policy actions that consider the multitude of direct and indirect drivers. The limitations of data and tools to unravel complex impact pathways to nutritional outcomes have constrained efficient policy actions in both developed and developing countries. Novel digital data sources and innovations in computational social science have resulted in new opportunities for understanding complex challenges and deriving policy outcomes. The current chapter discusses the major issues in the agriculture and nutrition data interface and provides a conceptual overview of analytical possibilities for deriving policy insights. The chapter also discusses emerging digital data sources, modelling approaches, machine learning and deep learning techniques that can potentially revolutionize the analysis and interpretation of nutritional outcomes in relation to food production, supply chains, food environment, individual behaviour and external drivers. An integrated data platform for digital diet data and nutritional information is required for realizing the presented possibilities.

11.1 Introduction

The global goal of ending hunger and malnutrition (Sustainable Development Goal-2) by 2030 is off track as the numbers of food insecure and malnourished people are increasing (Fanzo et al., 2020). The number of undernourished people climbed to 768 million in 2020 from 650 million in 2019 (FAO, 2021), belonging mainly to the Asian (>50%) and African continents (25%). This might be further increased in

T. S. Amjath-Babu (✉) · T. J. Krupnik
International Maize and Wheat Improvement Center (CIMMYT), Dhaka, Bangladesh
e-mail: t.amjath@cgiar.org; t.krupnik@cgiar.org

S. L. Riadura
International Maize and Wheat Improvement Center (CIMMYT), El Batan, Texcoco, Mexico
e-mail: s.l.ridaura@cgiar.org

E. Bertoni et al. (eds.), *Handbook of Computational Social Science for Policy*,
https://doi.org/10.1007/978-3-031-16624-2_11

the context of the economic disruption caused by COVID-19 pandemic and global price hikes due to recent Russia-Ukraine conflict.

Some may consider it ironic that a large proportion of the undernourished people, who cannot afford healthy diets, are those involved in the food production, including subsistence farmers and farm labourers (Fanzo et al., 2022). In addition, low- and middle-income (LMIC) as well as wealthy countries are burdened with overweight (BMI > 25), obesity (BMI > 30) and diet-related non-communicable diseases (Ferretti & Mariani, 2017; Global Panel on Agriculture and Food Systems for Nutrition, 2016). As such, there are increasing calls for agricultural and food system innovations and policies that can enhance diets and improve availability of quality foods for better nutrition and health outcomes (Fanzo et al., 2022; Global Panel on Agriculture and Food Systems for Nutrition, 2016).

Nevertheless, the linkages of global and national food systems to nutritional outcomes are complex and are influenced by diverse macro-level (trade, market access, climate change, technology, conflicts, wealth distribution, agricultural policies. etc.) and micro- and meso-level factors (farm types, income, gender considerations, diet preferences, attitude and beliefs, inter- and intra-household dynamics and power, cooking methods, sanitation, among others). A deeper understanding of these multi-scale (micro-, meso- and macro-level) drivers of nutritional outcomes is vital in devising agricultural policies and programmes and hence transforming the agri-food sector to meet the goal of ending hunger and malnutrition (Global Panel on Agriculture and Food Systems for Nutrition, 2016). There is a wide recognition of inadequate methods, data and metrics for understanding agri-food systems relationships to nutritional outcomes and dynamics (Marshall et al., 2021; Micha et al., 2018; Sparling et al., 2021). Towards this end, new sources of data and emerging computational social science methods may offer possibilities to test novel conceptual frameworks as well as empirical and experimental examination of the complex relationships and pathways. The current chapter focuses on how the availability of digital data and computational social science methods can support modelling and the analytics of a complex portfolio of factors (and their interactions) influencing food and nutritional outcomes. It also highlights the need of data-sharing protocols and platforms for fully utilizing the potential of emerging data and analytical tools for generating meaningful policy insights (Müller et al., 2020; Takeshima et al., 2020).

11.2 The Complex Pathways to Nutritional Outcomes: A Conceptual Note

Agricultural production and consequently nutrient availability and consumption are interrelated through complex pathways span across spatial and time scales. Nutritional outcomes (Sparling et al., 2021) are driven by a range of factors including food production, consumer purchasing power, trade and market systems as well as food transformation and consumer behaviour (Global Panel on Agriculture and Food Systems for Nutrition, 2016). The downturns in economies, climate stress and conflicts can contribute to changes in consumption practices that lead to malnutrition, while trade policies and supply chain infrastructure can impact food prices that influence the costs of food necessary for healthy diets that in part drive nutritional outcomes (FAO, 2021). Climate variability and extreme events can lead to losses in agricultural production and increased import demand from affected countries, leading to food price volatility (Willenbockel, 2012; Chatzopoulos et al., 2020). Recession or reduction in economic activity at country level can also lead to unemployment and reduction in wages and income, which may force households to shift to energy dense and cheaper foods (including 'junk food') instead of purchasing and consuming nutritious foods (Dave & Kelly, 2012). Income and social inequality amplify the impact of climate stresses or economic downturns in terms of access to nutritious diets (FAO, 2021). Lower productivities and low efficiency of supply chains can also lead to higher prices for diverse food groups needed for healthy diets. Conflicts or health crises like the COVID-19 pandemic disrupt the movement of goods, increase in prices of healthy foods and decrease in their availability (Amjath-Babu et al., 2020). Conflicts can also reduce access to capital, energy, labour or land and hence impact food production (FAO, 2021).

In the case of farm households, the raw nutrient availability for a household is determined by own production used for self-consumption, purchased food from market using the farm household income and food received through informal exchanges and social safety nets. Farm household income is determined by the yield (sold in market) of various farm enterprises (cereal crops, vegetables, cash crops, livestock, aquaculture, etc.), their farmgate price levels and the cost (inputs including land, labour, machinery, fertilizers, pest control, etc.) of production in addition to any available rental income, off-farm income and remittances. Farm production and income is further conditioned by environmental stresses and the state of natural resources (e.g. soil, water), agricultural policies and market infrastructure. Apart from these drivers, the technology available at farm level influences yield performance and post-harvest losses that impact food availability and access (Müller et al., 2020). The direct (self-consumption) and indirect (as a source of income for market purchase of food consumed) role of farm production in food nutrient availability depends on the strength and quality of market linkages (Bellon et al., 2020; Sibhatu et al., 2015).

The net energy and macro- and micronutrient availability to men, women and children are further conditioned by diet preferences, cooking methods, gender norms, nutritional knowledge, attitudes and beliefs (Monterrosa et al., 2020). A deficit in net availability and effective consumption of nutrients compared to requirement of individual members can potentially lead to malnutrition that can manifests in stunting, wasting of children, nutrition-deficit disorders as well as nutrition-related non-communicable diseases (N-NCD) in children and adults. Stunting, wasting and N-NCD are conditioned not only by nutritional deficiencies but also by nutrient utilization (determining bioavailability of nutrients through metabolic pathways) capacity of human bodies (Millward, 2017) and sanitary, hygiene and water quality conditions. Conversely overconsumption of high-calorie, low-nutrient foods can lead to overweight and obesity (Astrup & Bügel, 2019). Women's empowerment in terms of time, income and asset control can also positively influence the nutritional and health-related outcomes (Herforth & Ballard, 2016). The agriculture-nutrition pathway map by Kadiyala et al. (2014) includes health-care expenditure, health status as well as women's employment as additional determinants for nutritional outcomes. Figure 11.1 provides a comprehensive overview of the complex path of nutritional outcomes.

In the case of affordability of diets, nutrient adequate (e.g. the advised 'EAT-Lancet diet') diets is not affordable for 1.5 billion poor people globally (Hirvonen et al., 2020). Even in European Union, around 10% of population in 16 countries faces financial issues in affording healthy diets (Penne & Goedemé, 2021). Nutrient-rich food items are often costly to grow, store and transport compared to starchy food. Oil and sugar tend also to have a longer shelf life and is easier to transport (Fanzo et al., 2022). This calls for also further understanding on ways to make the nutritious food more affordable as high prices of nutritious food or its volatility can negatively impact consumption among the poor. Conversely, lower prices of sugar and sugar-rich food prices are related to higher prevalence of overweight and obesity (Headey & Alderman, 2019). These point to the importance of a better understanding of the macro-economic policies on nutritional outcomes as well as disconnects between agriculture-nutrition pathways.

The discussion so far highlights the requirement of deeper understanding the complex pathways linking nutritional security, health outcomes and public policies, especially for the most vulnerable groups (children and women). Below, we discuss existing modelling-based approaches as well as the role of emerging digital data and computational methods in opening new frontiers in quantifying the ex ante impacts of regional or national food and nutrition policies by unravelling the complex interactions of the macro-meso-micro-factors.

11.3   *Current* Ex Ante Analytical Models for Nutritional Policy Insights

In case of existing ex ante assessments (nutritional outcomes of agricultural policy), three studies are discussed here for documenting the current *state of the art* of methods employed. Lopez-Ridaura et al. (2018) took a nutrient-balancing approach where self-consumed farm products and the net annual farm income derived from all farm enterprises were converted to energy equivalents and compared with annual food energy requirement of households. Although the model focused on calories, the simplified relation allowed simulations of yield changes due to new technologies and their impact on potential household-level food availability ratios. The study provides a framework that could also be extended to macro and micronutrient availability and consumption (Bizimana & Richardson, 2019).

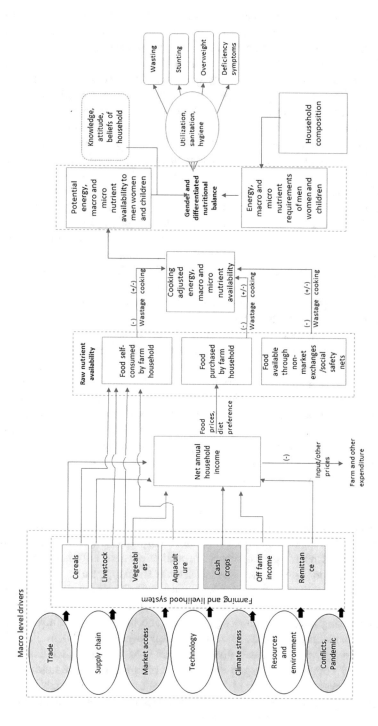

Fig. 11.1 Macro-level drivers and household-level nutrition outcome pathways of farm households [expanded framework based on an initial framework presented in Lopez-Ridaura et al. (2018)]

FARMSIM model is able to simulate the impact of net cash income from all farm enterprises on consumption of nutrients such as protein, calories, fat, calcium, iron and vitamin A. To represent the complex interrelations of macro- and micro-factors affecting food production and nutritional outcomes, current modelling approaches addressing nutritional security questions often use local-level proxies of macro-determinants (e.g. yield functions) or through representations of key variables (e.g. food prices or farm sizes) (Müller et al., 2020). In the case of the FARMSIM model, market prices are simulated (using Monte Carlo approaches) using probability distributions obtained from historical data or expert opinion, while yield distributions are generated by crop yields generated by the APEX (Agricultural Policy/Environmental eXtender) simulation model using historical weather data and plant growth parameters. These are matched to different technologies considered in simulation. Consideration of stochastic prices and yields allows modelling risk behaviour using stochastic efficiency with respect to a function (SERF).

FSSIM-Dev (European Commission. Joint Research Centre, 2020) is a farm-level model based on positive mathematical programming (PMP) which does not consider risk as the model is deterministic. FSSIM-Dev considers the non-separability of production and consumption decisions of farming households: It maximizes the utility from both the production and the consumption of food, and the decision to rely on home production or to go the market is governed by transaction costs. The model considers annual income beyond farm income by including subsidies, pensions, off farm income, remittances and other transfers as exogenous variables. Farm income is linked to the consumption by a linear expenditure function of uncompressible consumption below which consumption may not fall and supernumerary consumption, which is modelled as a fixed proportion (marginal budget share) of net income. FSSIM-Dev model is capable to generate food and nutrition security indicators as carbohydrates, proteins and lipids from the simulated food consumption. The simulation of micronutrients is not yet attempted by the existing model. Figure 11.1 shows an extended conceptual modelling frame that can offer wider insights to the questions related to nutritional impacts of agricultural policies, food environment, sanitation, etc.

The quoted policy simulation studies analysed policy impact on availability of macronutrients such as carbohydrates or proteins and had limited capacity in dealing with micronutrients. In addition, modelling efforts currently have limited ability to consider the access (income levels, impact of social safety nets, informal exchanges of food) and stability (seasonality and occasional shocks), gender roles (intra-household food allocation, women empowerment) and utilization (bioavailability) dimensions of the nutrient security question. Net nutritional availability is also affected by cooking methods, knowledge, attitude and beliefs that are not always integrated in modelling exercises to reduce complexity. In case of availability of nutrients for a given individual, distribution of food within households adds another layer of complexity. Despite the fact that ensuring adequate nutrition at an individual's level is at the heart of nutritional challenges, policy insights at this level are generally lacking. New sources of digital data and computational methods are expected to address the stated challenges.

11.4 The Data Scarcity for Nutritional Modelling and Analytics

The scarcity, within countries and among countries for harmonized data on food consumption and nutrition, is a major challenge for initiatives aimed at addressing global nutritional challenges. Currently, major data sources used for analytics are the specialized household surveys [demographic health surveys (DHS), multi-indicator cluster surveys (MICS), dietary intake surveys, consumption expenditure surveys, Living Standard Measurement Studies (LSMS), Food Security Monitoring System (FSMS) etc.]. These tend to include both economic-related variables and detailed diet data (Buckland et al., 2020). There is a need for efforts to make the datasets (cleaned data on food consumption and their nutritional equivalents) open through platforms with defined standards and efficient infrastructure to share data (de Beer, 2016). To make the open data sharing a reality, technological, legal and ethical challenges need to be considered. Communities like smallholder farmers hold data that, when subject to analytics, can be used to improve their well-being. But there is an absence of platforms and mechanisms that enable rapid and regular data acquisition and sharing (de Beer, 2016). Traka et al. (2020) suggested making food and nutrition data FAIR (findable, accessible, interoperable and reusable). An integrated data perspective of agriculture-food, nutrition and health is required for meaningful interventions ensuring sustainable production, shift in diets and reduction in non-communicable nutrition-related diseases (Traka et al., 2020). Even if data are available, they are often not available at sub-national levels, or may not be disaggregated across demographic groups, or may be out of date, in addition to a range of additional data quality challenges. Key complementary information required in preparing such databases is country-specific food composition tables (FCTs) that can be used to convert food products to their nutrient value, although often these tables are not comprehensive or may not even be available. As such, here is a need of coordinated effort to make sure that comprehensive FCTs are available (Ene-Obong et al., 2019).

The distribution of food within households adds another layer of difficulty, since information regarding intra-household distribution is lacking in many surveys. The disaggregation of household consumption data using adult male equivalent (AME) weights is a relevant disaggregation method (Coates et al., 2017) and is based on household members' relative caloric requirement. This deterministic method can be improved by adding error observed in dietary energy expenditure prediction models. Côté et al. (2022) compared traditional regression models against machine learning models in predicting individual vegetable and fruit consumption and did not observe a major improvement. Nevertheless, the scope of using machine learning or other innovative methods in disaggregation of consumption data is currently underexplored. Lager sets of food consumption data, disaggregated among household members, are required for validating and fine-tuning the methods.

11.5   Novel Digital Food and Nutrition Data for Computational Analytics

Global Individual Food consumption data Tool (GIFT-FAO/WHO) initiative aims to make harmonized data freely accessible online through an interactive web platform. This tool is based on FoodEx2, which is a food classification and description protocol (food items are coded with distinct hierarchy) developed by the European Food Safety Authority (EFSA) to standardize the 24 h recall data (Leclercq et al., 2019). There are also increasing attempts to collect nutritional data using telephonic surveys. Lamanna et al. (2019) reports that efforts to collect nutrition data from rural women in Kenya through telephone surveys result in a 0–25% increase in nutritional scores [minimum dietary diversity for women (MDD-W) and minimum acceptable diet for Infants and young children (MAD) estimates compared to face-to-face interviews]. This points to the potential of using digital tools and methods for data collection.

Nutrition apps for collecting diet information are increasingly available for diet recording and monitoring (Campbell & Porter, 2015). Hundreds of nutrition-related mobile apps are available, but their utility in tracking healthy food consumption and nutrition data generation is still limited. Fallaize et al. (2019) compared popular nutrition-related apps (Samsung Health, MyFitnessPal, FatSecret, Noom Coach and Lose It!) that assess macronutrients and micronutrients as well as energy from consumed food, against a reference method. They showed that apps are in general capable to assess macronutrient availability, while micronutrients estimates were inconsistent. Another similar study on nutritional apps (FatSecret, YAZIO, Fitatu, MyFitnessPal and Dine4Fit) showed inconsistent results on macronutrients and energy intake estimates (Bzikowska-Jura et al., 2021).

The efforts to use digital diet information collected by these or similar apps are so far very limited (Martinon et al., 2022), although they offer large potential. Smartphone image-based automatic food recognition and dietary assessment tools are currently emerging. These tools are attempting to identify, classify and estimate volume of food intake and nutrient content estimation. Machine learning and deep learning approaches are also now being used for classifying food items in a meal, which depends on generic and comprehensive food image datasets for training data. Deep learning approaches including convolutional neural network have been suggested as being more effective than machine learning algorithms such as support vector machines (SVM) and K-Nearest Neighbor (KNN) for this purpose due to their higher classification efficiency (Ciocca et al., 2020). Nevertheless, the quantification of the mass of food by visual assessment of volume and density is much more challenging. The estimation of calories and nutrients can be error prone due to poor classification and mass estimations. Further research and development of fully automated nutritional content detection applications using smartphones may transform it to a game changer for nutritional data (Subhi et al., 2019). An assessment of the meal *snapp* app by Keeney et al. (2016) showed that calorific values generated by such apps are comparable to standard application (Nutritionist

Pro™ used by dietitians). Nevertheless, user-documented diet data availability for analytical purposes tends to be constrained by lack of harmonization in collection of data and terms of use and privacy conditions (Maringer et al., 2018).

In case of developing nations, nutrition apps are mainly used at urban locations, and this may lead to a 'digital divide' if diet datasets from rural areas remain limited (Samoggia et al., 2021). Such a divide could be less pronounced in high-income countries. A novel digital tool that is being prototyped by CIMMYT (2022) in Bangladesh may address the possible rural-urban digital nutrition data divide in lower-income countries. This mobile app can be used by extension officers to assess diet data of smallholders to detect potential macro- and micronutrient deficiencies and suggest possible seasonal crops that can be grown in their homesteads to address the potential nutritional deficiencies in diets.

When used in larger scale, similar digital tools can also generate large-scale anonymised diet data that can be used for analytical purposes including modelling. The tool mentioned above helps extension officers to digitalize 7-day diet diaries recorded by farm household members. Digital tools that can lead to healthy food consumption while passively collecting food consumption data can prove useful for large-scale collection of diet datasets. Such datasets can be used for more comprehensive monitoring of national-level nutritional deficiencies and intervention targeting (Buckland et al., 2020) and policy simulations. In addition, if digital tools are used to source information on dietary supplements or additional constituents in foods (other than those meeting basic nutritional needs) that have health impacts, the role of bioactive dietary components (Yasmeen et al., 2017) in nutritional outcomes can also be explored using such datasets (Barnett & Ferguson, 2017).

Retailers' data on food purchases by consumers is another emerging source of nutrition-related data (Saarijärvi et al., 2016). Digital data generated at points of sale (POS) can be used for consumption pattern recognition and then mobilized to promote healthier consumption behaviour. Application such as 'NutriSavings', a healthy grocery shopping reward programme in the United States, attempted to influence consumers' decision to purchase healthy foods such as fruits and vegetables while reducing unhealthy fats, sugars and sodium using POS data (Nierenberg et al., 2019). Applications of point-of-sale nutrition data therefore could be advantageous in understanding nutrition consumption behaviour in developed countries and urban areas of developing nations.

Social media analytics (SMA) is also an emerging field for dietary data collection, behaviour analytics and population health assessments (Stirling et al., 2021). SMA is currently in data preparation and exploration phase, and future developments in quantitative data generation and analytics is expected to contribute towards nutritional surveillance and triangulation of other nutritional datasets. SMA applications in opinion mining, sentiment and content analysis and predictive analytics related to nutrition and health are emerging (Stirling et al., 2021). Initial studies on SMA show its high potential to yield policy insights similar to large surveys (Shah et al., 2020).

11.6   The Way Forward

The ideal (for computational social science applications) scenario (Fig. 11.2) is
that all kinds of diet and nutrition data sources, such as the harmonized diet data
from multi-indicator and other nutrition-related surveys, high frequency data from
telephone surveys, diet data from mobile applications aiming nutritional advise, diet
data from image-based diet detection applications, consumption information from
point-of-sale data, diet- and nutrition-related social media data getting aggregated to
a single data platform. These kinds of large datasets can be used for machine or deep
learning as well as for simulation and modelling studies (e.g. FARMSIM) or for
more conventional statistical analysis. Such national nutritional data-sharing platform can also include spatial data related to food supply chains, food environment
and external factors so that major drivers can be diagnosed for observed diet patterns
and for predictive analytics. The creation of unified platform for interoperable
digital data can facilitate analytics that could in turn result in more insightful
policy advice. The challenge is in the developing agreements with private firms
and public agencies who are data holders to ensure the data access to researchers
and the privacy of users and respondents. There is a need of policy innovations
that encourage the creation of centralized data-sharing platforms, especially on
nutritional data that allow researchers to analyse and derive policy insights that can
lead to achievement of sustainable development goals.

Several recent studies showed the viability of using satellite data and mobile
operator call detail records (CDR) to predict poverty levels at higher frequency and

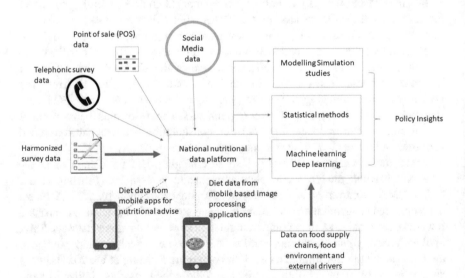

Fig. 11.2 Idealized national data platform for digital diet data and nutritional information. In case
of data flows, the thick arrow shows the existing data stream, the thinner arrows represent novel
data streams and dotted arrows represent upcoming and future data streams

spatial granularity (Pokhriyal & Jacques, 2017; Steele et al., 2017). There are also attempts to track poverty using e-commerce data (Wijaya et al., 2022) and mobile money transactions data (Engelmann et al., 2018). Once large-scale geographic location-specific diet and nutrition datasets are available, it may also be possible to make predictions regarding diet patterns and nutritional deficiencies using satellite, CDR data, mobile money or e-commerce transactions. Social media data can also complement such nutritional surveillance. This kind of near real-time monitoring of agri-food systems components and nutritional analytics could potentially results in important new insights for policy, given the current infrequent and inadequate food consumption and nutrition-related data and limited capacity of the modelling tools.

The chapter provides an overview of significant challenges in agriculture and nutrition policy development and presents emerging digital data sources and computational social science methods that can potentially address the stated challenges. Given the right analytical framework, data platforms and enabling conditions, computational social science techniques can unravel the complex impact pathways to nutritional outcomes and contribute significantly to addressing the global burden of overweight, obesity, malnutrition, hunger and nutrition-related non-communicable diseases.

References

Amjath-Babu, T. S., Krupnik, T. J., Thilsted, S. H., & McDonald, A. J. (2020). Key indicators for monitoring food system disruptions caused by the COVID-19 pandemic: Insights from Bangladesh towards effective response. *Food Security, 12*(4), 761–768. https://doi.org/10.1007/s12571-020-01083-2

Astrup, A., & Bügel, S. (2019). Overfed but undernourished: Recognizing nutritional inadequacies/deficiencies in patients with overweight or obesity. *International Journal of Obesity, 43*(2), 219–232. https://doi.org/10.1038/s41366-018-0143-9

Barnett, M. P. G., & Ferguson, L. R. (2017). Nutrigenomics. In *Molecular diagnostics* (pp. 305–326). Elsevier. https://doi.org/10.1016/B978-0-12-802971-8.00017-1

Bellon, M. R., Kotu, B. H., Azzarri, C., & Caracciolo, F. (2020). To diversify or not to diversify, that is the question. Pursuing agricultural development for smallholder farmers in marginal areas of Ghana. *World Development, 125*, 104682. https://doi.org/10.1016/j.worlddev.2019.104682

Bizimana, J.-C., & Richardson, J. W. (2019). Agricultural technology assessment for smallholder farms: An analysis using a farm simulation model (FARMSIM). *Computers and Electronics in Agriculture, 156*, 406–425. https://doi.org/10.1016/j.compag.2018.11.038

Buckland, A. J., Thorne-Lyman, A. L., Aung, T., King, S. E., Manorat, R., Becker, L., Piwoz, E., Rawat, R., & Heidkamp, R. (2020). Nutrition data use and needs: Findings from an online survey of global nutrition stakeholders. *Journal of Global Health, 10*(2), 020403. https://doi.org/10.7189/jogh.10.020403

Bzikowska-Jura, A., Sobieraj, P., & Raciborski, F. (2021). Low comparability of nutrition-related mobile apps against the polish reference method—A validity study. *Nutrients, 13*(8), 2868. https://doi.org/10.3390/nu13082868

Campbell, J., & Porter, J. (2015). Dietary mobile apps and their effect on nutritional indicators in chronic renal disease: A systematic review: Dietary apps in chronic renal disease. *Nephrology, 20*(10), 744–751. https://doi.org/10.1111/nep.12500

Chatzopoulos, T., Pérez Domínguez, I., Zampieri, M., & Toreti, A. (2020). Climate extremes and agricultural commodity markets: A global economic analysis of regionally simulated events. *Weather and Climate Extremes, 27*. https://doi.org/10.1016/j.wace.2019.100193

CIMMYT. (2022). *Nutrition smart homestead app.*

Ciocca, G., Micali, G., & Napoletano, P. (2020). State recognition of food images using deep features. *IEEE Access, 8*, 32003–32017. https://doi.org/10.1109/ACCESS.2020.2973704

Coates, J., Rogers, B. L., Blau, A., Lauer, J., & Roba, A. (2017). Filling a dietary data gap? Validation of the adult male equivalent method of estimating individual nutrient intakes from household-level data in Ethiopia and Bangladesh. *Food Policy, 72*, 27–42. https://doi.org/10.1016/j.foodpol.2017.08.010

Côté, M., Osseni, M. A., Brassard, D., Carbonneau, É., Robitaille, J., Vohl, M.-C., Lemieux, S., Laviolette, F., & Lamarche, B. (2022). Are machine learning algorithms more accurate in predicting vegetable and fruit consumption than traditional statistical models? An exploratory analysis. *Frontiers in Nutrition, 9*, 740898. https://doi.org/10.3389/fnut.2022.740898

Dave, D. M., Kelly, I. R. (2012). How does the business cycle affect eating habits?. *Social Science and Medicine,74*(2), 254–62. https://doi.org/10.1016/j.socscimed.2011.10.005

de Beer, J. (2016). *Ownership of open data: Governance options for agriculture and nutrition.* Wallingford: Global Open Data for Agriculture and Nutrition. de Beer, J., *Ownership of Open Data: Governance Options for Agriculture and Nutrition* (2016). (Wallingford: Global Open Data for Agriculture and Nutrition) [Report], 2016, Available at SSRN: https://ssrn.com/abstract=3015958

Ene-Obong, H., Schönfeldt, H. C., Campaore, E., Kimani, A., Mwaisaka, R., Vincent, A., El Ati, J., Kouebou, P., Presser, K., Finglas, P., & Charrondiere, U. R. (2019). Importance and use of reliable food composition data generation by nutrition/dietetic professionals towards solving Africa's nutrition problem: Constraints and the role of FAO/INFOODS/AFROFOODS and other stakeholders in future initiatives. *Proceedings of the Nutrition Society, 78*(4), 496–505. https://doi.org/10.1017/S0029665118002926

Engelmann, G., Smith, G., & Goulding, J. (2018). The Unbanked and Poverty: Predicting area-level socio-economic vulnerability from M-Money transactions. *2018 IEEE International Conference on Big Data (Big Data)*, 1357–1366. https://doi.org/10.1109/BigData.2018.8622268

European Commission. Joint Research Centre. (2020). *Modelling farm-household livelihoods in developing economies: Insights from three country case studies using LSMS ISA data.* Publications Office. https://data.europa.eu/doi/10.2760/185665

Fallaize, R., Zenun Franco, R., Pasang, J., Hwang, F., & Lovegrove, J. A. (2019). Popular nutrition-related mobile apps: An agreement assessment against a UK reference method. *JMIR MHealth and UHealth, 7*(2), e9838. https://doi.org/10.2196/mhealth.9838

Fanzo, J., Haddad, L., McLaren, R., Marshall, Q., Davis, C., Herforth, A., Jones, A., Beal, T., Tschirley, D., Bellows, A., Miachon, L., Gu, Y., Bloem, M., & Kapuria, A. (2020). The Food Systems Dashboard is a new tool to inform better food policy. *Nature Food, 1*(5), 243–246. https://doi.org/10.1038/s43016-020-0077-y

Fanzo, J., Rudie, C., Sigman, I., Grinspoon, S., Benton, T. G., Brown, M. E., Covic, N., Fitch, K., Golden, C. D., Grace, D., Hivert, M.-F., Huybers, P., Jaacks, L. M., Masters, W. A., Nisbett, N., Richardson, R. A., Singleton, C. R., Webb, P., & Willett, W. C. (2022). Sustainable food systems and nutrition in the 21st century: A report from the 22nd annual Harvard Nutrition Obesity Symposium. *The American Journal of Clinical Nutrition, 115*(1), 18–33. https://doi.org/10.1093/ajcn/nqab315

FAO. (2021). *The State of Food Security and Nutrition in the World 2021: Transforming food systems for food security, improved nutrition and affordable healthy diets for all.* Food & Agriculture Organization.

Ferretti, F., & Mariani, M. (2017). Simple vs. complex carbohydrate dietary patterns and the global overweight and obesity pandemic. *International Journal of Environmental Research and Public Health, 14*(10), 1174. https://doi.org/10.3390/ijerph14101174

Global Panel on Agriculture and Food Systems for Nutrition. (2016). *Food systems and diets: Facing the challenges of the 21st century.* http://glopan.org/sites/default/files/ForesightReport.pdf

Headey, D. D., & Alderman, H. H. (2019). The relative caloric prices of healthy and unhealthy foods differ systematically across income levels and continents. *The Journal of Nutrition, 149*(11), 2020–2033. https://doi.org/10.1093/jn/nxz158

Herforth, A., & Ballard, T. J. (2016). Nutrition indicators in agriculture projects: Current measurement, priorities, and gaps. *Global Food Security, 10*, 1–10. https://doi.org/10.1016/j.gfs.2016.07.004

Hirvonen, K., Bai, Y., Headey, D., & Masters, W. A. (2020). Affordability of the EAT–Lancet reference diet: A global analysis. *The Lancet Global Health, 8*(1), e59–e66. https://doi.org/10.1016/S2214-109X(19)30447-4

Kadiyala, S., Harris, J., Headey, D., Yosef, S., & Gillespie, S. (2014). Agriculture and nutrition in India: Mapping evidence to pathways: Agriculture-nutrition pathways in India. *Annals of the New York Academy of Sciences, 1331*(1), 43–56. https://doi.org/10.1111/nyas.12477

Keeney, M., Yeh, M.-C., Landman, R., May Leung, M., Gaba, A., & Navder, K. (2016). Exploring the use of an IPhone app: A novel approach to dietary assessment. *International Journal of Nutrition, 1*(4), 22–30. https://doi.org/10.14302/issn.2379-7835.ijn-14-566

Lamanna, C., Hachhethu, K., Chesterman, S., Singhal, G., Mwongela, B., Nge'ndo, M., Passeri, S., Farhikhtah, A., Kadiyala, S., Bauer, J.-M., & Rosenstock, T. S. (2019). Strengths and limitations of computer assisted telephone interviews (CATI) for nutrition data collection in rural Kenya. *PLoS One, 14*(1), e0210050. https://doi.org/10.1371/journal.pone.0210050

Leclercq, C., Allemand, P., Balcerzak, A., Branca, F., Sousa, R. F., Lartey, A., Lipp, M., Quadros, V. P., & Verger, P. (2019). FAO/WHO GIFT (Global Individual Food consumption data Tool): A global repository for harmonised individual quantitative food consumption studies. *Proceedings of the Nutrition Society, 78*(4), 484–495. https://doi.org/10.1017/S0029665119000491

Lopez-Ridaura, S., Frelat, R., van Wijk, M. T., Valbuena, D., Krupnik, T. J., & Jat, M. L. (2018). Climate smart agriculture, farm household typologies and food security. *Agricultural Systems, 159*, 57–68. https://doi.org/10.1016/j.agsy.2017.09.007

Maringer, M., van't Veer, P., Klepacz, N., Verain, M. C. D., Normann, A., Ekman, S., Timotijevic, L., Raats, M. M., & Geelen, A. (2018). User-documented food consumption data from publicly available apps: An analysis of opportunities and challenges for nutrition research. *Nutrition Journal, 17*(1), 59. https://doi.org/10.1186/s12937-018-0366-6

Marshall, Q., Bellows, A. L., McLaren, R., Jones, A. D., & Fanzo, J. (2021). You say you want a data revolution? Taking on food systems accountability. *Agriculture, 11*(5), 422. https://doi.org/10.3390/agriculture11050422

Martinon, P., Saliasi, I., Bourgeois, D., Smentek, C., Dussart, C., Fraticelli, L., & Carrouel, F. (2022). Nutrition-related mobile apps in the French app stores: Assessment of functionality and quality. *JMIR MHealth and UHealth, 10*(3), e35879. https://doi.org/10.2196/35879

Micha, R., Coates, J., Leclercq, C., Charrondiere, U. R., & Mozaffarian, D. (2018). Global dietary surveillance: Data gaps and challenges. *Food and Nutrition Bulletin, 39*(2), 175–205. https://doi.org/10.1177/0379572117752986

Millward, D. J. (2017). Nutrition, infection and stunting: The roles of deficiencies of individual nutrients and foods, and of inflammation, as determinants of reduced linear growth of children. *Nutrition Research Reviews, 30*(1), 50–72. https://doi.org/10.1017/S0954422416000238

Monterrosa, E. C., Frongillo, E. A., Drewnowski, A., de Pee, S., & Vandevijvere, S. (2020). Sociocultural influences on food choices and implications for sustainable healthy diets. *Food and Nutrition Bulletin, 41*(2_suppl), 59S–73S. https://doi.org/10.1177/0379572120975874

Müller, B., Hoffmann, F., Heckelei, T., Müller, C., Hertel, T. W., Polhill, J. G., van Wijk, M., Achterbosch, T., Alexander, P., Brown, C., Kreuer, D., Ewert, F., Ge, J., Millington, J. D. A., Seppelt, R., Verburg, P. H., & Webber, H. (2020). Modelling food security: Bridging the gap between the micro and the macro scale. *Global Environmental Change, 63*, 102085. https://doi.org/10.1016/j.gloenvcha.2020.102085

Nierenberg, D., Powers, A., & Papazoglakis, S. (2019). Data-driven nutrition in the digital age. *Sight and Life, 33*(1), 83–87.

Penne, T., & Goedemé, T. (2021). Can low-income households afford a healthy diet? Insufficient income as a driver of food insecurity in Europe. *Food Policy, 99,* 101978. https://doi.org/10.1016/j.foodpol.2020.101978

Pokhriyal, N., & Jacques, D. C. (2017). Combining disparate data sources for improved poverty prediction and mapping. *Proceedings of the National Academy of Sciences, 114*(46). https://doi.org/10.1073/pnas.1700319114

Saarijärvi, H., Kuusela, H., Kannan, P. K., Kulkarni, G., & Rintamäki, T. (2016). Unlocking the transformative potential of customer data in retailing. *The International Review of Retail, Distribution and Consumer Research, 26*(3), 225–241. https://doi.org/10.1080/09593969.2015.1105846

Samoggia, A., Monticone, F., & Bertazzoli, A. (2021). Innovative digital technologies for purchasing and consumption in urban and regional agro-food systems: A systematic review. *Foods, 10*(2), 208. https://doi.org/10.3390/foods10020208

Shah, N., Srivastava, G., Savage, D. W., & Mago, V. (2020). Assessing Canadians health activity and nutritional habits through social media. *Frontiers in Public Health, 7,* 400. https://doi.org/10.3389/fpubh.2019.00400

Sibhatu, K. T., Krishna, V. V., & Qaim, M. (2015). Production diversity and dietary diversity in smallholder farm households. *Proceedings of the National Academy of Sciences, 112*(34), 10657–10662. https://doi.org/10.1073/pnas.1510982112

Sparling, T. M., White, H., Boakye, S., John, D., & Kadiyala, S. (2021). Understanding pathways between agriculture, food systems, and nutrition: An evidence and gap map of research tools, metrics, and methods in the last 10 years. *Advances in Nutrition, 12*(4), 1122–1136. https://doi.org/10.1093/advances/nmaa158

Steele, J. E., Sundsøy, P. R., Pezzulo, C., Alegana, V. A., Bird, T. J., Blumenstock, J., Bjelland, J., Engø-Monsen, K., de Montjoye, Y.-A., Iqbal, A. M., Hadiuzzaman, K. N., Lu, X., Wetter, E., Tatem, A. J., & Bengtsson, L. (2017). Mapping poverty using mobile phone and satellite data. *Journal of The Royal Society Interface, 14*(127), 20160690. https://doi.org/10.1098/rsif.2016.0690

Stirling, E., Willcox, J., Ong, K.-L., & Forsyth, A. (2021). Social media analytics in nutrition research: A rapid review of current usage in investigation of dietary behaviours. *Public Health Nutrition, 24*(6), 1193–1209. https://doi.org/10.1017/S1368980020005248

Subhi, M. A., Ali, S. H., & Mohammed, M. A. (2019). Vision-based approaches for automatic food recognition and dietary assessment: A survey. *IEEE Access, 7,* 35370–35381. https://doi.org/10.1109/ACCESS.2019.2904519

Takeshima, H., Akramov, K., Park, A., Ilyasov, J., Liu, Y., & Ergasheva, T. (2020). Agriculture–nutrition linkages with heterogeneous, unobserved returns and costs: Insights from Tajikistan. *Agricultural Economics, 51*(4), 553–565. https://doi.org/10.1111/agec.12571

Traka, M. H., Plumb, J., Berry, R., Pinchen, H., & Finglas, P. M. (2020). Maintaining and updating food composition datasets for multiple users and novel technologies: Current challenges from a UK perspective. *Nutrition Bulletin, 45*(2), 230–240. https://doi.org/10.1111/nbu.12433

Wijaya, D. R., Paramita, N. L. P. S. P., Uluwiyah, A., Rheza, M., Zahara, A., & Puspita, D. R. (2022). Estimating city-level poverty rate based on e-commerce data with machine learning. *Electronic Commerce Research, 22*(1), 195–221. https://doi.org/10.1007/s10660-020-09424-1

Willenbockel, D. (2012). Extreme Weather Events and Crop Price Spikes in a Changing Climate: Illustrative Global Simulation Scenarios. *Research Report Oxfam International,* p. 60. http://hdl.handle.net/10546/241338

Yasmeen, R., Fukagawa, N. K., & Wang, T. T. (2017). Establishing health benefits of bioactive food components: A basic research scientist's perspective. *Current Opinion in Biotechnology, 44,* 109–114. https://doi.org/10.1016/j.copbio.2016.11.016

Chapter 12
Big Data and Computational Social Science for Economic Analysis and Policy

Sebastiano Manzan

Abstract The goal of this chapter is to survey the recent applications of big data in economics and finance. An important advantage of these large alternative datasets is that they provide very detailed information about economic behaviour and decisions which has spurred research aiming at answering long-standing economic questions. Another relevant characteristic of these datasets is that they might be available in real time, a property that can be used to construct economic indicators at high frequencies. Overall, big alternative datasets have the potential to make an impact on economic research and policy and to complement the information used by governmental agencies to produce the official statistics.

12.1 Introduction

Computational social science (CSS) can be broadly defined as the area of the social sciences that makes computing power an essential tool to conduct the analysis. The field has a long tradition in economics that goes back to the 1970s when economists started to use computers to solve numerically economic models. Since then, there has been an exponential growth in applications as documented by four *Handbooks of Computational Economics* published between 1996 and 2018 (see Amman et al., 1996; Hommes & LeBaron, 2018; Schmedders & Judd, 2013; Tesfatsion & Judd, 2006). Computational economics can be broadly characterized in three main areas of activity: numerical methods to solve economic models, agent-based models, and computationally intensive techniques to analyse and model big datasets. The limited goal of this chapter is to provide an overview that focuses on the analysis and modelling of large datasets, while I refer to Fontana and Guerzoni (2023) in this Handbook for a review of agent-based models (ABM) and their use in economic problems. The availability of big data offers the possibility to investigate

S. Manzan (✉)
Zicklin School of Business, Baruch College, CUNY, New York, NY, USA
e-mail: sebastiano.manzan@baruch.cuny.edu

E. Bertoni et al. (eds.), *Handbook of Computational Social Science for Policy*,
https://doi.org/10.1007/978-3-031-16624-2_12

long-lasting questions using more detailed information about economic behaviour. In addition, these datasets allow to uncover new empirical facts that were not previously known due to lack of information.

What exactly is "big data" in the context of economic applications? It can be defined as datasets that require advanced computing hardware and/or software tools to conduct the analysis. One such tool is distributed computing that shares the processing of a task across several machines, instead of a single machine as typically done by economists. Examples of large datasets used in economic analysis are administrative data (e.g. tax records for the whole population of a country), commercial datasets (e.g. consumer panels), and textual data (e.g. such as Twitter or news data) just to mention a few. In some cases, the datasets are structured and ready for analysis, while in other cases (e.g. text), the data is unstructured and requires a preliminary step to extract and organize the relevant information. As discussed in Einav and Levin (2014), economists are still in the early stages of analysing big data and are learning from developments in other disciplines. In particular, there is renewed interest in machine learning (ML) algorithms after the early applications of the 1990s (Kuan & White, 1994). Varian (2014) discusses techniques that can be used to analyse large datasets.

How can big data contribute to a better understanding of the economy and to support policy? In the highly aggregate context of macroeconomic analysis, big data offer the opportunity to bring to light the heterogeneity in consumers and firms that is typically neglected in official statistics. The high granularity of big datasets can be exploited to construct indicators that are better designed to explain certain phenomena, for example, along a geographic or demographic dimension. In addition, many economic models make assumptions about deep behavioural parameters that are difficult to estimate without detailed datasets. An example is represented by the work of Chetty et al. (2014b) where individual information about the school performance of a child is matched to his/her path of future earnings derived from tax data of the Internal Revenue Service (IRS). In other situations, big data allow to measure quantities that we could not measure until now. A field that is benefiting from these alternative sources of data is development economics. For instance, Storeygard (2016) uses night-light satellite data to estimate the income of sub-Saharan African cities.

Another important dimension in which big data can contribute to economic analysis is by offering information that is not only more granular but also more frequent in the time dimension. At times when economic conditions are rapidly changing, policy-makers need an accurate measure of the state of the economy to design the appropriate policy response. An example is provided by the early days of the Covid-19 pandemic in March 2020 when policy-makers felt the pressure to act in support of the economy despite the lack of official statistics to measure the extent of the slump, as discussed by Barbaglia et al. (2022). Many relevant economic indicators are observed infrequently, such as gross domestic product (GDP) at the quarterly frequency and the unemployment rate and the industrial production index at the monthly frequency. In addition, these variables are released with delays that range from a few days to several months. For these reasons, big data have the potential to produce indicators of business conditions that are more accurate and timely.

More generally, private companies are amassing significant amounts of data that could be used to complement official statistics and inform economic policy. As discussed by Bostic et al. (2016), the approach of governmental agencies to produce official statistics is based, to a large extent, on consumer and business surveys. The approach guarantees the accuracy and the representativeness of the sample, although it comes at the cost of being an expensive and time-consuming exercise. Hence, the availability of alternative datasets offers the possibility of extracting information that can complement the evidence obtained from the surveys (a deeper analysis on the issue of the use of digital trace data and unconventional data in official statistics can be found in Signorelli et al., 2023).

However, we are also faced with a new set of issues regarding data governance and ethics issues as discussed by Taylor (2023) in this Handbook.

The chapter is organized as follows. I first review some of the recent work in economics and finance that leverages large datasets and emphasize the role of big data in allowing the researcher to conduct the analysis. I then draw some conclusions and discuss areas of potential development of the field.

12.2 Big Data in Economics

In this sect. 12.1, I discuss the main findings of recent applications of big data in economics and finance. I organize the discussion by *data source* with the intention to provide a more consistent review of the results. The goal of this section is not to be exhaustive, but rather to offer a concise overview of some of the main applications of big data to economics.

12.2.1 Administrative Data

Administrative data refer to data collected by governmental agencies as part of their mandate. As discussed by Card et al. (2010), the main advantages of administrative data, relative to surveys, are their large samples, the low attrition and non-response rates, and the small measurement error. In addition, administrative datasets are very detailed in terms of the information available regarding individuals. However, the researcher is confronted with significant challenges in conducting the analysis given the restricted access to the data. Typically, the researcher is required to provide the code to the government agency that actually conducts the analysis, slowing down significantly the development of the research project.

An influential paper using tax record data is Chetty et al. (2014). The goal of the paper is to investigate intergenerational mobility in the USA. They use a sample of 40 million children born between 1980 and 1982 and relate their income at age

30 to the parents' income. This administrative dataset represents a unique setting to evaluate intergenerational mobility since it provides a large sample going back to the 1980 and allows to link children and parents with very high accuracy.

Information from the Social Security Administration (SSA) is used in Kopczuk et al. (2010) to investigate income inequality and social mobility in the USA starting in 1937. They find that inequality decreased up to the early 1950s and increased steadily since then. In terms of social mobility, they show that it has been relatively constant over time, including at the top end of the income distribution.

Big administrative datasets are also used to evaluate educational attainments and teaching effectiveness. Dobbie and Fryer (2011) uses administrative data from the New York City Department of Education to evaluate the effect of charter school programmes on students' achievement. The evidence suggests that charter schools have a significant positive effect on improving the academic performance of poor children across several metrics. One of the possible explanations for these improvements is that the schools employ high-quality teachers. The issue of measuring the quality of the teachers and their impact on student performance is investigated by Chetty et al. (2014a) and Chetty et al. (2014b). They use a sample of one million students and match data from the school districts and tax records to track the evolution of earnings for the children in the sample. They find that measures of teacher's value added (VA), such as student's test scores, do not show a significant bias as proxies of teacher's quality. In addition, by matching students to their subsequent tax record, Chetty et al. (2014b) find that elementary school teachers with higher VA have a positive effect on college attendance and average earnings, among other measures.

Another source of administrative data is the credit register used by Jiménez et al. (2014) to evaluate the effect of monetary policy on bank's lending behaviour. The credit register records all loans and contracts between the public and the banking sector in a country. They show that a lower interest rate has the effect of increasing bank's risk-taking behaviour which leads to an increase in the supply of credit, in particular to more risky borrowers.

12.2.2 Financial Data

Financial transaction data represent a prominent source of big data in economic analysis. An early application is represented by Gross and Souleles (2002) that use a random sample of 24 thousand credit card accounts to investigate the effect on debt of changes to credit limits. Their results show that individuals respond to an increase in credit limits by borrowing more, in particular for those that started near the limit. Another more recent application using credit card data is Gallagher and Hartley (2017) that use a random 5% sample of individuals with credit history. They use hurricane Katrina as a natural experiment and find that households that lived in areas most affected by the flood experienced large reductions in debt, mostly due to the decline in home loan obligations. Horvath et al. (2021) use credit card data to evaluate the behaviour of consumers during the 2020 pandemic. They find that credit card spending and balances declined rapidly during March/April 2020, in particular in areas with the highest incidence of cases. The recovery in spending started in May 2020 with riskier borrowers leading the way relative to those with high credit score. Dunn et al. (2020) use daily credit card data to assess the geographical and sectoral impact of the pandemic on consumer spending. They show that their measure of spending closely proxy for the monthly retail trade official statistic, which demonstrates the benefit of using big data to monitor the economy in real time. A similar analysis is provided in Bodas et al. (2019) and Carvalho et al. (2020) for Spain.

Calvet et al. (2009) use administrative data on the asset holdings and demographic information of all taxpayers in Sweden. The aim of the paper is to evaluate the financial sophistication of households in avoiding investment mistakes, such as under-diversification, inertia in risk taking, and holding losing stocks while selling winning stocks. They find that households with higher wealth and education levels are more sophisticated and less prone to investment mistakes.

12.2.3 Labour Markets

Labour market statistics have historically been data-rich due to the direct involvement of government agencies in the administration of unemployment benefits. Recently, private companies have started collecting information about the labour market. Naturally, the question is the representativeness of these private datasets for the overall labour market and the US economy. Horton and Tambe (2015) is a recent survey of the various sources of alternative labour market data that have emerged in recent years and provide a detailed discussion of the advantages and disadvantages of using such data. Napierala and Kvetan (2023) in this Handbook provide a complementary analysis of the role that big data can play in the analysis of the evolution of job skills.

An example of the use of alternative labour market data for policy is provided by Cajner et al. (2019). They use payroll data from the private company ADP to construct employment measures similar to those constructed by the Bureau of Labor Statistics (BLS) using the Current Employment Statistics (CES). They find that the two measures of employment complement each other and jointly they provide information about the dynamics of the labour market. This is a very important contribution since it shows that alternative data can provide information that is complementary and highly correlated with official statistics. The additional advantage of these private data sources is that they are available at higher frequencies and allow the researcher to segment the sample geographically and by demographic characteristics. This benefit is discussed in Cajner et al. (2020) that shows the real-time behaviour of the weekly employment measure during the Covid-19 pandemic relative to the monthly official statistic from CES. Similar results are also obtained by Gregory and Zhu (2014).

12.2.4 Textual Data

An alternative source of data that is gaining interest in economics and finance is textual data. In this case, the goal is to use text from newspapers, speeches, company reports, and Twitter, among others, to construct measures that help understand economic behaviour or predict economic variables. Gentzkow et al. (2019) provide a recent overview of the work done so far.

An important source of text data is newspaper articles that might be considered a proxy for the information set available to the public when making an economic decision. An early paper is Tetlock (2007) that extracts sentiment from a column of *The Wall Street Journal* and finds that it is useful to predict daily returns of the aggregate market. Baker et al. (2016) aim at measuring economic and political uncertainty by counting the number of articles that contain a set of keywords associated with uncertainty. They show that their measure is highly correlated with measures of uncertainty. Other recent applications analyse news to construct proxies for economic sentiment (see Barbaglia et al., forthcoming; Larsen & Thorsrud, 2019; Shapiro et al., 2020; Thorsrud, 2020). Monitoring the sentiment of consumers and businesses has a long tradition in economics, and it is typically based on surveys. The contribution of these papers is to show that sentiment based on newspaper articles has a similar behaviour to survey-based sentiment. These indicators are found to have forecasting power for several macroeconomic variables that is incremental relative to the typical macroeconomic predictors (Barbaglia et al., forthcoming). Larsen and Thorsrud (2019) investigate the relation between news and consumer expectations and find that the topics extracted from the news contribute to explain the consumers' decision to update their inflation expectations.

Another line of research has investigated the role of communication in the implementation of monetary policy. Hansen and McMahon (2016) use the text of verbal and written communication by the Federal Reserve to understand its role in predicting economic variables. They find that the forward guidance embedded in the central bank statements is more relevant relative to the communication of the state of the economy. Hansen et al. (2018) investigate the role of increasing transparency in the central bank communication by analysing the internal deliberation of the policy-makers. They find that their communication patterns changed significantly after transparency was introduced.

The GDELT project[1] is another source of textual data that has been used in several applications. Consoli et al. (2021) use sentiment analysis to understand the dynamics of sovereign yields in Europe. Acemoglu et al. (2018) use GDELT to identify events of political and social unrest in Egypt and to evaluate their effect on stock returns.

A data source that is gathering momentum in economic and financial analysis is Twitter. Baker et al. (2021) use Twitter messages to construct a Twitter Economic

---

[1] More information about GDELT is available at https://www.gdeltproject.org/.

Uncertainty (TEU) indicator similar to the EPU indicator proposed by Baker et al. (2021) that is based on newspaper articles. Their results show that there is a very high correlation between TEU and EPU.

12.2.5   Mobile Phone Data

Mobile phone data represents an additional source of big data for economic analysis. This type of data is potentially very high dimensional since it tracks the location of a user over time. An economic application is represented by Blumenstock et al. (2015) that use mobile phone data to measure the socio-economic status of the caller. This is a particularly useful initiative for developing countries where official statistics are not very reliable and well-developed. Milusheva (2020) uses mobile phone data to track the effect of the movement of people from high-disease areas to low-disease areas on malaria spreading. A similar idea is developed in Iacus et al. (2020) that investigate the effect of the containment measures on the spreading of the Covid-19 virus. Their findings suggest that a measure of mobility constructed from mobile phone data is a highly accurate predictor of the initial spread of the virus in Italy and France.

12.2.6   Internet Data

The emergence of the internet has created the opportunity for researchers to collect online data to proxy for economic variables of interest (see Edelman, 2012, for a detailed discussion). An example is provided by the emergence of eBay as a marketplace for the exchange of goods that allowed economists to test market design mechanisms and to investigate the behaviour of bidders and sellers. An early paper is Bajari and Hortacsu (2003) that examine the empirical regularities of eBay auctions and estimate a model of bidding.

An area of intense recent work has been measuring social ties based on online platform, such as Facebook. Bailey et al. (2018a) discuss the construction of the Social Connectedness Index (SCI) which measures the friendship connections between Facebook users living in different geographical areas of the USA and abroad. An application of the SCI to explain the housing market is provided in Bailey et al. (2018b). They find that social connections contribute to explain the surge in house prices which they argue to be the result of the similarity of experience and expectations about the housing market.

Cavallo and Rigobon (2016) uses price data that are scraped from online stores to construct measures of inflation. These measures are found to track well the official statistics and have the advantage that can be calculated at high frequencies. Goolsbee and Klenow (2018) use a large dataset of e-commerce transactions to calculate the inflation rate. They find that during the period 2014–2017, the inflation rate was 3% lower relative to the official Consumer Price Index (CPI).

Another big dataset that has recently gained interest among economists is Google Trends. It represents a measure of the intensity of queries in the Google search engine regarding a set of keywords in a certain geographic area. The big data feature of Google Trends is that the time series for the search terms is the outcome of the aggregation across millions of queries by Google users around the world. Google Trends can be interpreted as a sentiment measure since it captures the public interest on a specific topic at a certain point in time. An early contribution using Google Trends is Choi and Varian (2012) that finds that including appropriately selected trends improves the accuracy of nowcasts for several economic variables. D'Amuri and Marcucci (2017) use job search-related queries to forecast the unemployment rate in the USA. Their results show that using Google Trends improves accuracy also relative to professional forecasters and are particularly accurate during turning points that are difficult to predict in real time. Castelnuovo and Tran (2017) construct an indicator that they call Google Trends Uncertainty (GTU) that aims at capturing Economic and Political Uncertainty (EPU) in the spirit of Baker et al. (2016) using series from Google Trends.

12.2.7 Other Data

An interesting application of seismic data to economics is represented by Tiozzo Pezzoli and Tosetti (2021). They use seismic data to identify vibrations produced by human activity, such as air and road traffic and manufacturing activity among others. They find that the indicator they construct is strongly correlated with several official measures of economic activity.

Another source of alternative big data is obtained from satellite images that are used in a variety of CSS applications. However, only recently, economists realized the potential of satellite image data for economic analysis. Donaldson and Storeygard (2016) and Gibson et al. (2020) provide overviews of the application of satellite data in economics and a primer on remote sensing.

Chen and Nordhaus (2011) use night-light satellite data to improve GDP measures for developing countries, which is particularly relevant when official statistics are missing. The paper shows that luminosity provides informational value that can help improve the accuracy of output measures. Galimberti (2020) performs a similar exercise with the focus on the forecasting ability of the measures of economic activity based on the luminosity data. The results indicate that these measures are useful to improve the accuracy of simple forecasting models, although country-specific models deliver better forecast performance relative to the pooled model. In a similar context, Hu and Yao (2021) propose an econometric methodology to use luminosity to improve GDP measures. Henderson et al. (2011) provides a detailed discussion of applications of night lights to measure national income, in particular in the case of developing economies. Another application of night-light data is represented by Storeygard (2016) that evaluates whether the distance of cities from a port influenced their growth in sub-Saharan African countries. The role of the satellite data in this case is to provide a measure of economic activity at the city level that are not otherwise available from official statistics.

12.3   Conclusion

The discussion in this chapter demonstrates how big data can be valuable to answer long-standing questions and to test the validity of economic assumptions. An illustration is the work with administrative data discussed earlier that shows the great potential of providing economic researchers access to these data, but highlights also the severe limitations of scaling up the availability of these data to a wider audience of users. Another challenge is represented by the fact that many of these alternative datasets are collected by private companies that might have low incentives to share the data with researchers. However, big data have a significant public role to play which calls for a framework that facilitates sharing of the information. An example of the public relevance of using big data is to produce real-time indicators of business conditions. In this respect, the collaboration between the Federal Reserve and the payroll processor ADP (Cajner et al., 2019) indicates how the private big dataset can complement the existing information provided by statistical agencies to support economic policy in real time. This collaboration is likely to set the path for more extensive partnerships between the private sector and statistical agencies. As argued in Bostic et al. (2016), the current model of the production of economic data is the domain of governmental agencies that are funding and running the collection of data, typically in the form of consumer and business surveys. This model is likely to evolve in the future as companies collect increasing amounts of economic data that are valuable, and most likely cheaper, to the production of official statistics.

Acknowledgments The author is grateful to Eleonora Bertoni, Matteo Fontana, Lorenzo Gabrielli, Serena Signorelli, Michele Vespe, and Luca Barbaglia of the Joint Research Centre of the European Commission for the helpful comments that have improved the organization and clarity of the chapter.

References

Acemoglu, D., Hassan, T. A., & Tahoun, A. (2018). The power of the street: Evidence from Egypt's Arab Spring. *Review of Financial Studies, 31*(1), 1–42.

Amman, H. M., Tesfatsion, L., Kendrick, D. A., Rust, J., Judd, K. L., Schmedders, K., Hommes, C. H., & LeBaron, B. D. (1996). *Handbook of computational economics: Agent-based computational economics* (Vol. 2). Elsevier.

Bailey, M., Cao, R., Kuchler, T., Stroebel, J., & Wong, A. (2018a). Social connectedness: Measurement, determinants, and effects. *Journal of Economic Perspectives, 32*(3), 259–80.

Bailey, M., Cao, R., Kuchler, T., & Stroebel, J. (2018b). The economic effects of social networks: Evidence from the housing market. *Journal of Political Economy, 126*(6), 2224–2276.

Bajari, P., & Hortacsu, A. (2003). The winner's curse, reserve prices, and endogenous entry: Empirical insights from ebay auctions. *Rand Journal of Economics, 34*, 329–355.

Baker, S. R., Bloom, N., Davis, S., & Renault, T. (2021). Twitter-derived measures of economic uncertainty. *Working paper*.

Baker, S. R., Bloom, N., & Davis, S. J. (2016). Measuring economic policy uncertainty. *Quarterly Journal of Economics, 131*(4), 1593–1636.

Barbaglia, L., Consoli, S., & Manzan, S. (forthcoming). Forecasting with economic news. *Journal of Business and Economic Statistics*.

Barbaglia, L., Frattarolo, L., Onorante, L., Pericoli, F. M., Ratto, M., & Pezzoli, L. T. (2022). Testing big data in a big crisis: Nowcasting under COVID-19. *International Journal of Forecasting*, https://doi.org/10.1016/j.ijforecast.2022.10.005

Blumenstock, J., Cadamuro, G., & On, R. (2015). Predicting poverty and wealth from mobile phone metadata. *Science, 350*(6264), 1073–1076.

Bodas, D., Garcia Lopez, J. R., Murillo Arias, J., Pacce, M. J., Rodrigo López, T., Romero Palop, J. d. D., Ruiz de Aguirre, P., Ulloa Ariza, C. A., & Valero Lapaz, H. (2019). Measuring retail trade using card transactional data. *Documentos de trabajo/Banco de España, 1921*.

Bostic, W. G., Jarmin, R. S., & Moyer, B. (2016). Modernizing federal economic statistics. *American Economic Review, 106*(5), 161–64.

Cajner, T., Crane, L., Decker, R., Hamins-Puertolas, A., Kurz, C. J., et al. (2019). Tracking the labor market with "Big Data". *Working paper*.

Cajner, T., Crane, L. D., Decker, R., Hamins-Puertolas, A., & Kurz, C. J. (2020). Tracking labor market developments during the Covid-19 pandemic: A preliminary assessment. *Working paper*.

Calvet, L. E., Campbell, J. Y., & Sodini, P. (2009). Measuring the financial sophistication of households. *American Economic Review, 99*(2), 393–98.

Card, D., Chetty, R., Feldstein, M. S., & Saez, E. (2010). Expanding access to administrative data for research in the United States. *Working paper*.

Carvalho, V. M., Hansen, S., Ortiz, A., Garcia, J. R., Rodrigo, T., Rodriguez Mora, S., & Ruiz de Aguirre, P. (2020). Tracking the COVID-19 crisis with high-resolution transaction data. *Working paper*.

Castelnuovo, E., & Tran, T. D. (2017). Google it up! a Google Trends-based uncertainty index for the United States and Australia. *Economics Letters, 161*, 149–153.

Cavallo, A., & Rigobon, R. (2016). The billion prices project: Using online prices for measurement and research. *Journal of Economic Perspectives, 30*(2), 151–78.

Chen, X., & Nordhaus, W. D. (2011). Using luminosity data as a proxy for economic statistics. *Proceedings of the National Academy of Sciences, 108*(21), 8589–8594.

Chetty, R., Friedman, J. N., & Rockoff, J. E. (2014a). Measuring the impacts of teachers I: Evaluating bias in teacher value-added estimates. *American Economic Review, 104*(9), 2593–2632.

Chetty, R., Friedman, J. N., & Rockoff, J. E. (2014b). Measuring the impacts of teachers II: Teacher value-added and student outcomes in adulthood. *American Economic Review, 104*(9), 2633–79.

Chetty, R., Hendren, N., Kline, P., & Saez, E. (2014). Where is the land of opportunity? the geography of intergenerational mobility in the United States. *Quarterly Journal of Economics, 129*(4), 1553–1623.

Choi, H., & Varian, H. (2012). Predicting the present with Google Trends. *Economic Record, 88*, 2–9.

Consoli, S., Pezzoli, L. T., & Tosetti, E. (2021). Emotions in macroeconomic news and their impact on the european bond market. *Journal of International Money and Finance, 118*, 102472.

D'Amuri, F., & Marcucci, J. (2017). The predictive power of Google searches in forecasting US unemployment. *International Journal of Forecasting, 33*(4), 801–816.

Dobbie, W., & Fryer, R. G. (2011). Are high-quality schools enough to increase achievement among the poor? Evidence from the Harlem Children's Zone. *American Economic Journal: Applied Economics, 3*(3), 158–87.

Donaldson, D., & Storeygard, A. (2016). The view from above: Applications of satellite data in economics. *Journal of Economic Perspectives, 30*(4), 171–98.

Dunn, A., Hood, K., & Driessen, A. (2020). Measuring the effects of the COVID-19 pandemic on consumer spending using card transaction data. *Working paper.*

Edelman, B. (2012). Using internet data for economic research. *Journal of Economic Perspectives, 26*(2), 189–206.

Einav, L., & Levin, J. (2014). Economics in the age of big data. *Science, 346*(6210), 1243089.

Fontana, M., & Guerzoni, M. (2023). Modeling complexity with unconventional data: Foundational issues in computational social science. In Bertoni, E., Fontana, M., Gabrielli, L., Signorelli, S., & Vespe, M. (Eds.), *Handbook of computational social science for policy.* Springer.

Galimberti, J. K. (2020). Forecasting GDP growth from outer space. *Oxford Bulletin of Economics and Statistics, 82*(4), 697–722.

Gallagher, J., & Hartley, D. (2017). Household finance after a natural disaster: The case of hurricane Katrina. *American Economic Journal: Economic Policy, 9*(3), 199–228.

Gentzkow, M., Kelly, B., & Taddy, M. (2019). Text as data. *Journal of Economic Literature, 57*(3), 535–74.

Gibson, J., Olivia, S., & Boe-Gibson, G. (2020). Night lights in economics: Sources and uses 1. *Journal of Economic Surveys, 34*(5), 955–980.

Goolsbee, A. D., & Klenow, P. J. (2018). Internet rising, prices falling: Measuring inflation in a world of e-commerce. *Aea Papers and Proceedings, 108*, 488–92.

Gregory, A. W., & Zhu, H. (2014). Testing the value of lead information in forecasting monthly changes in employment from the bureau of labor statistics. *Applied Financial Economics, 24*(7), 505–514.

Gross, D. B., & Souleles, N. S. (2002). Do liquidity constraints and interest rates matter for consumer behavior? evidence from credit card data. *Quarterly Journal of Economics, 117*(1), 149–185.

Hansen, S., & McMahon, M. (2016). Shocking language: Understanding the macroeconomic effects of central bank communication. *Journal of International Economics, 99*, S114–S133.

Hansen, S., McMahon, M., & Prat, A. (2018). Transparency and deliberation within the FOMC: A computational linguistics approach. *The Quarterly Journal of Economics, 133*(2), 801–870.

Henderson, V., Storeygard, A., & Weil, D. N. (2011). A bright idea for measuring economic growth. *American Economic Review, 101*(3), 194–99.

Hommes, C., & LeBaron, B. (2018). *Computational economics: Heterogeneous agent modeling.* Elsevier.

Horton, J. J., & Tambe, P. (2015). Labor economists get their microscope: Big data and labor market analysis. *Big Data, 3*(3), 130–137.

Horvath, A., Kay, B. S., & Wix, C. (2021). The Covid-19 shock and consumer credit: Evidence from credit card data. *Working paper.*

Hu, Y., & Yao, J. (2021). Illuminating economic growth. *Journal of Econometrics, 228*(2), 359–378.

Iacus, S. M., Santamaria, C., Sermi, F., Spyratos, S., Tarchi, D., & Vespe, M. (2020). Human mobility and covid-19 initial dynamics. *Nonlinear Dynamics, 101*(3), 1901–1919.

Jiménez, G., Ongena, S., Peydró, J.-L., & Saurina, J. (2014). Hazardous times for monetary policy: What do twenty-three million bank loans say about the effects of monetary policy on credit risk-taking? *Econometrica, 82*(2), 463–505.

Kopczuk, W., Saez, E., & Song, J. (2010). Earnings inequality and mobility in the United States: Evidence from social security data since 1937. *Quarterly Journal of Economics, 125*(1), 91–128.

Kuan, C.-M., & White, H. (1994). Artificial neural networks: An econometric perspective. *Econometric Reviews, 13*(1), 1–91.

Larsen, V. H., & Thorsrud, L. A. (2019). The value of news for economic developments. *Journal of Econometrics, 210*(1), 203–218.

Milusheva, S. (2020). Managing the spread of disease with mobile phone data. *Journal of Development Economics, 147*, 102559.

Napierala, J., & Kvetan, V. (2023). Changing job skills in a changing world. In Bertoni, E., Fontana, M., Gabrielli, L., Signorelli, S., & Vespe, M. (Eds.), *Handbook of computational social science.* Springer.

Schmedders, K., & Judd, K. L. (2013). *Handbook of computational economics.* Newnes.

Shapiro, A. H., Sudhof, M., & Wilson, D. J. (2020). Measuring news sentiment. *Journal of Econometrics, 228*(2), 221–243.

Signorelli, S., Fontana, M., Gabrielli, L., & Vespe, M. (2023). Challenges for official statistics in the digital age. In Bertoni, E., Fontana, M., Gabrielli, L., Signorelli, S., & Vespe, M. (Eds.), *Handbook of computational social science for policy.* Springer.

Storeygard, A. (2016). Farther on down the road: Transport costs, trade and urban growth in Sub-Saharan Africa. *Review of Economic Studies, 83*(3), 1263–1295.

Taylor, L. (2023). Data justice, computational social science and policy. Bertoni, E., Fontana, M., Gabrielli, L., Signorelli, S., & Vespe, M. (Eds.), *Handbook of computational social science.* Springer.

Tesfatsion, L., & Judd, K. L. (2006). *Handbook of computational economics: Agent-based computational economics.* Elsevier.

Tetlock, P. C. (2007). Giving content to investor sentiment: The role of media in the stock market. *Journal of Finance, 62*(3), 1139–1168.

Thorsrud, L. A. (2020). Words are the new numbers: A newsy coincident index of the business cycle. *Journal of Business & Economic Statistics, 38*(2), 393–409.

Tiozzo Pezzoli, L., & Tosetti, E. (2021). Seismonomics: Listening to the heartbeat of the economy. *Working paper.*

Varian, H. R. (2014). Big data: New tricks for econometrics. *Journal of Economic Perspectives, 28*(2), 3–28.

Chapter 13
Changing Job Skills in a Changing World

Joanna Napierala and Vladimir Kvetan

Abstract Digitalization, automation, robotization and green transition are key current drivers changing the labour markets and the structure of skills needed to perform tasks within jobs. Mitigating skills shortages in this dynamic world requires an adequate response from key stakeholders. However, recommendations derived from the traditional data sources, which lack granularity or are available with a significant time lag, may not address the emerging issues rightly. At the same time, society's increasing reliance on the use of the Internet for day-to-day needs, including the way individuals search for a job and match with employers, generates a considerable amount of timely and high granularity data. Analysing such nontraditional data as content of online job advertisements may help understand emerging issues across sectors and regions and allow policy makers to act accordingly. In this chapter, we are drawing on experience setting the Cedefop project based on big data and presenting examples of other numerous research projects to confirm the potential of using nontraditional sources of information in addressing a variety of research questions related to the topic of changing skills in a changing world.

13.1 Introduction

We live in the world where huge amount of data on almost any aspect of our life is produced and collected. Capturing, understanding and fully exploiting nontraditional data, through advanced analytics, machine learning and artificial intelligence, might yield benefits for policy makers. For example, the dynamic changes on the labour markets driven by digitalization, automation, robotization and green transition require adequate response of key players to mitigate skills shortages. The timely understanding of emerging issues across sectors and regions

J. Napierala (✉) · V. Kvetan
The European Centre for the Development of Vocational Training (Cedefop), Thessaloniki, Greece
e-mail: joanna.NAPIERALA@cedefop.europa.eu; Vladimir.Kvetan@cedefop.europa.eu

would allow policy makers to better manage strands of education and training (E&T) policy and better design labour market policies. The relevance of this issues is underlined by the related policy questions in a recently published European Commission policy report (European Commission, Joint Research Centre, 2022).

Below we will discuss how the analysis of nontraditional data might allow to address various questions, for which the granularity of traditional data was not adequate, such as:

- Which are the most demanded skills brought about by the recent changes in labour markets?
- Is there a gap between the supply and demand side with respect to skills?
- What is the best possible way to reskill and upskill individuals building on their current skill sets?
- Are "new skills" concentrated in specific economic sectors or regions?

Against this background the European Centre for the Development of Vocational Training (Cedefop) has taken up the challenge to integrate its work on skills intelligence making use of big data collected via web sourcing of online job advertisements.[1] Drawing on Cedefop's work and expertise, in this chapter we will focus on the presentation of existing research that based analysis on nontraditional data or applied data science or AI-based analytical approaches to better understand ongoing changes in skills.

13.2   Existing Literature

Traditionally, the labour market intelligence (LMI) was based on information collected via well-established surveys or administrative data. In the rapidly changing labour market, job seekers, teachers and trainers search for timely and more fine-grained information to support their decisions. As the recent society more and more relies on the Internet for the day-to-day needs, it naturally changes also the way employer-employee job matching process occurs.

With the growing number of employers who use websites to reach out for potential candidates and also the increase of users searching for jobs online, the analysis based on the information extracted from online job advertisements (OJAs) became the most promising approach for addressing some of the most relevant questions that new labour market trends are posing (Colombo et al., 2018). Potential of labour market systems based on big data lays in giving access to a greater variety of data sources producing information beyond ability of traditional survey. This allows for labour market comparisons at regional level and for subpopulations, as well as at the level of skills [2] rather than occupations. Yet, these systems are not

---

[1] https://www.cedefop.europa.eu/en/projects/skills-online-job-advertisements

[2] In this chapter we will not make distinction between skills and competences or knowledge, but the interested reader could refer to the discussions on this topic that are summarized in, e.g. paper by Rodrigues, M., Fernandez Macias, E., and Sostero, M. (2021). A unified conceptual framework of

free from shortcomings, e.g. not suitable for long-term projections, having limits in representativeness related to coverage or completeness and subject of missing data as a result of inconsistencies in unstructured text [see Naughtin et al. (2017) and Cedefop et al. (2021)].

Although these examples come from mainly one-off and exploratory research projects, which very often are based on one data source, they still confirm the capacity of using online job advertisements in a variety of analysis. The potential of this source of information to draw conclusions about labour market trends across multiple dimensions as occupation, geography, level of education and the type of contract was confirmed in the study by Tkalec et al. (2020). Moreover, there were various examples of efforts to identify skills for emerging jobs, for example, skill requirements for business and data analytics positions (Verma et al., 2019), for ICT and statistician positions (Lovaglio et al., 2018) and for software engineering jobs (Gurcan & Cagiltay, 2019; Papoutsoglou et al., 2019). Few studies focused on skills identification in specific sectors—IT (Ternikov & Aleksandrova, 2020), tourism (Marrero-Rodríguez et al., 2020) and manufacturing (Leigh et al., 2020)—or requested for specific occupations: computer scientist positions (Grüger & Schneider, 2019), various types of analyst positions (Nasir et al., 2020) or skills requested in the public health job (Watts et al., 2019). Alekseeva et al. (2019) searched online job advertisements data for terms related to artificial intelligence (AI) to understand which professions are demanding these skills. The jobs information collected over time allows identification of trends in the skill set requirements for different industries as done by Prüfer and Prüfer (2019), who provided insights into the dynamics of demand for entrepreneurial skills in the Netherlands and also identified professions for which entrepreneurial skills are particularly important. Fabo et al. (2017) analysed how important are foreign language skills in the labour markets of Central and Eastern Europe. Pater et al. (2019) analysed demand for transversal skills on Polish labour market. Dawson et al. (2021a) used longitudinal job advertisements data to analyse changes in journalists' skills and understand changes in the situation of this occupation group on the Australian labour market.

Although a growing number of employers use the web to advertise job openings, this data is still being criticized for being more skewed toward employers seeking more highly skilled professions or those more exposed to the Internet (Carnevale et al., 2014; Kureková et al., 2015b). Nevertheless, Beblavý et al. (2016) using Slovak job portals or Kureková et al. (2015a) using Czech, Irish and Danish publicly administered cross-European job search portal data delivered evidence on skills requested specifically in low- and medium-skilled occupations. Wardrip et al. (2015) focused on understanding employers' educational preferences studying medium-skilled job advertisements.

---

tasks, skills and competences. *JRC Working Papers Series on Labour, education and Technology, 2021/02.* https://ec.europa.eu/jrc/sites/default/files/jrc121897.pdf

The extraction of information on skills level allows to calculate different types of skills required (e.g. soft, transversal, digital) and therefore better understand how soft and hard skills influence each other which was analysed by Borner et al. (2018).

The potential of extending studies on labour market polarization by including the information on relevance of specific skills and skill bundles was indicated in few studies (Alabdulkareem et al., 2018; Salvatori, 2018; Xu et al., 2021). The skills-based approach to study possible transitions from lower-wage into better-paying occupations based on online job advertisements data was explored by Demaria et al. (2020). The rich structure of neural language models encourages researchers to make attempts in building more sophisticated models, e.g. that predicts wages from job postings' text [see Bana (2021)].

Building online job advertisements databases over extended periods of time allows introducing longitudinal perspective into the analysis. For example, Adams-Prassl et al. (2020) used job advertisements data to get more insights into determinants of employers' demand for flexible work arrangements. Blair and Deming (2020) analysed changes in demand for skills in the USA indicating that increase in the demand for graduates with bachelor's degree is of structural rather than cyclical nature. Shandra (2020) analysed trends in skills requirements of internship positions. Das et al. (2020) explored how occupational task demands have changed over the past decade due to AI innovation. Acemoglu et al. (2020) studied AI effects on the labour market, indicating the increase in demand for AI skills after 2014. Recently, the job advertisements data were used to study the impact of introduced social distancing measures during Covid-19 pandemic on the labour market in the EU (Pouliakas & Branka, 2020). Similar analysis carried out on the Covid-19 impact on labour market demand in the USA gave insights about state-level, for essential and non-essential sectors, and teleworkable and non-teleworkable occupations (Forsythe et al., 2020).

Using real-time labour market data can also bring valuable insight on the reasons of the low employability of graduates and help learners make informed decisions about acquisition of skills requested by employers. Persaud (2020) combined information extracted from job postings and from programmes offered by universities and colleges and identified what skills employers are seeking for big data analytics professions and to what extent these competencies are acquired by students. Universities may use AI-based analytics for mapping of competences from job adverts and compare them with curricula and course descriptions to better design future education offer (Ketamo et al., 2019). Borner et al. (2018) systematically analysed the interplay of job advertisement contents, courses and degrees offered and publication records to understand skills gaps and proposed not only methodology but also visualizations to ease making data-driven decisions by less tech-savvy stakeholders' groups. Brüning and Mangeol (2020) using job posting data analysed geographical differences in demand for graduates' skills in the USA. They tried to find the answers on what skills employers look when searching for graduates that did not follow the vocational career pathway. They looked also how open are employers in need of ICT specialist to hire graduates from other study fields (ibidem).

Although there is little evidence that individuals when receiving supporting information from job recommending tools change their job search behaviour (Hensvik et al., 2020) and also that such recommending tools are effectively decreasing skills mismatches on the labour market, the recent advances in artificial intelligence has spurred research contributions in the areas of career pathway planning, curriculum planning, job transition tools and software supporting job search. The advancements in extraction of information on skills based on online job advertisement lead to the proposals of solutions of the job searching tools allowing to filter recommendations by skill set and company attribute (e.g. size, revenue) (Muthyala et al., 2017) or proposal of models that can be used in building job prediction applications based on descriptions of user knowledge and skills (Van Huynh et al., 2020). There are also solutions being developed (based on vacancies information) that given starting sets of skills recommend job options which are not only matched with individual skill sets (Giabelli et al., 2020b, 2021) but also aligned with career ambitions and personal interests of job seeker (Sadro & Klenk, 2021).

The aggregation of data sources on skills with job advertisements data and education offer (e.g. from local university, database of online courses) allows for building solutions with more personalized career information, advice and guidance. Such recommendation tools have a matching solution with in-built information from existing sources on education offer that provides job seekers with information about potential career opportunities together with information what courses to take to acquire missing skills. At the same time, these tools account for the available time to learn new skills a job seeker/person interested in changing profession has (Sadro & Klenk, 2021). Recommendation tools powered by labour market information could be personalized even further, for example, to allow job opportunities to be filtered to match individuals' health requirements (Sadro & Klenk, 2021) or commuting expectations of job seeker (Berg, 2018; Sadro & Klenk, 2021).

Networking websites for matching workers with employers could serve as source of information to give more insights about demand but also about supply of skills. From the demand perspective, such data allows to retrieve additional company metadata to investigate the relationship between company characteristics and workers' skills (Chang et al., 2019). Information extracted from workers' career profiles allows to extend the analysis to differentiate skills from entry- to middle- to top-level jobs. Such data was used to test the effectiveness of proposed framework to predict career trajectory with in-built time variable that allows to account for different length of workers' experience (Wang, 2021). The networking websites may allow also for checking which job advertisements were visited more frequently and which information could be used, e.g. to improve analysis on the tightness of the labour market (Adrjan & Lydon, 2019).

The information about users' skills from online CV profiles could also be used as an input in career guiding tools as proposed by Ghosh et al. (2020) to support people in their decisions on which skills to acquire to achieve their career goals. The analysis of big data of real changes in careers allows getting more understanding into possibilities for intersectoral mobility (International Labour Organization, 2020). Natural language processing (NLP) solutions were applied to find overlapping skills between occupations based on which potential job transition was established (Kanders et al., 2020). Dawson et al. (2021b) built job transitions recommender system getting more insights of similar sets of skills by combining information from longitudinal datasets of real-time job advertisements and occupational transitions from a household survey. In the ongoing project (Cedefop, 2020), the analysis based on data about labour market transitions extracted from more than 10 million anonymized CVs from across EU member states was carried out to feed the recommendation tool. [3] This tool will support job seekers by providing them with information on occupations alternative to their own. Allowing the worker for identification of skills which acquisition would yield the highest utility gains could translate into the improvement of his/her employment outcomes and increase his/her productivity. Sun (2021) presented a new data-driven skill recommendation tool based on deep reinforcement learning solution that also allows to account for learning difficulty. Stephany (2021) combining information about freelancers' skills and wages calculated the marginal gains of learning a new skill. Insights from this study could help designing individual reskilling pathways and help to increase individuals' employability. There is ongoing feasibility study [4] which aims is to explore the potential of information extracted from work platforms that play intermediary role on the labour market, in better understanding of interplay between workers' skills, tasks and occupations.

13.3 Computational Guidelines

The growing body of knowledge on labour market generated based on the online sources translated into the increasing interest in taking advantage of the skills intelligence for policy making. In 2014, Cedefop started building a pan-European system to collect and classify online job advertisements data. The initial phase included only five EU countries. Yet, with time the project was scaled up and extended to the whole EU, including all 27 Member States + UK and all 24 official languages of the EU (Cedefop, 2019). This positive experience led Cedefop to join efforts with Eurostat (and creation of its Web Intelligence Hub) in developing well-documented data production system that has big data element integrated into the production of official statistics (Descy et al., 2019). Yet, the retrieval of good-quality and robust information from online data sources to deliver labour market analysis in an efficient way is still a challenging task. The identified key challenges in using online job advertisements (OJAs) for skills and labour market analysis are

---

[3] The open source codes and libraries are shared on the project website: https://cran.r-project.org/web/packages/labourR/index.html.

[4] It is part of Cedefop/Eurostat project titled "Towards the European Web Intelligence Hub—European System for Collection and Analysis of Online Job Advertisement Data (WIH-OJA)" carried out in the 2020–2024 period under the contract reference number: AO/DSL/VKVET-JBRAN/WIH-OJA/002/20.

representativeness, completeness, maturity, simplification, duplication and status of vacancies (International Labour Organization, 2020). The computational challenges with building reliable time series data based on collecting information from online data sources can be grouped into four areas related to:

- Data ingestion
- Deduplication
- Classification of occupations and skills
- Representativeness

When the focus is on the data ingestion and landscaping part, then the source stability is one of the main technical problems, which has a direct impact on the representativeness of collected information and the reliability of further analysis. Firstly, some sources of information might be blocked from data collection not allowing for extraction of information, and prior agreements with the website's owners will be needed to access the information. Secondly, some websites may not be available during the data extraction because of technical problems. Thirdly, there is also a natural lifecycle of the online sources as some new websites may appear while existing ones can close or rebrand. It has been shown that inclusion of the website that contained a large volume of spurious and anonymous job postings could lead to the discrepancy with the official vacancy statistics (International Labour Organization, 2020). In order to ensure stability of data sources, the added value of using tools like analytic hierarchy process to help in ranking of the online sources based on various dimensions, including information coverage, update frequency, popularity and expert assessment and validation, is explored. [5]

The challenges with deduplication relate to the fact that it is common that the same job advertisement appears in various sources on the Internet. This can happen either intentionally, when employer publishes OJA on more than one portal, or unintentionally due to activities of aggregators—portals that automatically crawl other websites with the view of republishing OJAs. Very often, the content of such job advertisements is almost identical differing only in a small portion of the text (e.g. date of release). There are several ways to allow for identification of near-duplicate job advertisement to avoid counting the same information multiple times, e.g. using bag of words, shingling and hashing techniques (Lecocq, 2015). In the process of deduplication, the comparison of several fields in the job advertisement (e.g. job title, name of employer, sector) is done to determine whether it is a duplicate or not. [6] Metadata derived from job portals is another way to help identifying duplicate advertisements (e.g. reference ID, page URL). In addition, machine learning algorithms could be used to remove irrelevant content, e.g. training offers.

---

[5] This is ongoing work carried out by Cedefop and Eurostat under WIH-OJA project mentioned earlier.

[6] It should be mentioned that information from deduplicated advertisements is used to enrich the final observation as additional information from across all sources is merged into one.

In the next phase of the data processing, the challenges relate to the classification of occupations, and skills emerge from the fact that the information is extracted from unstructured fields of job advertisements. For example, employers might have a tendency to conceal tacitly expected requirements by explicitly mentioning only a few skills from the list of required ones in online job advertisements. Similarly, the candidates building their online career profiles may signal only selected skills they have, for example, indication of "Hadoop" and "Java" could infer workers' expertise as well as for "MapReduce" (Muthyala et al., 2017). Sometimes the same word may have different meaning depending on the context, e.g. philosophy as the field of study or as the company philosophy, informal written guidelines on how people should perform and conduct themselves at work; Java could either come from the job advertisement searching for IT or coffee making person.

In general, two approaches are used in the information extraction from unstructured text: cluster analysis and classification (Ternikov & Aleksandrova, 2020). For example, Zhao et al. (2015) developed a system for skill entity recognition and normalization based on information from resumes, while Djumalieva and Sleeman (2018) used online job advertisements data and employed machine learning methods, such as word embeddings, network community detection algorithms and consensus clustering to build general skills taxonomy. In a similar way, Khaouja et al. (2019) created a taxonomy of soft skills applying combination of DBpedia and word embeddings and evaluated similarity of concepts with cosine distance. Moreover, a social network analysis was used to build a hierarchy of terms.

The unavailability of high-quality training datasets was believed to constrain advancements in the use of AI in extraction of information from unstructured text. Yet, it is observed that solutions based on structured and fully semantic ontological approaches or taxonomies proved to work better allowing to extract meaningful information from online data compared to applications exclusively based on machine learning approaches (International Labour Organization, 2020; Sadro & Klenk, 2021). Nevertheless, the taxonomy-based extraction processes are not free from deficiencies, as the quality of extracted information tends to be as good as the underlying taxonomies used for this purpose (Cedefop et al., 2021). Plaimauer (2018) studying matches between taxonomy terms and language used in vacancies published on Austrian labour market shows that 56% of the terms from taxonomy never appeared in job advertisements. She also observed that longer terms were identified with less frequency in the vacancies' descriptions. Grammatical cases in some language seem challenging for natural language processing tools, which often leads to misinterpretation of recognized skills (Ketamo et al., 2019).

The mapping of the unstructured text (e.g. of job titles, skills) to existing taxonomies (e.g. ISCO—International Standard Classification of Occupations) is usually done in a few steps, and pipelines are built for separate languages (Boselli et al., 2017). First the text needs to be extracted from the body of the job adverts;

this could be done by bag of word or Word2Vec approach [7] (Boselli et al., 2017). In both cases the usual steps were applied to preprocess the text. [8] The bag of word extraction leads to creation of sets of *n*-consecutive words (so called n-grams); usually unigrams or bigrams are analysed (ibidem). The Word2Vec extraction is based on replacement of each word in a title by a corresponding vector of n-dimensional space. This approach requires huge text corpora for producing meaningful vectors (ibidem). The corpuses with specific domain can significantly improve the quality of obtained word embeddings. In the next step of the classification pipeline, machine learning techniques (e.g. decision trees, naïve Bayes, K-nearest neighbour (k-NN), support vector machines (SVM), convolutional neural network) are applied to match with the "closest" code. The similarity is judged based on the value of one of the existing indexes of similarity (e.g. Cosine, Motyka, Ruzicka, Jaccard, Levenshtein distance, Sørensen-Dice index).

The evaluation of the quality of obtained matches (e.g. between job titles and occupation classifier) is not an easy task, although the problem related to matching itself is not a new one as previously some AI solutions were developed for coding of open answers on job titles provided by respondents in survey data (Schierholz & Schonlau, 2020).

Yet, the main difference between the information on job titles provided by individual worker and information originating from job titles mentioned in online job advertisements is that the latter includes more extraneous information (e.g. "ideal candidate", "involve regular travel") and tends to be more difficult to parse (Turrell et al., 2019). One way to validate that the occupation classifier is generating meaningful predictions is to check the implied occupational hierarchies (Bana et al., 2021). For example, a classifier that misclassifies a high-skilled profession with a low-skilled one would be judged as performing worse than the one that would categorize such occupation as belonging to more general category but within adequate hierarchical occupational group. Nevertheless, Malandri et al. (2021a) who applied word embeddings approach to job advertisements data identified existing mismatches in the taxonomy compared to real market examples. In particular, analysing the market of ICT occupations, they showed that although in ESCO taxonomy data engineer and a data scientist belonged to the same occupation group, these are not similar occupations in the real labour market (ibidem). The previous studies show that the level of accuracy of extracted information depends from field to field and also on the level of detail, as the accuracy rate of six-digit occupation coding was about 10 percentage points lower than when done for major groups at two-digit ISCO level (Carnevale et al., 2014). A similar trade-off between more granularity and less accuracy was observed by Turrell et al. (2019)

---

[7] Word2Vec is a technique that uses words' proximity to each other within the corpus as an indicator of relatedness. It creates a co-occurrence matrix that shows how often each word in the corpus is found within a "window" of other adjacent words.

[8] The usual steps were applied as *(i)* HTML tag removal, *(ii)* tokenization, *(iii)* lowercase reduction, *(iv)* stop words removal and *(v)* stemming, which will not be discussed in detail here.

who decided to use three-digit occupation classification. Yet, using supervised algorithms it was proven to be possible at least for English language to achieve good performances (over 80%) in classifying textual job vacancies gathered from the online advertisements with respect to the fourth-level ISCO taxonomy (Boselli et al., 2017). Nevertheless, less than 85% of titles were correctly classified in the matching exercise of job titles advertised on Dutch websites with ISCO-08 ontology (Tijdens & Kaandorp, 2019). The manual check of the unclassified terms showed that job titles in vacancies could be either more specialized compared to the terms in ontology or vice versa. However, some wrong classifications also occurred despite the high reliability score of classifications for these titles that included some similar words, e.g. *campaign manager* versus *camping manager* (ibidem). Another challenge with finding matching occupation classifier is that sometimes job advertisements can have generic, meaningless job titles or no title at all. Therefore, it is also important to design and train the classifiers that, e.g. could suggest a job title acknowledging the content of entire job description as, for example, the proposed Job-Oriented Asymmetrical Pairing System (JOAPS) by Bernard et al. (2020).

Overall, the main disadvantage of classifying unstructured information with use of taxonomies is that they are not forward-looking, and the frequency of revisions that oftentimes lean on expert panels and surveys allows updating them with the information on the emerging skills and/or occupations only with substantial delays. The AI solutions were introduced to update ESCO taxonomy with information on occupations; however the detailed information on the applied procedure was not provided in the official reports (European Commission, 2021a, b). A tool with capacity to automatically enrich the standard occupation and skills taxonomy with terms that represent new occupations was proposed by Giabelli et al. (2020b, 2021). This tool identifies the most suitable terms to be added to the taxonomy on the basis of four measures, namely, Generality, Adequacy Specificity and Comparability (GASC) (for formal definitions of these measures, see Giabelli et al. (2020a)). Very often inconsistencies in terminology used by job seekers and the jargon of employers when describing the same skills are the reasons for which the solution developers struggle when matching information from different data holders (Sadro & Klenk, 2021). One way to overcome the problem is to apply the AI and advanced linguistic understanding and build a platform which "translates" jargon of job advertisements to a simpler language for job seekers (Sadro & Klenk, 2021). The revealed comparative advantage (RCA) [9] was used as a measure of the importance of a skill for an individual job by Anna Giabelli et al. (2020a) to enrich ESCO taxonomy with real labour market-derived information about skills relevance and skills similarity. [10] Another AI-based methodology to refine taxonomy was proposed by Malandri et al. (2021b). The novelty of this approach is based on the

---

[9] For formula see Alabdulkareem, A., Frank, M. R., Sun, L., AlShebli, B., Hidalgo, C., and Rahwan, I. (2018, Jul). Unpacking the polarization of workplace skills. *Sci Adv, 4*(7), eaao6030. https://doi.org/10.1126/sciadv.aao6030

[10] Skills similarity was measured by Jaccard index.

automation of the process, which is to be carried out without involvement of experts. It is based on the implementation of domain-independent metric called hierarchical semantic similarity applied to judge the semantic similarity between new terms and taxonomic elements, which value is later used to evaluate the embeddings obtained from domain-specific corpus and, eventually, the suggestions on which new terms should be assigned to a different concept are made based on comparison of these evaluations. Chiarello et al. (2021) proposed a methodology that can be used to improve taxonomy. The innovation of this approach lays in the use of the natural language processing tools for knowledge extraction from scientific papers. The extracted terms are later linked with the existing ones allowing for identification of these which were not included in the taxonomy before.

As the final step before starting to analyse online data, it is crucial to explore its representativeness. The sources of potential bias in online job advertisements are multiple (B6ręsewicz & Pater, 2021). Moreover, the population of job vacancies and its structure are practically unknown, and for non-probability samples the traditional weighting cannot be used as an adjusting method (Kureková et al., 2015b). Researchers who recognize the problem of online data representativeness very often provide results of their analysis together with information from other data sources, i.e. representative surveys and registry data [e.g. Colombo et al. (2019)]. Beresewicz et al. (2021) suggest applying a combined traditional calibration with the LASSO-assisted approach to correct representation error in the online data.

13.4 The Way Forward

The aim of this chapter was to map the diversity of existing research projects that used big data and artificial intelligence approaches to research the topic of changing job skills in the changing world. It also tried to summarize the computational challenges related to extraction of information from online, unstructured data and the other issues that analysts using such data may struggle with. Based on the existing evidence, suggestions were made for the design of future research projects, which in this very vivid research area may already be addressed but were not identified by us in our mapping exercise.

Having said that, firstly, one needs to focus more on the understanding of the quality of applied classification methods. Although one ongoing project that investigates the quality of job title classifiers was identified [see (Bana et al., 2021)], the projects focusing on design and testing of some alternative approaches with outputs allowing to understand and improve the quality of existing solutions will be welcomed.

Secondly, the projects which aim at delivering comparable information across countries (e.g. Skills Online Vacancy Analysis Tool for Europe—OVATE [11]) would

---

[11] https://www.cedefop.europa.eu/en/data-visualisations/skills-online-vacancies

benefit from further research aiming at understanding the language characteristics' role in the extraction capacity of taxonomies or quality of these extractions. For example, the analysis on the number of extracted skills obtained in each OVATE language extraction pipeline shows huge variation across language pipelines. In general, translated version of ESCO taxonomy [12] from English to any other language used in EU countries brings lower number of extracted skills, but the reasons behind this are not known. Research projects with approach presented in Sostero and Fernández-Macías (2021) or similar approaches with other existing ontologies used as a benchmark or applied to other than English languages would be highly welcomed.

Thirdly, the use of artificial intelligence approaches in identification of new/emerging skills, which are not included in taxonomies, is another research area that requires more investment and knowledge building. The ongoing research tendered by Cedefop/Eurostat may bring some more understanding to this discussion, but other possibilities should also be explored, e.g. identifying gaps by merging taxonomy terms with the information extracted from academic journals [see Chiarello et al. (2021)].

Furthermore, the researchers using results of online data analysis to inform policy makers to be transparent about the potential biases should also include an explanatory methodological note on the representativeness of their data. The AI-based approaches to correct representation error in the online data are also a developing field in the researchers' discussions [see Beresewicz et al. (2021)].

Lastly, the various recommendation tools that appear on the market to offer help to job seekers are based on the similarities or overlap between skills of two occupations, and these similarities are very often calculated with use of similar techniques to classifying unstructured text with existing taxonomies [see Amdur et al. (2016) and Domeniconi et al. (2016)]. It would be worth researching and evaluating the quality of existing solutions and the suggested transitions offered to job seekers, especially that some recommendation tools do not account for hierarchical structure of skills or duration of learning time.

References

Acemoglu, D., Autor, D., Hazell, J., & Restrepo, P. (2020). *AI and jobs: Evidence from online vacancies*. NBER working paper series, 28257. https://www.nber.org/papers/w28257

Adams-Prassl, A., Balgova, M., & Qian, M. (2020). *Flexible work arrangements in low wage jobs: Evidence from job vacancy data*. IZA Institute of Labour Economics, Discussion Paper Series No. 13691. https://ftp.iza.org/dp13691.pdf

Adrjan, P., & Lydon, R. (2019). Clicks and jobs: measuring labour market tightness using online data. *Economic Letter, Vol. 2019*(No. 6). https://doi.org/https://www.centralbank.ie/docs/default-source/publications/economic-letters/vol-2019-no-6-clicks-and-jobs-measuring-labour-market-tightness-using-online-data-(adrjan-and-lydon).pdf?sfvrsn=6

---

[12] Version 1.0.

Alabdulkareem, A., Frank, M. R., Sun, L., AlShebli, B., Hidalgo, C., & Rahwan, I. (2018). Unpacking the polarization of workplace skills. *Science Advances, 4*(7), eaao6030. https://doi.org/10.1126/sciadv.aao6030

Alekseeva, L., Azar, J., Gine, M., Samila, S., & Taska, B. (2019). The demand for AI skills in the labor market. *Labour Economics.* https://doi.org/10.2139/ssrn.3470610

Amdur, B., Redino, C., & Ma, A. Y. (2016). *Using machine learning to measure job skill similarities.* https://www.datasciencecentral.com/profiles/blogs/using-machine-learning-to-measure-job-skill-similarities

Bana, S., Brynjolfsson, E., Rock, D., & Steffen, S. (2021). *job2vec: Learning a representation of jobs 4th IDSC of IZA workshop: Matching workers and jobs online - New developments and opportunities for social science and practice,* Online event.

Bana, S. H. (2021). *job2vec: Using language models to understand wage premia.* https://www.chapman.edu/research/institutes-and-centers/economic-science-institute/_files/ifree-papers-and-photos/sarah-bana-job2vec-wage-premia.pdf

Beblavý, M., Kureková, L. M., & Haita, C. (2016). The surprisingly exclusive nature of medium- and low-skilled jobs. *Personnel Review, 45*(2), 255–273. https://doi.org/10.1108/pr-12-2014-0276

Bereęsewicz, M., & Pater, R. (2021). *Inferring job vacancies from online job advertisements.* https://data.europa.eu/doi/10.2785/963837.

Beresewicz, M., Białkowska, G., Marcinkowski, K., Maslak, M., Opiela, P., Pater, R., & Zadroga, K. (2021). Enhancing the demand for labour survey by including skills from online job advertisements using model-assisted calibration. *Survey Research Methods, 15*(2), 147–167. https://doi.org/10.18148/srm/2021.v15i2.7670

Berg, A. M. (2018). *Combining learning analytics with job market intelligence to support learning at the workplace.* https://doi.org/10.1007/978-3-319-46215-8_8.

Bernard, T., Moreau, T., Viricel, C., Mougel, P., Gravier, C., & Laforest, F. (2020). Learning joint job embeddings using a job-oriented asymmetrical pairing system. *24th European Conference on Artificial Intelligence - ECAI 2020,* Santiago de Compostela. http://ecai2020.eu/papers/705_paper.pdf

Blair, P. Q., & Deming, D. J. (2020). *Structural increases in skill demand after the great recession.* NBER working paper series, 26680. https://www.nber.org/papers/w26680

Borner, K., Scrivner, O., Gallant, M., Ma, S., Liu, X., Chewning, K., Wu, L., & Evans, J. A. (2018). Skill discrepancies between research, education, and jobs reveal the critical need to supply soft skills for the data economy. *Proceedings of the National Academy of Sciences of the United States of America, 115*(50), 12630–12637. https://doi.org/10.1073/pnas.1804247115

Boselli, R., Cesarini, M., Marrara, S., Mercorio, F., Mezzanzanica, M., Pasi, G., & Viviani, M. (2017). WoLMIS: A labor market intelligence system for classifying web job vacancies. *Journal of Intelligent Information Systems, 51*(3), 477–502. https://doi.org/10.1007/s10844-017-0488-x

Brüning, N., & Mangeol, P. (2020). *What skills do employers seek in graduates? Using online job posting data to support policy and practice in higher education.* OECD Education Working Papers, 231. doi:https://doi.org/10.1787/bf533d35-en.

Carnevale, A. P., Jayasundera, T., & Repnikov, D. (2014). *Understanding online job ads data.* A technical report. M. S. o. P. P. Center on Education and the Workforce. https://cew.georgetown.edu/wp-content/uploads/2014/11/OCLM.Tech_.Web_.pdf

Cedefop. (2019). *Online job vacancies and skills analysis: A Cedefop pan-European approach.* http://data.europa.eu/doi/10.2801/097022

Cedefop. (2020). *Ex-ante publicity notice 'Curriculum Vitae (CV) data analytics & intelligence'.* https://www.cedefop.europa.eu/files/1._ex-ante_publicity_notice_cv_data_analytics_intelligence.pdf

Cedefop, European Commission, ETF, ILO, OECD, & UNESCO. (2021). *Perspectives on policy and practice: tapping into the potential of big data for skills policy.* Publications Office. https://doi.org/10.2801/25160

Chang, H.-C., Wang, C.-Y., & Hawamdeh, S. (2019). Emerging trends in data analytics and knowledge management job market: extending KSA framework. *Journal of Knowledge Management, 23*(4), 664–686. https://doi.org/10.1108/jkm-02-2018-0088

Chiarello, F., Fantoni, G., Hogarth, T., Giordano, V., Baltina, L., & Spada, I. (2021, 2021/12/01/). Towards ESCO 4.0 – Is the European classification of skills in line with Industry 4.0? A text mining approach. *Technological Forecasting and Social Change, 173*, 121177. https://doi.org/10.1016/j.techfore.2021.121177.

Colombo, E., Mercorio, F., & Mezzanzanica, M. (2018). Applying machine learning tools on web vacancies for labour market and skill analysis [Conference paper]. https://doi.org/https://techpolicyinstitute.org/wp-content/uploads/2018/02/Colombo_paper.pdf

Colombo, E., Mercorio, F., & Mezzanzanica, M. (2019). AI meets labor market: Exploring the link between automation and skills. *Information Economics and Policy, 47*, 27–37. https://doi.org/10.1016/j.infoecopol.2019.05.003

European Commission, Joint Research Centre. (2022). In M. Fontana, E. Bertoni, M. Vespe, L. Gabrielli, & S. Signorelli (Eds.), *Mapping the demand side of computational social science for policy: Harnessing digital trace data and computational methods to address societal challenges*. Publications Office. https://doi.org/10.2760/901622

Das, S., Steffen, S., Clarke, W., Reddy, P., Brynjolfsson, E., & Fleming, M. (2020). Learning occupational task-shares dynamics for the future of work. *Proceedings of the AAAI/ACM Conference on AI, Ethics, and Society*, New York, NY. doi:https://doi.org/10.1145/3375627.3375826.

Dawson, N., Molitorisz, S., Rizoiu, M.-A., & Fray, P. (2021a). Layoffs, inequity and COVID-19: A longitudinal study of the journalism jobs crisis in Australia from 2012 to 2020. *Journalism*. https://doi.org/10.1177/1464884921996286

Dawson, N., Williams, M. A., & Rizoiu, M. A. (2021b). Skill-driven recommendations for job transition pathways. *PLoS One, 16*(8), e0254722. https://doi.org/10.1371/journal.pone.0254722

Demaria, K., Fee, K., & Wardrip, K. (2020). *Exploring a skills-based approach to occupational mobility*. F. R. B. o. P. a. https://www.philadelphiafed.org/-/media/frbp/assets/community-development/reports/skills-based-mobility.pdf?la=en

Descy, P., Kvetan, V., Wirthmann, A., & Reis, F. (2019). Towards a shared infrastructure for online job advertisement data. *Statistical Journal of the IAOS, 35*, 669–675. https://doi.org/10.3233/SJI-190547

Djumalieva, J., & Sleeman, C. (2018). *An open and data-driven taxonomy of skills extracted from online job adverts* (ESCoE DP-2018-13). https://EconPapers.repec.org/RePEc:nsr:escoed:escoe-dp-2018-13

Domeniconi, G., Moro, G., Pagliarani, A., Pasini, K., & Pasolini, R. (2016). *Job recommendation from semantic similarity of LinkedIn users' skills*. ICPRAM.

European Commission. (2021a). *Leveraging Artificial Intelligence to update the ESCO Occupations Pillar. Report* – May 2021. S. A. a. I. Employment. https://ec.europa.eu/esco/portal/document/en/15388500-824d-4af6-8126-ab33c9495cc0

European Commission. (2021b). *Leveraging Artificial Intelligence to maintain the ESCO Occupations Pillar. Report* - April 2021. S. A. a. I. Employment. https://ec.europa.eu/esco/portal/document/en/ccc8d633-bfe1-4052-826e-613cd1202b60

Fabo, B., Beblavý, M., & Lenaerts, K. (2017). The importance of foreign language skills in the labour markets of Central and Eastern Europe: Assessment based on data from online job portals. *Empirica, 44*(3), 487–508. https://doi.org/10.1007/s10663-017-9374-6

Forsythe, E., Kahn, L. B., Lange, F., & Wiczer, D. (2020). Labor demand in the time of COVID-19: Evidence from vacancy postings and UI claims. *Journal of Public Economics, 189*, 104238. https://doi.org/10.1016/j.jpubeco.2020.104238

Ghosh, A., Woolf, B., Zilberstein, S., & Lan, A. (2020). Skill-based career path modeling and recommendation. *2020 IEEE International Conference on Big Data (Big Data)*.

Giabelli, A., Malandri, L., Mercorio, F., & Mezzanzanica, M. (2020a). GraphLMI: A data driven system for exploring labor market information through graph databases. *Multimedia Tools and Applications*. https://doi.org/10.1007/s11042-020-09115-x

Giabelli, A., Malandri, L., Mercorio, F., Mezzanzanica, M., & Seveso, A. (2020b). NEO: A tool for taxonomy enrichment with new emerging occupations. *International Semantic Web Conference.*

Giabelli, A., Malandri, L., Mercorio, F., Mezzanzanica, M., & Seveso, A. (2021). NEO: A system for identifying new emerging occupation from job ads. *The Thirty-Fifth AAAI Conference on Artificial Intelligence (AAAI-21).*

Grüger, J., & Schneider, G. (2019). Automated analysis of job requirements for computer scientists in online job advertisements. *Proceedings of the 15th International Conference on Web Information Systems and Technologies.*

Gurcan, F., & Cagiltay, N. E. (2019). *Big data software engineering: Analysis of knowledge domains and skill sets using LDA-based topic modeling.*

Hensvik, L., Le Barbanchon, T., & Rathelot, R. (2020). *Do algorithmic job recommendations improve search and matching? Evidence from a large-scale randomised field experiment in Sweden. 4th IDSC of IZA Workshop: Matching Workers and Jobs Online - New Developments and Opportunities for Social Science and Practice*, Online event.

International Labour Organization. (2020). *The feasibility of using big data in anticipating and matching skills needs* (978-92-2-032855-2).

Kanders, K., Djumalieva, J., Sleeman, C., & Orlik, J. (2020). Mapping career cause-ways: Supporting workers at risk. *A new system for supporting job transitions and informing skills policy in a changing labour market.* https://media.nesta.org.uk/documents/Mapping_Career_Causeways_01_G2XA7Sl.pdf

Ketamo, H., Moisio, M., Passi-Rauste, A., & Alamäki, A. (2019). Mapping the future curriculum: Adopting artificial intelligence and analytics in forecasting competence needs. *Proceedings of the 10th European Conference on Intangibles and Intellectual Capital ECIIC 2019*, Chieti-Pescara.

Khaouja, I., Mezzour, G., Carley, K. M., & Kassou, I. (2019). Building a soft skill taxonomy from job openings. *Social Network Analysis and Mining, 9*(1), 43. https://doi.org/10.1007/s13278-019-0583-9

Kureková, L. M., Beblavý, M., Haita, C., & Thum, A.-E. (2015a). Employers' skill preferences across Europe: Between cognitive and non-cognitive skills. *Journal of Education and Work, 29*(6), 662–687. https://doi.org/10.1080/13639080.2015.1024641

Kureková, L. M., Beblavý, M., & Thum-Thysen, A. (2015b). Using online vacancies and web surveys to analyse the labour market: A methodological inquiry. *IZA Journal of Labor Economics, 4*(1). https://doi.org/10.1186/s40172-015-0034-4

Lecocq, D. (2015). *Near-duplicate detection.* Moz Developer Blog. https://moz.com/devblog/near-duplicate-detection

Leigh, N. G., Lee, H., & Kraft, B. (2020). Robots, skill demand and manufacturing in US regional labour markets. *Cambridge Journal of Regions, Economy and Society, 13*(1), 77–97. https://doi.org/10.1093/cjres/rsz019

Lovaglio, P. G., Cesarini, M., Mercorio, F., & Mezzanzanica, M. (2018). Skills in demand for ICT and statistical occupations: Evidence from web-based job vacancies. *Statistical Analysis and Data Mining: The ASA Data Science Journal, 11*(2), 78–91. https://doi.org/10.1002/sam.11372

Malandri, L., Mercorio, F., Mezzanzanica, M., & Nobani, N. (2021a). MEET-LM: A method for embeddings evaluation for taxonomic data in the labour market. *Computers in Industry, 124.* https://doi.org/10.1016/j.compind.2020.103341

Malandri, L., Mercorio, F., Mezzanzanica, M., & Nobani, N. (2021b). *TaxoRef: Embeddings evaluation for AI-driven Taxonomy Refinement [Proceedings]. The European Conference on Machine Learning and Principles and Practice of Knowledge Discovery in Databases.* https://2021.ecmlpkdd.org/wp-content/uploads/2021/07/sub_917.pdf

Marrero-Rodríguez, R., Morini-Marrero, S., & Ramos-Henriquez, J. M. (2020). Tourism jobs in demand: Where the best contracts and high salaries go at online offers. *Tourism Management Perspectives, 35*, 100721. https://doi.org/10.1016/j.tmp.2020.100721

Muthyala, R., Wood, S., Jin, Y., Qin, Y., Gao, H., & Rai, A. (2017). *Data-driven job search engine using skills and company attribute filters [Proceedings]. 2017 IEEE International*

Conference on Data Mining Workshops (ICDMW) (pp. 199–206). doi:https://doi.org/10.1109/ICDMW.2017.33.

Nasir, S. A. M., Wan Yaacob, W. F., & Wan Aziz, W. A. H. (2020). Analysing online vacancy and skills demand using text mining. *Journal of Physics: Conference Series* (1496). doi:https://doi.org/10.1088/1742-6596/1496/1/012011.

Naughtin, C., Reeson, A., Mason, C., Sanderson, T., Bratanova, A., Singh, J., McLaughlin, J., & Hajkowicz, S. (2017). *Employment data ecosystem: Equipping Australians with the information they need to navigate the future labour market [Technical Report].* http://hdl.voced.edu.au/10707/523489.

Papoutsoglou, M., Ampatzoglou, A., Mittas, N., & Angelis, L. (2019). Extracting knowledge from on-line sources for software engineering labor market: A mapping study. *IEEE Access, 7,* 157595–157613. https://doi.org/10.1109/access.2019.2949905

Pater, R., Szkola, J., & Kozak, M. (2019). A method for measuring detailed demand for workers' competences. *Economics, 13*(1). https://doi.org/10.5018/economics-ejournal.ja.2019-27

Persaud, A. (2020). Key competencies for big data analytics professions: A multimethod study. *Information Technology and People, 34*(1), 178–203. https://doi.org/10.1108/itp-06-2019-0290

Plaimauer, C. (2018). Using vacancy mining for validating and supplementing labour market taxonomies. *Semantics conference*, Vienna.

Pouliakas, K., & Branka, J. (2020). *EU jobs at highest risk of Covid-19 social distancing. Is the pandemic exacerbating the labour market divide?* Cedefop Working Paper, no 1. doi:https://doi.org/10.2801/968483.

Prüfer, J., & Prüfer, P. (2019). Data science for entrepreneurship research: Studying demand dynamics for entrepreneurial skills in the Netherlands. *Small Business Economics, 55*(3), 651–672. https://doi.org/10.1007/s11187-019-00208-y

Rodrigues, M., Fernandez Macias, E., & Sostero, M. (2021). A unified conceptual framework of tasks, skills and competences. *JRC Working Papers Series on Labour, education and Technology, 2021/02.* https://ec.europa.eu/jrc/sites/default/files/jrc121897.pdf

Sadro, F., & Klenk, H. (2021). *Using labour market data to support adults to plan for their future career: Experience from the CareerTech Challenge.* https://learningandwork.org.uk/wp-content/uploads/2021/06/Using-Labour-Market-Data-to-Support-Adults-to-Plan-for-their-Future-Career.pdf

Salvatori, A. (2018). The anatomy of job polarisation in the UK. *Journal of Labour Market Research, 52*(1), 8. https://doi.org/10.1186/s12651-018-0242-z

Schierholz, M., & Schonlau, M. (2020). Machine learning for occupation coding—A comparison study. *Journal of Survey Statistics and Methodology.* https://doi.org/10.1093/jssam/smaa023

Shandra, C. (2020). What employers want from interns: Demand-side trends in the internship market. *SocArXiv.* https://doi.org/10.31235/osf.io/4mzbv

Sostero, M., & Fernández-Macías, E. (2021). The professional lens: What online job advertisements can say about occupational task profiles. *JRC Working Papers Series on Labour, education and Technology,* 2021/13. https://ec.europa.eu/jrc/sites/default/files/jrc125917.pdf

Stephany, F. (2021). When does it pay off to learn a new skill? Revealing the Complementary Benefit of Cross-Skilling. *SocArXiv.* https://doi.org/10.31219/osf.io/sv9de

Sun, Y. (2021). *Cost-effective and interpretable job skill recommendation with deep reinforcement learning*, Institute of Computing Technology, Chinese Academy of Sciences. https://www2021.thewebconf.org/

Ternikov, A., & Aleksandrova, E. (2020). Demand for skills on the labor market in the IT sector. *Business Informatics, 14*(2), 64–83. https://doi.org/10.17323/2587-814x.2020.2.64.83

Tijdens, K., & Kaandorp, C. (2019). *Classifying job titles from job vacancies into ISCO-08 and related job features - the Netherlands.* doi: https://doi.org/10.13140/RG.2.2.27133.72164.

Tkalec, M., Tomić, I., & Žilić, I. (2020). Potražnja za radom u Hrvatskoj: Indeks online slobodnih radnih mjesta [Labor Demand in Croatia: Online Vacancy Index]. *Ekonomski Pregled, 71*(5), 433–462. https://doi.org/10.32910/ep.71.5.1

Turrell, A., Speigner, B. J., Djumalieva, J., Copple, D., & Thurgood, J. (2019). *Transforming naturally occurring text data into economic statistics: The case of online job vacancy postings.* NBER working paper series, 25837. doi:https://doi.org/10.3386/w25837.

Van Huynh, T., Van Nguyen, K., Nguyen, N. L., & Nguyen, A. G. (2020). Job prediction: From deep neural network models to applications. *RIVF International Conference on Computing and Communication Technologies (RIVF).*

Verma, A., Yurov, K. M., Lane, P. L., & Yurova, Y. V. (2019). An investigation of skill requirements for business and data analytics positions: A content analysis of job advertisements. *Journal of Education for Business, 94*(4), 243–250. https://doi.org/10.1080/08832323.2018.1520685

Wang, C. (2021). *Variable interval time sequence modeling for career trajectory prediction: Deep collaborative perspective.* http://videolectures.net/www2021_wang_trajectory_prediction/

Wardrip, K., Fee, K., Nelson, L., & Andreason, S. T. (2015). *Identifying opportunity occupations in the nation's largest metropolitan economies.* https://www.clevelandfed.org/newsroom-and-events/publications/special-reports/sr-20150909-identifying-opportunity-occupations.aspx

Watts, R. D., Bowles, D. C., Fisher, C., & Li, I. W. (2019). Public health job advertisements in Australia and New Zealand: A changing landscape. *Australian and New Zealand Journal of Public Health, 43*(6), 522–528. https://doi.org/10.1111/1753-6405.12931

Xu, W., Qin, X., Li, X., Chen, H., Frank, M., Rutherford, A., Reeson, A., & Rahwan, I. (2021). Developing China's workforce skill taxonomy reveals extent of labor market polarization. *Humanities and Social Sciences Communications, 8*(1). https://doi.org/10.1057/s41599-021-00862-2

Zhao, M., Javed, F., Jacob, F., & McNair, M. (2015). *SKILL: A system for skill identification and normalization.* AAAI.

Chapter 14
Computational Climate Change: How Data Science and Numerical Models Can Help Build Good Climate Policies and Practices

Massimo Tavoni

Abstract Computational social science can help advance climate policy and help solve the climate crises. To do so, several steps need to be overcome to make the best use of the wealth of data and variety of models available to evaluate climate change policies. Here, we review the state of the art of numerical modelling and data science methods applied to policy evaluation. We emphasize that significant progress has been made but that critical social and economic phenomena—especially related to climate justice—are not yet fully captured and thus limit the predictivity and usefulness of computational approaches. We posit that the integration of statistical and numerical approaches is key to developing a new impact evaluation science that overcomes the traditional divide between ex ante and ex post approaches.

14.1 Introduction

Climate change is one of the defining societal and policy issues of our time. As climate impacts are already felt in societies and economies, governments around the world are mobilizing enormous resources to help reduce greenhouse gases and to adapt to climate impacts. Given the complexity of the mitigation and adaptation strategies, which involve a variety of different constituencies and span a broad technology spectrum, advanced research methods can help guide policies to be most effective, efficient, and inclusive.

Computational social science has already made significant contributions to climate policy evaluation and climate impact research. This is an area of high inter-disciplinarity that combines climate physical information with data representing human systems. Broadly speaking, methodologies have focused on either retro-

M. Tavoni (✉)
Politecnico di Milano and RFF-CMCC European Institute on Economics and the Environment, Milan, Italy
e-mail: massimo.tavoni@eiee.org

© The Author(s) 2023
E. Bertoni et al. (eds.), *Handbook of Computational Social Science for Policy*,
https://doi.org/10.1007/978-3-031-16624-2_14

spective or prospective assessments, though mixed approaches have also emerged. Retrospective, ex-post analysis has focused on evaluating the impact of climate policies and global warming on a variety of social and economic outcomes, such as economic productivity, inequality, labour market participation, social acceptance, etc. This strand of research has employed a variety of statistical approaches such as econometrics and machine learning to data either historically observed or purposefully generated (e.g. via surveys, or experimental trials). Prospective, ex ante approaches have tackled the issue of projecting the consequences of climate change and climate strategies into the future, often far distant ones given the inertia in the climate and economic systems. Here, methodologies have focused on numerical approaches such as optimization and simulation models, often integrating different components of the human and climate systems. Prominent examples include integrated assessment models (IAMs), energy system models, computable general equilibrium models, and agent-based models.

Both approaches have had a significant policy influence in Europe and elsewhere. The increased empirical recognition of the social and economic risks of climate change has helped to make climate a policy priority. Ex-post policy evaluation has helped improve our understanding of the functioning of public interventions and to improve them. Future scenarios of emissions and the energy and economy transformations compatible with decarbonization have had a major influence in determining the outcomes of international climate negotiations such as the Paris Agreement and informing society of the possible course of actions via international panels such as the Intergovernmental Panel on Climate Change (IPCC).

This chapter provides a succinct review of the role played by both statistical and numerical methods in climate change mitigation and adaptation research. It then discussed the policy implications of the research done so far on specific policy-relevant issues. Finally, it maps possible evolutions and future contributions of computational social sciences to help the impending fight against climate change.

14.2 Modelling the Climate Economy

14.2.1 Model Paradigms

Understanding the complex relationship between climate change, social and economic factors, and the needed transition in the emitting sectors such as agriculture and energy cannot be done without complex tools. Indeed, computational approaches have become the dominant paradigm for generating scenarios of future climate and of climate-resilient strategies. One class of models that is prominent in this field goes under the name of IAMs. As evident from the title, integrated assessment modelling is a general term that captures a variety of paradigms, often of a very different nature.

A general distinction has been made between benefit-cost and detailed process models (Weyant, 2017). Both model paradigms include greenhouse gas (GHG) emissions, compute their climate consequences, and feature technologies to mitigate and adapt to climate change. They have been used for decades to inform the design of climate policies. However, they have some fundamental differences which have set them apart, despite often being equated. Benefit-cost models have a relatively aggregated representation of the mitigation component but include the feedback of climate change on the economic system. The closed-loop formulation allows doing what the name suggests: to compute costs and benefits of climate action and to optimize the trade-off between the two, suggesting courses of action which are therefore economically optimal. This class of models originates from economists' role in early climate debates, such as the US National Academy of Science climate committee established in the early 1980s which included two future Nobel prize winners in economics, Tom Schelling and William Nordhaus. Nordhaus developed back then what has become a standard benefit-cost model, the DICE model (Nordhaus, 1994, 2008; Nordhaus & Boyer, 2000), for which he eventually won the prestigious award in 2018. DICE is a dynamic, non-linear optimization model based on the optimal growth framework of Ramsey-Cass-Koopmans, coupled with a simplified climate model and a very simple representation of emission reduction technologies. Despite the simplicity and reliance on standard, neo-classical approaches, the model has been used extensively by many scholars in different fields, thus becoming a classic still in use today even in regulatory work. Other benefit-cost IAMs are FUND (Tol 1997, www.fund-model.org) and PAGE (Hope, 2006).

In parallel to the development of benefit-cost models, a different approach emerged. This built on the work done in the 1970s to model energy systems in response to the oil price shocks, for example, by the Energy Modeling Forum, as well as by the establishment in the late 1980s of the IPCC. By the early 1990s, several detailed process models had been developed, and even a model comparison project on the economic costs of climate control was completed (Gaskins & Weyant, 1993), at the same time when structured model comparisons emerged in climate science (Smith et al., 2015). Disaggregated process-based models represent the underlying processes more explicitly than aggregate models: for example, mitigation technologies are represented in much greater detail, the climate components are based on intermediate complexity models calibrated upon large-scale climate models, and the economic sectors might be represented at higher granularity. This class of models includes simulation and optimization approaches but tends to focus on the evaluation of policies such as emission reduction ones rather than finding the optimal climate conditions for the economy. Over time, dozens of such models have been developed, whether global tools for understanding international climate policies such as the ones envisioned in the Paris Agreement or national or subnational tools to simulate local policies. An association has been established 14 years ago with the purpose of (Emmerling & Tavoni, 2021; Gambhir et al., 2019; Weyant, 2017) creating a community of scholars and practitioners focused on integrated modelling for climate change.[1]

In addition to these two broad classes of integrated assessment models, additional numerical approaches have been developed over the years. A large number of computable general equilibrium (CGE) models are now available, though not always classified as IAMs (Parrado, 2010; Rausch et al., 2011). These models, alongside dynamic stochastic general equilibrium (DSGE) ones, have a detailed representation of economic sectors and of their interaction. They are used for policy evaluation and not optimization, thus belonging to the detailed process category. For example, the European Commission regularly employs a CGE and a DSGE for the impact assessment of its policy proposals, including the ambitious Fit-for-55 policy package. Model paradigms that do not enforce equilibrium are also available and also used for policy appraisal. These include macro-econometric approaches as well as agent-based models applied to climate change (Keppo et al., 2021; Lamperti et al., 2018; Ma & Nakamori, 2009).

14.2.2   Modelling Relevance for Climate Policy

It is hard to underestimate the contribution of computational models to the climate change policy debate, whether it is about policy impact assessment or international negotiations. The reliance of scientific bodies such as the IPCC on model-generated scenarios and numbers is clear evidence of this process: from the less than 200 scenarios in the 4th assessment report of the IPCC, scenarios have grown to well over 1000 in most recent ones. The Paris Agreement agreed upon in 2015 was, for example, heavily influenced by the fifth assessment report and in particular by the results of integrated assessment models which simulated the implications of stabilizing temperature below 2 °C. The climate neutrality pledges recently announced by several major economies can be partly attributed to a sentence in the IPCC 1.5 special report which is the outcome of model-based evaluations: 'In model pathways with no or limited overshoot of 1.5°C, global net anthropogenic CO_2 emissions decline by about 45% from 2010 levels by 2030 (40–60% interquartile range), reaching net zero around 2050 (2045–2055 interquartile range)'.

Models have not just provided the timing of climate neutrality, which has become such a focal point for international climate policies. They have also depicted transformation pathways for the economic, energy, and land systems compatible with climate stabilization: most of the scenario work has indeed focused on cost-effective pathways meeting given climate targets. These constraints have been taken as given by policy, including temperature targets but progressively carbon budgets, which have emerged as a reliable climate metric from the climate science

---

[1] https://www.iamconsortium.org/

community (Allen et al., 2009). The integration of process-based models with climate science speaks of the multidisciplinary nature of mathematical modelling.

Models have also laid out different technological and behavioural pathways to net-zero emissions: though in all climate stabilization scenarios fossil fuels are phased out quite rapidly and replaced by renewable sources and energy demand measures, the combination of different technologies and behavioural changes can change substantially across pathways consistent with the Paris Agreement. For example, the same temperature goal can be achieved with different usage of CO_2 removal strategies. Although all scenarios compatible with 1.5 °C envisage some negative emission technologies, the timing and extent of removals vary across models and scenarios and have been the subject of intense academic debate (Fuss et al., 2014; Tavoni & Socolow, 2013). The extent of CO_2 removals is driven as much by techno-economic assumptions made in the models about the technologies as by normative hypothesis and scenario design. For example, the choice of the intertemporal discount rate is a well-known key parameter in integrated assessment modelling but mostly for benefit-cost optimization rather than cost-effective analysis of a given temperature target. The introduction into IAMs of negative emission strategies, however, has made this normative assumption relevant also for cost-effectiveness: by sharing the burden towards future generations, scenarios with high discount rates are characterized by a higher reliance on CO_2 removals (Emmerling et al., 2019). This example shows how normative judgments, often implicit in model formulations, matter for climate change (Saltelli et al., 2020). The consequences of these choices matter not just for academic purposes but also for policy design: the extent and need of negative emission technologies are now discussed in international policy such as in the revision of the Nationally Determined Contributions, as well as in national policies such as the EU Green Deal where a separate accounting of removals from standard emission reductions and their management into the Emission Trading Scheme is now debated (Rickels et al., 2021).

Computational impact assessments have also examined the social and economic consequences of climate policies. For example, the European Commission legislative proposals—including the Fit-for-55 and the mid-century strategies—have been vetted by a series of climate-energy-economy models, which have computed the repercussions for economic activity, employment, and other social dimensions. Table 14.1 (*EC, 'Policy scenarios for delivering the European Green Deal'*) reports the latest estimates for the increased emission reduction ambition recently announced by the European Commission for the GDP of Europe, as computed by three climate-energy-economy models (JRC-GEM-E3, E3ME, and E-QUEST). Although all models tend to agree on relatively small macroeconomic impacts of decarbonizing the European economy, it is worth noting that different models produce estimates of different signs, as well as that the results will depend on the details of the policy formulation. For example, the GEM-E3 model of the JRC suggests that the economy will slightly contract, whereas the E3ME by Cambridge Econometrics foresees a policy-induced economic expansion. The reason for this discrepancy is the underlying economic framework assumed in each model: GEM-E3 is a computational general equilibrium model which embeds assumptions about

Table 14.1 Macroeconomic implications in terms of EU GDP variations of implementing an emission reduction of 55% by 2030. Source: EC, 'Policy scenarios for delivering the European Green Deal'

Policy setup	• Lump sum transfers • Imperfect labour market • Free allocation ETS • Scope extension ETS • No carbon pricing non-ETS	• Tax recycling • Imperfect labour market • Free allocation ETS • Scope extension ETS • No carbon pricing non-ETS	• Tax recycling • Imperfect labour market • Free allocation ETS • Scope extension ETS • Carbon pricing non-ETS
JRC-GEM-E3	−0.39	−0.27	−0.27
Policy setup	• Lump sum transfers • Free allocation ETS • No carbon pricing non-ETS	• Tax recycling • Free allocation ETS • Carbon pricing non-ETS	• Tax recycling • Auctioning ETS • Carbon pricing non-ETS
E3ME	0.19	0.42	0.50
Policy setup	Lump sum transfers	Lower taxation low-skilled labour	Support green investing
E-QUEST	−0.29	0.00	0.13

relatively well-functioning markets. E3ME is a macro-econometric model which does not assume optimizing behaviour and full utilization of resources but is rather based on a simulation approach based on economic accounting matrices and historical relations which include, for example, voluntary and involuntary unemployment. E-QUEST is a micro-founded dynamic stochastic general equilibrium model. The choice of the European Commission to employ three models of different nature highlights the fact that when it comes to economic and social repercussions, the model paradigm choice is essential and it is hard to discriminate between good and bad models, contrary to physical models such as Global Circulation Models. Importantly for policy evaluation, different models can simulate different policy provisions: Table 14.1 shows just how relevant is the type of policies which will be implemented, for example, how carbon tax revenues will be used.

If economic policy consequences are hard to predict, social impacts are even more complicated and yet increasingly important in the objective of achieving a just transition. Economic and social inequalities, for example, are a major driver of policy acceptance and a crucial policy objective. Traditionally, integrated assessment models have not focused on inequality (Emmerling & Tavoni, 2021): however, this has now become a policy focus, and new work to expand models in order to address this request is ongoing (Gazzotti et al., 2021). The use of computational models for understanding behavioural responses has proven more difficult. As a result, models have prioritized technological, supply-side solutions over demand-side ones (Creutzig et al., 2018). This is because of the difficulties of portraying human behaviour into tractable mathematical formulations, and the traditional paucity of empirical evidence on how households respond to economic and behavioural interventions. As we will discuss later on, the empirical evidence has accumulated in recent years thanks to more robust statistical approaches and data with higher resolutions. This has opened up the possibility of using models, such as agent-based models, to better capture the behavioural responses to climate policies. Even standard integrated assessment models have developed and can now account for lifestyle changes necessary to achieve the low carbon transformation (van den Berg et al., 2019).

Besides climate mitigation policies, computational models have been used to compute the impacts of climate change and to design adaptation strategies. One policy-relevant application of IAMs, for example, has been to compute the social cost of carbon (SCC)—the monetized damages associated with an incremental increase in carbon emissions. The SCC is used to evaluate the cost-effectiveness of climate policies in the USA and has been traditionally computed using three benefit-cost IAMs. The economic valuation of climate impacts is important also in climate negotiations on the discussion of loss and damages. One major driving factor in the wide range of estimates of the SCC is the formulation and parametrization of the damage functions. The damage function originally used in simple benefit-cost models such as DICE has been criticized for lack of empirical basis and for recommending insufficient climate ambition (or equivalently for producing too low SCC). This highlights the importance of better integration of empirical and modelling approaches, a point on which we will return further in the chapter.

One area where computational models have indirectly contributed to policy assessment, both for mitigation and adaptation, is the generation of counterfactual emission scenarios. Evaluating policies ex ante requires first defining a world in which those policies are absent, as a reference over which to calculate the policy repercussions. This is a notoriously difficult task, given the uncertainties of predicting future outcomes but also the challenges of defining which trends and policies to include. For global climate issues, the workhorse of counterfactual emission scenarios is that of the Shared Socio-economic Pathways [SSPs (Riahi et al., 2017)]. The SSPs depict five possible scenarios of the future, with different demographic, economic, and technological trajectories, and consequent challenges for climate mitigation and adaptation challenges (O'Neill et al., 2013). The narratives span different evolutions of prosperity, inequality, and environmental degradation: they are characterized by both quantitative elements, such as population and GDP growth, and qualitative elements such as technological narratives. The SSPs have been simulated by five IAMs, which have produced the resulting emission trajectories, and consequent climate outcomes. These have been used by several other scientific communities, most notably the climate science and climate impact ones. They also have had policy repercussions, for example, on the social cost of carbon.

14.2.3 Challenges in Using Integrated Assessment Models to Inform Societal Change

So far, we have highlighted the growing relevance of computational mathematical models for the prospective evaluation of climate policies. Climate strategies' repercussions for societies, households, and businesses are now routinely quantified using numerical models. Although this speaks of the growing importance of computational sciences in the climate domain, the increased reliance on structured approaches has not come without problems. For example, quantitative approaches have been condemned for exploring only a narrow set of possible futures and not keeping track of the rapid evolution of the climate technology and policy context. Most models implicitly represent value judgments and social preference and have been criticized for not exploring these normative assumptions, which are also plagued by uncertainties (MacAskill, 2016).

The modelling community has responded in various ways to the challenges of underplaying or not representing the full range of uncertainties. For example, global sensitivity analysis (Razavi et al., 2021) is a well-proven approach to ensure model-generated results are robust. It has been used in the context of large-scale climate-energy models only to a limited extent (Butler et al., 2014; Marangoni et al., 2017). An additional strategy to deal with model uncertainty has been a coordinated community response based on multi-model ensembles. Multi-model ensembles provide a range of plausible outcomes from a set of harmonized assumptions (i.e. given carbon budgets or temperature targets, still within this century). The ensemble spread is typically used to quantify model uncertainty. Although this process appears unambiguous, it can deceptively be so, since it is based on the assumption that models constitute independent estimates (Abramowitz et al., 2019; Merrifield et al., 2020). Selection and availability biases determine the typology and number of models involved in the comparison project and model dependencies inherent in the community work. Uncertainty is influenced by choices made in the model comparison project construction (Knutti et al., 2010). For example, ensemble members are not independent: they have historically shared code, use similar parametrization, and—an issue that is especially important for models examining socio-techno-economic transitions—belong to similar paradigms. The implicit normativity of climate-energy-economy models further contributes to compounding the sources of uncertainties and model relations.

The challenges in formulating policy recommendation from a vast number of scenarios and models, which often disagree, have led the policy community to often embrace simpler approaches based on few, representative scenarios. This, for example, has become the standard approach of IPCC in presenting its scenario space in an accessible way. International organizations, such as the International Energy Agency, typically present very few scenarios that become standard ones. However, the problem of the plausibility of scenarios and of the associated uncertainties is not solved by reducing the scenario space arbitrarily, unless statistical valid approaches are employed, which is typically not the case. This has important policy ramifications: for example, impact studies typically take the SSP5-RCP8.5 as a benchmark scenario, despite this being a relatively extreme one which was judged as a low probability by the scenario community itself (Ho et al., 2019). Advancements in statistical approaches and in behavioural science can be used towards not only more robust empirical evidence of policy effectiveness but also to make scenarios credible and insightful, something we turn to in the next section.

14.3   Data Science for Climate Impacts and Policy

The contribution of computational social sciences to climate change is not limited to numerical modelling. Actually, some of the most important contributions of the literature with direct or indirect policy repercussions have come from empirical and statistical approaches.

14.3.1   Data-Driven Approaches for Climate Economics

The fact that climate change has been changing, in addition to the natural and large variability in weather patterns, and the fact that climate and energy policies have been slowly but gradually deployed have provided previous information to scholars interested in the causal relationship between climate and its solutions and high-stakes social and economic issues. The growth of observations not just on climate outcomes but also on social and economic ones at high spatial resolution has provided sufficient statistical power to conduct innovative empirical research.

Several approaches have been used to infer causal climate-socio-economic linkages. Panel data econometrics, for example, has been applied to understand the impacts of climate change on a large set of outcomes, as discussed below. Other econometric approaches such as difference in difference, matching, and regression discontinuity designs have been used to infer causal relationships in the absence of an exogenous variation to be exploited. Standard regression approaches have been used where counterfactual randomization was ensured, such as in randomized controlled trials. Finally, machine learning methods are increasingly used to understand and promote sustainable policies: for example, machine learning has been applied to satellite imagery, whose increased abundance and resolution can provide crucial information on sustainability in areas of the world where data is scarce (Burke et al., 2020) and where climate change impacts are also more likely to occur. Novel algorithms have also been used to better understand energy usage patterns, for example, in the residential and transportation sectors where high-frequency information is now available, and to study policies to motivate behavioural and technological changes towards a greener society.

14.3.2   Relevance of Empirical Methods for Climate Policy

Empirical methods are key for understanding policy effectiveness and environmental social and economic disruptions. They are also needed in order to calibrate prospective impact assessments. Over the past few years, empirical studies have greatly advanced the understanding of climate change impacts and of the policies meant to address them.

One major area has been in the quantification of climate social and economic impacts. Traditionally, the climate impact functions used in benefit-cost analysis and for the social cost of carbon were based on prospective studies which raised issues of replicability and transparency. Over the course of the past 10 years, a wealth of data-driven approaches have highlighted the relationship between historical weather variability and many outcomes (Carleton & Hsiang, 2016). For example, temperature heat induces mortality and has been connected to aggression and violence. Agriculture and crop yields are related to temperature in a strongly non-linear way, with yields dropping when temperature exceeds certain thresholds. This non-linear relationship has been documented also for energy demand (Auffhammer et al., 2017), with peak demand rising when temperatures are high.

On the economic side, temperature variability has been associated with significant macroeconomic repercussions. The identification of a non-linear relationship between temperature and economic growth has highlighted how climate impacts can persistently slow economic progress (Burke et al., 2015; Dell et al., 2012; Kalkuhl & Wenz, 2020). This view is in stark contrast to the previously assumed relations which were based on the levels and not the growth of the economy. The consequences of this new empirical evidence have been particularly prominent in the benefit-cost assessments of climate policies and in the calculations of the social cost of carbon. Once the empirically derived damage functions were plugged into the IAMs, the recommendations for policy stringency changed dramatically, and for the first time, it appeared that stabilizing climate change within the goals of the Paris Agreement made global economic sense (Gazzotti et al., 2021; Glanemann et al., 2020; Hänsel et al., 2020). Similarly, the social cost of carbon—a policy-relevant metric for setting policy in the USA—increased substantially over previously available estimates (Ricke et al., 2018).

The data science advanced on the economics of climate change impact also highlighted the major economic inequalities brought about by climate change. These economic inequalities are detectable already today (Diffenbaugh & Burke, 2019) and are forecasted to persist even in case of ambitious emission reductions, and even more in the absence of cooperation, as shown in Fig. 14.1. Although the extent of persistency of climate economic impacts is a subject of intense academic debate (Piontek et al., 2021), the accumulated evidence has shown the importance of reducing emissions as fast as possible, preparing adaptation systems, and considering additional climate interventions such as CO_2 removals, to avoid temperature overshoots and consequent social and economic repercussions.

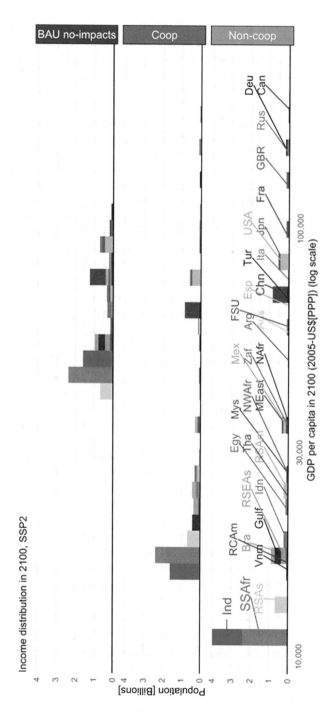

Fig. 14.1 GDP per capita, net of costs and impacts, population-weighted distribution in 2100 for three climate scenarios (BAU without climate impacts, global cooperation and noncooperation). From Gazzotti et al. (2021)

Another area of data-driven research which has important ramifications for policy design is that of behavioural science and of the experimental economics literature quantifying traditional and behavioural interventions. Behavioural sciences have consistently shown how human behaviour is fraught by biases, but also that several of these can be predicted, and thus partly addressed (Ariely, 2010). Many governments around the world have promoted the use of behavioural informed public policies, including but going beyond the use of 'nudges' (Banerjee et al., 2021). Methodologically, disentangling the impact of policy interventions, including behavioural ones, on outcome variables is difficult. Confounding factors such as exogenous trends (e.g. in energy prices, preferences, etc.) and self-selection (e.g. environmentally sensitive households more likely to enrol in pro-environmental programmes) have traditionally made it difficult to quantify the causal impact of policies. However, the embracing of statistical approaches based on counterfactual randomization, such as laboratory, online, and field experiments, has opened up the possibility to test for causality of policy interventions. Randomized controlled trials have been now done on millions of households and have helped evaluate a variety of interventions, such as information provision, message framing, social comparisons, monetary, and symbolic incentives (Allcott, 2011; Allcott & Mullainathan, 2010; Bonan et al., 2020, 2021; Ferraro & Price, 2013; Fowlie et al., 2015). These interventions have been assessed using reliable metrics of energy usage (and consequent GHG emissions) such as actual metered electricity consumption, thus providing a reliable line of evidence. The main results of this stream of computational social science have shown that behavioural interventions, if properly designed and implemented, can lead to small but significant energy and emission reductions. However, their effectiveness is context-dependent and varies significantly across population subgroups. As such, these policy instruments should complement but not substitute traditional interventions, including infrastructural and incentive-based ones.

The potential of data science to inform climate policymaking is enormous, but it has not yet realized its full capacity and should anticipate possible critiques which might emerge in the future. In terms of potential applications, data-driven approaches can help inform local decisions and design climate-resilient infrastructures at the local level. Cities are places that abound both in data and emissions, and where well-designed infrastructural policies can promote lifestyles that are both sustainable and inclusive (Creutzig et al., 2018). Data can be used to transform mobility services and reduce congestion and pollution. Some local institutions have already begun using high-resolution data for public purposes related to sustainable planning, but only limited potential has concretized. One concern with such an extension of data-driven urban policies regards the question of data equity and privacy. The way data is handled when it comes to policymaking is as crucial as the actual policies which will derive from it: data availability is often skewed towards certain sociodemographic areas and population subgroups, and resulting policies need to ensure to go beyond pre-existing social arrangements. Furthermore, the question of privacy which has become a central element of regulatory design needs to be accounted for when relying on data-driven impact evaluation.

14.4 Towards an Integrated Computational Approach

Overall, we have highlighted that computational approaches—both model-based and data-driven—have played an increasingly important role in climate change policies, both for mitigation and adaptation. Computational approaches have become ubiquitous, and policymaking is now heavily dependent on them, whether it is for determining the impacts of proposed legislation or of already implemented one. However, in order to serve society well, mathematical and statistical modelling should be accompanied by an epistemic strengthening of the underlying theoretical basis, empirical validity, and scientific practices (Saltelli et al., 2020).

One focus area for climate-purposed computational approaches is that of integrating data and model-driven approaches. Traditionally, these two approaches have been used to look at retrospective and prospective policy assessment, respectively. This rigid division of labour needs to be overcome if we want to have policy appraisal which can be learned from actual experiences: the growing number of energy and climate policies being tested in real-world conditions can now provide important information for calibrating models and made them more policy-relevant. Furthermore, the growing availability of high-resolution data such as those from satellite imagery, social media, and high-frequency metered energy and environmental indicators can be harnessed to understand behavioural policy responses at a high level of granularity. Machine learning approaches can then be combined with model inputs and output to increase the understanding of the model-embedded processes and to better predict policy responses. Finally, model validation and adequate exploration of the uncertainties should become scientific practices fully integrated in mathematical modelling. Computational approaches to do that effectively are now available, and their properties are well known (Razavi et al., 2021), and yet they are often not done (Saltelli et al., 2019). This speaks of the importance of a tighter and regulated relationship between researchers and policymakers, with clear guidance from policy evaluation agencies on scientific practices and robust methodological approaches. If crafted properly in a coordinated and co-designed manner, computational social science can be of tremendous value to climate policy-making and help accelerate the climate transition and ensure it is carried out in a just and inclusive way.

References

Abramowitz, G., Herger, N., Gutmann, E., Hammerling, D., Knutti, R., Leduc, M., Lorenz, R., Pincus, R., & Schmidt, G. A. (2019). ESD reviews: Model dependence in multi-model climate ensembles: Weighting, sub-selection and out-of-sample testing. *Earth System Dynamics, 10*(1), 91–105. https://doi.org/10.5194/esd-10-91-2019

Allcott, H. (2011). Social norms and energy conservation. *Journal of Public Economics, 95*(9–10), 1082–1095. https://doi.org/10.1016/j.jpubeco.2011.03.003

Allcott, H., & Mullainathan, S. (2010). Behavior and energy policy. *Science, 327*(5970), 1204–1205.

Allen, M. R., Frame, D. J., Huntingford, C., Jones, C. D., Lowe, J. A., Meinshausen, M., & Meinshausen, N. (2009). Warming caused by cumulative carbon emissions towards the trillionth tonne. *Nature, 458*(7242), 1163–1166. https://doi.org/10.1038/nature08019

Ariely, D. D. (2010). *Predictably irrational, revised and expanded edition: The hidden forces that shape our decisions* (Revised and Expanded ed. edition). Harper Perennial.

Auffhammer, M., Baylis, P., & Hausman, C. H. (2017). Climate change is projected to have severe impacts on the frequency and intensity of peak electricity demand across the United States. *Proceedings of the National Academy of Sciences of the United States of America, 114*(8), 1886–1891. https://doi.org/10.1073/pnas.1613193114

Banerjee, A., Chandrasekhar, A. G., Dalpath, S., Duflo, E., Floretta, J., Jackson, M. O., Kannan, H., Loza, F. N., Sankar, A., Schrimpf, A., & Shrestha, M. (2021). *Selecting the most effective nudge: Evidence from a large-scale experiment on immunization* (Working Paper No. 28726; Working Paper Series). National Bureau of Economic Research. doi: 10.3386/w28726.

Bonan, J., Cattaneo, C., d'Adda, G., & Tavoni, M. (2020). The interaction of descriptive and injunctive social norms in promoting energy conservation. *Nature Energy, 5*(11), 900–909. https://doi.org/10.1038/s41560-020-00719-z

Bonan, J., Cattaneo, C., d'Adda, G., & Tavoni, M. (2021). Can social information programs be more effective? The role of environmental identity for energy conservation. *Journal of Environmental Economics and Management, 108*, 102467. https://doi.org/10.1016/j.jeem.2021.102467

Burke, M., Hsiang, S. M., & Miguel, E. (2015). Global non-linear effect of temperature on economic production. *Nature, 527*(7577), 235–239. https://doi.org/10.1038/nature15725

Burke, M., Driscoll, A., Lobell, D., & Ermon, S. (2020). *Using satellite imagery to understand and promote sustainable development* (Working Paper No. 27879; Working Paper Series). National Bureau of Economic Research. doi: https://doi.org/10.3386/w27879.

Butler, M. P., Reed, P. M., Fisher-Vanden, K., Keller, K., & Wagener, T. (2014). Identifying parametric controls and dependencies in integrated assessment models using global sensitivity analysis. *Environmental Modelling and Software, 59*, 10–29. https://doi.org/10.1016/j.envsoft.2014.05.001

Carleton, T. A., & Hsiang, S. M. (2016). Social and economic impacts of climate. *Science, 353*(6304), aad9837. https://doi.org/10.1126/science.aad9837

Creutzig, F., Roy, J., Lamb, W. F., Azevedo, I. M. L., de Bruin, W. B., Dalkmann, H., Edelenbosch, O. Y., Geels, F. W., Grubler, A., Hepburn, C., Hertwich, E. G., Khosla, R., Mattauch, L., Minx, J. C., Ramakrishnan, A., Rao, N. D., Steinberger, J. K., Tavoni, M., Ürge-Vorsatz, D., & Weber, E. U. (2018). Towards demand-side solutions for mitigating climate change. *Nature Climate Change, 8*(4), 268–271. https://doi.org/10.1038/s41558-018-0121-1

Dell, M., Jones, B. F., & Olken, B. A. (2012). Temperature shocks and economic growth: Evidence from the last half century. *American Economic Journal: Macroeconomics, 4*(3), 66–95.

Diffenbaugh, N. S., & Burke, M. (2019). Global warming has increased global economic inequality. *Proceedings of the National Academy of Sciences, 116*, 201816020. https://doi.org/10.1073/pnas.1816020116

Emmerling, J., & Tavoni, M. (2021). Representing inequalities in integrated assessment modeling of climate change. *One Earth, 4*(2), 177–180. https://doi.org/10.1016/j.oneear.2021.01.013

Emmerling, J., Drouet, L., van der Wijst, K.-I., van Vuuren, D., Bosetti, V., & Tavoni, M. (2019). The role of the discount rate for emission pathways and negative emissions. *Environmental Research Letters, 14*(10), 104008. https://doi.org/10.1088/1748-9326/ab3cc9

Ferraro, P. J., & Price, M. K. (2013). Using nonpecuniary strategies to influence behavior: Evidence from a large-scale field experiment. *Review of Economics and Statistics, 95*(1), 64–73.

Fowlie, M., Greenstone, M., & Wolfram, C. (2015). Are the non-monetary costs of energy efficiency investments large? Understanding low take-up of a free energy efficiency program. *American Economic Review, 105*(5), 201–204. https://doi.org/10.1257/aer.p20151011

Fuss, S., Canadell, J. G., Peters, G. P., Tavoni, M., Andrew, R. M., Ciais, P., Jackson, R. B., Jones, C. D., Kraxner, F., Nakicenovic, N., Le Quéré, C., Raupach, M. R., Sharifi, A., Smith, P., & Yamagata, Y. (2014). Betting on negative emissions. *Nature Climate Change, 4*(10), 850–853. https://doi.org/10.1038/nclimate2392

Gambhir, A., Butnar, I., Li, P.-H., Smith, P., & Strachan, N. (2019). A review of criticisms of integrated assessment models and proposed approaches to address these, through the lens of BECCS. *Energies, 12*(9), 1747. https://doi.org/10.3390/en12091747

Gaskins, D. W., & Weyant, J. P. (1993). Model comparisons of the costs of reducing CO2 emissions. *American Economic Review, 83*(2), 318–323.

Gazzotti, P., Emmerling, J., Marangoni, G., Castelletti, A., van der Wijst, K.-I., Hof, A., & Tavoni, M. (2021). Persistent inequality in economically optimal climate policies. *Nature Communications, 12*(1), 3421. https://doi.org/10.1038/s41467-021-23613-y

Glanemann, N., Willner, S. N., & Levermann, A. (2020). Paris Climate Agreement passes the cost-benefit test. *Nature Communications, 11*(1), 110. https://doi.org/10.1038/s41467-019-13961-1

Hänsel, M. C., Drupp, M. A., Johansson, D. J. A., Nesje, F., Azar, C., Freeman, M. C., Groom, B., & Sterner, T. (2020). Climate economics support for the UN climate targets. *Nature Climate Change, 10*(8), 781–789. https://doi.org/10.1038/s41558-020-0833-x

Ho, E., Budescu, D. V., Bosetti, V., van Vuuren, D. P., & Keller, K. (2019). Not all carbon dioxide emission scenarios are equally likely: A subjective expert assessment. *Climatic Change, 155*(4), 545–561. https://doi.org/10.1007/s10584-019-02500-y

Hope, C. (2006). The marginal impact of CO2 from PAGE2002: An integrated assessment model incorporating the IPCC's five reasons for concern. *Integrated Assessment Journal, 6*(1), 19–56.

Kalkuhl, M., & Wenz, L. (2020). The impact of climate conditions on economic production. Evidence from a global panel of regions. *Journal of Environmental Economics and Management, 103*, 102360. https://doi.org/10.1016/j.jeem.2020.102360

Keppo, I., Butnar, I., Bauer, N., Caspani, M., Edelenbosch, O., Emmerling, J., Fragkos, P., Guivarch, C., Harmsen, M., Lefèvre, J., Gallic, T. L., Leimbach, M., McDowall, W., Mercure, J.-F., Schaeffer, R., Trutnevyte, E., & Wagner, F. (2021). Exploring the possibility space: Taking stock of the diverse capabilities and gaps in integrated assessment models. *Environmental Research Letters, 16*(5), 053006. https://doi.org/10.1088/1748-9326/abe5d8

Knutti, R., Furrer, R., Tebaldi, C., Cermak, J., & Meehl, G. A. (2010). Challenges in combining projections from multiple climate models. *Journal of Climate, 23*(10), 2739–2758. https://doi.org/10.1175/2009JCLI3361.1

Lamperti, F., Dosi, G., Napoletano, M., Roventini, A., & Sapio, A. (2018). Faraway, so close: Coupled climate and economic dynamics in an agent-based integrated assessment model. *Ecological Economics, 150*, 315–339. https://doi.org/10.1016/j.ecolecon.2018.03.023

Ma, T., & Nakamori, Y. (2009). Modeling technological change in energy systems – From optimization to agent-based modeling. *Energy, 34*(7), 873–879. https://doi.org/10.1016/j.energy.2009.03.005

MacAskill, W. (2016). Normative Uncertainty as a Voting Problem. *Mind, 125*(500), 967–1004. https://doi.org/10.1093/mind/fzv169

Marangoni, G., Tavoni, M., Bosetti, V., Borgonovo, E., Capros, P., Fricko, O., Gernaat, D. E. H. J., Guivarch, C., Havlik, P., Huppmann, D., Johnson, N., Karkatsoulis, P., Keppo, I., Krey, V., Broin, Ó., & E., Price, J., & Van Vuuren, D. P. (2017). Sensitivity of projected long-term CO 2 emissions across the Shared Socioeconomic Pathways. *Nature Climate Change, 7*(2), 113–117. https://doi.org/10.1038/nclimate3199

Merrifield, A. L., Brunner, L., Lorenz, R., Medhaug, I., & Knutti, R. (2020). An investigation of weighting schemes suitable for incorporating large ensembles into multi-model ensembles. *Earth System Dynamics, 11*(3), 807–834. https://doi.org/10.5194/esd-11-807-2020

Nordhaus, W. D. (1994). *Managing the global commons: The economics of climate change.* MIT Press.

Nordhaus, W. D. (2008). *A question of balance.* Yale University Press.

Nordhaus, W. D., & Boyer, J. (2000). *Warming the world.* MIT Press. http://books.google.com/books?id=GbcZHGQliwC&printsec=frontcover&dq=Warming+the+world+book&hl=en&ei=mvDUTdGzJZCq8AOa94zgDA&sa=X&oi=book_result&ct=result&resnum=1&ved=0CCkQ6AEwAA#v=onepage&q=Warming%20the%20world%20book&f=false

O'Neill, B. C., Kriegler, E., Riahi, K., Ebi, K. L., Hallegatte, S., Carter, T. R., Mathur, R., & van Vuuren, D. P. (2013). A new scenario framework for climate change research: The concept of shared socioeconomic pathways. *Climatic Change, 122*(3), 387–400. https://doi.org/10.1007/s10584-013-0905-2

Parrado, R. (2010). Climate-change feedback on economic growth: Explorations with a dynamic general equilibrium model. *Environment and Development Economics, 15*, 515–533.

Piontek, F., Drouet, L., Emmerling, J., Kompas, T., Méjean, A., Otto, C., Rising, J., Soergel, B., Taconet, N., & Tavoni, M. (2021). Integrated perspective on translating biophysical to economic impacts of climate change. *Nature Climate Change, 11*(7), 563–572. https://doi.org/10.1038/s41558-021-01065-y

Rausch, S., Metcalf, G. E., & Reilly, J. (2011). Distributional impacts of carbon pricing: A general equilibrium approach with micro-data for households. *Energy Economics.*

Razavi, S., Jakeman, A., Saltelli, A., Prieur, C., Iooss, B., Borgonovo, E., Plischke, E., Lo Piano, S., Iwanaga, T., Becker, W., Tarantola, S., Guillaume, J. H. A., Jakeman, J., Gupta, H., Melillo, N., Rabitti, G., Chabridon, V., Duan, Q., Sun, X., et al. (2021). The Future of Sensitivity Analysis: An essential discipline for systems modeling and policy support. *Environmental Modelling and Software, 137*, 104954. https://doi.org/10.1016/j.envsoft.2020.104954

Riahi, K., van Vuuren, D. P., Kriegler, E., Edmonds, J., O'Neill, B. C., Fujimori, S., Bauer, N., Calvin, K., Dellink, R., Fricko, O., Lutz, W., Popp, A., Cuaresma, J. C., Samir, K. C., Leimbach, M., Jiang, L., Kram, T., Rao, S., Emmerling, J., et al. (2017). The Shared Socioeconomic Pathways and their energy, land use, and greenhouse gas emissions implications: An overview. *Global Environmental Change, 42*, 153–168. https://doi.org/10.1016/j.gloenvcha.2016.05.009

Ricke, K., Drouet, L., Caldeira, K., & Tavoni, M. (2018). Country-level social cost of carbon. *Nature Climate Change, 8*(10), 895–900. https://doi.org/10.1038/s41558-018-0282-y

Rickels, W., Proelß, A., Geden, O., Burhenne, J., & Fridahl, M. (2021). Integrating carbon dioxide removal into European emissions trading. *Frontiers in Climate, 3*, 62. https://doi.org/10.3389/fclim.2021.690023

Saltelli, A., Aleksankina, K., Becker, W., Fennell, P., Ferretti, F., Holst, N., Li, S., & Wu, Q. (2019). Why so many published sensitivity analyses are false: A systematic review of sensitivity analysis practices. *Environmental Modelling and Software, 114*, 29–39. https://doi.org/10.1016/j.envsoft.2019.01.012

Saltelli, A., Bammer, G., Bruno, I., Charters, E., Fiore, M. D., Didier, E., Espeland, W. N., Kay, J., Piano, S. L., Mayo, D., Jr., Pielke, R., Portaluri, T., Porter, T. M., Puy, A., Rafols, I., Ravetz, J. R., Reinert, E., Sarewitz, D., Stark, P. B., et al. (2020). Five ways to ensure that models serve society: A manifesto. *Nature, 582*(7813), 482–484. https://doi.org/10.1038/d41586-020-01812-9

Smith, S. J., Clarke, L. E., Edmonds, J. A., Kejun, J., Kriegler, E., Masui, T., Riahi, K., Shukla, P. R., Tavoni, M., van Vuuren, D. P., & Weyant, J. P. (2015). Long history of IAM comparisons. *Nature Climate Change, 5*(5), 391–391. https://doi.org/10.1038/nclimate2576

Tavoni, M., & Socolow, R. (2013). Modeling meets science and technology: An introduction to a special issue on negative emissions. *Climatic Change, 118*(1), 1–14. https://doi.org/10.1007/s10584-013-0757-9

Tol, R. S. J. (1997). On the optimal control of carbon dioxide emissions: An application of FUND. *Environmental Modeling and Assessment, 2*(3), 151–163. https://doi.org/10.1023/A:1019017529030

van den Berg, N. J., Hof, A. F., Akenji, L., Edelenbosch, O. Y., van Sluisveld, M. A. E., Timmer, V. J., & van Vuuren, D. P. (2019). Improved modelling of lifestyle changes in Integrated Assessment Models: Cross-disciplinary insights from methodologies and theories. *Energy Strategy Reviews, 26*, 100420. https://doi.org/10.1016/j.esr.2019.100420

Weyant, J. (2017). Some contributions of integrated assessment models of global climate change. *Review of Environmental Economics and Policy, 11*(1), 115–137. https://doi.org/10.1093/reep/rew018

Chapter 15
Digital Epidemiology

Yelena Mejova

Abstract Computational social science has had a profound impact on the study of health and disease, mainly by providing new data sources for all of the primary Ws—what, who, when, and where—in order to understand the final "why" of disease. Anonymized digital trace data bring a new level of detail to contact networks, search engine and social media logs allow for the now-casting of symptoms and behaviours, and media sharing informs the formation of attitudes pivotal in health decision-making. Advances in computational methods in network analysis, agent-based modelling, as well as natural language processing, data mining, and time series analysis allow both the extraction of fine-grained insights and the construction of abstractions over the new data sources. Meanwhile, numerous challenges around bias, privacy, and ethics are being negotiated between data providers, academia, the public, and policymakers in order to ensure the legitimacy of the resulting insights and their responsible incorporation into the public health decision-making. This chapter outlines the latest research on the application of computational social science to epidemiology and the data sources and computational methods involved and spotlights ongoing efforts to address the challenges in its integration into policymaking.

15.1 Introduction

From the beginnings of epidemiology, the importance of data has been central. Often considered fathers of the field, John Graunt analysed London's bills of mortality to measure the mortality of certain diseases in 1663, and later in 1854, John Snow mapped the cholera cases to identify its sources. Although since those early days in London the medical and mathematical understanding of disease have greatly advanced, one of the primary roles of the epidemiologist is still to prepare and

Y. Mejova (✉)
ISI Foundation, Turin, Italy
e-mail: yelenamejova@acm.org

E. Bertoni et al. (eds.), *Handbook of Computational Social Science for Policy*,
https://doi.org/10.1007/978-3-031-16624-2_15

organize the collection of relevant and useful data and to use it to model disease (Obi et al., 2020). This data includes the fundamental *W*s that are necessary to understand disease: health event (what), people involved (who), place (where), time (when), and causes, risk factors, and modes of transmission (why/how) (Dicker et al., 2006). Thus, some of the main tasks of an epidemiologist are disease surveillance, field investigation, contact tracing, evaluation of interventions, and public communication—all of which have been transformed by the digital and computing revolutions.

Scientifically, the field is highly multidisciplinary, first measuring the basics of the *W*s—identifying the people, places, and time frames of the health events—and then introducing higher-order considerations, the biology of disease, behaviour of its carriers, and ecological influences on the transmission. By building models around this knowledge, it attempts to recommend possible interventions, which then require additional measurement and modelling of complex feedback effects and the psychological and behavioural factors. Advances in disparate fields like genetics, behavioural economics, and ecology on the one hand and more recent strides in computing methods and digitization on the other are making it possible for epidemiology to develop a systems conceptualization of the fields it connects. Computational social science (CSS) in particular adds new tools via large-scale detection, tracking, and contextualizing of disease. As we will see below, digital traces such as mobility and cellphone data have been used to better understand human networks, user-generated content on social media and the web has been employed to now-cast symptoms and disease, and social interactions have been monitored to understand the impact of social contact and new information on health-related behaviour change. Capturing the latest modelling and computing techniques, the umbrella terms of *digital* or *computational epidemiology* encompass these new methodological developments.

A string of epidemics in the early twenty-first century—H1N1 (swine flu) in 2009, Ebola in 2014 and 2019, and Zika in 2016—has brought epidemiology to the forefront of public awareness, culminating in the COVID-19 pandemic (at the time of this writing, in any case). Meanwhile, public health policy and interventions are being increasingly informed by telecommunications and other digital data (Budd et al., 2020; Oliver et al., 2020; Rich & Miah, 2017). Governments are collaborating with major cellphone companies to perform privacy-preserving contact tracing, internet companies are releasing aggregated mobility data for contagion modelling, and social media giants are partnering with public health organizations to tackle health misinformation and to support public health messaging campaigns. Throughout, a constant negotiation is at play between the needs of public health researchers and the release of commercially valuable information by the companies.

Moreover, a less publicized, but nevertheless critical, battle is being waged against non-communicable diseases including cardiovascular diseases, cancer, diabetes, and mental health disorders. Daily digital traces, such as social media posts and location check-ins, are being used to understand the lifestyle choices of large cohorts, as an alternative to surveys and diaries. Discussions around mental health, disordered eating, illicit drugs, and other topics that are difficult to capture using traditional surveillance methods are now presenting a window into vulnerable populations, even before they register in medical records.

Despite the great promise of new data sources and methodologies, big data approaches are a subject to a slew of challenges that the field needs to overcome in order to establish fruitful collaboration with policymakers. Although big, the datasets often present biased view of the population, which is more tech-savvy and affluent, while excluding those who may have a more urgent need of monitoring and assistance. However, integration of this new data into existing datasets allows for the reduction of overall bias and helps in extending analyses performed on traditional data sources. Encompassing many disciplines who have their own organization, research frameworks, peculiar jargon, and publication venues, digital epidemiology is still in the processing of bridging the siloes to encourage truly multidisciplinary insight. Standardizing the reporting and transparency among these disciplines aims to reduce the number of isolated studies which may suffer from the lack of reproducibility due to the peculiar nature of the available data, application domain, or poorly documented methodology. The legal and ethical standards of using digital data are still being decided through a dialogue between data owners, public health researchers, academics in various disciplines, and representatives of the users of the digital platforms. Thus, the field is still building the structures of cooperation, trust, and legitimacy that are necessary to provide impactful insights for policymakers. Nevertheless, COVID-19 has accelerated the integration of digital epidemiology into its decision-making process. Below, we outline the major accomplishments in the application of computational social science to epidemiology, the accompanying challenges, and the possible ways forward to greater legitimacy and impact.

15.2 Existing Literature

The explosion in the utilization of computational methods for epidemiology has been spurred by the combination of new computational techniques and the availability of new sources of data. The immense volume of available data has encouraged further development and integration into the scientific toolkit of distributed computing frameworks, as well as data-intense deep learning algorithms, with frameworks such as Apache Spark and TensorFlow that allow the ingestion and processing of terabytes of data (Kleppmann, 2017; Weidman, 2019). The rise of infrastructure as a service (IaaS) business model from giants of industry including Amazon Web Services, Microsoft Azure, and cloud services from Oracle, Google, and IBM has allowed the researchers to access sophisticated infrastructure without purchasing the hardware and support staff within their institutions (on a similar topic, see also Fontana & Guerzoni, 2023).

Much of the data that has accompanied these developments in the computing field has been put to use by epidemiologists, opening new scientific ground. The ongoing digitization of medical records, insurance claims, and governmental public health data continues to provide large-scale, high-quality view of individuals within the medical system. Ongoing efforts, such as the European Health Data Space,[1] aggregate such datasets, handle privacy concerns, and make it available for research and policymaking (European Commission., 2021). Moreover, the communication revolution has enabled researchers to better understand these individuals even before they enter the public health system. Digital traces of people's daily activities, including the apps they use, web searches they make, social media posts they publish, as well as the signals from the wearables they keep on their bodies, can help create a view of health-related activities with an unprecedented resolution and reach. One of the earliest attempts to track influenza-like illness (ILI) using user-generated data was proposed by a team of Google researchers who tracked the occurrence of specific keywords in the company's search query logs (Ginsberg et al., 2009). Although highly criticized by subsequent researchers (Lazer et al., 2014) (we will discuss these concerns below), research on web logs continues to produce encouraging results, including detecting adverse reactions (Yom-Tov & Lev-Ran, 2017), predicting diagnosis of diabetes (Hochberg et al., 2019), and understanding the information needs around medical topics (Rosenblum & Yom-Tov, 2017). Specialized application use has been used to understand the effects of gamification (Althoff et al., 2016) and social contagion (Aral & Nicolaides, 2017) on exercise and the characteristics of (un-)successful diets (Weber & Achananuparp, 2016). The text posted by thousands of users on social media platforms has been used to identify and track depression (De Choudhury et al., 2013), eating disorders (Stewart et al., 2017), attitudes toward vaccination (Cossard et al., 2020), and other health interventions. The networked nature of the data often allows the study of the way in which information (Johnson et al., 2020), behaviours, and diseases propagate. Finally, anonymized mobility data, often coming from telephone and transportation companies, has allowed a more fine-grained transmission modelling of the disease (Vespe et al., 2021), as well as the impact of mobility-related interventions (Jeffrey et al., 2020). These data sources add immense value to the traditional ones by increasing the population coverage (some into millions of people), temporal resolution (allowing "now-casting"), and qualitative depth that are impossible or prohibitively expensive to reach outside the digital domain.

One of the earliest examples of the application of computational models to infectious diseases was human influenza, which is an ongoing public health battle. It is continuously analysed via viral phylodynamics in order to better understand its transmission dynamics. Computational phylogenetics methods are applied to datasets of genetic sequences sampled over time and sub-populations in order to assemble a phylogenetic tree and estimate various dynamics of the process (Volz et al., 2013). Fitness models also help in selecting the vaccines year over year (Łuksza & Lässig, 2014). Beyond the study of the virus itself, CSS has introduced several behavioural aspects to the models, many of which have been used during the COVID-19 epidemic. Mobility data (including that provided publicly by large corporations during the pandemic) has been used to monitor the compliance with

---

[1] https://ec.europa.eu/health/ehealth-digital-health-and-care/european-health-data-space_en

interventions, such as the stay-at-home orders during COVID-19, revealing the role of awareness and fatigue in modelling risky behaviours (Weitz et al., 2020). Large-scale online surveys and crowdsourcing have been used to gauge psychological and behavioural responses to the pandemic around the world (Yamada et al., 2021). Even larger efforts, such as InfluenzaNet, recruit thousands of volunteers across Europe to regularly report ILI symptoms, allowing researchers to identify risk factors and gauge influenza vaccine effectiveness (Koppeschaar et al., 2017). Travel records have been used to track the international transmission of disease (Azad & Devi, 2020), whereas a machine-learned anonymized smartphone mobility map has been used to forecast influenza within and across countries (Venkatramanan et al., 2021). For instance, the Global Epidemic and Mobility (GLEaM) framework uses local and international mobility data to build epidemic models, allowing for the simulation of worldwide pandemics, including estimating the impact of interventions during the COVID-19 epidemic (Chinazzi et al., 2020; Van den Broeck et al., 2011). To better understand the reasons behind risky behaviours and non-compliance with public health advice, researchers utilized discussions on social media, often finding misunderstandings and downright misinformation (Betti et al., 2021; Keller et al., 2021). Finally, public health communication campaigns have been evaluated using outreach online by influencers (Bonnevie et al., 2020) as well as news websites and popular social media sites (Carlson et al., 2020).

Unlike in the beginning of epidemiology's development as a science, the infectious diseases have these days given way to non-communicable diseases as the cause of illness and death, especially in the developed countries. The daily behaviours captured in digital trace data, especially social media, have been extensively used to study non-communicable diseases including obesity and diabetes type 2, mental illness, and even suicide. At population level, diabetes has been tracked using store purchase data (Aiello et al., 2019), as well as social media posts (Abbar et al., 2015), and some environmental causes have been tracked in the USA, with a focus on "food deserts" where access to healthy food is limited (De Choudhury, Sharma, et al., 2016). Attempts to inform potential interventions have been made by measuring the importance of community support during a weight loss journey (Cunha et al., 2016) and the effect of intervention messaging on those affected by anorexia (Yom-Tov et al., 2012). Observational studies of exercise in particular through specialized exercise applications have shown that information about other people's routine may affect one's own (Aral & Nicolaides, 2017) and that gender plays an important role in the continued use of such apps (Mejova & Kalimeri, 2019). Further, a combination of web search and wearables data has been used to show the health impact of applications not necessarily meant for exercise, such as Pokémon Go, which resulted in potentially years worth of life spans added to the fans of the game (Althoff et al., 2016). The anonymous and connected nature of social media and specialized forums have also allowed a better understanding of depression, anxiety, eating disorders, and other mental health issues (for an overview, see Chancellor & De Choudhury, 2020). The text of the posts has been used to predict suicidal ideation (Cheng et al., 2017), psychotic relapses (Birnbaum et al., 2019), and PTSD (Coppersmith et al., 2014). More specialized data sources have been used to track recreational drug use (Deluca et al., 2012), as well as the use of "dark web" as a marketplace for such activities (Aldridge & Décary-Hétu, 2016). In combination with screening questionnaires which use validated scales such as Center for Epidemiologic Studies Depression Scale (CES-D) and Beck Depression Inventory (BDI), the daily self-expressions of those dealing with mental health issues provide an unintrusive record of the condition's progression and reactions to potential interventions.

These encouraging developments have been accompanied by a vigorous discussion of their limitations. The privacy concerns regarding secondary use of personal data, even if originally posted on public platforms, demand a critical evaluation of the balance between potential benefits of public health research, compared to the privacy risks to the individuals captured in the data (see, e.g. Taylor, 2023). Other critiques are more unique to the field of epidemiology. For instance, the machine learning framework of classification, as well as most deterministic compartmental models (such as Susceptible-Infected-Recovered (SIR), more on which later), makes necessary simplifying assumptions about the natural progression of a disease, its behaviour, as well as the pharmaceutical and non-pharmaceutical interventions introduced to slow its spread, although more sophisticated models with more complex representations are continuously being proposed.

The separation between traditional epidemiology and computing disciplines in the research teams often results in the failure to take into consideration the established theories in clinical science, using operationalization that is most convenient technically, but not as well matched to the medical condition tracked, while a vague communication of the technical aspects of computing pipelines makes it difficult to integrate the results into clinical practice (Chancellor & De Choudhury, 2020). Observational studies have also lacked the rigor of causal analysis, often stopping at correlational observations. Despite capturing multitudes of people, each data source has substantial biases that must be not only acknowledged by the researchers but accounted for in the analytical pipeline (Yom-Tov, 2019). Finally, data ownership, global justice, and ethical oversight are all important problems that need to be addressed for digital epidemiology to gain legitimacy on the scientific and policy stage (Vayena et al., 2015). We will touch on these and other peculiarities of using computational social science for epidemiology in the next section.

15.3   Computational Guidelines

The abovementioned literature not only pushes the boundaries of traditional epidemiology and the purview of computing but addresses multiple important policy questions regarding public health. The third goal of the UN Sustainable Development Goals (SDGs) is to "Ensure healthy lives and promote well-being for all at all ages".[2] For instance, the goal encompasses the work on alleviating communicable and non-communicable diseases, prevention and treatment of substance abuse, ensuring access to sexual and reproductive services, and increasing the healthcare capacity in all countries, but especially in the developing ones. Although CSS cannot build the necessary infrastructure, it can measure, on both community and individual scale, the utilization of healthcare services, the barriers experienced by the populous, and the expression of unfulfilled needs. Furthermore, it can help in tracking and forecasting disease, again at the scales including individuals, thus measuring the impact of potential ongoing interventions. In fact, CSS can help to craft, deploy, and monitor epidemiological interventions by providing detailed profiling of the target audience, individualized message delivery, and fine-grained behavioural feedback. In order to bring these promises to fruition, a slew of challenges remain to be fully addressed by the research and policy community, including data access and privacy, construct validity, methodological transparency, sampling bias, accounting for confounders, and finally sufficiently clear communication to ensure real-world application. Below, we discuss several policy questions that CSS may address and outline technical and organizational best practices.

15.3.1 Infectious Diseases

The modelling and predicting of infectious diseases is perhaps the most well-known purview of digital epidemiology. Some of the simplest models of disease spread use a system of states as a basis, such as the Susceptible-Infected-Recovered (SIR) model wherein the population can be put into one of these three states (Bjørnstad et al., 2020). Other compartmental models exist which describe the progression of disease with more states ("compartments"), including Asymptomatic infectious, Hospitalized, etc. (Blackwood & Childs, 2018). Such states may also include behaviours of the population segments, including those produced via interventions such as quarantining (Maier & Brockmann, 2020) and wearing masks (Ngonghala et al., 2020). The SIR model has also been extended to incorporate the age structure in the contact matrices (Walker et al., 2020). Compartmental models are popular because they can be designed to frame the essential parts of a question and to work with reduced amounts of data for calibration. By varying parameters such as time between cases, average rate an individual can infect another, and the time infected individual can recover, researchers can estimate the case increase, as well as other properties of the epidemic. For instance, during the COVID-19 epidemic, the effective reproduction number R, or average number of secondary cases per infectious case in a population made up of both susceptible and non-susceptible hosts, has been closely watched and estimated in different affected countries, providing an important characterization of the disease's spread (D'Arienzo &

---

[2] https://sdgs.un.org/goals/goal3

Coniglio, 2020). This classic model has been recently challenged and improvements have been proposed. For instance, the assumption that any individual may contact and thus infect any other in a population (*homogeneous mixing*) has been shown to be oversimplification of the way people interact in reality; instead, considering other information, such as differential susceptibility by age, may improve the models models (Q.-H. Liu et al., 2018).

Further, the availability of large-scale data has allowed scholars to model the real-world networks more accurately. The effect of network structure has been studied in the context of epidemic spreading velocity (Cui et al., 2014) and size (Y. Liu et al., 2016; Wu et al., 2015) and thresholds (Silva et al., 2019). Pandemic outbreaks have been found to be supported in networks with high assortativity (Moreno et al., 2003) and those having community structures (Z. Liu & Hu, 2005). The plethora of data has also allowed the application of agent-based models (ABMs) which attempt to capture empirical socio-demographic characteristics such as household's sizes and compositions, however at a larger computational cost. Such models have been used to incorporate empirical knowledge about contact rates within and between age groups (Ogden et al., 2020) and comorbidities (Wilder et al., 2020). Most such models are built using known population statistics, such as the ABM built to simulate disease evolution in France in order to evaluate the effectiveness of COVID-19 lockdowns, physical distancing, and mask-wearing (Hoertel et al., 2020). Alternatively, contact tracing data has been used to build detailed community network approximations, such as one built for Boston, by considering anonymized GDPR-compliant mobile location data in combination with 83,000 places from Foursquare (Aleta et al., 2020). To make sure data sparsity does not result in individual privacy violations, the authors use a probabilistic approach to measure co-presence. Thus, ABMs have been useful in furthering our understanding of the changes to contact networks and their impact on disease transmission.

Fine-grained mobile phone data has been used to estimate population movements affecting the spread of influenza-like illness (ILI) predating COVID-19. In Tizzoni et al. (2014), the data comes as a set of phone calls georeferenced to the cellphone tower. The authors estimate that a user's most frequent location in the data is their residence and second-most frequent is the place of employment. Usually obtained via extensive (and expensive) surveys, such information is revolutionizing disease modelling on both local and global scales. Beyond phone records, internet data has also been used to monitor mobility. These works show the possibility for large corporations to surface anonymized, aggregated, and differentially private data in order to assist public health researchers and decision-makers. These include Google COVID-19 Community Mobility Reports (Google, 2021a), Apple Mobility Trends Reports (Apple., 2021), and Facebook Disease Prevention Maps (Facebook, 2021b), all of which aggregate the massive amounts of information their platforms collect about the location of their users. All three resources have been used to gauge the changes in mobility of during the COVID-19 lockdowns (Mejova & Kourtellis, 2021; Shepherd et al., 2021; Woskie et al., 2021). However, if one wants to obtain a more nuanced understanding of contact networks, wearable technologies can be used to detect face-to-face interactions within, say, an organization or a building.

Unobtrusive sensors have been used to detect close proximity interactions at 1.5 m in order to reveal the interaction patterns among healthcare workers and patients in a hospital (Vanhems et al., 2013), as well as at an academic conference (Smieszek et al., 2016) and within several households in Kenya (Kiti et al., 2016). Large-scale proximity sensors were later used by many governments during the COVID-19 epidemic through passive contact tracing apps, which use anonymous identifiers to remember devices which were in a close proximity of a person and which can notify their users in case somebody within their contact history has been found to be COVID-positive (Barrat et al., 2020).

But before the disease can be tracked, its very presence needs to be detected. Computational social science presents several unprecedented data sources that enable researchers to "now-cast" disease as it moves through the population. As mentioned, web search data has been used to monitor ILI symptoms (Ginsberg et al., 2009) and is still used for many others. However, one does not need to be a Google employee to perform such research, as aggregated search data is surfaced by the company via Google Search Trends (Google, 2021b), which has been used to track anything from Lyme disease (Kapitány-Fövény et al., 2019) to type 2 diabetes (Tkachenko et al., 2017). Of course, other dynamic social media have been used to track disease, including Twitter, Reddit, and Sina Weibo, all of which have been used to track non-communicable diseases as well. Beyond observation, self-reported data can be obtained from participatory surveillance systems, such as InfluenzaNet (Koppeschaar et al., 2017), which collects influenza-related information from thousands of volunteers from countries around the EU.

Both algorithmic and data advances described above come with many caveats which both the scientific and policy communities are yet to tackle effectively. As machine learning and other modelling algorithms become more complex, difficulties in communicating their benefits and—more importantly—limitations to those outside the initiated trained practitioners result in misunderstandings about the certainty of the predictions and limits of their applications, leading to a limited deployment in the field. However, the solution may not lie in a more detailed description of the algorithms, but in the clarification of their merits, such that we can be determined whether their performance warrants their integration in the decision-making process of policymaking. One could take a page from the social science "reproducibility crisis" (Camerer et al., 2018) which illustrated the bias toward significant, positive, and theoretically neat results at the cost of valid, generalizable insights. Several actions, including the Social Sciences Replication Project (SSRP), the Reproducibility Project: Psychology (RPP), and the Experimental Economics Replication Project (EERP), have been organized to provide increased rigor to the insights on important theories and results in each field. Beyond reproducibility, integration of new methodologies should be tested in prediction competitions, such as CDC's FluSight, a competition that brings together researchers and industry leaders to forecast the timing, peak, and intensity of the flu season (Centers for Disease Control and Prevention., 2021). Another ongoing effort is the ECDC's European Covid-19 Forecast Hub which collates and combines short-term forecasts of COVID-19 generated by different independent modelling teams across Europe and makes available a near-term future trajectory of the pandemic (European Centre for Disease Control and Prevention (ECDC), 2021). The legitimacy afforded by such efforts would encourage the data owners (e.g. internet/technology companies including social media websites and phone companies) to contribute datasets that would level the playing field between well-funded and smaller players. It is especially important to solicit both algorithmic and expert (human) predictions in order to provide a baseline for comparison, as it has been shown that people tend to distrust algorithms faster when they make mistakes, compared to when humans do the same (Dietvorst et al., 2015). Increased transparency in the way epidemiological studies are designed, the kind of data they use, and—crucially—their predictions ahead of the target date are all likely not only to clarify the potential impact of the new methods on public health but also to unify the field under a set of common goals (Miguel et al., 2014).

This proposal will hopefully address several other critiques. Legitimizing and clearly describing the uses of data would give a greater transparency to the secondary use of data, greater oversight over anonymization standards, and aggregate statistics of its biases. Biases in data collection have been a constant critique of scientific endeavours; however, it may be even easier to gloss over biases in big datasets, but it has been shown that even large datasets of internet or technology users have substantial biases in terms of demographics, wealth, and technological access (Hargittai, 2020; Yom-Tov, 2019). Sampling biases limit the generalizability of the scientific studies. As such biases tend to underrepresent those coming from more disadvantaged backgrounds and locales, systematic testing of the algorithms on different populations would provide a quantifiable measure of the change in performance across groups of interest (Olteanu et al., 2019). The peculiarities of the digital platforms provide another constraint, including the affordances provided by each website, as well as the peculiar user base and culture. For instance, the privacy and identification limitations on Facebook distinguish it from more open platforms, like Twitter, or community-oriented ones, like Reddit, resulting in differences of information disclosure and propagation. The very timing of the studies imposes biases specific to the time period selected for the analysis (for instance, 2020 will likely be a special year in many datasets), making some observations unique to the contemporary societal, technological, and public health situation. To address some of these problems, scientists must be encouraged to publish replication studies, as well as to extend them into long-term projects, in order to test the models initially proposed on different data and time spans. Further, establishing data partnerships addressing important public health concerns will insure the infrastructure is in place in case a crisis, such as the COVID-19 epidemic, strikes.

15.3.2 Non-communicable Diseases

As medicine advanced against infectious diseases, non-communicable diseases have become the leading causes of death and illness throughout developed and developing world. Many of such conditions, including obesity and the overweight, diabetes, and cardiovascular complications, have a strong "lifestyle" component, wherein the daily activities of the population accumulate to contribute to worsening outcomes. CSS provides a unique view of such behaviours, using the digital traces left through these daily activities such as social media posts, business check-ins, web searches, use of applications, and many others. Behaviours around food consumption and nutrition have been studied using Twitter (Abbar et al., 2015), Instagram (Mejova et al., 2015), as well as large datasets of grocery purchases (Aiello et al., 2019). Often, natural language processing (NLP) tools are used to process the text obtained from many internet users or deep machine learning (ML) models to "recognize" relevant objects in the shared images in order to understand the daily behaviours of the internet users. Crucially, these activities can be put into a cultural context to better understand the societal, economic, and psychological forces shaping these daily decisions, much as proposed by Weiss as "cultural epidemiology" (Weiss, 2001) that combines quantitative and qualitative methodologies. For instance, large datasets of recipes have been examined in order to establish a network of flavours and ingredients across countries and relate it to the health outcomes of different locales (Sajadmanesh et al., 2017). The relationship between economic deprivation on diet in the USA has shown that those living in "food deserts" mention food that is higher in fat, cholesterol, and sugar than otherwise (De Choudhury, Sharma, et al., 2016). Further, specialized apps and wearables are used to monitor physical activity. For example, a study of running tracking app data (Aral & Nicolaides, 2017) aimed to understand the role of social interaction and comparison on the duration of one's run. However, some researchers aim to go beyond behavioural profiling and use internet search data to detect those potentially having serious illness. A team used search query logs to first identify users who mentioned having a diabetes diagnosis and compare them to a control group (Hochberg et al., 2019). Researchers were able to predict whether a user will be searching for diabetes-related words from their previous queries with a positive predictive value of 56% at a false-positive rate of 1% at up to 240 days before they mention the diagnosis. In general, it was found that people tend to search about symptoms some time before they are diagnosed with the underlying condition (Hochberg et al., 2020), especially if the symptoms are serious. Yet more data is available to monitor disease on a population level via information surfaced by the advertising systems of large social media platforms. For instance, Facebook allows potential advertisers to run detailed queries on their target audience, specifying their demographics, precise location, language, and interests (which span health concerns, activities, hobbies, worldviews, and many more categories) (Facebook, 2021a). These can then be used as a kind of "digital census" to quantify awareness of health-related topics and behaviours related to non-communicable diseases within well-defined demographic groups across fine and broad geographies (Mejova, Weber, et al., 2018). Compared to traditional survey-based monitoring, the above studies provide unobtrusive, real-time, and extremely rich sources of behavioural observation. Especially on social media, the users are self-motivated to share their meals and activities, to annotate them with geographic and other metadata, and to interact with other posts. Although suffering from social desirability bias, in combination with other consumption statistics, social media and app use data provide important signals about the social and psychological context of health-related behaviours.

Further, non-communicable disease interventions can be studied on a personal level while delivered through a myriad of technologies. Integration of smartphones with user-generated content is leading to sophisticated personalized interventions aiming at motivating the users to increase their physical activity level (Harrington et al., 2018; op den Akker et al., 2014). Different messaging strategies have been explored including personalized exercise recommendation (Tseng et al., 2015), also employing machine learning via supervised learning (Hales et al., 2016; Marsaux et al., 2016) and reinforcement learning (Rabbi et al., 2015; Yom-Tov et al., 2017). Others help users find exercise partners (Hales et al., 2016) and provide educational materials (Short et al., 2017) and emotional support (Vandelanotte et al., 2015). The applications have been embraced by the governments and businesses worldwide. For instance, UK's National Health Service promotes an *Active 10* app that encourages everyone to have a brisk walk and for those ready for a bigger challenge has *Couch to 5K* app for beginner runners (National Health Service., 2021). India's Ministry of Youth Affairs and Sports launched its *Fit India* app to help its populous keep track of their fitness goals, water intake, and sleep (Play Store., 2021). Social media is, of course, another popular outlet for public health outreach. Many associations, such as the National Eating Disorders Association in the USA, run annual health awareness campaigns on different social media channels, making it possible to measure the impact of their campaigns on the sustainability of the attention to the topic and other subsequent behaviours expressed by their audience (Mejova & Suarez-Lledó, 2020). To assist in the efforts, some researchers focus on which influencers and content (especially contagious "memes") are particularly successful in attracting an audience (Kostygina et al., 2020) or how to better identify the relevant users to target (Chu et al., 2019).

Although the above studies provide a valuable context to the ongoing epidemics of non-communicable diseases, and potential avenues to communicate about them, mostly observational studies usually fail to reach the threshold for causal insight. Often large datasets lack the information on important confounders that may affect the outcome of the study. For instance, while comparing health-related interests expressed by Facebook users to rates of obesity, diabetes, and alcoholism, researchers have found that unrelated (or "placebo") interests, such as those in entertainment or technology, also had substantial correlation with the rates of disease (Mejova, Weber, et al., 2018). Some attempt to improve the quality of their models by employing *instrumental variables*, especially when the explanatory variable of interest is correlated with the error term. Weather is a popular instrumental variable, as it is often not related to the dependent variable, but may have some relationship with the independent ones. In their study of social contagion in a community of runners, the authors used the weather at one person's location as an instrumental variable when modelling the running behaviour of another (Aral & Nicolaides, 2017). They show that without the corrections, the effect would have been overestimated by 71–82%. The inability to acquire multi-dimensional data that has important confounders (which are often demographics, protected by numerous privacy regulations) has an additional effect of hiding the unequal relevance of the ongoing work to those less represented in these datasets. Inferring sensitive information, including age, gender, and location, may be possible from some sources of data, but such activity may both break the privacy of the platform and violate the protections imposed by the EU General Data Protection Regulation (GDPR). It is thus imperative to engage legitimate stakeholders who will negotiate controlled releases of highly detailed data for research on pressing topics and especially provide input during policy changes when a "natural experiment" may take place. Policymakers may also want to explicitly outline the under-served populations they would like to focus on, thus encouraging the creation of datasets around groups that are not yet captured in currently available data. For instance, India's efforts in the National Mission for Empowerment of Women (NMEW) may be augmented by encouraging the monitoring of technology use through available data (Mejova, Gandhi et al., 2018). Alternatively, access to care can be monitored using online tools, such as those for women's health services (Dodge et al., 2018) across the USA.

15.3.3 Mental Illness and Suicide

An especially vulnerable population that has been extensively studied by CSS in the context of epidemiology is people with diagnosed mental illness, or those simply expressing mental distress, alongside those who vocally contemplate suicide. The anonymity and social support provided by the internet forums and websites allow many to express feelings and thoughts which may be difficult to evoke using standard public health methods like surveys and medical records. The pervasive use of social media, including on mobile devices, allows users to post instantly during the moments of mental distress and for some to integrate digital platforms into their coping mechanisms. Communities around eating disorders (anorexia, bulimia, etc.) (Stewart et al., 2017; Yom-Tov et al., 2012), depression (De Choudhury et al., 2013; Reece & Danforth, 2017), and drug abuse (Kazemi et al., 2017) and recovery (Chancellor et al., 2019) are providing valuable insights in the way people experience these conditions, seek and provide support, and even provide practical advice. For instance, by combining automated machine learning classification and text processing techniques with clinical expertise, researchers have used the Reddit opioid addiction recovery forums to discover alternative treatments that the users share and discuss (Chancellor et al., 2019). It is also possible to monitor the progression of mental illness to serious suicide ideation by examining suicide prevention forums (De Choudhury, Kiciman et al., 2016), as well as studying web search patterns (Adler et al., 2019). Studies of search engine usage have been able to confirm behavioural signs of people with autism, for instance, finding that users who have self-stated that they have autism spend less time examining image results (Yechiam, Yom-Tov et al., 2021). Whereas most studies rely on self-declaration of diagnosis, some studies use social media to better understand those who have been confirmed to be clinically diagnosed. Facebook posts of patients diagnosed with a primary psychotic disorder have been analysed to find predictors of a future psychotic relapse (Birnbaum et al., 2019).

However, the very fact that self-expression of mental distress may come before official diagnosis makes such research struggle with construct validity, that is, what exactly is being measured, and how robust it is in clinical terms. Reviews of literature on mental health status on social media show that few use the definitions and theories developed in the clinical setting to define, for instance, the conditions of "anxiety" or "depression" that are being tracked (Chancellor & De Choudhury, 2020). Whether mentions of disorders on social media capture users who are struggling with them, merely interested in the topic, or even misusing the terms is an important question to answer before these methods can be applied to the clinical setting. It is imperative to foster a closer collaboration between the medical establishment and researchers attempting to contribute to the epidemiology of conditions possibly discussed in user-generated data. From the CSS research community's side, it is important to rigorously define the cohorts of interest and follow clinically validated diagnostic procedures (Ernala et al., 2019) when studying new sources of data and methods for identifying those potentially struggling with mental illness. However, it is also desirable to have the medical community to acknowledge these new sources of information as an additional signal that should be clinically studied and which may play a role in official diagnostic (and possibly treatment) frameworks. As mentioned earlier, methods based on alternative data sources may play a role in the profiling of future recruits for studies, potentially expanding their reach beyond those already in the medical system.

15.3.4 Beliefs, Information, and Misinformation

User-generated data provides yet another unique context around health and disease: the dynamics of individual's knowledge, opinion, and belief and their interactions with various information sources that shape these important precursors to behaviour. The quality of medical information available to people on social media and through web search can be evaluated using big data NLP tools and in collaboration with area experts. YouTube videos have been found to be some of the worst offenders in terms of advocating methods proven to be harmful or having no scientific basis (Madathil et al., 2015). Twitter (Rosenberg et al., 2020), Reddit (Jang et al., 2019), and Pinterest (Guidry & Messner, 2017) all have been examined for links to potentially harmful health advice. One of the most serious problems is the anti-vaccination movement that has been strengthening in both developed and developing countries. Twitter data has proven to be useful in explaining some variation in the vaccine coverage rates, as reported by the immunization monitoring system of WHO (Bello-Orgaz et al., 2017). Classifying whether social media users support or oppose vaccinations has been shown to be feasible, both using deep learning on the posted text and images (Wang et al., 2020) and using network algorithms on the conversation network (Cossard et al., 2020). However, it is in the more specialized websites, such as the discussion forums for parents, that give space to those who are hesitant and are in the process of making healthcare decisions for themselves or their family. There, researchers can find lists of concerns, previous experiences, and information seeking, as well as testimonials about the experiences with the medical establishment (Betti et al., 2021).

Further, internet captures myriad interactions with medical services and consequences of health interventions. Social media has been used extensively for *pharmacovigilance*, discovering drug side effects (Alvaro et al., 2015), drug interactions (Correia et al., 2016), and recreational drug use (Deluca et al., 2012) and even uncovering illicit online pharmacies (Katsuki et al., 2015). Patient experiences can be found on business review websites (Rastegar-Mojarad et al., 2015), as well as general-purpose social media, where communities can discuss their perceptions of treatment (Booth et al., 2019; Hswen et al., 2020). Super-utilizers of healthcare services have also been studied on social media in order to inform online social support interventions and complement offline community care services (Guntuku et al., 2021), and efforts have been made to integrate patient experiences in online discussions into customer satisfaction and service quality measures (Albarrak & Li, 2018).

As more and more people use internet and social media as a source of medical knowledge and advice, as well as social support, understanding how this information is translated into behaviours and life choices is an increasingly urgent research direction. Although the detection of cyberbullying and other negative speech on social media is an active research direction (Chatzakou et al., 2019), ethical concerns prevent the integration of user profiling and targeting in mental health interventions. However, health misinformation has been acknowledged to be a parallel pandemic in the COVID-19 era, and concerted efforts are ongoing in monitoring and tackling potentially harmful information (World Health Organization, 2021a). In this sphere, CSS will continue to play an important role by providing the tools for the analysis of new social and information sharing platforms that are increasingly permeating the information landscape.

15.4 The Way Forward

Epidemiology was one of the first of the sciences to use large datasets, and thus, it is in a natural position to take advantage of the latest developments in digitization, big data, and computing methods. The year 2020 has forced the field to mobilize its best resources to address the COVID-19 pandemic and put in stark light the challenges facing the field. The silver lining of this dark cloud could be an understanding of the necessary steps in bringing digital epidemiology into the policy sphere, making it agile and relevant in a fast-moving globalized world.

The COVID-19 pandemic has imparted an important component to the epidemiological field—a clarity of vision. It has shown in a stark contrast the cost of indecision and the global repercussions on the lives and economies and forced the realities of a global pandemic to the public and the governmental attention. It has also revealed weaknesses in the current health policy structures, the slow response of the governments to the WHO's messaging, and disarray in the case tracking and reporting standards. Already, actions are in place to remedy these weaknesses. Attempts are being made to formalize the government responses through treaties and international agreements (though enforcing such agreements remains a struggle) (Maxmen, 2021). Partnerships are being forged, and large companies released detailed datasets of user activity and mobility to aid in monitoring and modelling (National Institutes of Health., 2021).

Such clarity of vision is necessary to improve the impact of digital epidemiology also in other spheres. The UN Sustainable Development Goals (SDGs)[3] provide a general prioritization for the health and well-being challenges, but these must be defined clearly in order to encourage the building of tools and partnerships. One such effort is the European Data Space, which aims to legitimize and operationalize the data usage across the member states while complying with its established privacy regulations.[4] Another is the WHO Hub for Pandemic and Epidemic Intelligence which aims to build a "global trust architecture" that will encourage greater sharing of data through addressing numerous aspects: "governance, legal frameworks and data-sharing agreements; data solidarity, fairness and benefits sharing; transparency about how pandemic and epidemic intelligence outputs are used; openness of technology solutions and artificial intelligence applications; security of data; combating misinformation and addressing infodemics; privacy by design principles; and public participation and people's data literacy" (World Health Organization, 2021b,c). Additionally, the One Health movement, supported by the WHO, emphasizes the collaboration between disparate domains to accomplish a systems-level perspective on problems such as antibiotic resistance (World Health Organization., 2017). These ambitious projects are a response to a complex problem that involves many parties, some of which only recently began weighing the benefits and dangers of massive surveillance for the greater good.

Several important steps need to be taken in order to engage all major parties involved. First, civil society must be educated in the basics of digital literacy, data privacy, and its governance in order to ensure the users of technologies contributing to the big data revolution provide truly informed consent. For instance, the EU has proposed the Digital Competence Framework that comprises not only information literacy but also skills in communication, digital content creation, safety, and problem-solving (EU Science Hub., 2021). Second, the professionals coming from different civic, academic, and policy silos must be brought together and upskilled to legibly communicate about the role of data in public health. For instance, efforts

---

[3] https://sdgs.un.org/goals
[4] https://ec.europa.eu/health/ehealth/dataspace_en

such as the Lagrange Fellowships in Italy (Fondazione CRT., 2019); the Data Science for Social Good Fellowship in Chicago, USA (University of Chicago., 2021); and the Data Fellowship at the OCHA Centre for Humanitarian Data (Centre for Humanitarian Data, 2021) are excellent efforts to impart data science skills in the next generation of humanitarians, epidemiologists, and academics. Institutionally, the normalization of building teams that incorporate data literacy (and analytics skills, if possible) is an ongoing process that is only recently being supported by educational resources. Third, the governance of technology giants that own much of the data necessary for monitoring and modelling disease must be kept clear and up-to-date considering the latest technological developments. Interestingly, during the COVID-19 efforts to build contact tracing apps, it was the corporations (Apple and Google) that refused to implement features that would threaten the privacy of their users (privacy being an important feature of their services) (Meyer, 2021). However, one must not rely on the businesses to maintain ethical standards of data use, which must be carefully negotiated before the next disaster strikes.

Much of this chapter describes the impressive accomplishments by the academic researchers in the fields of disease monitoring, modelling, prediction, and contextualization. However, to bring these tools to the policymakers' table, they must be robust, vetted, and available on demand. Additional organization is necessary to establish a well-defined set of problems for the community to tackle and to provide legitimacy in order to foster data exchange to support research. Standardizing the tasks (such as flu season prediction), metrics, available data, and benchmarks will allow for an increased accountability and reproducibility of academic endeavours that go beyond publication peer review. Such tasks should be defined in collaboration with the policymakers in order to align the priorities with the societal needs and system outputs with the information needs. The way Netflix Prize has invigorated the recommender systems community (Netflix, 2009) and Google Flu Trends spurred interest in the digital disease tracking (Google., 2014), ambitious competitions not only would provide clarity of vision for the field but would also be able to direct the research agenda to under-served areas and communities. It would be beneficial if the collaborative efforts described above would include a space for the academics and researchers to tackle specific problems within an evaluation framework that produces benchmark datasets and reproducible methods, beyond scientific publications.

Finally, the technological development will continue revolutionizing the field, spurring debate on additional policy considerations. The advances in deep machine learning are allowing to process speech, images, and video at scale and are already being used for plant (Ferentinos, 2018) and human (Li et al., 2020) disease detection. The rise of confidential computing, wherein user data is isolated and protected on the user's device, and only trusted operations can be run on it, eliminates the need to transfer the data for processing elsewhere (Rashid, 2020). The negotiation between the new potential insights and the cost to the society will require thoughtful, informed, and urgent consideration.

References

Abbar, S., Mejova, Y., & Weber, I. (2015). You tweet what you eat: Studying food consumption through twitter. In *Proceedings of the 33rd Annual ACM Conference on Human Factors in Computing Systems* (pp. 3197–3206).

Adler, N., Cattuto, C., Kalimeri, K., Paolotti, D., Tizzoni, M., Verhulst, S., Yom-Tov, E., & Young, A. (2019). How search engine data enhance the understanding of determinants of suicide in india and inform prevention: Observational study. *Journal of medical internet research, 21*(1), e10179.

Aiello, L. M., Schifanella, R., Quercia, D., & Del Prete, L. (2019). Large-scale and high-resolution analysis of food purchases and health outcomes. *EPJ Data Science, 8*(1), 1–22.

Albarrak, A., & Li, Y. (2018). Quality and customer satisfaction health accessibility framework using social media platform. In *Proceedings of the 51st Hawaii International Conference on System Sciences.*

Aldridge, J., & Décary-Hétu, D. (2016). Hidden wholesale: The drug diffusing capacity of online drug cryptomarkets. *International Journal of Drug Policy, 35*, 7–15.

Aleta, A., Martin-Corral, D., Pastore y Piontti, A., Ajelli, M., Litvinova, M., Chinazzi, M., Dean, N. E., Halloran, M. E., Longini Jr, I. M., Merler, S., et al. (2020). Modelling the impact of testing, contact tracing and household quarantine on second waves of covid-19. *Nature Human Behaviour, 4*(9), 964–971.

Althoff, T., White, R. W., & Horvitz, E. (2016). Influence of pokémon go on physical activity: Study and implications. *Journal of Medical Internet Research, 18*(12), e315.

Alvaro, N., Conway, M., Doan, S., Lofi, C., Overington, J., & Collier, N. (2015). Crowdsourcing twitter annotations to identify first-hand experiences of prescription drug use. *Journal of Biomedical Informatics, 58*, 280–287.

Apple. (2021). *Mobility trends reports.* Accessed 1 Sep 2021.

Aral, S., & Nicolaides, C. (2017). Exercise contagion in a global social network. *Nature Communications, 8*(1), 1–8.

Azad, S., & Devi, S. (2020). Tracking the spread of covid-19 in india via social networks in the early phase of the pandemic. *Journal of Travel Medicine, 27*(8), taaa130.

Barrat, A., Cattuto, C., Kivelä, M., Lehmann, S., & Saramäki, J. (2020). Effect of manual and digital contact tracing on covid-19 outbreaks: A study on empirical contact data. *Journal of the Royal Society Interface, 18*(178), 20201000.

Bello-Orgaz, G., Hernandez-Castro, J., & Camacho, D. (2017). Detecting discussion communities on vaccination in twitter. *Future Generation Computer Systems, 66*, 125–136.

Betti, L., De Francisci Morales, G., Gauvin, L., Kalimeri, K., Mejova, Y., Paolotti, D., & Starnini, M. (2021). Detecting adherence to the recommended childhood vaccination schedule from user-generated content in a us parenting forum. *PLoS Computational Biology, 17*(4), e1008919.

Birnbaum, M. L., Ernala, S. K., Rizvi, A., Arenare, E., Van Meter, A., De Choudhury, M., & Kane, J. M. (2019). Detecting relapse in youth with psychotic disorders utilizing patient-generated and patient-contributed digital data from facebook. *NPJ Schizophrenia, 5*(1), 1–9.

Bjørnstad, O. N., Shea, K., Krzywinski, M., & Altman, N. (2020). Modeling infectious epidemics. *Nature Methods, 17*(5), 455–456.

Blackwood, J. C., & Childs, L. M. (2018). An introduction to compartmental modeling for the budding infectious disease modeler. *Letters in Biomathematics 5*, 195–221.

Bonnevie, E., Rosenberg, S. D., Kummeth, C., Goldbarg, J., Wartella, E., & Smyser, J. (2020). Using social media influencers to increase knowledge and positive attitudes toward the flu vaccine. *Plos One, 15*(10), e0240828.

Booth, A., Bell, T., Halhol, S., Pan, S., Welch, V., Merinopoulou, E., Lambrelli, D., & Cox, A. (2019). Using social media to uncover treatment experiences and decisions in patients with acute myeloid leukemia or myelodysplastic syndrome who are ineligible for intensive chemotherapy: Patient-centric qualitative data analysis. *Journal of Medical Internet Research, 21*(11), e14285.

Budd, J., Miller, B. S., Manning, E. M., Lampos, V., Zhuang, M., Edelstein, M., Rees, G., Emery, V. C., Stevens, M. M., Keegan, N., Short, M. J., Pillay, D., Manley, E., Cox, I. J., Heymann, D., Johnson, A. M., & McKendry, R. A. (2020). Digital technologies in the public-health response to covid-19. *Nature medicine, 26*(8), 1183–1192.

Camerer, C. F., Dreber, A., Holzmeister, F., Ho, T.-H., Huber, J., Johannesson, M., Kirchler, M., Nave, G., Nosek, B. A., Pfeiffer, T., Altmejd, A., Buttrick, N., Chan, T., Chen, Y., Forsell, E., Gampa, A., Heikensten, E., Hummer, L., Imai, T., ... Wu, H. (2018). Evaluating the replicability of social science experiments in nature and science between 2010 and 2015. *Nature Human Behaviour, 2*(9), 637–644.

Carlson, S., Dey, A., & Beard, F. (2020). An evaluation of the 2016 influenza vaccination in pregnancy campaign in nsw, australia. *Public Health Res Pract, 30*(1), pii–29121908.

Centers for Disease Control and Prevention. (2021). *Flusight: Flu forecasting.* Accessed 1 Sep 2021.

Centre for Humanitarian Data. (2021). *Data fellows programme.* Accessed 18 Sep 2021.

Chancellor, S., & De Choudhury, M. (2020). Methods in predictive techniques for mental health status on social media: A critical review. *NPJ Digital Medicine, 3*(1), 1–11.

Chancellor, S., Nitzburg, G., Hu, A., Zampieri, F., & De Choudhury, M. (2019). Discovering alternative treatments for opioid use recovery using social media. In *Proceedings of the 2019 CHI Conference on Human Factors in Computing Systems* (pp. 1–15).

Chatzakou, D., Leontiadis, I., Blackburn, J., Cristofaro, E. D., Stringhini, G., Vakali, A., & Kourtellis, N. (2019). Detecting cyberbullying and cyberaggression in social media. *ACM Transactions on the Web (TWEB), 13*(3), 1–51.

Cheng, Q., Li, T. M., Kwok, C.-L., Zhu, T., & Yip, P. S. (2017). Assessing suicide risk and emotional distress in chinese social media: A text mining and machine learning study. *Journal of Medical Internet Research, 19*(7), e243.

Chinazzi, M., Davis, J. T., Ajelli, M., Gioannini, C., Litvinova, M., Merler, S., y Piontti, A. P., Mu, K., Rossi, L., Sun, K., et al. (2020). The effect of travel restrictions on the spread of the 2019 novel coronavirus (covid-19) outbreak. *Science, 368*(6489), 395–400.

Chu, K.-H., Colditz, J., Malik, M., Yates, T., & Primack, B. (2019). Identifying key target audiences for public health campaigns: Leveraging machine learning in the case of hookah tobacco smoking. *Journal of Medical Internet Research, 21*(7), e12443.

Coppersmith, G., Harman, C.,& Dredze, M. (2014). Measuring post traumatic stress disorder in twitter. In *Eighth International AAAI Conference on Weblogs and Social Media.*

Correia, R. B., Li, L., & Rocha, L. M. (2016). Monitoring potential drug interactions and reactions via network analysis of instagram user timelines. In *Biocomputing 2016: Proceedings of the Pacific Symposium* (pp. 492–503)

Cossard, A., Morales, G. D. F., Kalimeri, K., Mejova, Y., Paolotti, D., & Starnini, M. (2020). Falling into the echo chamber: The italian vaccination debate on twitter. In *Proceedings of the International AAAI Conference on Web and Social Media* (Vol. 14, pp. 130–140).

Cui, A.-X., Wang, W., Tang, M., Fu, Y., Liang, X., & Do, Y. (2014). Efficient allocation of heterogeneous response times in information spreading process. *Chaos: An Interdisciplinary Journal of Nonlinear Science, 24*(3), 033113.

Cunha, T. O., Weber, I., Haddadi, H., & Pappa, G. L. (2016). The effect of social feedback in a reddit weight loss community. In *Proceedings of the 6th International Conference on Digital Health Conference* (pp. 99–103).

D'Arienzo, M., & Coniglio, A. (2020). Assessment of the sars-cov-2 basic reproduction number, r0, based on the early phase of covid-19 outbreak in italy. *Biosafety and Health, 2*(2), 57–59.

De Choudhury, M., Gamon, M., Counts, S., & Horvitz, E. (2013). Predicting depression via social media. In *Seventh International AAAI Conference on Weblogs and Social Media.*

De Choudhury, M., Kiciman, E., Dredze, M., Coppersmith, G.,&Kumar, M. (2016). Discovering shifts to suicidal ideation from mental health content in social media. In *Proceedings of the 2016 CHI Conference on Human Factors in Computing Systems* (pp. 2098–2110).

De Choudhury, M., Sharma, S., & Kiciman, E. (2016). Characterizing dietary choices, nutrition, and language in food deserts via social media. In *Proceedings of the 19th Acm Conference on Computer-Supported Cooperative Work & Social Computing* (pp. 1157–1170).

Deluca, P., Davey, Z., Corazza, O., Di Furia, L., Farre, M., Flesland, L. H., Mannonen, M., Majava, A., Peltoniemi, T., Pasinetti, M., et al. (2012). Identifying emerging trends in recreational drug use; outcomes from the psychonaut web mapping project. *Progress in Neuro-Psychopharmacology and Biological Psychiatry, 39*(2), 221–226.

Dicker, R. C., Coronado, F., Koo, D., & Parrish, R. G. (2006). Principles of epidemiology in public health practice; an introduction to applied epidemiology and biostatistics. Self-study course. Stephen B. Thacker CDC Library collection.

Dietvorst, B. J., Simmons, J. P., & Massey, C. (2015). Algorithm aversion: People erroneously avoid algorithms after seeing them err. *Journal of Experimental Psychology: General, 144*(1), 114.

Dodge, L. E., Phillips, S. J., Neo, D. T., Nippita, S., Paul, M. E., & Hacker, M. R. (2018). Quality of information available online for abortion self-referral. *Obstetrics and Gynecology, 132*(6), 1443.

Ernala, S. K., Birnbaum, M. L., Candan, K. A., Rizvi, A. F., Sterling, W. A., Kane, J. M., & De Choudhury, M. (2019). Methodological gaps in predicting mental health states from social media: Triangulating diagnostic signals. In *Proceedings of the 2019 Chi Conference on Human Factors in Computing Systems* (pp. 1–16).

EU Science Hub. (2021). *The digital competence framework 2.0.* Accessed 19 Sep 2021.

European Centre for Disease Control and Prevention (ECDC). (2021). *European covid-19 forecast hub.* Accessed 1 Sep 2021.

European Commission. (2021). *European health data space.* Accessed 1 Sep 2021.

Facebook. (2021a). *Ads manager.* Accessed 1 Sep 2021.

Facebook. (2021b). *Disease prevention maps.* Accessed 1 Sep 2021.

Ferentinos, K. P. (2018). Deep learning models for plant disease detection and diagnosis. *Computers and Electronics in Agriculture, 145,* 311–318.

Fondazione CRT. (2019). *Borse lagrange.* Accessed 18 Sep 2021.

Fontana, M., & Guerzoni, M. (2023). Modeling complexity with unconventional data: Foundational issues in computational social science. In Bertoni, E., Fontana, M., Gabrielli, L., Signorelli, S., Vespe, M. (Eds.), *Handbook of computational social science for policy.* Springer.

Ginsberg, J., Mohebbi, M. H., Patel, R. S., Brammer, L., Smolinski, M. S., & Brilliant, L. (2009). Detecting influenza epidemics using search engine query data. *Nature, 457*(7232), 1012–1014.

Google. (2014). *Google flu trends.* Accessed 18 Sep 2021.

Google. (2021a). *Covid-19 community mobility reports.* Accessed 1 Sep 2021.

Google. (2021b). *Trends.* Accessed 1 Sep 2021.

Guidry, J., & Messner, M. (2017). Health misinformation via social media: The case of vaccine safety on pinterest. In *Social media and crisis communication* (pp. 267–279). Routledge.

Guntuku, S. C., Klinger, E. V., McCalpin, H. J., Ungar, L. H., Asch, D. A., & Merchant, R. M. (2021). Social media language of healthcare super-utilizers. *NPJ Digital Medicine, 4*(1), 1–6.

Hales, S., Turner-McGrievy, G., Fahim, A., Freix, A., Wilcox, S., Davis, R. E., Huhns, M., & Valafar, H. (2016). A mixed-methods approach to the development, refinement, and pilot testing of social networks for improving healthy behaviors. *JMIR Human Factors, 3*(1), e4512.

Hargittai, E. (2020). Potential biases in big data: Omitted voices on social media. *Social Science Computer Review, 38*(1), 10–24.

Harrington, C. N., Wilcox, L., Connelly, K., Rogers, W., & Sanford, J. (2018). Designing health and fitness apps with older adults: Examining the value of experience-based co-design. In *Proceedings of the 12th EAI International Conference on Pervasive Computing Technologies for Healthcare* (pp. 15–24).

Hochberg, I., Allon, R., & Yom-Tov, E. (2020). Assessment of the frequency of online searches for symptoms before diagnosis: Analysis of archival data. *Journal of Medical Internet Research, 22*(3), e15065.

Hochberg, I., Daoud, D., Shehadeh, N., & Yom-Tov, E. (2019). Can internet search engine queries be used to diagnose diabetes? Analysis of archival search data. *Acta Diabetologica, 56*(10), 1149–1154.

Hoertel, N., Blachier, M., Blanco, C., Olfson, M., Massetti, M., Rico, M. S., Limosin, F., & Leleu, H. (2020). A stochastic agent-based model of the SARS-CoV-2 epidemic in France. *Nature Medicine, 26*(9), 1417–1421.

Hswen, Y., Zhang, A., Sewalk, K. C., Tuli, G., Brownstein, J. S., & Hawkins, J. B. (2020). Investigation of geographic and macrolevel variations in LGBTQ patient experiences: Longitudinal social media analysis. *Journal of Medical Internet Research,, 22*, e17087.

Jang, S. M., Mckeever, B. W., Mckeever, R., & Kim, J. K. (2019). From social media to mainstream news: The information flow of the vaccine-autism controversy in the US, Canada, and the UK. *Health Communication, 34*(1), 110–117.

Jeffrey, B., Walters, C. E., Ainslie, K. E., Eales, O., Ciavarella, C., Bhatia, S., Hayes, S., Baguelin, M., Boonyasiri, A., Brazeau, N. F., Cuomo-Dannenburg, G., FitzJohn, R. G., Gaythorpe, K., Green, W., Imai, N., Mellan, T. A., Mishra, S., Nouvellet, P., Juliette, H., . . . Riley, S. (2020). Anonymised and aggregated crowd level mobility data from mobile phones suggests that initial compliance with COVID-19 social distancing interventions was high and geographically consistent across the UK. *Wellcome Open Research, 5*, 170.

Johnson, N. F., Velásquez, N., Restrepo, N. J., Leahy, R., Gabriel, N., El Oud, S., Zheng, M., Manrique, P., Wuchty, S., & Lupu, Y. (2020). The online competition between pro-and anti-vaccination views. *Nature, 582*(7811), 230–233.

Kapitány-Fövény, M., Ferenci, T., Sulyok, Z., Kegele, J., Richter, H., Vályi-Nagy, I., & Sulyok, M. (2019). Can google trends data improve forecasting of lyme disease incidence? *Zoonoses and Public Health, 66*(1), 101–107.

Katsuki, T., Mackey, T. K., & Cuomo, R. (2015). Establishing a link between prescription drug abuse and illicit online pharmacies: Analysis of twitter data. *Journal of Medical Internet Research, 17*(12), e280.

Kazemi, D. M., Borsari, B., Levine, M. J., & Dooley, B. (2017). Systematic review of surveillance by social media platforms for illicit drug use. *Journal of Public Health, 39*(4), 763–776.

Keller, S. N., Honea, J. C., & Ollivant, R. (2021). How social media comments inform the promotion of mask-wearing and other covid-19 prevention strategies. *International Journal of Environmental Research and Public Health, 18*(11), 5624.

Kiti, M. C., Tizzoni, M., Kinyanjui, T. M., Koech, D. C., Munywoki, P. K., Meriac, M., Cappa, L., Panisson, A., Barrat, A., Cattuto, C., et al. (2016). Quantifying social contacts in a household setting of rural kenya using wearable proximity sensors. *EPJ Data Science, 5*(1), 1–21.

Kleppmann, M. (2017). *Designing data-intensive applications: The big ideas behind reliable, scalable, and maintainable systems.* O'Reilly Media.

Koppeschaar, C. E., Colizza, V., Guerrisi, C., Turbelin, C., Duggan, J., Edmunds, W. J., Kjelsø, C., Mexia, R., Moreno, Y., Meloni, S., Paolotti, D., Perrotta, D., van Straten, E., Franco, A. O. (2017). Influenzanet: Citizens among 10 countries collaborating to monitor influenza in Europe. *JMIR Public Health and Surveillance, 3*(3), e7429.

Kostygina, G., Tran, H., Binns, S., Szczypka, G., Emery, S., Vallone, D., & Hair, E. (2020). Boosting health campaign reach and engagement through use of social media influencers and memes. *Social Media+ Society, 6*(2), 2056305120912475.

Lazer, D., Kennedy, R., King, G., & Vespignani, A. (2014). The parable of google flu: Traps in big data analysis. *Science, 343*(6176), 1203–1205.

Li, L.-F., Wang, X., Hu, W.-J., Xiong, N. N., Du, Y.-X., & Li, B.-S. (2020). Deep learning in skin disease image recognition: A review. *IEEE Access, 8*, 208264–208280.

Liu, Q.-H., Ajelli, M., Aleta, A., Merler, S., Moreno, Y., & Vespignani, A. (2018). Measurability of the epidemic reproduction number in data-driven contact networks. *Proceedings of the National Academy of Sciences, 115*(50), 12680–12685.

Liu, Y., Deng, Y., Jusup, M., & Wang, Z. (2016). A biologically inspired immunization strategy for network epidemiology. *Journal of Theoretical Biology, 400*, 92–102.

Liu, Z., & Hu, B. (2005). Epidemic spreading in community networks. *EPL (Europhysics Letters)*, *72*(2), 315.

Łuksza, M., & Lässig, M. (2014). A predictive fitness model for influenza. *Nature, 507*(7490), 57–61.

Madathil, K. C., Rivera-Rodriguez, A. J., Greenstein, J. S., & Gramopadhye, A. K. (2015). Healthcare information on youtube: A systematic review. *Health Informatics Journal, 21*(3), 173–194.

Maier, B. F., & Brockmann, D. (2020). Effective containment explains subexponential growth in recent confirmed covid-19 cases in China. *Science, 368*(6492), 742–746.

Marsaux, C. F., Celis-Morales, C., Livingstone, K. M., Fallaize, R., Kolossa, S., Hallmann, J., San-Cristobal, R., Navas-Carretero, S., O'Donovan, C. B., Woolhead, C., Forster, H., Moschonis, G., Lambrinou, C.-P., Surwillo, A., Godlewska, M., Hoonhout, J., Goris, A., Macready, A. L., Walsh, M. C., . . . Saris, W. H. M. (2016). Changes in physical activity following a geneticbased internet-delivered personalized intervention: Randomized controlled trial (food4me). *Journal of medical Internet research, 18*(2), e30.

Maxmen, A. (2021). Why did the world's pandemic warning system fail when covid hit? *Nature, 589*, 499–500.

Mejova, Y., Gandhi, H. R., Rafaliya, T. J., Sitapara, M. R., Kashyap, R., & Weber, I. (2018). Measuring subnational digital gender inequality in India through gender gaps in facebook use. In *Proceedings of the 1st ACM SIGCAS Conference on Computing and Sustainable Societies* (pp. 1–5).

Mejova, Y., Haddadi, H., Noulas, A., & Weber, I. (2015). # Foodporn: Obesity patterns in culinary interactions. In *Proceedings of the 5th International Conference on Digital Health 2015* (pp. 51–58).

Mejova, Y., & Kalimeri, K. (2019). Effect of values and technology use on exercise: Implications for personalized behavior change interventions. In *Proceedings of the 27th ACM Conference on User Modeling, Adaptation and Personalization* (pp. 36–45).

Mejova, Y., & Kourtellis, N. (2021). Youtubing at home: Media sharing behavior change as proxy for mobility around covid-19 lockdowns. In *13th Acm Web Science Conference 2021* (pp. 272–281). Association for Computing Machinery. https://doi.org/10.1145/3447535.3462494

Mejova, Y., & Suarez-Lledó, V. (2020). Impact of online health awareness campaign: Case of national eating disorders association. In *International Conference on Social Informatics* (pp. 192–205).

Mejova, Y., Weber, I., & Fernandez-Luque, L. (2018). Online health monitoring using facebook advertisement audience estimates in the united states: Evaluation study. *JMIR Public Health and Surveillance, 4*(1), e7217.

Meyer, D. (2021). *Apple and google flex privacy muscles with blockage of english covid contact-tracing app update.* Accessed 18 Sep 2021.

Miguel, E., Camerer, C., Casey, K., Cohen, J., Esterling, K. M., Gerber, A., Glennerster, R., Green, D. P., Humphreys, M., Imbens, G., et al. (2014). Promoting transparency in social science research. *Science, 343*(6166), 30–31.

Moreno, Y., Gómez, J. B., &Pacheco, A. F. (2003). Epidemic incidence in correlated complex networks. *Physical Review E, 68*(3), 035103.

National Health Service. (2021). *Better health.* Accessed 1 Sep 2021.

National Institutes of Health. (2021). *Open-access data and computational resources to address covid-19.* Accessed 1 Sep 2021.

Netflix. (2009). *Netflix prize.* Accessed 18 Sep 2021.

Ngonghala, C. N., Iboi, E., Eikenberry, S., Scotch, M., MacIntyre, C. R., Bonds, M. H., & Gumel, A. B. (2020). Mathematical assessment of the impact of non-pharmaceutical interventions on curtailing the 2019 novel coronavirus. *Mathematical biosciences, 325*, 108364.

Obi, C. G., Ezaka, E. I., Nwankwo, J. I., & Onuigbo, I. I. (2020). Role of the epidemiologist in the containment of COVID-19 pandemic. *AIJR Preprints.* https://doi.org/10.21467/preprints.183

Ogden, N. H., Fazil, A., Arino, J., Berthiaume, P., Fisman, D. N., Greer, A. L., Ludwig, A., Ng, V., Tuite, A. R., Turgeon, P., Waddell, L. A., & Wu, J. (2020). Artificial intelligence in public health: Modelling scenarios of the epidemic of COVID-19 in Canada. *Canada Communicable Disease Report, 46*(8), 198.

Oliver, N., Lepri, B., Sterly, H., Lambiotte, R., Deletaille, S., De Nadai, M., Letouzé, E., Salah, A. A., Benjamins, R., Cattuto, C., Colizza, V., de Cordes, N., Fraiberger, S. P., Koebe, T., Lehmann, S., Murillo, J., Pentland, A., Pham, P. N., Pivetta, F., ...Vinck, P. (2020). Mobile phone data for informing public health actions across the COVID-19 pandemic life cycle. *Science Advance, 6*(23), eabc0764.

Olteanu, A., Castillo, C., Diaz, F., & Kýcýman, E. (2019). Social data: Biases, methodological pitfalls, and ethical boundaries. *Frontiers in Big Data, 2*, 13.

op den Akker, H., Jones,V. M.,& Hermens, H. J. (2014). Tailoring real-time physical activity coaching systems: A literature survey and model. *User Modeling and User-Adapted Interaction, 24*(5), 351–392. https://doi.org/10.1007/s11257-014-9146-y

Play Store. (2021). *Fit India*, Accessed 1 Sep 2021.

Rabbi, M., Aung, M. H., Zhang, M., & Choudhury, T. (2015). Mybehavior: Automatic personalized health feedback from user behaviors and preferences using smartphones. In *Proceedings of the 2015 ACM International Joint Conference on Pervasive and Ubiquitous Computing* (pp. 707–718).

Rashid, F. Y. (2020). The rise of confidential computing: Big tech companies are adopting a new security model to protect data while it's in use-[news]. *IEEE Spectrum, 57*(6), 8–9.

Rastegar-Mojarad, M., Ye, Z., Wall, D., Murali, N., & Lin, S. (2015). Collecting and analyzing patient experiences of health care from social media. *JMIR Research Protocols, 4*(3), e3433.

Reece, A. G., & Danforth, C. M. (2017). Instagram photos reveal predictive markers of depression. *EPJ Data Science, 6*, 1–12.

Rich, E., & Miah, A. (2017). Mobile, wearable and ingestible health technologies: Towards a critical research agenda. *Health Sociology Review, 26*(1), 84–97.

Rosenberg, H., Syed, S., & Rezaie, S. (2020). The twitter pandemic: The critical role of twitter in the dissemination of medical information and misinformation during the covid-19 pandemic. *Canadian Journal of Emergency Medicine, 22*(4), 418–421.

Rosenblum, S., & Yom-Tov, E. (2017). Seeking web-based information about attention deficit hyperactivity disorder: Where, what, and when. *Journal of Medical Internet Research, 19*(4), e6579.

Sajadmanesh, S., Jafarzadeh, S., Ossia, S. A., Rabiee, H. R., Haddadi, H., Mejova, Y., Musolesi, M., Cristofaro, E. D., & Stringhini, G. (2017). Kissing cuisines: Exploring worldwide culinary habits on the web. In *Proceedings of the 26th International Conference on World Wide Web Companion* (pp. 1013–1021).

Shepherd, H. E., Atherden, F. S., Chan, H. M. T., Loveridge, A., & Tatem, A. J. (2021). Domestic and international mobility trends in the united kingdom during the covid-19 pandemic: An analysis of facebook data. *International Journal of Health Geographics 20*, 46 (2021). https://doi.org/10.1186/s12942-021-00299-5

Short, C., Rebar, A., James, E., Duncan, M., Courneya, K., Plotnikoff, R., Crutzen, R., & Vandelanotte, C. (2017). How do different delivery schedules of tailored web-based physical activity advice for breast cancer survivors influence intervention use and efficacy? *Journal of Cancer Survivorship, 11*(1), 80–91.

Silva, D. H., Ferreira, S. C., Cota, W., Pastor-Satorras, R., & Castellano, C. (2019). Spectral properties and the accuracy of mean-field approaches for epidemics on correlated power-lawnetworks. *Physical Review Research, 1*(3), 033024.

Smieszek, T., Castell, S., Barrat, A., Cattuto, C., White, P. J., & Krause, G. (2016). Contact diaries versus wearable proximity sensors in measuring contact patterns at a conference: Method comparison and participants' attitudes. *BMC Infectious Diseases, 16*(1), 1–14.

Stewart, I., Chancellor, S., De Choudhury, M., & Eisenstein, J. (2017). # Anorexia, # anarexia, # anarexyia: Characterizing online community practices with orthographic variation. In *2017 IEEE International Conference on Big Data (Big Data)* (pp. 4353–4361).

Taylor, L. (2023). Data justice, computational social science and policy. In Bertoni, E., Fontana, M., Gabrielli, L., Signorelli, S., Vespe, M. (Eds.), *Handbook of computational social science for policy*. Springer.

Tizzoni, M., Bajardi, P., Decuyper, A., Kon Kam King, G., Schneider, C. M., Blondel, V., Smoreda, Z., González, M. C., & Colizza, V. (2014). On the use of human mobility proxies for modeling epidemics. *PLoS Computational Biology, 10*(7), e1003716.

Tkachenko, N., Chotvijit, S., Gupta, N., Bradley, E., Gilks, C., Guo, W., Crosby, H., Shore, E., Thiarai, M., Procter, R., et al. (2017). Google trends can improve surveillance of type 2 diabetes. *Scientific Reports, 7*(1), 1–10.

Tseng, J. C., Lin, B.-H., Lin, Y.-F., Tseng, V. S., Day, M.-L., Wang, S.-C., Lo, K.-R.,& Yang, Y.-C. (2015). An interactive healthcare system with personalized diet and exercise guideline recommendation. In *2015 Conference on Technologies and Applications of Artificial Intelligence (TAAI)* (pp. 525–532).

University of Chicago. (2021). *Data science for social good summer fellowship*. Accessed 18 Sep 2021.

Van den Broeck, W., Gioannini, C., Gonçalves, B., Quaggiotto, M., Colizza, V., & Vespignani, A. (2011). The gleamviz computational tool, a publicly available software to explore realistic epidemic spreading scenarios at the global scale. *BMC Infectious Diseases, 11*(1), 1–14.

Vandelanotte, C., Short, C., Plotnikoff, R. C., Hooker, C., Canoy, D., Rebar, A., Alley, S., Schoeppe, S., Mummery, W. K., & Duncan, M. J. (2015). Tayloractive–examining the effectiveness of web-based personally-tailored videos to increase physical activity: A randomised controlled trial protocol. *BMC Public Health, 15*(1), 1020.

Vanhems, P., Barrat, A., Cattuto, C., Pinton, J.-F., Khanafer, N., Régis, C., Kim, B.-a., Comte, B., & Voirin, N. (2013). Estimating potential infection transmission routes in hospital wards using wearable proximity sensors. *PloS One, 8*(9), e73970.

Vayena, E., Salathé, M., Madoff, L. C., & Brownstein, J. S. (2015). Ethical challenges of big data in public health. *PLOS Computational Biology, 11*, e1003904.

Venkatramanan, S., Sadilek, A., Fadikar, A., Barrett, C. L., Biggerstaff, M., Chen, J., Dotiwalla, X., Eastham, P., Gipson, B., Higdon, D., Kucuktunc, O., Lieber, A., Lewis, B. L., Reynolds, Z., Vullikanti, A. K., Wang, L., & Marathe, M. (2021). Forecasting influenza activity using machine-learned mobility map. *Nature communications, 12*(1), 1–12.

Vespe, M., Iacus, S. M., Santamaria, C., Sermi, F., & Spyratos, S. (2021). On the use of data from multiple mobile network operators in europe to fight COVID-19. *Data & Policy, 3*, E9.

Volz, E. M., Koelle, K., & Bedford, T. (2013). Viral phylodynamics. *PLoS Computational Biology, 9*(3), e1002947.

Walker, P. G., Whittaker, C., Watson, O. J., Baguelin, M., Winskill, P., Hamlet, A., Djafaara, B. A., Cucunubá, Z., Olivera Mesa, D., Green, W., Thompson, H., Nayagam, S., Ainslie, K. E. C., Bhatia, S., Bhatt, S., Boonyasiri, A., Boyd, O., Brazeau, N. F., Cattarino, L., ... Ghani, A. C. (2020). The impact of COVID-19 and strategies for mitigation and suppression in low-and middle-income countries. *Science, 369*(6502), 413–422.

Wang, Z., Yin, Z., & Argyris, Y. A. (2020). Detecting medical misinformation on social media using multimodal deep learning. *IEEE Journal of Biomedical and Health Informatics, 25*(6), 2193–2203.

Weber, I., & Achananuparp, P. (2016). Insights from machine-learned diet success prediction. In *Biocomputing 2016: Proceedings of the Pacific Symposium* (pp. 540–551).

Weidman, S. (2019). *Deep learning from scratch: Building with python from first principles*. O;Reilly Media.

Weiss, M. G. (2001). Cultural epidemiology: An introduction and overview. *Anthropology & Medicine, 8*(1), 5–29.

Weitz, J. S., Park, S.W., Eksin, C., & Dushoff, J. (2020). Awareness-driven behavior changes can shift the shape of epidemics away from peaks and toward plateaus, shoulders, and oscillations. *Proceedings of the National Academy of Sciences, 117*(51), 32764–32771.

Wilder, B., Charpignon, M., Killian, J. A., Ou, H.-C., Mate, A., Jabbari, S., Perrault, A., Desai, A.N., Tambe, M., & Majumder, M. S. (2020). Modeling between population variation in

COVID-19 dynamics in Hubei, Lombardy, and New York City. *Proceedings of the National Academy of Sciences, 117*(41), 25904–25910.

World Health Organization. (2017). *One health*. Accessed 21 Sep 2021.

World Health Organization. (2021a). *Fighting misinformation in the time of COVID-19, one click at a time*. Accessed 21 Sep 2021.

World Health Organization. (2021b). *Who hub for pandemic and epidemic intelligence*. Accessed 21 Sep 2021.

World Health Organization. (2021c). *Who, germany open hub for pandemic and epidemic intelligence in Berlin*. Accessed 21 Sep 2021.

Woskie, L. R., Hennessy, J., Espinosa, V., Tsai, T. C., Vispute, S., Jacobson, B. H., Cattuto, C., Gauvin, L., Tizzoni, M., Fabrikant, A., Gadepalli, K., Boulanger, A., Pearce, A., Kamath, C., Schlosberg, A., Stanton, C., Bavadekar, S., Abueg, M., Hogue, M., . . . , Gabrilovich, E. (2021). Early social distancing policies in Europe, changes in mobility & COVID-19 case trajectories: Insights from spring 2020. *Plos one, 16*(6), e0253071.

Wu, Q., Fu, X., Jin, Z., & Small, M. (2015). Influence of dynamic immunization on epidemic spreading in networks. *Physica A: Statistical Mechanics and its Applications, 419*, 566–574.

Yamada, Y., Ćepulić, D.-B., Coll-Martın, T., Debove, S., Gautreau, G., Han, H., Rasmussen, J., Tran, T. P., Travaglino, G. A., & Lieberoth, A. (2021). Covidistress global survey dataset on psychological and behavioural consequences of the covid-19 outbreak. *Scientific Data, 8*(1), 1–23.

Yechiam, E., Yom-Tov, E. (2021). Unique internet search strategies of individuals with self-stated autism: Quantitative analysis of search engine users' investigative behaviors. *Journal of Medical Internet Research, 23*(7), e23829.

Yom-Tov, E. (2019). Demographic differences in search engine use with implications for cohort selection. *Information Retrieval Journal, 22*(6), 570–580.

Yom-Tov, E., Feraru, G., Kozdoba, M., Mannor, S., Tennenholtz, M., & Hochberg, I. (2017). Encouraging physical activity in patients with diabetes: Intervention using a reinforcement learning system. *Journal of Medical Internet Research, 19*(10), e338.

Yom-Tov, E., Fernandez-Luque, L., Weber, I., & Crain, S. P. (2012). Pro-anorexia and pro-recovery photo sharing: A tale of two warring tribes. *Journal of Medical Internet Research, 14*(6), e151.

Yom-Tov, E., & Lev-Ran, S. (2017). Adverse reactions associated with cannabis consumption as evident from search engine queries. *JMIR Public Health and Surveillance, 3*(4), e77.

Chapter 16
Learning Analytics in Education for the Twenty-First Century

Kristof De Witte and Marc-André Chénier

Abstract The online traces that students leave on electronic learning platforms; the improved integration of educational, administrative and online data sources; and the increasing accessibility of hands-on software allow the domain of learning analytics to flourish. Learning analytics, as in interdisciplinary domain borrowing from statistics, computer sciences and education, exploits the increased accessibility of technology to foster an optimal learning environment that is both transparent and cost-effective. This chapter illustrates the potential of learning analytics to stimulate learning outcomes and to contribute to educational quality management. Moreover, it discusses the increasing emergence of large and accessible data sets in education and compares the cost-effectiveness of learning analytics to that of costly and unreliable retrospective studies and surveys. The chapter showcases the potential of methods that permit savvy users to make insightful predictions about student types, performance and the potential of reforms. The chapter concludes with recommendations, challenges to the implementation and growth of learning analytics.

16.1 Introduction

Education stakeholders are currently working within an environment where vast quantities of data can be leveraged to have a deeper understanding of the educational attainment of learners. A growing pool of data is generated through software with

K. De Witte (✉)
Leuven Economics of Education Research (LEER), KU Leuven, Leuven, Belgium

Maastricht Economic and Social Research Institute on Innovation and Technology (UNU-MERIT), United Nations University, Maastricht, The Netherlands
e-mail: kristof.dewitte@kuleuven.be; k.dewitte@maastrichtuniversity.nl

M.-A. Chénier
Leuven Economics of Education Research (LEER), KU Leuven, Leuven, Belgium
e-mail: marcandre.chenier@kuleuven.be

© The Author(s) 2023
E. Bertoni et al. (eds.), *Handbook of Computational Social Science for Policy*,
https://doi.org/10.1007/978-3-031-16624-2_16

which students, teachers and administrators interact (Kassab et al., 2020), through apps, social networking and the collection of user behaviour on aggregators such as *YouTube* and *Google* (De Wit & Broucker, 2017). Moreover, thanks to the Internet of Everything phenomenon, stakeholders in the education domain have access to data in which people, processes, data and things connect to the internet and to each other (Langedijk et al., 2019). That data takes on non-traditional formats and retains language, location, movement, networks, images and video information (Lazer et al., 2020). Such non-traditional data sets require cutting-edge analytical techniques in order to be effectively used for learning purposes and to be translated into succinct policy recommendations.

Learning analytics, as an interdisciplinary domain borrowing from statistics, computer sciences and education (Leitner et al., 2017), exploits this new data-rich landscape to improve the learning process and outcomes of current and future citizens (De Wit & Broucker, 2017). In education, learning analytics is set squarely within the new computational social sciences, which consist in the "development and application of computational methods to complex, typically large-scale human behavioral data" (Lazer et al., 2009). Learning analytics directs these advances towards the creation of actionable information in education. It applies data analytics to the field of education, and it attempts to propose ways to explore, analyse and visualize data from any relevant data source (Vanthienen & De Witte, 2017). An important role of learning analytics is the exploitation of the traces left by students on electronic learning platforms (Greller & Drachsler, 2012). As such, learning analytics allows teachers to maximize the cognitive and non-cognitive education outcomes of students (Long & Siemens, 2011). In an optimal learning environment, one would maximally leverage the potential of students to increase their welfare and performance not only during schooling but also afterwards, across civil society.

As the COVID-19 pandemic induced shifts towards online and home education, there is an increased opportunity for data analytics in general and to mitigate the crisis' effects both on learning outcomes (Maldonado & De Witte, 2021) and on the well-being of students (Iterbeke & De Witte, 2020) in particular. The online traces that students leave on electronic learning platforms allow teachers, schools and policy-makers to better tailor targeted remedial teaching interventions to the most needy students. The closures of schools also showed how unequally digital devices are spread among students, with significant groups of disadvantaged students without access to basic digital instruments such as stable broadband access and computer. Similarly, the school closures revealed significant differences between countries in their readiness for online teaching and in the availability of high-quality digital instruction. Still, thanks to the unprecedented crisis, multiple countries made significant investments in the educational ICT infrastructure (De Witte & Smet, 2021). If this coincides with improved training of teachers and school managers; an improved integration of educational, administrative and online data sources; and the improved accessibility of hands-on software, we expect to see the domain of learning analytics to further flourish in the next decades.

The following chapter aims to contribute to this accelerated use of learning analytics by picturing its potential in multiple educational domains. We first discuss the increasing emergence of large and accessible data sets in education and the associated growth in expertise in educational data collection and analysis. This is sustained by real-time streamed data and increasingly autonomous administrative data sets. Section 16.2 compares the cost-effectiveness of learning analytics to that of costly and unreliable retrospective studies and surveys. Learning analytics may also contribute to the improvement in the quality of the currently dispensed education through fraud detection and student performance prediction, for example. In Sect. 16.3, three tools of growing popularity and potential for learning analytics are presented: the Bayesian Additive Regression Trees (BART), the Social Network Analysis (SNA) and the Natural Language Processing (NLP). These tools permit savvy users to make insightful predictions about student types, performance and the potential of reforms. The brief description of these techniques aims to familiarize practitioners and decision-makers with their potential. Finally, alongside recommendations, technical and non-technical challenges to the implementation and growth of learning analytics and empirically based education in general are discussed. As the growing possibilities of learning analytics result in sensitive options regarding data usage and linkages, we discuss in the conclusion section the related ethical and legal concerns.

16.2   Potential for Educators and Citizens

16.2.1   Growing Opportunities for Data-Driven Policies in Education

"Students and teachers are leaving large amounts of digital footprints and traces in various educational apps and learning management platforms, and education administrators register various processes and outcomes in digital administrative systems" (Nouri et al., 2019). In this section, we discuss three trends that allow for growing opportunities in fomenting creative data-driven policies in education: (1) the development of online teaching platforms, (2) software-oriented administrative data collection with links between heterogeneous data sets (Langedijk et al., 2019) and (3) the Internet of Things (Langedijk et al., 2019).

First, consider the online teaching platforms. A prime example of the latter are massive open online courses (i.e. MOOCs, De Smedt et al., 2017). Institutional MOOC initiatives have been contributing to making high-quality educational material accessible to a wide range of students and to maintaining the prestige of the participating institutions (Dalipi et al., 2018). For adults, MOOC completion has also been associated with increased resilience to unemployment (Castaño-Muñoz & Rodrigues, 2021). From a learning analytics perspective, it is interesting to observe that all student activities can be tracked within the MOOC. This information has been studied to give empirical grounding to suggestions to reduce course dropout by fostering peer engagements on online forums, team homeworks and peer evaluations

(Dalipi et al., 2018). From a methodological perspective, some of the innovative methodologies exploiting MOOC's large data sets include K-means clustering, support vector machines and hidden Markov models.[1]

A second trend in data-driven policies in education arise from software-oriented administrative data collections. These refer to the digital warehousing of administrative data such that this data can be relatively easily linked with other data sets and easily transformed through, for example, the inclusion of a large quantity of new observations (e.g. student files) and the ad hoc addition of new variables of interest (Agasisti et al., 2017). Administrative data sets are built around procedures whose aims are not primarily to foster data-driven policies (Barra et al., 2017; Rettore & Trivellato, 2019). In that sense, they can provide rich information about students and other educational stakeholders while being quicker to gather and significantly cheaper than retrospective surveys (Figlio et al., 2016).

As a major advantage, software-oriented administrative data collections can be easily linked to other data sources, such as the wide array of information surveyed by local governments in their interactions with citizens. Through software integration, data regarding such diverse domains as public health and agriculture may be seamlessly captured. To conceptualize the diversity of potential data sources, Langedijk et al. (2019) describe those data as divided into thematic silos. Each silo represents an important civil concern, health or education, for example, and within each silo, stakeholders can define sub-themes onto which interesting data sets are attached. For example, in the case of education, some proposed sub-themes are standardized test results, textbook quality and teacher quality[2] (Langedijk et al., 2019). Through the development of electronic networks, links cannot only be established within silos, where policy-makers may, for instance, be interested in the relation between teacher quality and test scores, but also across silos, where improvement in learning outcomes can be associated with changes in the health of citizens (Langedijk et al., 2019). The analyses required to measure such associations can take advantage of the typically long-run collection of administrative data (Figlio et al., 2016). As an additional advantage of the electronic networks, whereas data has traditionally been transmitted in batches, in order to produce descriptive reports at set time intervals, for example, electronic networks now permit event registration in real time (De Wit & Broucker, 2017; Mukala et al., 2015). The real-time extraction

---

[1] K-means clustering divides the observations (e.g. a sample of teachers) into a quantity K of groups that share similar measured characteristics. That similarity is defined as the squared distance to the mean of the group's characteristics (Bishop, 2006). Support vector machines construct a porous hyperplane that maximally separate the observations closest to it. They are particularly useful to solve classification problems with high-dimensional data (e.g. registered student activity during multiple lecture) (Bishop, 2006). Finally, hidden Markov models assume that measurements are generated by underlying hidden states. These hidden states are modelled as a Markov process (Bishop, 2006). That approach is particularly suited to the analysis of sequential data such as the quantity of attempts in an educational game (Tadayon, 2020).

[2] Teacher quality is a multi-dimensional concept that is often proxied by teacher value-added scores.

of data benefits teachers and students who can rely, for example, on automated assignments and online dashboards in order to improve their learning experience and their learning outcomes (De Smedt et al., 2017).

A good example of data set linkages in education are studies with population data that aim to explore education outcomes in specific subgroups. A recent study by Mazrekaj et al. (2020) made use of the rich micro-data sets made available to researchers by the Dutch Central Bureau of Statistics (CBS). These micro-data cover many themes of social life (e.g. financial, educational, health, environmental, professional silos and more) and are, though of limited access because of privacy issues, easy to link together with standard analytics software.

Third, consider the Internet of Things. The Internet of Things denotes the numerous physical devices with integrated internet connectivity (DeNardis, 2020). In educational settings, these devices are the computers, SNS services, mobile devices, camera, sensors and software with which students, teachers and administrators interact (Kassab et al., 2020). They are used to monitor student attendance and class behaviour and their interactions with online teaching services and laboratories. On online platforms, but also through mobile apps and logging platforms (e.g. library access, blogs, electronic learning environment), students' and tutors' behaviours and opinions can be monitored in real time and passed through automatic analytics platforms or saved to solve future policy issues (De Smedt et al., 2017; De Wit & Broucker, 2017). Similarly, RFID (radio-frequency identification) sensors track the locations and availability of educational appliances such as laboratory equipment and projectors. Students and tutors can communicate with each other regardless of location, and assessment feedback can be delivered instantaneously, resulting in higher-quality education.

16.2.2 Learning Analytics as a Toolset

The toolset of learning analytics can be used for several purposes. We first provide some examples on how it can contribute to improve the cost-effectiveness of education and next how it can foster education outcomes on cognitive and non-cognitive scales. Finally, we provide examples of how learning analytics can assist in educational quality management.

16.2.2.1 Improving Cost-Effectiveness of Education

The increasing public scrutiny and tighter budgets, which are an ever-present reality of the educational landscape, motivate a double goal for data-driven solutions. These must improve efficiency and performance with regard to learning outcomes while also proposing solutions that are competitive in terms of cost (Barra et al., 2017). There are two poles through which cost-effective learning analytics solutions can be proposed.

The first pole stands at the level of data collection. Administrative data sets suffer from their high cost of data cleaning and collection. Indeed, although data extraction is usually native to recent administrative software (King, 2016), administrative data sets typically require ad hoc linkages and research designs (Agasisti et al., 2017). In the sense that their inclusion in data-driven decision-making is not their primary purpose, they constitute an opportunistic data source and thus may occasionally demand more resource investments than deliberate data collection procedures. Meanwhile, the omnipresent network of computing devices and the associated online educational platforms permit data extraction at every step of the learning process (De Smedt et al., 2017). As previously indicated, this type of unstructured data can be saved, but the real-time data stream can also be designed in such a way to permit automatic analyses. This deliberate pipeline associating the collected data to useful analyses can insure cost-effectiveness through economies of scales. It can also serve as a baseline to future improvements in summarizing data for students, teachers and stakeholders in general. In short, rich data sets and insightful analyses can be produced without requiring punctual organizational involvement. In that sense, the environment in which learning analytics is embedded permits professionals and stakeholders to benefit from opportunistic analyses and from insights that are delivered efficiently (Barra et al., 2017). For example, during the COVID-19 crisis, learning analytics was used to monitor how students were reached by online teaching.

The second pole to achieve cost-effectiveness in the establishment of data-driven policy-making for education is that of data analytics. Up until now, technologically able and creative teams have been achieving parity with the expanding volume, variety and velocity of data by developing and applying advanced analytical methods (De Wit & Broucker, 2017; King, 2016). One such method is Data Envelopment Analysis (DEA). It permits the employment of administrative and learning data in order to directly fulfil goals related to cost minimization (Barra et al., 2017; De Witte & López-Torres, 2017; Mergoni & De Witte, 2021). The result of such analyses may be useful in promoting efficient investments in educational resources (see, e.g. the report by the European Commission Expert Group on Quality Investment in Education and Training). Additional spending brings to the forefront its paradoxical effect of increasing cost-effectiveness in the long run. Advances in social sciences have already demonstrated the consequences of poor learning outcomes, the principal of which are "lower incomes and economic growth, lower tax revenues, and higher costs of such public services as health, criminal justice, and public assistance" (Groot & van den Brink, 2017). Hence, learning outcomes deserve an important place in discussions around the cost-effectiveness of education (De Witte & Smet, 2021).

16.2.2.2 Improving Learning Outcomes

In terms of directly improving educational quality, three ambitions can be distinguished for learning analytics: making improvements in (non-)cognitive learning outcomes, reducing learning support frictions and a wide deployment and long-term maintenance for each teaching tool (Viberg et al., 2018). These ambitions are now discussed.

First, learning outcomes can be interpreted as the academic performance of students, as measured by quizzes and examinations (Viberg et al., 2018). Learning outcomes can also be defined in a broader way than similar testable outcomes, for example, by being related to interpersonal skills and civic qualities. However widely defined, it is important that the set of criteria identifying educational success is well-defined by stakeholders and that it is clearly communicated to and open to the contributions of citizens. In that way, educational policy discussions can be centred around transparent and recognized aims.

Although there is a rich literature evaluating learning analytics in higher education, the contributions of learning analytics tools to improving the (non-)cognitive learning outcomes of secondary school students have received relatively little attention in the empirical literature (Bruno et al., 2021). Nevertheless, clear improvements in writing and argumentative quality have been associated with the use of automatic text evaluation softwares (Lee et al., 2019; Palermo & Wilson, 2020). These softwares use Natural Language Processing (NLP) to analyse data extracted from online learning platforms. Automatic text evaluation has also shown promising results at higher education levels and with non-traditional adult students (Whitelock et al., 2015b). There is thus flexibility in terms of the type of students or teachers to whom learning analytics approaches apply.

Another interesting contribution of learning analytics to the outcomes of secondary school students has been in improving their computer programming abilities. This has been accomplished through another advanced data analysis technique, process mining, which helped teachers in pairing students based on captured behavioural traces during programming exercises (Berland et al., 2015).

Second, with respect to learning support frictions, there is often a lag between the assumptions behind the design of learning platforms and the observed behaviours of students (Nguyen et al., 2018). An example of this lag is that students tend to spend less time studying than recommended by their instructors. Less involved students also tend to spend less time preparing assignments (Nguyen et al., 2018). By reducing their ability to receive feedback in a timely manner, a similar lag can negatively affect both students' and teachers' involvement in the learning process. Thanks to learning analytics tools, students will receive tailored feedback, will rehearse exercises that are particularly difficult for them and will receive stimulating examples that fit their interest (Iterbeke et al., 2020). This reduces the learning support frictions and consequently improves learning outcomes.

Yet, the lag between the desired learning outcomes and student behaviour cannot be corrected simply through the implementation of electronic platforms or through a gamification of the learning process. It is critical that the digital tools being implemented and those implementing them take students' feedback into account. Many students are now used to accessing information without having to pass through much in the manner of physical and social barriers. For those students, the interactivity and the practicality of the digital learning tools are particularly important (Pardo et al., 2018; Selwyn, 2019). Other students may not have the same familiarity with online computing devices. For these, accessibility has to be negotiated into the tools.

Many authors warn of a transfer from magisterial education to learning platforms in which feedback and exercises may be too numerous, superficial or ill-adapted to students' capabilities or learning ambitions (Lonn et al., 2015; Pardo et al., 2018; Topolovec, 2018). Hence, a hybrid approach to learning support is suggested wherein technologies, such as those just touched upon of automatic text analyses and process mining, are combined with personalized feedback from teachers and tutors. Indeed, classroom teaching is often characterized by a lack of personalization and biases in the dispensation of feedback and exercises. For example, low-performing students are over-represented among the receiver of teacher feedback. Additionally, given the same learning objectives, feedback may be administrated differently to students of different genders and origins. Teachers may find learning analytics tools useful in helping their students attain the desired learning outcomes while fostering their personal learning ambitions and their self-confidence (Evans, 2013; Hattie & Timperley, 2007).

Third, learning analytics can provide additional value to students and teachers. In that sense, we observe several clear advantageous applications of learning analytics.

- Learning analytics could contribute to non-cognitive skills, as collaboration is an area where non-cognitive skills play an important role. Identifying collaboration and the factors that incite it can improve learning outcomes and even help in preventing fraud. The implementation of analytics methods such as Social Network Analysis (SNA) in learning platforms may allow teachers to prevent or foster such collaborations (De Smedt et al., 2017). Simple indicators like the time of assignment submission can be treated as proxies for collaboration. We discuss SNA more into depth in Sect. 16.3.
- Another computational approach, process mining, can exploit real-time behavioural data to summarize the interactions of students with a given course's material. Students can then be distinguished based on their mode of involvement in the course (Mukala et al., 2015). It allows teachers to learn how the teaching method results in behavioural actions. These insights can be incorporated in the course design, and on the detection of inefficient behaviour, allowing fast and personalized intervention (De Smedt et al., 2017).
- A conjoint method to generate value from learning analytics is by implementing text analyses directly on the learning platforms. Natural Language Processing (NLP) is a text analysis method that has been shown to greatly improve the performance of students with regard to assignments such as the writing of essays (Whitelock et al., 2015a). Generally, text analysis can provide automated feedback shared with the students and their teachers (De Smedt et al., 2017). Providing automated feedback makes another argument for the cost-effectiveness of learning analytics. By giving course providers the ability to score large student bodies, it allows teachers to put more focus onto providing adapted support

to their students (De Smedt et al., 2017). We discuss NLP into more depth in Sect. 16.3.

- Not the least advantage of online learning is that it allows asynchronous and synchronous interactions and communications between the participants to a course (Broadbent & Poon, 2015). These interactions can be logged as unstructured data and incorporated into useful text, process and social network analyses.

16.2.2.3 Educational Quality Management

A key component of quality improvement in education is the creation of quality and performance indicators related to teachers and schools (Vanthienen & De Witte, 2017). Learning analytics' contribution to educational quality improvement is in providing data sources and computational methods and combining them in order to produce actionable summaries of teaching and schooling quality (Barra et al., 2017). Whereas, traditionally, data analyses have required punctual involvement and costly (time) investments from stakeholders, learning analytics can rely on computational power and dense networks of computational devices to automatically propose real-time reports to policy-makers. Below, contributions in terms of quality measurement and predictions are introduced.

16.2.2.4 Underlying Data for Quality Measurement

Through the exploitation of unstructured, streamed, behavioural data and pre-existing administrative data sets, analytical reports can be updated in real time to reflect the state of education at any desired level, from the individual student and classroom to the country as a whole. That information is commonly ordered in online dashboards (De Smedt et al., 2017). Analysts and programmers can even allow the user to customize the presented summary in real time, by applying filters on maps and subgroups of students, for example.

16.2.2.5 Efficiency Measurement

An aspect of the quality measurements provided by learning analytics is efficiency research, in which inputs and outputs are compared against a best practice frontier (see the earlier discussed Data Envelopment Analysis model). In this branch of literature, schools are, for instance, compared based on their ability to maximize learning outcomes given a set of educational inputs (De Witte & López-Torres, 2017; e Silva & Camanho, 2017; Mergoni & De Witte, 2021). The outcome of a similar analysis might be used for quality assessment purposes.

16.2.2.6 Predictions

When discussing the potential of learning analytics for educators and stakeholders, the ability to make predictions about learning outcomes is an unavoidable point of interest. In quantitative analyses, predictions are generated by translating latent patterns in historical data, be it structured or unstructured, in order to identify likely future outcomes (De Witte & Vanthienen, 2017).

Predictions can be produced using, for example, the Bayesian Additive Regression Trees (BART) model (see Sect. 16.3), as applied in Stoffi et al. (2021). There, linked administrative and PISA data available only in Flanders is used to distinguish a group of overwhelmingly under-performing Walloon students and explain their situation. Typically, such a technique uses administrative data that is available for both endowment groups in order to make a sensible generalization from one to the other.

Alternatively, process mining can be used to identify clusters of students and distinguish successful interaction patterns with a course's material (Mukala et al., 2015). Similar applications can be imagined for Social Network Analysis (De Smedt et al., 2017), through the evaluation of collaborative behaviour, and Natural Language Processing. These techniques are usually perceived as descriptive, but their output may very well be included in a predictive framework by education professionals and researchers.

Learning analytics has initiated a shift from using purely predictive analytics as a mean to identify student retention probabilities and grades towards the application of a wider set of methods (Viberg et al., 2018). In return, cutting-edge exploratory and descriptive methods can improve traditional predictive pipelines.

16.3   An Array of Policy-Driving Tools

It is one thing to comb over the numerous contributions and potential of learning analytics to data-informed decision-making; it is yet another to actually take the plunge and settle on tools for problem-solving in education. In what follows, a brief introduction to distinct methods from the field of computational social sciences is provided. In that way, the reader can get acquainted with the intuition of the methods and how they can be used to improve learning outcomes and quality measurement in education. To set the scene, we also illustrate how the approaches open up the range of innovative educational questions that can be answered through learning analytics.

16.3.1   *Bayesian Additive Regression Trees*

The Bayesian Additive Regression Trees (BART) stems from machine learning and probabilistic programming. It is a predictive and classifying algorithm that makes solving complex prediction problems simple by relying on a set of sane parameter configurations. Earlier comparable algorithms such as the Gradient Boosting Machine (GBM) and the Random Forest (RF) require repeated adjustments that hinge the quality of their predictions on an analyst's programming ability and limited computational resources. By contrast, the BART incorporates prior knowledge about educational science problems in order to produce competitive predictions and measures of uncertainty after a single estimation run (Dorie et al., 2019). This contributes to the accessibility of knowledge discovery and the credibility of policy statements in education.

As with the GBM and the RF, the essential and most basic component of the BART algorithm is the decision or prediction tree. The prediction tree is a classic predictive method that, unlike traditional regression methods, does not assume linear associations between sets of variable. It is robust to outlying variable values, such as those due to measurement error, and can accommodate a large quantity of data and high-dimensional data sets.

Their accuracy and relative simplicity have made regression trees popular diagnostic and prediction tools in medicine and public health (Lemon et al., 2003; Podgorelec et al., 2002). In education, a recent application of regression trees has been to explore dropout motivations and predictors in tertiary education (Alfermann et al., 2021). The regression tree algorithm (i.e. CART or classification and regression trees, Breiman et al., 2017) does variable selection automatically, so researchers are able to distinguish a few salient motivations, such as the perceived usefulness of the work, from a vast endowment of possible predictors.

To predict quantities such as test scores or dropout risk, regression trees separate the observations into boxes associating a set of characteristics with an outcome. The trees are created in multiple steps. In each of these steps, all observations comprised in a box of characteristics are split in two new boxes. Each split is selected by the algorithm to maximize the accuracy of the desired predictions. The end result of this division of observations into smaller and smaller boxes are branches through which each individual observation descends into a leaf. That leaf is the final box that assigns a single prediction value (e.g. a student's well-being score) to the set of observations sharing its branch. Graphically, the end result is a binary decision tree where each split is illustrated by a programmatic *if* statement leading onto either the next binary split or a leaf.

The Bayesian Additive Regression Trees (BART) algorithm is the combination of many such small regression trees (Kapelner & Bleich, 2016). Each regression tree adds to the predictive performance of the algorithm by picking up on the mistakes and leftover information from the previously estimated trees. After hundreds or possibly thousands of such trees are estimated, complex and subtle associations can be detected in the data. This makes the BART algorithm particularly competitive in areas of learning analytics where a large quantity of data are collected and there is little existing theory as to how interesting variables may be related to the outcome of interest, be it some aspect of the well-being of students or their learning outcomes.

The specific characteristic of the BART algorithm is its underlying Bayesian probability model (Kapelner & Bleich, 2016). By using prior probabilistic knowledge to restrict estimation possibilities to realistic prediction scenario, the algorithm can avoid detecting spurious association between variables. Each data set, unless it constitutes a perfect survey of the entire population of interest, contains variable associations that are present purely due to chance. Such coincidental associations reduce the ability to predict true outcomes when they are included in predictive models. Thus, each regression tree estimated by the BART algorithm is kept relatively small. Because each tree tends to assign predictions to larger sets of observations (i.e. large boxes), the predictive ability of individual trees is bad. This is why analysts call them weak learners. However, by combining many such weak learners, a flexible, precise and accurate prediction function can be generated (Hill et al., 2020).

The BART algorithm has already been presented earlier in this chapter as a flexible technique to detect and explain learning outcome inequalities (Stoffi et al., 2021). A refinement of the algorithm also permits the detection of heterogeneous policy effects on the learning outcomes of students. This is showed in Bargagli-Stoffi et al. (2019), where it is found that Flemish schools with a young and less experienced school director benefit most from a certain public funding policy. The large administrative data sets provided by educational institutions and governments are well fit for the application of rewarding but computationally demanding techniques such as the BART (Bargagli-Stoffi et al., 2019).

16.3.2   Social Network Analysis

The aim of Social Network Analysis (SNA) is to study the relations between individuals or organizations belonging to the same social networks (Wasserman, Faust, et al., 1994). Relations between these actors are defined by nodes and ties. The nodes are points of observations, which can be students, schools, administrations and more. The ties indicate a relationship between nodes and can contain additional information about the intensity of various components of that relationship (e.g. the time spent collaborating, the type of communication; Grunspan et al., 2014). Specifically for education, SNA aims to describe the networks of students and staff and make that information actionable to stakeholders. Applications of SNA include the optimization of learning design, the reorganization of student groups and the identification of at-risk clusters of students (Cela et al., 2015). Through text analysis and other advanced analytics methods, SNA can handle unstructured data from school blogs, wikis, forums, etc. (Cela et al., 2015). We discuss five examples more in detail next and refer the interested readers to the review by Cela et al. (2015), who provides many other concrete applications of SNA in education.

As a first example, the recognized importance of peer effects, both within and outside the classroom, makes Social Network Analysis (SNA) a particularly useful tool in education (Agasisti et al., 2017; Cela et al., 2015; Iterbeke et al., 2020). Applications of SNA model peer effects indirectly as a component of unobserved school or classroom effects that influence the (non-)cognitive skills (Cooc & Kim,

2017). As a second example, SNA has been applied to describe and explain a multiplicity of phenomena in schools. In a study of second and third primary school graders from 41 schools in North Carolina, Cooc and Kim (2017) found that pupils with a low reading ability who associated with higher ability peers for guidance significantly improved their reading scores over a summer. Third, other relevant applications of SNA have been in assessing the participation of peers in the well-being, be it mental or physical, of students. Surveying 1458 Belgian teenagers, Wegge et al. (2014) showed that the authors of cyber-bullying were often also responsible for physically bullying a student. Additionally, it was observed that a majority of bullies were in the same class as the bullied students. Moreover, a map of bullying networks isolated some students as being perpetrators of the bullying of multiple students. In cases of intimidation and bullying, a clear advantage of SNA over the usual approaches is that the data does not depend on isolated denunciations from victims and peers. The analysis of Wegge et al. (2014) simultaneously identifies culprits and victims, suggesting a course of action that does not focus attention on an isolated victim of bullying. A fourth example application of SNA is in improving the managerial efficacy and the performance of employees within educational organizations. One way to do this is by identifying bottlenecks in the transmission of information through the mapping of social networks. This can take two forms in the language of SNA: brokerage and structural holes (Everton, 2012). In a brokerage situation, a single agent or node controls the passing of information from one organizational sub-unit to the other. Meanwhile, structural holes identify absent ties between sub-units in the network. In a school, an important broker may be the principal's secretary, whereas structural holes may be present if teachers or staffs do not communicate well with one another (Hawe & Ghali, 2008). As a fifth illustration, the SNA method has been used to propose a typology of teachers based on the nature of their ties with students and to identify clusters of students more likely to be plagiarising with each other (Chang et al., 2010; Merlo et al., 2010; Ryymin et al., 2008). The ability to cluster students based on the intensity of their collaborations in a course has also been distinguished as a way to prevent fraud. Detecting cooperation between students is one of the key application of SNA in learning analytics (De Smedt et al., 2017).

16.3.3 Natural Language Processing

Natural Language Processing (NLP) is an illustration of the ability of computing machines to communicate with human languages (Smith et al., 2020). NLP applications can be achieved with relatively simple sets of rules or heuristics (e.g. word counts, word matching) or without applying cutting-edge machine learning techniques (Smith et al., 2020). Given NLP relies on machine learning techniques, it is better able to understand the context and reveal hidden meanings in communications (e.g. irony) (Smith et al., 2020).

In education, the use of NLP has been shown to improving students' learning outcomes (Whitelock et al., 2015a) and promoting student engagement. Moreover, NLP systems have the potential to provide one-on-one tutoring and personalized study material (Litman, 2016). The automatic grading of complex assignments is a precious feature of NLP models in education. These may eventually become a cost-effective solution that facilitate the evaluation of deeper learning skills than those evaluated through answers to multiple-choice questions (Smith et al., 2020). By efficiently adjusting the evaluation of knowledge to the learning outcomes desired by stakeholders, NLPs can contribute to educational performance. External and open data sets have allowed NLP solutions to achieve better accuracy in tasks such as grading. Such data sets can situate words within commonly invoked themes or contexts, for example, allowing the NLP model to make a more nuanced analysis of language data (Smith et al., 2020). Access to rich language data sets and algorithmic improvements may even allow NLP solutions to produce course assessment material automatically (Litman, 2016). However, an open issue with machine learning implementations of NLP is that the features used in grading by the computer may not provide useful feedback to the student or the teacher (e.g. by basing the grade on word counts) (Litman, 2016). Reasonable feedback may still require human input.

16.4   Issues and Recommendations

Despite the outlined benefits and contributions of learning analytics, there are, however, still some issues and limitations. A clear distinction can be made between issues belonging to the technical and non-technical parts of learning analytics (De Wit & Broucker, 2017). In the first case, there are the issues related to platform and analytics implementations, data warehousing, device networking, etc. With regard to the non-technical issues, there are concerns over the public acceptance and involvement in learning analytics, private and public regulations, human resources acquisition and the enthusiasm of stakeholders as to the technical potential of learning analytics. We summarize these challenges and propose a nuanced policy pathway to learning analytics implementation and promotion.

16.4.1   Non-technical Issues

Few learning analytics papers mention ethical and legal issues linked to the applications of their recommendations (Viberg et al., 2018). Clearly, developments in learning analytics participate to and benefit from the expansion of behavioural data collection. The spread and depth of data collection are generating new controversies around data privacy and security. These have an important place in public discourse and, if mishandled by stakeholders, could contribute to further limiting the potential of data availability and computational power in learning analytics and similar disciplines (Langedijk et al., 2019). Scientists are currently complaining about the restrictions put upon their research by rules and accountability procedures. Such rules curtail data-driven enterprises and may be detrimental to learning outcome's improvements (Groot & van den Brink, 2017). To facilitate collaboration between decision-makers, it is important that the administrative procedures related to learning analytics been seen by researchers as contributing to a healthy professional environment (Groot & van den Brink, 2017).

Additionally, public accountability and policies promoting organizational transparency may be a proper counter-balance to privacy concerns among citizens (e Silva & Camanho, 2017). The transparency and accessibility of information, by making relevant educational data sets public, for example, can involve citizens in the knowledge discovery related to education and foster enthusiasm for data-driven inference in that domain (De Smedt et al., 2017). It is also important that the concerned parties, including civil society, are interested in applying data-driven decision-making (Agasisti et al., 2017). It can be difficult to convince leaders in education to shift to data-driven policies since, for them, "experience and gut-instinct have a stronger pull" (Long & Siemens, 2011).

Just as necessary as political commitment, the acquisition of a skilled workforce is another sizeable non-technical issue (Agasisti et al., 2017). The growth of data-driven decision-making has yielded an increase in the demand for higher-educated workers while reducing the employment of unskilled workers (Groot & van den Brink, 2017). In other words, there is a gap between the growing availability of large, complex data sets and the pool of human resources that is necessary to clean and analyse those data (De Smedt et al., 2017). This invokes the problem, shared across the computational social sciences, of the double requirement of technical and analytical skills. Often, even domain-specific knowledge is an unavoidable component of useful policy insights (De Smedt et al., 2017). That multiplicity of professional requirements has made certain authors talk of the desirable modern data analyst as a scholar-practitioner (Streitwieser & Ogden, 2016).

16.4.2 Technical Issues

Many technical problems must be tackled before data-driven educational policies become a gold standard. Generally, there is a need for additional research regarding the effects of online educational softwares and of digital data collection pipelines on student and teacher outcomes. Additionally, inequalities in terms of the access to online education and its usage are an ever-present challenge (Jacob et al., 2016; Robinson et al., 2015).

There is yet relatively little evidence indicating that learning analytics improve the learning outcomes of students (Alpert et al., 2016; Bettinger et al., 2017; Jacob et al., 2016; Viberg et al., 2018). For example, less sophisticated correction algorithms may be exploited by students who will tailor their solution to obtain maximal scores without obtaining the desired knowledge (De Wit & Broucker, 2017). This is a question of adjustment between the spirit and the letter of the learning process.

Additionally, although the combination of administrative and streamed data is in many ways advantageous compared to survey data (Langedijk et al., 2019), the fast collection and analysis of data create issues of data accuracy. With real-time data analyses and reorientations of the learning process, accessible computing power becomes an issue.

Meanwhile, the unequal access to online resources and devices plainly removes a section of the student and teacher population from being reached by the digital tools of education. In part, this creates issues of under-representation in educational studies that increasingly rely on data obtained online (Robinson et al., 2015). It also creates a divide between those stakeholders that can make an informed choice between using and developing digital tools and face-to-face education and those that cannot access it or to whom digital education has a prohibitive cost (Bettinger et al., 2017; Di Pietro et al., 2020; Robinson et al., 2015).

Lack of access to digital or hybrid learning tools (i.e. a mix of face-to-face and digital education) may directly impede the learning and well-being of students. Indeed, students with access to online and hybrid education can access resources independently to enhance their educational pathway (Di Pietro et al., 2020). In a sense, a larger range of choices makes better educational outcomes attainable. For example, students at a school within a neighbourhood of low socio-economic standing may access a diverse network of students and teachers on electronic platforms (Jacob et al., 2016). In times of crisis such as with the COVID-19 school lockdowns, ready access to online educational platform also reduces the opportunity cost of education (Chakraborty et al., 2021; Di Pietro et al., 2020).

However, access is not a purely technical challenge. There are also noted gaps between populations in terms of the usage that is made of educational platforms and internet resources more generally (Di Pietro et al., 2020; Jacob et al., 2016). Students participating to MOOC, for example, are overwhelmingly highly educated professionals (Zafras et al., 2020). Online education may also leave more discretion to students. This discretion has proven to be a disadvantage to those who perform less well and are less motivated in face-to-face classes (Di Pietro et al., 2020).

16.4.3 Recommendations

Data-driven policies will require vast investments in information technology systems towards both data centres and highly skilled human resources. Therefore, additional data warehouses need to be built and maintained. Those require strong engineering capabilities (De Smedt et al., 2017). The integration of teaching and peer collaborations within computer systems promises to accelerate innovations in education. One can imagine that, in the future, administrative and real-time learning data will be updated and analysed in real time. The analyses will also benefit from combining data from other areas of interest such as health or finance. Additionally, the reach of analytics programs could be international, allowing for the shared integration and advancement of knowledge systems across countries (Langedijk et al., 2019).

Although there is a large practical potential of data-driven policies and educational tools, it is important that an educational data strategy not be developed in and of itself. Unlike what some *big data* enthusiasts have claimed, the data does not "speaks for itself" in education (Anderson, 2008). Those teachers, administrators and policy-makers, who are working to better educate our children, will still face complicated dilemma appealing to their professional expertise regardless of the level of integration of data analytics in education.

Furthermore, to insure political willingness, it is critical that work teams and stakeholders profit from the collected and analysed data (De Smedt et al., 2017). This contributes to the transparency of data use. Finally, although the evidence is still quite thin regarding the benefits of learning analytics, it must be noted that only a small quantity of validated instruments are actually being used to measure the quality and transmission of knowledge through learning platforms (Jivet et al., 2018).

Despite this scarcity of evidence pertaining to education, the exploitation of data through learning analytics can be linked to the recognized advantages of *big data* in driving public policy. Namely, it can facilitate a differentiation of services, increased decisional transparency, needs identification and organizational efficiency (Broucker, 2016). Generally, the lack of available data backing a decision is an indication of a lack of information and, thus, sub-optimal decision-making (Broucker, 2016).

Policies can be better implemented through quick and vast access to information about students and other educational stakeholders. In other words, the needs of students and other educational stakeholders can be more efficiently satisfied with evidence obtained from data collection (e.g. lower cost, higher speed of implementation). Such evidence-based education is a rational response to the so-called *fetishization* of change that has been plaguing educational reforms (Furedi, 2010; Groot & van den Brink, 2017).

It follows that data analytics should not become a new object for the *fetishization* of change in educational reforms. Indeed, quantitative goals (e.g. quantity of sensors in a classroom) should not be confounded with educational attainments (Long & Siemens, 2011; Mandl et al., 2008). Rather, data analytics should be developed and motivated as an approach that ensures that there are opportunities to use data in order to sustain mutually agreeable educational objectives.

These objectives may pertain to the lifetime health, job satisfaction, time allocation and creativity of current students (Oreopoulos & Salvanes, 2011). In other words, learning analytics pipelines must be carefully implemented in order to ensure that they are a rational response to contemporary challenges in education.

Acknowledgments The authors are grateful for valuable comments and suggestions from the participants of the Education panel of the CSS4P workshop, particularly Federico Biagi and Zsuzsa

Blaskó. Moreover, they wish to thank Alexandre Leroux of GERME, Francisco do Nascimento Pitthan, Willem De Cort, Silvia Palmaccio and the members of the LEER and CSS4P team for the rewarding discussions and suggestions.

References

Agasisti, T., Ieva, F., Masci, C., Paganoni, A. M., & Soncin, M. (2017). *Data analytics applications in education*. Auerbach Publications. https://doi.org/10.4324/9781315154145-8

Alfermann, D., Holl, C., & Reimann, S. (2021). Should i stay or should i go? indicators of dropout thoughts of doctoral students in computer science. *International Journal of Higher Education, 10*(3), 246–258. https://doi.org/10.5430/ijhe.v10n3p246

Alpert, W. T., Couch, K. A., & Harmon, O. R. (2016). A randomized assessment of online learning. *American Economic Review, 106*(5), 378–82. https://doi.org/10.1257/aer.p20161057

Anderson, C. (2008). The end of theory: The data deluge makes the scientific method obsolete. *Wired Magazine, 16*(7), 16–07.

Bargagli-Stoffi, F. J., De Witte, K., & Gnecco, G. (2019). Heterogeneous causal effects with imperfect compliance: A novel bayesian machine learning approach. *Preprint arXiv:1905.12707*.

Barra, C., Destefanis, S., Sena, V., & Zotti, R. (2017). Disentangling faculty efficiency from students' effort. *Data Analytics Applications in Education* (pp. 105–128). Auerbach Publications. https://doi.org/10.4324/9781315154145-5

Berland, M., Davis, D., & Smith, C. P. (2015). Amoeba: Designing for collaboration in computer science classrooms through live learning analytics. *International Journal of Computer-Supported Collaborative Learning, 10*(4), 425–447. https://doi.org/10.1007/s11412-015-9217-z

Bettinger, E. P., Fox, L., Loeb, S., & Taylor, E. S. (2017). Virtual classrooms: How online college courses affect student success. *American Economic Review, 107*(9), 2855–75. https://doi.org/10.1257/aer.20151193

Bishop, C. M. (2006). *Pattern recognition and machine learning*. Springer.

Breiman, L., Friedman, J. H., Olshen R. A., & Stone, C. J. (2017). *Classification and regression trees*. Routledge.

Broadbent, J., & Poon, W. L. (2015). Self-regulated learning strategies & academic achievement in online higher education learning environments: A systematic review. *The Internet and Higher Education, 27*, 1–13. https://doi.org/10.1016/j.iheduc.2015.04.007

Broucker, B. (2016). Big data governance; een analytisch kader. *Bestuurskunde, 25*(1), 24–28.

Bruno, E., Alexandre, B., Ferreira Mello, R., Falcão, T. P., Vesin, B., & Gašević, D. (2021). Applications of learning analytics in high schools: A systematic literature review. *Frontiers in Artificial Intelligence, 4*, 132.

Castaño-Muñoz, J., & Rodrigues, M. (2021). Open to moocs? Evidence of their impact on labour market outcomes. *Computers & Education, 173*, 104289. https://doi.org/10.1016/j.compedu.2021.104289

Cela, K. L., Sicilia, M. Á., & Sánchez, S. (2015). Social network analysis in e-learning environments: A preliminary systematic review. *Educational Psychology Review, 27*(1), 219–246. https://doi.org/10.1007/s10648-014-9276-0

Chakraborty P., Mittal, P., Gupta, M. S., Yadav, S., & Arora, A. (2021). Opinion of students on online education during the COVID-19 pandemic. *Human Behavior and Emerging Technologies, 3*(3), 357–365. https://doi.org/10.1002/hbe2.240

Chang, W.-C., Lin, H.-W., & Wu, L.-C. (2010). Applied social network anaysis to project curriculum. In *The 6th International Conference on Networked Computing and Advanced Information Management* (pp. 710–715).

Cooc, N., & Kim, J. S. (2017). Peer influence on children's reading skills: A social network analysis of elementary school classrooms. *Journal of Educational Psychology, 109*(5), 727. https://doi. org/10.1037/edu0000166

Dalipi, F., Imran, A. S., & Kastrati, Z. (2018). Mooc dropout prediction using machine learning techniques: Review and research challenges. In *2018 IEEE Global Engineering Education Conference (EDUCON)* (pp. 1007–1014).

De Smedt, J., vanden Broucke, S. K., Vanthienen, J., & De Witte, K. (2017). Improved student feedback with process and data analytics. In *Data analytics applications in education* (pp. 11–36). Auerbach Publications. https://doi.org/10.4324/9781315154145-2

De Wit, K., & Broucker, B. (2017). The governance of big data in higher education. In *Data analytics applications in education* (pp. 213–234). Auerbach Publications. https://doi.org/10. 4324/9781315154145-9

De Witte, K., & Vanthienen, J. (2017). *Data analytics applications in education.* Auerbach Publications. https://doi.org/10.1201/b20438

De Witte, K., & López-Torres, L. (2017). Efficiency in education: A review of literature and a way forward. *Journal of the Operational Research Society, 68*(4), 339–363. https://doi.org/10.1057/ jors.2015.92

De Witte, K., & Smet, M. (2021). *Financing Education in the Context of COVID-19* (Ad hoc report No. 3/2021). European Expert Network on Economics of Education (EENEE).

DeNardis, L. (2020). The cyber-physical disruption. In *The internet in everything* (pp. 25–56). Yale University Press.

Di Pietro, G., Biagi, F., Costa, P., Karpiński, Z., & Mazza, J. (2020). *The likely impact of covid-19 on education: Reflections based on the existing literature and recent international datasets* (Vol. 30275). Publications Office of the European Union.

Dorie, V., Hill, J., Shalit, U., Scott, M., & Cervone, D. (2019). Automated versus do-it-yourself methods for causal inference: Lessons learned from a data analysis competition. *Statistical Science, 34*(1), 43–68. https://doi.org/10.1214/18-STS667

e Silva, M. C. A., & Camanho, A. S. (2017). Using data analytics to benchmark schools: The case of Portugal. In *Data analytics applications in education* (pp. 129–162). Auerbach Publications. https://doi.org/10.4324/9781315154145-6

Evans, C. (2013). Making sense of assessment feedback in higher education. *Review of Educational Research, 83*(1), 70–120. https://doi.org/10.3102/0034654312474350

Everton, S. F. (2012). *Disrupting dark networks.* Cambridge University Press. https://doi.org/10. 1017/CBO9781139136877

Figlio, D., Karbownik, K., & Salvanes, K. G. (2016). Education research and administrative data. In *Handbook of the economics of education* (pp. 75–138). Elsevier.

Furedi, F. (2010). *Wasted: Why education isn't educating.* Bloomsbury Publishing.

Greller, W., & Drachsler, H. (2012). Translating learning into numbers: A generic framework for learning analytics. *Educational Technology & Society, 15*(3), 42–57.

Groot, W., & van den Brink, H. M. (2017). Evidence-based education and its implications for research and data analytics with an application to the overeducation literature. In *Data analytics applications in education* (pp. 235–260). Auerbach Publications. https://doi.org/10. 4324/9781315154145-10

Grunspan, D. Z., Wiggins, B. L., & Goodreau, S. M. (2014). Understanding classrooms through social network analysis: A primer for social network analysis in education research. *CBE—Life Sciences Education, 13*(2), 167–178. https://doi.org/10.1187/cbe.13-08-0162

Hattie, J., & Timperley, H. (2007). The power of feedback. *Review of Educational Research, 77*(1), 81–112. https://doi.org/10.3102/003465430298487

Hawe, P., & Ghali, L. (2008). Use of social network analysis to map the social relationships of staff and teachers at school. *Health Education Research, 23*(1), 62–69. https://doi.org/10.1093/her/ cyl162

Hill, J., Linero, A., & Murray J. (2020). Bayesian additive regression trees: A review and look forward. *Annual Review of Statistics and Its Application, 7*, 251–278. https://doi.org/10.1146/ annurev-statistics-031219-041110

Iterbeke, K., & De Witte, K. (2020). Helpful or harmful? The role of personality traits in student experiences of the covid-19 crisis and school closure. *FEB Research Report Department of Economics.* https://doi.org/10.1177/01461672211050515

Iterbeke, K., De Witte, K., Declercq, K., & Schelfhout, W. (2020). The effect of ability matching and differentiated instruction in financial literacy education. evidence from two randomised control trials. *Economics of Education Review, 78,* 101949. https://doi.org/10.1016/j.econedurev.2019.101949

Jacob, B., Berger, D., Hart, C., & Loeb, S. (2016). Can technology help promote equality of educational opportunities? *RSF: The Russell Sage Foundation Journal of the Social Sciences, 2*(5), 242–271. https://doi.org/10.7758/rsf.2016.2.5.12

Jivet, I., Scheffel, M., Specht, M., & Drachsler, H. (2018). License to evaluate: Preparing learning analytics dashboards for educational practice. In *Proceedings of the 8th International Conference on Learning Analytics and Knowledge* (pp. 31–40). https://doi.org/10.1145/3170358.3170421

Kapelner, A., & Bleich, J. (2016). Bartmachine: Machine learning with bayesian additive regression trees. *Journal of Statistical Software, Articles, 70*(4), 1–40. https://doi.org/10.18637/jss.v070.i04

Kassab, M., DeFranco, J., & Laplante, P. (2020). A systematic literature review on internet of things in education: Benefits and challenges. *Journal of Computer Assisted Learning, 36*(2), 115–127. https://doi.org/10.1111/jcal.12383

King, G. (2016). *Big data is not about the data! computational social science: Discovery and prediction.*

Langedijk, S., Vollbracht, I., & Paruolo, P. (2019). The potential of administrative microdata for better policy-making in Europe. In *Data-driven policy impact evaluation,* (p. 333). https://doi.org/10.1007/978-3-319-78461-8_20

Lazer, D., Pentland, A., Adamic, L., Aral, S., Barabasi, A.-L., Brewer, D., Christakis, N., Contractor, N., Fowler, J., Gutmann, M., Jebara, T., King, G., Macy M., Roy D., & Van Alstyne, M. (2009). Social Science: Computational social science. *Science, 323*(5915), 721–723. https://doi.org/10.1126/science.1167742

Lazer, D., Pentland, A., Watts, D. J., Aral, S., Athey S., Contractor, N., Freelon, D., Gonzalez-Bailon, S., King, G., Margetts, H., Nelson, A., Salganik, M. J., Strohmaier, M., Vespignani, A., & Wagner, C. (2020). Computational social science: Obstacles and opportunities. *Science, 369*(6507), 1060–1062. https://doi.org/10.1126/science.aaz8170

Lee, H.-S., Pallant, A., Pryputniewicz, S., Lord, T., Mulholland, M., & Liu, O. L. (2019). Automated text scoring and real-time adjustable feedback: Supporting revision of scientific arguments involving uncertainty *Science Education, 103*(3), 590–622. https://doi.org/10.1002/sce.21504

Leitner, P., Khalil, M., & Ebner, M. (2017). Learning analytics in higher education—a literature review. In *Learning analytics: Fundaments, applications, and trends* (pp. 1–23). https://doi.org/10.1007/978-3-319-52977-6_1

Lemon, S. C., Roy J., Clark, M. A., Friedmann, P. D., & Rakowski, W. (2003). Classification and regression tree analysis in public health: Methodological review and comparison with logistic regression. *Annals of Behavioral Medicine, 26*(3), 172–181. https://doi.org/10.1207/S15324796ABM2603_02

Litman, D. (2016). Natural language processing for enhancing teaching and learning. In *Thirtieth AAAI Conference on Artificial Intelligence.*

Long, P., & Siemens, G. (2011). Penetrating the fog: Analytics in learning and education. *EDUCAUSE Review, 46*(5), 30.

Lonn, S., Aguilar, S. J., & Teasley, S. D. (2015). Investigating student motivation in the context of a learning analytics intervention during a summer bridge program. *Computers in Human Behavior, 47,* 90–97. https://doi.org/10.1016/j.chb.2014.07.013

Maldonado, J., & De Witte, K. (2021). The effect of school closures on standardised student test. *British Educational Research Journal, 48*(1), 49–94. https://doi.org/10.1002/berj.3754

Mandl, U., Dierx, A., & Ilzkovitz, F. (2008). *The effectiveness and efficiency of public spending* (Technical Report). Directorate General Economic and Financial Affairs (DG ECFIN).

Mazrekaj, D., De Witte, K., & Cabus, S. (2020). School outcomes of children raised by same-sex parents: Evidence from administrative panel data. *American Sociological Review, 85*(5), 830–856. https://doi.org/10.1177/0003122420957249

Mergoni, A., & De Witte, K. (2021). Policy evaluation and efficiency: A systematic literature review. *International Transactions in Operational Research.* https://doi.org/10.1111/itor.13012

Merlo, E., Ríos, S. A., Álvarez, H., L'Huillier, G., & Velásquez, J. D. (2010). Finding inner copy communities using social network analysis. In *International Conference on Knowledge-Based and Intelligent Information and Engineering Systems* (pp. 581–590).

Mukala, P., Buijs, J. C., Leemans, M., & van der Aalst, W. M. (2015). Learning analytics on coursera event data: A process mining approach. In *SIMPDA* (pp. 18–32).

Nguyen, Q., Huptych, M., & Rienties, B. (2018). Linking students' timing of engagement to learning design and academic performance. In *Proceedings of the 8th International Conference on Learning Analytics and Knowledge* (pp. 141–150). https://doi.org/10.1145/3170358.3170398

Nouri, J., Ebner, M., Ifenthaler, D., Saqr, M., Malmberg, J., Khalil, M., Bruun, J., Viberg, O., Conde González, M. Á., Papamitsiou, Z., & Berthelsen, U. D. (2019). Efforts in Europe for Data-Driven Improvement of Education–A Review of Learning Analytics Research in Seven Countries. *International Journal of Learning Analytics and Artificial Intelligence for Education (iJAI), 1*(1), 8–27. https://doi.org/10.3991/ijai.v1i1.11053

Oreopoulos, P., & Salvanes, K. G. (2011). Priceless: The nonpecuniary benefits of schooling. *Journal of Economic Perspectives, 25*(1), 159–84. https://doi.org/10.1257/jep.25.1.159

Palermo, C., & Wilson, J. (2020). Implementing automated writing evaluation in different instructional contexts: A mixed-methods study. *Journal of Writing Research, 12*(1), 63–108.

Pardo, A., Bartimote, K., Shum, S. B., Dawson, S., Gao, J., Gašević, D., Leichtweis, S., Liu, D., Martínez-Maldonado, R., Mirriahi, N., Moskal, A. C. M., Schulte, J., Siemens, G., & Vigentini, L. (2018). Ontask: Delivering data-informed, personalized learning support actions. *Journal of Learning Analytics, 5*(3), 235–249.

Podgorelec, V., Kokol, P., Stiglic, B., & Rozman, I. (2002). Decision trees: An overview and their use in medicine. *Journal of Medical Systems, 26*(5), 445–463. https://doi.org/10.1023/A:1016409317640

Rettore, E., & Trivellato, U. (2019). The use of administrative data to evaluate the impact of active labor market policies: The case of the italian liste di mobilità. In *Data-driven policy impact evaluation* (pp. 165–182). Springer. https://doi.org/10.1007/978-3-319-78461-8_11

Robinson, L., Cotten, S. R., Ono, H., Quan-Haase, A., Mesch, G., Chen, W., Schulz, J., Hale, T. M., & Stern, M. J. (2015). Digital inequalities and why they matter. *Information, Communication & Society, 18*(5), 569–582. https://doi.org/10.1080/1369118X.2015.1012532

Ryymin, E., Palonen, T., & Hakkarainen, K. (2008). Networking relations of using ict within a teacher community. *Computers & Education, 51*(3), 1264–1282. https://doi.org/10.1016/j.compedu.2007.12.001

Selwyn, N. (2019). What's the problem with learning analytics? *Journal of Learning Analytics, 6*(3), 11–19.

Smith, G. G., Haworth, R., & Žitnik, S. (2020). Computer science meets education: Natural language processing for automatic grading of open-ended questions in ebooks. *Journal of Educational Computing Research, 58*(7), 1227–1255. https://doi.org/10.1177/0735633120927486

Stoffi, F. J. B., De Beckker, K., Maldonado, J. E., & De Witte, K. (2021). Assessing sensitivity of machine learning predictions. a novel toolbox with an application to financial literacy. *Preprint arXiv:2102.04382.*

Streitwieser, B., & Ogden, A. C. (2016). *International higher education's scholar-practitioners: Bridging research and practice* , Books, S., (Ed.).

Tadayon, M., & Pottie, G. J. (2020). Predicting student performance in an educational game using a hidden markov model. *IEEE Transactions on Education, 63*(4), 299–304. https://doi.org/10.1109/TE.2020.2984900

Topolovec, S. (2018). *A comparison of self-paced and instructor-paced online courses: The interactive effects of course delivery mode and student characteristics.*

Vanthienen, J., & De Witte, K. (2017). *Data analytics applications in education.* Auerbach Publications. https://doi.org/10.4324/9781315154145

Viberg, O., Hatakka, M., Bälter, O., & Mavroudi, A. (2018). The current landscape of learning analytics in higher education. *Computers in Human Behavior, 89*, 98–110. https://doi.org/10.1016/j.chb.2018.07.027

Wasserman, S., Faust, K. (1994). *Social network analysis: Methods and applications.* Cambridge University Press.

Wegge, D., Vandebosch, H., & Eggermont, S. (2014). Who bullies whom online: A social network analysis of cyberbullying in a school context. *Communications, 39*(4), 415–433. https://doi.org/10.1515/commun-2014-0019

Whitelock, D., Twiner, A., Richardson, J. T., Field, D., & Pulman, S. (2015a). Feedback on academic essay writing through pre-emptive hints: Moving towards. *European Journal of Open, Distance and E-learning, 18*(1), 1–15.

Whitelock, D., Twiner, A., Richardson, J. T., Field, D., & Pulman, S. (2015b). Openessayist: A supply and demand learning analytics tool for drafting academic essays. In *Proceedings of the Fifth International Conference on Learning Analytics and Knowledge* (pp. 208–212).

Zafras, I., Kostas, A., & Sofos, A. (2020). Moocs & participation inequalities in distance education: A systematic literature review 2009-2019. *European Journal of Open Education and E-learning Studies, 5*(1), 68–89.

Chapter 17
Leveraging Digital and Computational Demography for Policy Insights

Ridhi Kashyap and Emilio Zagheni

Abstract Situated at the intersection of the computational and demographic sciences, digital and computational demography explores how new digital data streams and computational methods advance the understanding of population dynamics, along with the impacts of digital technologies on population outcomes, e.g. linked to health, fertility and migration. Encompassing the data, methodological and social impacts of digital technologies, we outline key opportunities provided by digital and computational demography for generating policy insights. Within methodological opportunities, individual-level simulation approaches, such as microsimulation and agent-based modelling, infused with different data, provide tools to create empirically informed synthetic populations that can serve as virtual laboratories to test the impact of different social policies (e.g. fertility policies, support for the elderly or bereaved people). Individual-level simulation approaches allow also to assess policy-relevant questions about the impacts of demographic changes linked to ageing, climate change and migration. Within data opportunities, digital trace data provide a system for early warning with detailed spatial and temporal granularity, which are useful to monitor demographic quantities in real time or for understanding societal responses to demographic change. The demographic perspective highlights the importance of understanding population heterogeneity in the use and impacts of different types of digital technologies, which is crucial towards building more inclusive digital spaces.

R. Kashyap (✉)
Department of Sociology, Leverhulme Centre for Demographic Science, and Nuffield College, University of Oxford, Oxford, UK
e-mail: ridhi.kashyap@nuffield.ox.ac.uk

E. Zagheni
Max Planck Institute for Demographic Research, Rostock, Germany
e-mail: zagheni@demogr.mpg.de

E. Bertoni et al. (eds.), *Handbook of Computational Social Science for Policy*,
https://doi.org/10.1007/978-3-031-16624-2_17

17.1 Introduction

Demography is the scientific study of populations, including the three fundamental forces that shape population dynamics—mortality, fertility and migration. While these three forces produce the essential events for which demographers have developed a range of measurement methods, each of these processes is also the result of complex individual behaviours that are shaped by multiple forces. Thus, in addition to measuring demographic phenomena and describing macrolevel population patterns, demographers examine how and why specific population-level outcomes emerge, seek to explain them and understand their consequences. While pursuing science-driven discovery, demographers also inevitably address several interrelated, policy-relevant themes, such as ageing, family change, the ethnic diversification of societies, spatial segregation and related outcomes and the relationship between environmental and population change. These and related policy-relevant topics are intimately connected with the three core demographic processes of mortality, fertility and migration. For example, significant reductions in mortality rates over the course of the twentieth and twenty-first centuries imply that individuals across Europe can expect to lead long lives, with an increasing overlap of generations within populations. How does the ageing of populations impact on key social institutions linked to the labour market, pension systems and provision of care? Moreover, how can societies better prepare for these changes?

Demography has historically been a data-driven discipline and one that has developed tools to repurpose different kinds of data—often not originally intended or collected for research—for measuring and understanding population change (Billari & Zagheni, 2017; Kashyap, 2021). Demography is thus uniquely positioned to take advantage of the opportunities enabled by the broader development of the computational social sciences, both in terms of new data streams and computational methods. A growing interest in this interface between demography and computational social science has led to the emergence of digital and computational demography (Kashyap et al., 2022). This chapter describes how insights from digital and computational demography can help augment the policy relevance of demographic research.

Demographic research is relevant for policy makers in several ways. At its most basic level, understanding the current as well as anticipated future size, composition and geographical distribution of a population—whether a national, regional or local population—is essential for planning for the provision of services, for identifying targets of aid and for setting policy priorities. For example, the needs for specific public services are closely tied to the age structure of a population—populations that have more young people have very different needs than those with a larger share of older people. The impacts of these age structures are also felt in economic and social domains. This shapes not only what services are needed, e.g. schools versus social care for the elderly, but also which issues require priority at a given time. Demographic analyses can also help identify population changes and trends for the future, to identify areas that will emerge in the future as relevant for policy making.

For instance, subnational areas where population is growing quickly have very different needs, and require a different type of planning, compared to those areas that experience depopulation. More broadly, demography sheds light on population heterogeneity along various dimensions and offers insights into the heterogeneous impact of policy interventions on different segments of the population. For example, when considering key demographic trends like ageing, or the impact of climate change on health and population dynamics, questions of the inequality in these impacts across different regions and socioeconomic groups are critical from a policy perspective for identifying vulnerable communities and for supporting them appropriately. Policy makers may also try to favour certain demographic trends, such as through fertility policies or migration policies, in a way that leads to co-benefits at the individual and societal level. For example, policy makers may pursue fertility policies oriented towards helping individuals achieve the desired number of children, which may in turn affect the long-term sustainability of social security systems.

17.2 The Digital Turn in Demography: An Overview

Demographers have conventionally relied on data sources such as government administrative registers, censuses and nationally representative surveys to describe and understand population trends. A key strength of these data sources that makes them well-suited for demographic research is their representativeness and population generalizability. Censuses and population registers target complete coverage and enumeration of populations. In contrast, the types of surveys conducted by and used by demographers draw on high-quality, probability samples to provide a richer, in-depth source of data with a view to testing specific theories, understanding individual behaviours and attitudes that underpin demographic patterns. While these data sources are critical for demographic research, they also have a number of limitations. These data sources are often slow (e.g. censuses are mostly decennial), resource- and time-intensive and often reactive (e.g. surveys that require asking individuals for information), although in some cases these data are generated as by-products of administrative transactions where individuals interact with state institutions (e.g. birth registration, tax registration). Demographers have developed and applied mathematical and statistical techniques to use quantitative data sources to carefully measure and describe macrolevel (aggregate) population patterns, understand the relationships between different demographic variables and decompose changes in population indicators into different underlying processes. Growing bodies of individual-level and linked datasets have also enabled demographers to address individual-level causal questions about how specific social policies or social changes affect demographic behaviours.

The growing use of digital technologies such as the internet and mobile phones, as well as advances in computational power for processing, storing and analysing data, has led to a digital and computational turn in demography (Kashyap et al., 2022). This digital turn has affected demographic research along three dimensions:

1. Advancements in data opportunities
2. Applications of computational methods for demographic questions
3. Growing interest in the impacts of digitalization for demographic behaviours

17.2.1   Advances in Data Opportunities

Technological changes in digitized information storage and processing have improved access and granularity of traditional demographic data sources, while also generating new types of data streams and new opportunities for data collection, thereby enriching the demographic data ecosystem (Kashyap, 2021). Some of these new data streams are opened up by the widespread use of digital technologies such as the internet, mobile phones and social media. However, the digitization of information more broadly means that diverse types of digital data sources can now be repurposed for demographic research, ranging from detailed administrative data to bibliometric and crowdsourced genealogical databases, many of which were not intentionally collected for the purpose of research (Alburez-Gutierrez et al., 2019). These new data sources offer novel possibilities, but also come with their own unique ethical and methodological challenges, as we describe in the next section on computational guidelines.

In terms of their opportunities, these new data streams can help fill data gaps in areas where conventional data may be lacking and can provide higher-frequency and real-time measurement than conventional sources of demographic data to capture events as they occur. In addition, they provide better temporal and/or spatial resolution that can help 'nowcast' and understand local patterns and indicators in a timely way. For example, a growing body of research has used digital trace data from the web, mobile and social media to measure international or internal migration (e.g. Zagheni & Weber, 2012; Deville et al., 2014; Gabrielli et al., 2019; Alexander et al., 2020; Fiorio et al., 2021; Rampazzo et al., 2021). Different types of digital traces have been used to capture mobility processes. Some widely used examples include aggregated social media audience counts from Facebook's marketing platform (Rampazzo et al., 2021; Alexander et al., 2020) and timestamped call detail records from mobile phones that provide changing spatiotemporal distributions of mobile users (e.g. Deville et al., 2014). Vehicle detection with machine learning (ML) techniques applied to satellite images obtained via remote sensing have also been used to track mobility processes (e.g. Chen et al., 2014). Conventional data on migration are often lacking, and these studies identify ways in which these nontraditional data can help fill gaps and complement traditional sources of demographic statistics. Digital traces of behaviours, such as those from aggregate web search queries or social media posts, can further provide non-elicited forms of measurement of contexts, norms and behaviours that are relevant for understanding demographic shifts (Kashyap, 2021). For example, aggregated web search queries have been shown to capture fertility intentions that are predictive of fertility rates (Billari et al., 2016; Wilde et al., 2020) or information-seeking about abortion (Reis & Brownstein, 2010; Leone et al., 2021). Social media posts have also been used to study sentiments surrounding parenthood (Mencarini et al., 2019), while satellite images have been used to assess the socioeconomic characteristics of geographical areas (Elvidge et al., 2009; Gebru et al., 2017; Jochem et al., 2021).

Beyond passive measurement from already existing digital traces, internet- and mobile-based technologies can also provide cost-effective opportunities for data collection. Targeted recruitment of survey respondents, based on social and demographic attributes such as those provided by the social media advertisement platforms (e.g. Facebook), has enabled research on hard-to-reach groups, e.g. migrant populations (Pötzschke & Braun, 2017), or those working in specific service sector jobs/occupations (Schneider & Harknett, 2019a, b). Digital modes of data collection also proved invaluable during the COVID-19 pandemic, when rapid understanding of social and behavioural responses to the pandemic and associated lockdowns was needed but traditional face-to-face forms of data collection were impossible (Grow et al., 2020). Combining passively collected information (e.g. from social media or mobile phones) with accurate surveys is an active area of research, with great promise in the context of monitoring indicators of sustainable development on a global scale (Kashyap et al., 2020; Aiken et al., 2022; Chi et al., 2022).

17.2.2 Computational Methods for Demographic Questions

Second, improvements in computational power have facilitated the adoption of computational methodologies, such as microsimulation and agent-based simulation, as well as ML techniques, for demographic applications. Microsimulation techniques, which take empirical transition rates of mortality, fertility and migration as their input to generate a synthetic population that has a realistic genealogical structure, have been used to study the evolution of population dynamics. Microsimulation techniques have been used to examine kinship dynamics and intergenerational processes, such as the availability and potential support of kin and extended family across the life course (Zagheni, 2010; Verdery & Margolis, 2017; Verdery, 2015) or the extent of generational overlap (Alburez-Gutierrez et al., 2021), as well as the impact of macrolevel changes, like technological changes (Kashyap & Villavicencio, 2016) or educational change (Potančoková & Marois, 2020) that affect demographic rates, on population dynamics. Agent-based simulation techniques build on microsimulation by incorporating individual-level behavioural rules, social interaction and feedback mechanisms to test behavioural theories for how macrolevel population phenomena emerge from individual-level behaviours. Agent-based simulation approaches have been used within the demographic literature to model migration decision-making (Klabunde & Willekens, 2016; Entwisle et al., 2016) as well as family and marriage formation processes (Billari et al., 2007;

Diaz et al., 2011; Grow & Van Bavel, 2015). Both these types of individual-based simulation techniques that model individual-level probabilities of experiencing events—when infused with different types of real demographic data—offer ways of building what Bijak et al. describe as 'semi-artificial' population models that are empirically informed (Bijak et al., 2013). Such semi-artificial models are useful for generating scenarios to examine social interaction and feedback effects or assess the likely consequences of policies given a set of theoretical expectations. These approaches can be used to generate synthetic counterfactual scenarios and are useful to identify causal relationships, especially for social and demographic questions for which experimental approaches like randomized trials are not possible nor ethically desirable. Given that policy making is often concerned with causal relationships, these tools are of quintessential importance, especially when used in combination with data-driven approaches.

Improvements in computational power combined with an increasingly data-rich environment have also opened up opportunities for the use of ML approaches in demographic research. The focus on the discovery of macrolevel regularities in population dynamics, its interest in exploring different dimensions of population heterogeneity and the discipline's orientation towards projection of unseen (future) trends based on seen (past) trends lends itself well to the applications of ML techniques (Kashyap et al., 2022). An emerging body of work has used supervised ML approaches that find predictive models that link some explanatory variables to some outcome to individual-level longitudinal survey data to assess the predictability of demographic and life course outcomes (Salganik et al., 2020; Arpino et al., 2022). ML approaches have also been used for demographic forecasting (Nigri et al., 2019; Levantesi et al., 2022) and for population estimation using geospatial data (Stevens et al., 2015; Lloyd et al., 2017). While demographic research has been broadly concerned with prediction of risk for population groups or sub-groups, ML techniques offer the opportunity to generate more accurate predictions at the individual level and to better quantify heterogeneity in outcomes or responses.

17.2.3 Demographic Impacts of Digitalization

Digitalization has implications for demographic processes as digital tools are used for information-seeking, social interaction and communication and accessing vital services. The importance of digital technologies as a lifeline for different domains was powerfully illuminated during the COVID-19 pandemic. Demographic research has highlighted how the use of internet and mobile technologies can directly impact on demographic outcomes linked to health (Rotondi et al., 2020), marriage (Bellou, 2015; Sironi & Kashyap, 2021), fertility (Billari et al., 2019, 2020) and migration (Pesando et al., 2021), by enabling access to information, promoting new paths for social learning and interaction and providing flexibility in reconciling work and family (e.g. through remote working). This research suggests that access to digital resources (e.g. broadband connectivity, mobile apps) may, for example, enhance the health, wellbeing and quality of life in sparsely populated areas, by enabling better connectivity, access to services and economic opportunities in those regions. This may contribute to reduce depopulation in certain rural areas of Europe, by making them more attractive places to live and work. At the same time, not everyone may have the same level of access or skills necessary to take full advantage of the digital revolution (van Deursen & van Dijk, 2011; Alvarez-Galvez et al., 2020), and a deeper examination of the heterogeneity of these impacts is necessary to understand who and under what conditions digital technologies can empower. In addition to understanding the social impacts of digital technologies, there is value in understanding the demographic characteristics of digital divides also from the perspective of using new streams of digital data for population generalizable measurement. This is an area where demographers have also begun to make contributions through exploring demographic dimensions of social media and internet use (Feehan & Cobb, 2019; Gil-Clavel & Zagheni, 2019; Kashyap et al., 2020).

17.3 Computational Guidelines

Digital and computational demography, which bridges computational social science with demography, offers several opportunities for addressing policy-relevant questions. We provide guidelines for leveraging these opportunities along three dimensions: methodological opportunities, data opportunities and understanding demographic heterogeneity in the impacts of digital technologies.

17.3.1 Methodological Opportunities

Policy makers frequently need to understand the impacts of specific policies or a basket of policies (e.g. fertility policies that seek to promote the realization of desired fertility), examine multiple scenarios and counterfactuals and assess the heterogeneity in the impacts of specific policies or social and environmental changes (e.g. climate change) on populations. Computational simulation techniques such as microsimulation and agent-based simulation, which have been increasingly adopted within digital and computational demography, are particularly useful for addressing these types of questions. By incorporating different types of data and forms of population heterogeneity (e.g. differences by educational groups) within simulation models, these approaches can be used to create synthetic populations where individual decisions and behaviours are guided by empirical survey data and/or observed demographic rates (e.g. birth, death or migration rates).

Agent-based simulation approaches are especially useful when the focus is on understanding non-linear feedback effects or social influence effects on behaviours, such as those linked to whether or not to have a child given a wider set of contextual conditions. Microsimulation approaches can help understand the broader implications of a current set of demographic rates for population composition and change, as well as for kinship and intergenerational processes. Microsimulation techniques, for example, can help understand the evolution of kinship availability and support as a consequence of changing demographic rates. By incorporating rates that vary by different population sub-groups (e.g. ethnic groups), microsimulation approaches can help explore questions about the future size and composition of the availability of kin support for different population groups, which is a central question for understanding and adapting in the context of population ageing. These approaches provide the necessary flexibility to create counterfactual scenarios and for an opportunity to link different types of data to understand how different parts of a population system respond—e.g. individual-level changes affect macrolevel patterns, or macrolevel shocks affect individuals. A central challenge when building simulation models is the trade-off between parsimony and complexity. On the one hand, while simulation models allow for flexibility to incorporate different parameters to model complex systems, the inclusion of too many parameters can be counterproductive for interpretability, i.e. for understanding which parameters directly affect the outcome of interest. Another separate concern is that of how best to understand model uncertainty and draw statistical inferences from model outcomes. To this end, different approaches for computationally intensive calibration of simulation models have been applied within the demographic literature. These approaches combine the tools of statistics (including Bayesian statistics) with simulation approaches to help assess model sensitivity and uncertainty (Poole & Raftery, 2000; Bijak et al., 2013).

As noted in the previous section, the data ecosystem of demography has been significantly enriched with the digital revolution. The availability of a greater variety of data sources and the ability to link them, either at individual or aggregate levels, offer an opportunity to apply tools of causal analysis for observational data, such as quasi-experimental techniques. These techniques can be especially powerful for analysing the impacts of climate shocks (e.g. temperature changes, natural disasters). Such research designs are enabled by the availability of georeferenced data and the ability to link these to other data, e.g. survey or census datasets, thereby facilitating analysis of the impacts of environmental contexts on demographic outcomes (e.g. Andriano & Behrman, 2020; Hauer et al., 2020; Thiede et al., 2022).

Computational methods like ML further provide new approaches to harness an enriched data ecosystem. While a lot of social demographic research has been guided by a theoretical perspective focused on analysing the specific relationship between a theoretical predictor and outcome of interest, ML techniques allow for ways in which a wider range of potential predictors (or features) that are increasingly available in our data sources as well as different functional forms can inform analyses such that new patterns can be learned from data. From a policy perspective, these approaches have the potential to help identify new types of regularities and relationships between variables (e.g. social factors and health outcomes), detect vulnerable population sub-groups and help guide new questions to identify new social mechanisms that can help streamline the targeting and delivery of public services and social policies (e.g. Wang et al., 2013; Mhasawade et al., 2021; Aiken et al., 2022). The deployment of algorithmic decision-making processes however also raises significant social and ethical challenges, such as those about bias and discrimination, whereby algorithms can amplify existing patterns of social disadvantage, as well as transparency and accountability, particularly given concerns about the opacity of complex algorithms (Lepri et al., 2018). Insights from the demographic literature further emphasize the importance of proceeding carefully when deploying these tools. Social demographic research that has applied ML techniques to long-standing survey datasets to predict life course outcomes such as educational performance or material hardship has shown that these outcomes are often challenging to predict at the individual level (Salganik et al., 2020). More work is needed to understand the conditions under which ML approaches can help improve predictive accuracy with different types of social data but also to better evaluate the social and ethical implications and trade-offs in the use of predictive approaches for policy making.

17.3.2 Data Opportunities

Policy makers are interested in knowing about real-time developments as they unfold. A key challenge with traditional sources of demographic data, as noted in the previous section, has often been their slower timeliness and lags between data collection, processing and publication. Digital trace data, which are generated as by-products of the use of web, social media and mobile technologies, are often able to more effectively capture real-time processes. The widespread use of different types of digital technologies in different domains of life implies that aggregated forms of these data can provide meaningful signals of population behaviours. For example, the reliance on search engines such as Google for information-seeking means that aggregated web search queries, such as those provided via Google Trends, can help us understand health concerns or behaviours, or fertility intentions within a population. When calibrated to 'ground truth' demographic data sources, these real-time data have the potential to help predict future changes and 'nowcast' patterns before they appear in official statistics.

More generally, new data opportunities provide a system for early warning with detailed spatial and temporal granularity. This can be useful in cases where demographic quantities, like migration flows, need to be monitored in response to a crisis, or for understanding the societal responses to demographic change, e.g. misinformation related to migration or media portrayals of immigrant populations. The value of nontraditional, digital trace datasets for monitoring mobility was highlighted during the COVID-19 pandemic, where Google mobility data was used to track the impacts of lockdowns and for other forms of public health surveillance (Google, 2022). These data proved useful to assess the potential impact of policy decisions related to partial or full lockdowns, and related reductions in mobility, on lives saved (Basellini et al., 2021).

Different types of digital trace data can also provide complementary measures of sentiments, attitudes, norms and current conversations in different formats (e.g. images, text) that are useful for capturing social responses to events, as might be required for policy makers. Online spaces have become salient spaces for social interaction and exchange, information-seeking and collective expression and mobilization. For example, in the area of fertility and family formation, online platforms and forums, such as Mumsnet or fertility apps, can provide a view on prevailing sentiments, concerns and aspirations surrounding parenthood. For other domains, such as for the labour market, when understanding supply or demand in specific sectors may be necessary (e.g. long-term care), online job search forums can provide insights into these dynamics (e.g. Buchmann et al., 2022). Social media can also provide a useful barometer to track sentiments surrounding immigration or policy changes surrounding immigration (e.g. Flores, 2017) while also providing novel ways to measure the integration of immigrant groups (e.g. Dubois et al., 2018).

While digital trace data provide unique opportunities, it is important to ensure that appropriate ethical, measurement and theoretical frameworks guide the use of the data for policy purposes and, where feasible, the data be triangulated and contextualized against traditional data sources. In many cases, aggregated data are sufficient to address a policy-relevant research question, whereas in others more fine-grained, individual-level information may be needed. In cases where aggregated data are insufficient, creating ways to appropriately anonymize the data and safeguard against any risk of harming respondents should remain priority. Given that digital trace data are often not expressly collected for research and collected with informed consent, which is a fundamental principle for survey research, higher standards of privacy protection should be adhered to when using these data. A central challenge with digital trace data remains data access. These data come from and are often owned by private companies, which implies that both their access can often be limited and important details of the proprietary algorithms that shape them may not be known. The landscape of access to digital trace data, via more democratic modes of access such as public application programming interfaces (APIs), has become increasingly more constrained, and in many cases platform terms of use have become more stringent. Policy initiatives to support the development of transparent frameworks for enabling ethically guided and privacy-preserving modes of data sharing between research institutions and private companies are urgently needed to ensure that the potential of these data is realized.

When analysing digital traces, it is important to consider demographic biases to better understand who is represented in them and the broader generalizability of the data. These biases may reflect broader digital divides in internet access or platform-specific patterns of use. Triangulation against high-quality traditional data, e.g. from probability surveys, can be valuable in assessing these biases. A separate, but equally important, consideration is that of algorithmic bias, i.e. whereby algorithms implemented on online platforms shape behaviours, such that it is difficult to assess whether observed patterns detected in the data reflect actual behaviours or the algorithms. One way to address algorithmic bias is to move beyond passively collected digital traces towards data collection that involves surveying respondents directly, as we describe next.

The increasing adoption of digital technologies has also facilitated online and mobile modes for primary data collection. For example, even in the case of traditional data sources such as censuses, respondents can fill in questionnaires online, although no census so far has shifted completely online as the exclusive mode of data collection. Digital technologies provide cost-efficient modes for survey data collection, although mode and demographic biases of these platforms need to be addressed when using these approaches. A significant opportunity for online recruitment of specific population groups, e.g. migrants or new parents, is provided by social media-targeted advertisement platforms. These are relevant from a policy perspective as they offer new opportunities for data collection that are cost-efficient, timely, and can help overcome some of the limitations of only passively collected digital traces. For example, Facebook allows ads that are targeted towards migrants from specific countries or language speakers, although the algorithms used to determine whether a user is a migrant are unclear. By conducting surveys on migrant groups where respondents are recruited using these algorithmic targeting capabilities of social media ad platforms, researchers can help audit the algorithms that are used in designing the targeting features of these platforms. While such online surveys offer advantages, they are not high-quality probability samples. Drawing population-level inferences from them requires users to collect demographic information within them followed by the application of de-biasing techniques such as post-stratification weighting, where population weights come from a source such as a census or a high-quality probability survey (Zagheni & Weber, 2015).

An important direction for extracting greater value from digital behavioural data is to integrate these with surveys—for example, mobile app-based modes of data collection may enable both the collection of self-reported information combined with data on location or movement (e.g. via an accelerometer) or time use. More broadly, data linkage of different types of data—e.g. survey and geospatial data, administrative data with survey data—can help bolster the value that can be derived from data for policy purposes. Linked administrative data, such as that from population registers, are a key resource for demographic research. The Nordic countries (Thomsen & Holmøy, 1998; Blom & Carlsson, 1999), but also others such as the Netherlands (Bakker et al., 2014), have led the way in creating robust data infrastructures and access to these data, and greater policy efforts across Europe to improve linkage of and access to administrative data are highly desirable.

17.3.3   Understanding Demographic Heterogeneity in the Impacts of Digital Technologies

Research suggests that digital technologies, by providing cost-effective ways of accessing information, enabling communication and exchange and providing access to vital services, can help empower individuals in different domains of life, including their health, wellbeing and family life, among others. Digital technologies have the potential to provide valuable tools, for example, for mitigating isolation and exclusion of rural or ageing populations, or providing modes for flexible working. While technology has the potential to make significant positive impacts, the internet is also not a singular technology, and one where content is often deregulated and user-generated and where the risk of misinformation is also present. From a policy perspective to ensure that the full potential of digital technologies is realized effectively and equitably, it is essential to understand who is using digital technologies and tools (or not), how they use them and who benefits from them. The demographic perspective can be especially valuable for understanding this with the aim of clarifying who and under what conditions technology can empower and when it does not.

For understanding demographic differences in the use of digital technologies and functionalities, different data sources are needed. First, a deeper assessment of these differences requires more detailed questions, moving beyond simple measures of internet use within traditional data sources, e.g. large-scale social and demographic survey data infrastructures, to understand how individuals are leveraging technologies for various life domains. Second, administrative data from governments, but also private companies (e.g. mobile phone operators), can provide important insights on the use of digital services by demographic groups. Policy makers should seek to incorporate demographic information (e.g. age, gender, education, ethnicity) where possible in identifying the uptake and impacts of digital tools. Third, digital traces from different platforms can themselves be useful for understanding demographic differences in the use of different platforms in some cases. For example, data from the social media marketing platforms can provide insights on the demographic composition of their user base, although the aforementioned limitations about potential algorithmic bias affecting these data should be carefully considered when interpreting these data.

17.4   Discussion

Demography is a highly policy-relevant discipline. As this chapter has highlighted, the new data sources and computational tools available to demographers enable us to provide sharper images of our societies and of sociodemographic mechanisms. This, in turn, amplifies our intuition of the implications of alternative policy choices. While the use of computational approaches, such as those outlined in this chapter, is clearly valuable, we emphasize that these are best thought as providing complementary and synergistic potential. The most fruitful use cases are likely to be those where both traditional and nontraditional data can be integrated for policy making purposes.

Computational modelling approaches that we have described, such as individual-level simulation models, will further benefit from integrating different types of data to help build 'semi-artificial' societies (Bijak et al., 2013), or in other words empirically informed synthetic models, that can serve as virtual laboratories to assess the potential social impacts of different policies. These provide useful tools to assess policy-relevant questions about the impacts of the future course of key demographic trends, such as ageing, climate change and immigration.

A distinct opportunity offered by the demographic perspective is the importance of understanding demographic differences in the use of different types of digital technologies and platforms. This is crucial both from the perspective of understanding their social impacts and also for more careful use, analysis and interpretation of the data generated by the use of technologies (e.g. digital trace data). The internet is not a singular technology, yet the digital revolution has affected nearly all domains of life. Understanding population-level heterogeneity in digital access and skills, as well as identifying pathways through which digital tools can empower different marginalized populations (e.g. rural populations, older populations), is crucial for addressing population inequalities. Ensuring that no one is left behind in digital spaces is something that needs to be addressed by policy makers, as presently significant digital divides in digital infrastructure, as well as digital skills, persist, such as between Eastern and Western Europe (OECD, 2019). Closing these divides will require policy efforts targeting both infrastructure and also digital (up)-skilling to facilitate the digital inclusion of communities.

Policy efforts that push for frameworks for data sharing and access between researchers and proprietary datasets to facilitate their scientific use are crucial for realizing the opportunities offered by new types of data. The involvement of researchers, not only at point of access but also in the process of coproduction of proprietary datasets and for algorithmic transparency, is desirable, to ensure constructive use for scientific and policy insights. Beyond proprietary data, the data revolution also encompasses administrative data held by governments, which is now increasingly digitized, and streamlined access to these data as well as frameworks to facilitate more effective data linkage between different governmental agencies is crucial. While the data ecosystem has diversified and become enriched, we stress that more and bigger datasets do not necessarily mean better data. The proper assessment of data quality and reliance on proper measurement should remain core principles when collecting, producing, using and analysing data, which are areas where demographic research has much to contribute. Lastly, it is useful to remember that while better data when used in an ethical way can provide better images of our societies, data itself can only help us identify problems, but does not solve them.

References

Aiken, E., Bellue, S., Karlan, D., Udry, C., & Blumenstock, J. E. (2022). Machine learning and phone data can improve targeting of humanitarian aid. *Nature, 603*(7903), 864–870. https://doi.org/10.1038/s41586-022-04484-9

Alburez-Gutierrez, D., Zagheni, E., Aref, S., Gil-Clavel, S., Grow, A., & Negraia, D. V. (2019). Demography in the digital era: New data sources for population research. Preprint. *SocArXiv.*https://doi.org/10.31235/osf.io/24jp7.

Alburez-Gutierrez, D., Mason, C., & Zagheni, E. 2021. The "sandwich generation" revisited: Global demographic drivers of care time demands. *Population and Development Review.* Advanced Publication. doi:https://doi.org/10.1111/padr.12436.

Alexander, M., Polimis, K., & Zagheni, E. (2020). Combining social media and survey data to nowcast migrant stocks in the United States. *Population Research and Policy Review*, August. doi:https://doi.org/10.1007/s11113-020-09599-3.

Alvarez-Galvez, J., Salinas-Perez, J. A., Montagni, I., & Salvador-Carulla, L. (2020). The persistence of digital divides in the use of health information: A comparative study in 28 European countries. *International Journal of Public Health, 65*(3), 325–333. https://doi.org/10.1007/s00038-020-01363-w

Andriano, L., & Behrman, J. (2020). The effects of growing-season drought on young women's life course transitions in a Sub-Saharan context. *Population Studies, 74*(3), 331–350. https://doi.org/10.1080/00324728.2020.1819551

Arpino, B., Le Moglie, M., & Mencarini, L. (2022). What tears couples apart: A machine learning analysis of union dissolution in Germany. *Demography, 59*(1), 161–186. https://doi.org/10.1215/00703370-9648346

Bakker, B. F. M., van Rooijen, J., & van Toor, L. (2014). The system of social statistical datasets of statistics Netherlands: An integral approach to the production of register-based social statistics. *Statistical Journal of the IAOS, 30*(4), 411–424. https://doi.org/10.3233/SJI-140803

Basellini, U., Alburez-Gutierrez, D., Del Fava, E., Perrotta, D., Bonetti, M., Camarda, C. G., & Zagheni, E. (2021). Linking excess mortality to mobility data during the first wave of COVID-19 in England and Wales. *SSM - Population Health, 14*, 100799. https://doi.org/10.1016/j.ssmph.2021.100799

Bellou, A. (2015). The impact of internet diffusion on marriage rates: Evidence from the broadband market. *Journal of Population Economics, 28*(2), 265–297. https://doi.org/10.1007/s00148-014-0527-7

Bijak, J., Hilton, J., Silverman, E., & Cao, V. D. (2013). Reforging the wedding ring: Exploring a semi-artificial model of population for the United Kingdom with Gaussian process. *Demographic Research, 29*, 729–766. https://doi.org/10.4054/DemRes.2013.29.27

Billari, F. C., & Zagheni, E. (2017). *Big data and population processes: A revolution?*, July. doi:https://doi.org/10.31235/osf.io/f9vzp.

Billari, F. C., Prskawetz, A., Diaz, B. A., & Fent, T. (2007). The "wedding-ring": An agent-based marriage model based on social interaction. *Demographic Research, 17*, 59–82.

Billari, F. C., D'Amuri, F., & Marcucci, J. (2016). Forecasting births using Google. In *CARMA 2016: 1st International Conference on Advanced Research Methods in Analytics* (pp. 119–119). Editorial Universitat Politècnica de València. doi:https://doi.org/10.4995/CARMA2016.2015.4301.

Billari, F. C., Giuntella, O., & Stella, L. (2019). Does broadband internet affect fertility? *Population Studies, 73*(3), 297–316. https://doi.org/10.1080/00324728.2019.1584327

Billari, F. C., Rotondi, V., & Trinitapoli, J. (2020). Mobile phones, digital inequality, and fertility: Longitudinal evidence from Malawi. *Demographic Research, 42*, 1057–1096.

Blom, E., & Carlsson, F. (1999). Integration of administrative registers in a statistical system: A Swedish perspective. *Statistical Journal of the United Nations Economic Commission for Europe, 16*(2–3), 181–196. https://doi.org/10.3233/SJU-1999-162-307

Buchmann, M., Buchs, H., Busch, F., Clematide, S., Gnehm, A-S., & Müller, J. (2022). Swiss job market monitor: A rich source of demand-side micro data of the labour market. *European Sociological Review*, January, jcac002. doi: https://doi.org/10.1093/esr/jcac002.

Chen, X., Xiang, S., Liu, C.-L., & Pan, C.-H. (2014). Vehicle detection in satellite images by hybrid deep convolutional neural networks. *IEEE Geoscience and Remote Sensing Letters, 11*(10), 1797–1801. https://doi.org/10.1109/LGRS.2014.2309695

Chi, G., Fang, H., Chatterjee, S., & Blumenstock, J. E. (2022). Microestimates of wealth for all low- and middle-income countries. *Proceedings of the National Academy of Sciences, 119*(3), e2113658119. https://doi.org/10.1073/pnas.2113658119

Deville, P., Linard, C., Martin, S., Gilbert, M., Stevens, F. R., Gaughan, A. E., Blondel, V. D., & Tatem, A. J. (2014). Dynamic population mapping using mobile phone data. *Proceedings of the National Academy of Sciences, 111*(45), 15888–15893. https://doi.org/10.1073/pnas.1408439111

Diaz, B. A., Fent, T., Prskawetz, A., & Bernardi, L. (2011). Transition to parenthood: The role of social interaction and endogenous networks. *Demography, 48*(2), 559–579. https://doi.org/10.1007/s13524-011-0023-6

Dubois, A., Zagheni, E., Garimella, K., & Weber, I. (2018). Studying migrant assimilation through Facebook interests. In S. Staab, O. Koltsova, & D. I. Ignatov (Eds.), *Social Informatics* (pp. 51–60). Lecture Notes in Computer Science. Springer International Publishing. doi:https://doi.org/10.1007/978-3-030-01159-8_5.

Elvidge, C. D., Sutton, P. C., Ghosh, T., Tuttle, B. T., Baugh, K. E., Bhaduri, B., & Bright, E. (2009). A global poverty map derived from satellite data. *Computers and Geosciences, 35*(8), 1652–1660. https://doi.org/10.1016/j.cageo.2009.01.009

Entwisle, B., Williams, N. E., Verdery, A. M., Rindfuss, R. R., Walsh, S. J., Malanson, G. P., Mucha, P. J., et al. (2016). Climate shocks and migration: An agent-based modeling approach. *Population and Environment, 38*(1), 47–71. https://doi.org/10.1007/s11111-016-0254-y

Feehan, D. M., & Cobb, C. (2019). Using an online sample to estimate the size of an offline population. *Demography, 56*(6), 2377–2392. https://doi.org/10.1007/s13524-019-00840-z

Fiorio, L., Zagheni, E., Abel, G., Hill, J., Pestre, G., Letouzé, E., & Cai, J. (2021). Analyzing the effect of time in migration measurement using georeferenced digital trace data. *Demography, 58*(1), 51–74. https://doi.org/10.1215/00703370-8917630

Flores, R. D. (2017). Do anti-immigrant laws shape public sentiment? A study of Arizona's SB 1070 using Twitter data. *American Journal of Sociology, 123*(2), 333–384. https://doi.org/10.1086/692983

Gabrielli, L., Deutschmann, E., Natale, F., Recchi, E., & Vespe, M. (2019). Dissecting global air traffic data to discern different types and trends of transnational human mobility. *EPJ Data Science, 8*(1), 26. https://doi.org/10.1140/epjds/s13688-019-0204-x

Gebru, T., Krause, J., Wang, Y., Chen, D., Deng, J., Aiden, E. L., & Fei-Fei, L. (2017). Using deep learning and Google Street View to estimate the demographic makeup of neighborhoods across the United States. *Proceedings of the National Academy of Sciences, 114*(50), 13108–13113.

Gil-Clavel, S., & Zagheni, E. (2019). Demographic differentials in Facebook usage around the world. *Proceedings of the International AAAI Conference on Web and Social Media, 13*(July), 647–650.

Google. (2022). *COVID-19 community mobility report.* 2022. https://www.google.com/covid19/mobility?hl=en

Grow, A., & Van Bavel, J. (2015). Assortative mating and the reversal of gender inequality in education in Europe: An agent-based model. Edited by Hemachandra Reddy. *PLoS One, 10*(6), e0127806. https://doi.org/10.1371/journal.pone.0127806

Grow, A., Perrotta, D., Del Fava, E., Cimentada, J., Rampazzo, F., Gil-Clavel, S., & Zagheni, E. (2020). Addressing public health emergencies via Facebook surveys: Advantages, challenges, and practical considerations. *Journal of Medical Internet Research, 22*(12), e20653. https://doi.org/10.2196/20653

Hauer, M. E., Holloway, S. R., & Oda, T. (2020). Evacuees and migrants exhibit different migration systems after the Great East Japan earthquake and tsunami. *Demography, 57*(4), 1437–1457. https://doi.org/10.1007/s13524-020-00883-7

Jochem, W. C., Leasure, D. R., Pannell, O., Chamberlain, H. R., Jones, P., & Tatem, A. J. (2021). Classifying settlement types from multi-scale spatial patterns of building footprints. *Environment and Planning B: Urban Analytics and City Science, 48*(5), 1161–1179. https://doi.org/10.1177/2399808320921208

Kashyap, R. (2021). Has demography witnessed a data revolution? Promises and pitfalls of a changing data ecosystem. *Population Studies, 75*(sup1), 47–75. https://doi.org/10.1080/00324728.2021.1969031

Kashyap, R., & Villavicencio, F. (2016). The dynamics of son preference, technology diffusion, and fertility decline underlying distorted sex ratios at birth: A simulation approach. *Demography, 53*(5), 1261–1281. https://doi.org/10.1007/s13524-016-0500-z

Kashyap, R., Fatehkia, M., Al Tamime, R., & Weber, I. (2020). Monitoring global digital gender inequality using the online populations of Facebook and Google. *Demographic Research, 43*, 779–816.

Kashyap, R., Gordon Rinderknecht, R., Akbaritabar, A., Alburez-Gutierrez, D., Gil-Clavel, S., Grow, A., Kim, J., et al. (2022). Digital and computational demography. *SocArXiv*. https://doi.org/10.31235/osf.io/7bvpt

Klabunde, A., & Willekens, F. (2016). Decision-making in agent-based models of migration: State of the art and challenges. *European Journal of Population, 32*(1), 73–97. https://doi.org/10.1007/s10680-015-9362-0

Leone, T., Coast, E., Correa, S., & Wenham, C. (2021). Web-based searching for abortion information during health emergencies: A case study of Brazil during the 2015/2016 Zika Outbreak. *Sexual and Reproductive Health Matters, 29*(1), 1883804. https://doi.org/10.1080/26410397.2021.1883804

Lepri, B., Oliver, N., Letouzé, E., Pentland, A., & Vinck, P. (2018). Fair, transparent, and accountable algorithmic decision-making processes. *Philosophy and Technology, 31*(4), 611–627. https://doi.org/10.1007/s13347-017-0279-x

Levantesi, S., Nigri, A., & Piscopo, G. (2022). Clustering-based simultaneous forecasting of life expectancy time series through long-short term memory neural networks. *International Journal of Approximate Reasoning, 140*(January), 282–297. https://doi.org/10.1016/j.ijar.2021.10.008

Lloyd, C. T., Sorichetta, A., & Tatem, A. J. (2017). High resolution global gridded data for use in population studies. *Scientific Data, 4*(1), 1–17. https://doi.org/10.1038/sdata.2017.1

Mencarini, L., Hernández-Farías, D. I., Lai, M., Patti, V., Sulis, E., & Vignoli, D. (2019). Happy parents' tweets: An exploration of Italian twitter data using sentiment analysis. *Demographic Research, 40*, 693–724.

Mhsawade, V., Zhao, Y., & Chunara, R. (2021). Machine learning and algorithmic fairness in public and population health. *Nature Machine Intelligence, 3*(8), 659–666. https://doi.org/10.1038/s42256-021-00373-4

Nigri, A., Levantesi, S., Marino, M., Scognamiglio, S., & Perla, F. (2019). A deep learning integrated Lee–Carter model. *Risks, 7*(1), 33. https://doi.org/10.3390/risks7010033

OECD. (2019). Skills for a digital society. In *OECD skills outlook 2019: Thriving in a digital world*. Organisation for Economic Co-operation and Development. https://www.oecd-ilibrary.org/education/oecd-skills-outlook-2019_df80bc12-en.

Pesando, L. M., Rotondi, V., Stranges, M., Kashyap, R., & Billari, F. C. (2021). The internetization of international migration. *Population and Development Review, 47*(1), 79–111. https://doi.org/10.1111/padr.12371

Poole, D., & Raftery, A. E. (2000). Inference for deterministic simulation models: The Bayesian melding approach. *Journal of the American Statistical Association, 95*(452), 1244–1255.

Potančoková, M., & Marois, G. (2020). Projecting future births with fertility differentials reflecting women's educational and migrant characteristics. *Vienna Yearbook of Population Research, 18*, 141–166.

Pötzschke, S., & Braun, M. (2017). Migrant sampling using Facebook advertisements: A case study of polish migrants in four European countries. *Social Science Computer Review, 35*(5), 633–653. https://doi.org/10.1177/0894439316666262

Rampazzo, F., Bijak, J., Vitali, A., Weber, I., & Zagheni, E. (2021). A framework for estimating migrant stocks using digital traces and survey data: An application in the United Kingdom. *Demography.*

Reis, B. Y., & Brownstein, J. S. (2010). Measuring the impact of health policies using internet search patterns: The case of abortion. *BMC Public Health, 10*(1), 514. https://doi.org/10.1186/1471-2458-10-514

Rotondi, V., Kashyap, R., Pesando, L. M., Spinelli, S., & Billari, F. C. (2020). Leveraging mobile phones to attain sustainable development. *Proceedings of the National Academy of Sciences, 117*(24), 13413–13420. https://doi.org/10.1073/pnas.1909326117

Salganik, M. J., Lundberg, I., Kindel, A. T., Ahearn, C. E., Al-Ghoneim, K., Almaatouq, A., Altschul, D. M., et al. (2020). Measuring the predictability of life outcomes with a scientific mass collaboration. *Proceedings of the National Academy of Sciences, 117*(15), 8398–8403. https://doi.org/10.1073/pnas.1915006117

Schneider, D., & Harknett, K. (2019a). Consequences of routine work-schedule instability for worker health and well-being. *American Sociological Review, 84*(1), 82–114. https://doi.org/10.1177/0003122418823184

Schneider, D., & Harknett, K. (2019b). What's to like? Facebook as a tool for survey data collection. *Sociological Methods and Research*, November, 0049124119882477. doi:https://doi.org/10.1177/0049124119882477.

Sironi, M., & Kashyap, R. (2021). Internet access and partnership formation in the United States. *Population Studies, November*, 1–19. https://doi.org/10.1080/00324728.2021.1999485

Stevens, F. R., Gaughan, A. E., Linard, C., & Tatem, A. J. (2015). Disaggregating census data for population mapping using random forests with remotely-sensed and ancillary data. *PLoS One, 10*(2), e0107042. https://doi.org/10.1371/journal.pone.0107042

Thiede, B. C., Randell, H., & Gray, C. (2022). The childhood origins of climate-induced mobility and immobility. *Population and Development Review.* https://doi.org/10.1111/padr.12482

Thomsen, I., & Holmøy, A. M. K. (1998). Combining data from surveys and administrative record systems. The Norwegian experience. *International Statistical Review, 66*(2), 201–221. https://doi.org/10.1111/j.1751-5823.1998.tb00414.x

van Deursen, A., & van Dijk, J. (2011). Internet skills and the digital divide. *New Media and Society, 13*(6), 893–911. https://doi.org/10.1177/1461444810386774

Verdery, A. M. (2015). Links between demographic and kinship transitions. *Population and Development Review, 41*(3), 465–484. https://doi.org/10.1111/j.1728-4457.2015.00068.x

Verdery, A. M., & Margolis, R. (2017). Projections of white and black older adults without living kin in the United States, 2015 to 2060. *Proceedings of the National Academy of Sciences, 114*(42), 11109–11114. https://doi.org/10.1073/pnas.1710341114

Wang, T., Rudin, C., Wagner, D., & Sevieri, R. (2013). Learning to detect patterns of crime. In H. Blockeel, K. Kersting, S. Nijssen, & F. Železný (Eds.), *Machine learning and knowledge discovery in databases* (pp. 515–530). Lecture Notes in Computer Science. Springer. doi:https://doi.org/10.1007/978-3-642-40994-3_33.

Wilde, J., Chen, W., & Lohmann, S. (2020). COVID-19 and the future of US fertility: What can we learn from Google? *Working Paper 13776. IZA Discussion Papers*. https://www.econstor.eu/handle/10419/227303.

Zagheni, E. (2010). *The impact of the HIV/AIDS epidemic on orphanhood probabilities and kinship structure in Zimbabwe.* UC Berkeley. https://portal.demogr.mpg.de/uc/item/,DanaInfo=escholarship.org,SSL+7xp9m970.

Zagheni, E., & Weber, I. (2012). You are where you E-Mail: Using e-Mail data to estimate international migration rates. In *Proceedings of the 4th Annual ACM Web Science Conference* (pp. 348–351). WebSci '12. Association for Computing Machinery. doi:https://doi.org/10.1145/2380718.2380764.

Zagheni, E., & Weber, I. (2015). Demographic research with non-representative internet data. *International Journal of Manpower, 36*(1), 13–25.

Chapter 18
New Migration Data: Challenges and Opportunities

Francesco Rampazzo, Marzia Rango, and Ingmar Weber

Abstract Migration is hard to measure due to the complexity of the phenomenon and the limitations of traditional data sources. The Digital Revolution has brought opportunities in terms of new data and new methodologies for migration research. Social scientists have started to leverage data from multiple digital data sources, which have huge potential given their timeliness and wide geographic availability. Novel digital data might help in estimating migrant stocks and flows, infer intentions to migrate, and investigate the integration and cultural assimilation of migrants. Moreover, innovative methodologies can help make sense of new and diverse streams of data. For example, Bayesian methods, natural language processing, high-intensity time series, and computational methods might be relevant to study different aspects of migration. Importantly, researchers should consider the ethical implications of using these data sources, as well as the repercussions of their results.

18.1 Introduction

Migration has become one of the most salient issues confronting policymakers around the world. The historic adoption of the Global Compact for Safe, Orderly and Regular Migration (GCM)—the first-ever intergovernmental agreement on international migration—and the Global Compact for Refugees in December 2018 and the inclusion of migration-related targets in the 2030 Agenda for Sustainable

F. Rampazzo (✉)
Leverhulme Centre for Demographic Science, Department of Sociology, Nuffield College, University of Oxford, Oxford, UK
e-mail: Francesco.rampazzo@demography.ox.ac.uk

M. Rango
UN Operations and Crisis Centre (UNOCC), New York, NY, USA
e-mail: marzia.rango@un.org

I. Weber
Universität des Saarlandes, Saarbrücken, Germany
e-mail: ingmar.weber@uni-saarland.de

© The Author(s) 2023
E. Bertoni et al. (eds.), *Handbook of Computational Social Science for Policy*,
https://doi.org/10.1007/978-3-031-16624-2_18

345

Development are a clear testament to this. These frameworks have also provided a renewed push to calls from the international community to improve migration statistics globally. The first of the 23 objectives of the GCM is about improving data for evidence-based policy and a more informed public discourse about migration. As a matter of fact, many countries still struggle to report basic facts and figures about migration, which limits their ability to make informed policy decisions and communicate those to the public, but also limits the ability of researchers to contribute to the production of evidence and knowledge on migration.

Migration is a complex phenomenon to measure. Population changes generally happen slowly as fertility and mortality tend to impact population dynamics gradually. However, a country's population structure might change more rapidly due to migration (Billari, 2022). Migration, and in particular international migration, has become increasingly important in shaping population change, especially in higher-income countries, where fertility is decreasing (Bijak, 2010). The study of migration is affected by many challenges (i.e. availability of data, measurement problems, harmonisation of definitions) (Bilsborrow et al., 1997). Above all, there is a lack of timely and comprehensive data about migrants, combined with the varying measures and definitions of migration used by different countries, which are barriers to accurately estimating international migration (Bijak, 2010; Willekens, 1994, 2019). Despite the best efforts of many researchers and official statistics offices, international migration estimates lack quality due to the limited data available in many countries (Kupiszewska & Nowok, 2008; Poulain et al., 2006; Zlotnik, 1987). Migration is a topic widely discussed in several research fields including demography (Lee, 1966), sociology (Petersen, 1958), political science (Boswell et al., 2011), and economics (Kennan & Walker, 2011). Insufficient availability of quality data on migration can have a high social and political impact, because these inaccuracies might limit the capacity to take evidence-based decisions.

The main data sources used to measure migration are censuses, administrative records, and household surveys, collectively referred to as 'traditional data sources'. These data sources have limitations related to the definition of migrants (i.e. the discrepancy between internationally recommended definition and applied definitions in each country), coverage of the entire migrant population, and the quality of the estimates (especially for admin records) (Azose & Raftery, 2019; Willekens, 2019). Moreover, traditional data on migration are not promptly and regularly available. There might be a gap of several months or even years between the time the data are collected and statistics are released to the public. Timely and granular migration data are needed not only for research purposes but also for informed policy and programmatic decisions related to migration. In times of global crisis, such as the COVID-19 pandemic or the Russian invasion of Ukraine, the need for accurate and timely data becomes particularly urgent, but the capacity to collect data from traditional sources can be significantly reduced (Stielike, 2022).

In the last 25 years, the world has experienced a data revolution (Kashyap, 2021). New data created by human digital interactions increased dramatically in volume, speed, and availability. The data revolution did come not only with the advent of new data sources but also with increased computational power. This, in turn, helped to create more sophisticated models to study social phenomena such as migration. New 'ready-made' data from digital sources, commonly referred to as 'digital trace data' (Salganik, 2019), have started to be repurposed to answer social science questions.

Cesare et al. (2018) addressed the challenges faced by social scientists when using digital traces. One of the main challenges is related to bias and non-representativeness, as users of social media platforms, for instance, are not representative of the broader population and might not necessarily reveal their true opinions or personal details. Correspondingly, understanding how to measure the bias of these online non-representative sources is critical to infer demographic trends for the wider population (Zagheni & Weber, 2015). Once the biases are quantified, one possible next step is to combine different data sources to extract more information and enhance the existing data. This is an ongoing process in which social scientists have started to combine survey data with digital traces, originally created for marketing, and repurposing them for scientific research (Alexander, Polimis and Zagheni, 2020; Gendronneau et al., 2019; Rampazzo et al., 2021; Zagheni et al., 2017). The idea of repurposing data is not new to the social sciences (Billari & Zagheni, 2017; Sutherland, 1963; Zagheni & Weber, 2015). For example, John Graunt's first Life Table (1662) was in fact a reworking of public health data from the *Bills of Mortality* to infer the size of the population of London at the time (Sutherland, 1963).

New data sources are a gold mine for migration studies because they offer an opportunity to address the lack of information which hinders this field of research. Digital traces (especially social media data) are quick to collect using, for example, Twitter's or Facebook's application programming interface (API)[1] (for a comprehensive overview of digital trace data for migration and mobility, check Bosco et al., 2022). This allows to know in close to real time how many of the users are in a specific location and have recently changed their country of residence or are foreign-born, contributing to 'nowcasting' migration (e.g. monitoring trends almost in real time). However, digital traces are not always available to academics and practitioners, as they are mostly owned by businesses and may not be fully and publicly accessible.

This chapter has two objectives. First, it aims to bring examples of how new data sources and methodologies have been used for studying migration and migrant characteristics. Second, it highlights advantages, limitations, and challenges of digital trace data in migration research.

---

[1] An API is a kind of middleman between data held by a company and a user requesting this data. While the actual database storing the data is protected and not exposed to the outside world, an API provides a link between the requesting user and the server where the data are stored in a database (Cooksey, 2014; Sloan & Quan-Haase, 2017). To be able to connect to an API, a key authentication is usually needed, which is a long series of letters and numbers that identifies the account querying the API (Cooksey, 2014).

18.2   New Data in Migration Research

As a statistical concept, international migration has been historically characterised by five building blocks:[2] (i) legal nationality, (ii) residence, (iii) place of birth, (iv) time, and (v) purpose of stay (Zlotnik, 1987). As these blocks are complexly entwined with each other, statistical systems use one or a combination of them to gather data on international migrants. The United Nations recommends a definition of international migration which explicitly focuses on residence and time (UN, 1998), defining a migrant as a 'person who moves from their country of usual residence for a period of at least 12 months'. Migrants that stay between 3 and 12 months are considered to be short-term migrants. The intended purpose of the UN's definition of international migrants is to harmonise data sources worldwide. However, current definitions of migrants vary between countries. While they all depend on the time of stay outside of the country of usual residence, definitions applied at the national level differ (i.e. 'minimum duration of stay in the destination country required for the change of residence in the origin country' Kupiszewska and Nowok, 2008, p. 58) (Kupiszewska & Nowok, 2008; Willekens, 1994).

It has been suggested that digital traces can help refine migration theory and modelling. Fiorio et al. (2017) and Fiorio et al. (2021) highlight the potential of using geotagged Twitter data to investigate short-term mobility and long-term migration. Indeed, the definition of an international migrant has become tied up with the increase in the number of individuals living transnational lives (Carling et al., 2021). Digital trace data might help broaden or qualify the distinction between short-term and long-term migrants, adding nuances. However, we need to consider that digital trace data do not follow the same definition as traditional data sources. For example, on Twitter, migrants can be identified through changes in their location over a period of time, while Facebook provides on their Advertising Marketing Platform a variable that can be used to characterise migrants. The Facebook variable is defined as 'People that used to live in country x and now live in country y' (Rampazzo et al., 2021), which refers to the concept of residence and usage of the social media. The Facebook migrant definition does not account for the time aspect, which creates problems when comparing official migration statistics and Facebook estimates. In Zagheni et al. (2017), the description of the Facebook migrant variable was 'Expat from country x', which highlights that the definition behind this variable may be subject to change.

The information on the categorisation of migrant users on social media is limited. In the case of Facebook, the evidence comes from internal and external research. Migrant users might be identified not only through self-declared public information (e.g. 'hometown') but also through inferred information based on their use of the

---

[2] The UN Expert Group on Migration Statistics is updating and revising concepts and definitions on international migration: https://unstats.un.org/unsd/demographic-social/migration-expert-group/task-forces/TF2-ConceptualFramework-Final.pdf and https://unstats.un.org/unsd/demographic-social/migration-expert-group/task-forces/taskforce-2.

social media (e.g. user's IP address) (US SEC Commision, 2018, 2019, 2020). Spyratos et al. (2018) conducted a survey of 114 Facebook users asking them to check whether they were classified by the Facebook Advertising Platform as migrants. The majority of the non-representative sample was classified correctly as an 'expat' despite not having self-reported country of birth or of previous residence on Facebook. Moreover, Facebook's researchers declared to use 'hometown' as a feature for characterising migrants (Herdağdelen et al., 2016). On Twitter, migrants are typically identified through geo-targeting for research studies. However, the number of geo-tagged tweets is limited: only 2/3% of the tweets are provided with a geo-location (Halford et al., 2018; Leetaru et al., 2013). Fake and duplicate accounts might also be a challenge when studying migrants on social media. For Facebook, the percentages of fake and duplicated accounts are reported every year on the US Securities and Exchange Commission documents and are stable at a 11% duplicate accounts and 5% fake accounts (US SEC Commision, 2018, 2019, 2020). Therefore, possible algorithm changes on the measure provided may affect continuity of data from these sources. Case in point, previous work (Palotti et al., 2020; Rampazzo et al., 2021) identified discontinuities in the Facebook data in March 2019 leading to a drop in the global estimates of the number of migrants active on the platform.

Although migrants are not clearly defined in digital trace data, stock estimates of migrant populations seem to be proportionally comparable to traditional data estimates. Zagheni et al. (2017) showed that Facebook Advertising data and American Community Survey data are highly correlated. Moreover, Facebook Advertising data has proved to be faster in capturing out-migration from Puerto Rico in the aftermath of Hurricane Maria. Alexander et al. (2020) show how Facebook Advertising data allowed to provide monthly estimates of the relocation of Puerto Ricans to mainland USA, and subsequent return migration, which traditional data sources were not able to register. The same result is supported by the use of Twitter data (Martín et al., 2020), as well as by monthly Airline Passenger Traffic data used by the US Census Bureau.[3] Facebook Advertising Platform could also be used to monitor out-migration from a country experiencing political turbulence, such as Venezuela (Palotti et al., 2020). These examples highlight another important feature of digital trace data: their broad geographic availability. These data can be widely available also in contexts of poor traditional statistics (e.g. low- and middle-income countries); for example, the Facebook migrant variable is available for 17 of the 54 African countries (Rampazzo & Weber, 2020).

Facebook Advertising data has also provided insights on migrant integration in Germany and the USA (Dubois et al., 2018; Stewart et al., 2019). Cultural assimilation was studied through the comparison of interests expressed online by the German population and Arabic-speaking migrants in Germany (Dubois et al., 2018). Results shows that Arabic-speaking migrants in Germany are less culturally similar compared to other European migrants in Germany, but the divide is less

---

[3] https://www.census.gov/library/stories/2020/08/estimating-puerto-rico-population-after-hurricane-maria.html

pronounced for younger and more educated men. Similarly, cultural integration in the USA was investigated through self-reported musical interests between Mexican first- and second-generation migrants and Anglo and African Americans (Stewart et al., 2019). The comparison between self-reported musical interests highlights that education and language spoken (e.g. English versus Spanish) are key characteristics determining assimilation. However, these studies are affected by limitations linked to self-reported information and 'black box' algorithms estimating interests on social media platforms.

Analysis of digital traces can do more than help with estimation of current migration stocks. Non-traditional data sources can also provide insights into migration intentions, migration flows, and more. For example, Google Trends data going back to 2004 has been used to estimate migration intentions and subsequently predict flows to selected destination countries (Böhme et al., 2020). Böhme et al. (2020) complemented Google Trends with survey data to predict migration flows and intentions. Their results are robust, but the authors highlight as a limitation that the predictive power of words chosen might change over time. Moreover, the models had higher performance when focusing on countries where internet usage is high (Böhme et al., 2020).

Wanner (2021) used a similar approach with Google Trends data to study migration flows to Switzerland from France, Italy, Germany, and Spain. They found that Google Trends data can anticipate migration flows to a certain extent when actual migration is decreasing in volume. Avramescu and Wiśniowski (2021) focused on Google Trends searches related to employment and education from Romania directed to the UK, creating a composite indicator in a time series model. They obtained mixed results in terms of predictive power, stressing that knowing the context of the origin and destination countries is important to increase accuracy of the predictions. Despite the challenges, all the authors agree that Google Trends is a powerful source for estimating potential migration.

New opportunities might arise also from consumer data from the retail sector (e.g. from basket analysis). For instance, some studies show how food consumption patterns can shed light on integration aspects (Guidotti et al., 2020; Sîrbu et al., 2021). Moreover, companies such as LinkedIn, Indeed, and Duolingo provide reports on their users that might reflect migration dynamics. LinkedIn[4] and Indeed[5] reports focus on economic migration, providing insights on the international job market, while Duolingo[6] featuring the most studied language per country shows, for example, how Swedish is the most popular language in Sweden or that German is the top language studied in the Balkans.

---

[4] https://www.ecb.europa.eu/pub/economic-bulletin/articles/2021/html/ecb.ebart202105_02~c429c01d24.en.html#toc4

[5] https://www.hiringlab.org/uk/blog/2021/10/05/foreign-interest-in-driving-jobs-rises-on-visa-announcement/

[6] https://blog.duolingo.com/2021-duolingo-language-report/

This section has looked at multiple digital data sources and what they can bring to the field of migration studies. Clearly, digital trace data have huge potential given their timeliness and wide geographic availability. However, calibrating new data sources with and validating them against traditional data are essential to use novel sources effectively for migration analysis and policy. New digital data offer possibilities to study a diverse range of topics, including the scale of migration, intentions to migrate, and integration and cultural assimilation of migrants. Given their wide applicability to often politically sensitive topics, such as migration and human displacement, social scientists should critically reflect on the risks of results being misinterpreted, or, worse, misused, and how unethical uses of the data could harm individuals, particularly those in vulnerable situations, and infringe upon their fundamental rights (Beduschi, 2017). While many of the applications of computational social science to study are motivated by a potential positive impact on both migrants and the wider society, similar methods could be used to limit freedom and rights of migrants (for a comprehensive analysis of ethical considerations, see Taylor, 2023).

18.3 New Opportunities in Migration Research

The Digital Revolution has brought not only new data sources but also opportunities to apply new methodologies or augment research possibilities. Modelling migration is necessary because of the lack of quality in migration data from both traditional and digital sources. Digital trace data needs to be calibrated with traditional data. A natural way of combining data sources is through Bayesian models; indeed, Alexander et al. (2020) suggest a framework to combine migration data from multiple sources over time through a Bayesian hierarchical model. One level of the model focuses on adjusting the bias related to non-representative data (e.g. digital trace data) for a 'gold standard' given by survey data (e.g. the American Community Survey). Rampazzo et al. (2021) proposed a Bayesian hierarchical model as well. Their model combines traditional and digital data considering both data sources to be biased. Both frameworks stress that digital trace data cannot be a substitute for traditional data sources and that more accurate results can be obtained through their combination, rather than replacement.

Moreover, social media could also be actively used to recruit survey respondents. Advertisements on social media can be repurposed to recruit survey participants to answer a questionnaire. Facebook and Instagram have been used to recruit survey respondents during the COVID-19 pandemic (Grow et al., 2020), LGBTQ+ minorities (Kühne & Zindel, 2020), but also migrants (Pötzschke & Braun, 2017; Pötzschke & Weiß, 2021). Recruiting migrant respondents for traditional sampling strategies is notoriously challenging. However, social media advertising platforms such as that offered by Facebook provide the opportunity for non-probabilistic sampling of migrants, through the use of the migration variable.[7] Pötzschke and Braun (2017) used Facebook to sample Polish migrants in four European countries—Austria, Ireland, Switzerland, and the UK. In the 4 weeks during which the ads were running, a total of 1100 respondents were recruited with a budget of 500 euro. Moreover, Pötzschke and Weiß (2021) used a similar design on Facebook and Instagram to recruit German migrants worldwide. 3800 individuals completed the questionnaire from 148 countries. The advantage of this strategy is to recruit migrant respondents worldwide in a timely manner and with modest budgets. However, it is challenging to produce representative results as there is no control over who opts in to the survey. This necessitates techniques such as post-stratification to make the results more representative of the specific migrant population. It may be worth noting that similar techniques are also used in traditional surveys (e.g. re-weighting, re-calibration), though with surveys on social media, the lack of a probability sampling results in a necessity to post-stratify.

Narratives around migration are usually investigated through qualitative interviews (Flores, 2017; Rowe et al., 2021). The proliferation of social media has also increased the volume of publicly available text that can be analysed to study general perceptions, narratives, and sentiments on a variety of topics. For instance, Twitter can also be used to analyse sentiments towards migrants and migration (Flores, 2017; Rowe et al., 2021). In 2010, the state of Arizona implemented an anti-immigrant law, the effect of which was studied using 250,000 tweets with natural language processing (NLP) techniques and a difference-in-difference design (Flores, 2017). Analysing the content of the tweets, the author stressed that policies have an effect on the perception of migrants, proving that micro-blogging data are an alternative source for public opinion on migrants (Flores, 2017). In Europe as well, analysis of Twitter text data delivered insights on sentiment towards migrants, describing a situation of polarisation of opinion (Rowe et al., 2021). The data provide an opportunity to track population sentiment towards migration in close to real time and monitor shifts over time. Moreover, focusing on the language used on social media, NLP might be useful to identify migrants and study migration flows (Kim et al., 2020).

High-intensity (e.g. weekly or monthly) time series are an opportunity to monitor change and create early alert systems for shifting migration patterns. Napierała et al. (2022) proposed a cumulative sum model to detect changes in trend of asylum applications. The use of flow data and early warning systems could help policymakers in anticipating refugee movements and improve preparedness and management capacities, if handled ethically and responsibly. However, these data and models can be used to make it more difficult for individuals to exercise their rights under the International Human Rights Law. Administrative data sources hold great potential for the study of migration patterns but present specific issues: for instance, their coverage is limited to the extent that people officially register or de-register from countries' administrative systems; also, administrative records track

---

[7] On Facebook Advertising Platform, it is possible to also create advertisements on Instagram.

events (e.g. asylum applications), not individuals, and are affected by issues of double-counting and biases that may affect their usability for official migration statistics. Eurostat data on number of applications lodged (which might also be biased) in EU countries could be augmented by including digital trace data in the model, increasing the ability to potentially anticipate future trends. This approach is suggested by Carammia et al. (2022) through an adaptive machine learning algorithm which combines data from Google Trends and traditional data sources. Given their frequency, data from social media platforms and Google Trends could indeed contribute to the early identification of shifting trends and, if managed responsibly, to greater capacities of migration policymakers and practitioners to inform adequate and timely measures (Alexander, Polimis and Zagheni, 2020; Martín et al., 2020).

Projects like Refugee.Ai and GeoMatch[8] propose to use data-driven algorithms to assign refugees across countries and improve their integration prospects (Bansak et al., 2018). Providing examples for the USA and Switzerland, Bansak et al. (2018) describe an algorithm based on supervised machine learning and optimal matching which takes into account the refugee characteristics (e.g. age, gender, language, education) and local site characteristics. The authors bring evidence of an improvement in subsequent refugee employment outcomes (from 34 to 48%). Moreover, they suggest that the model is flexible and can focus on different integration metrics to optimise for. The matching system is described also in the context of the UK (Jones & Teytelboym, 2018). Similar systems have been suggested also in Sweden to match refugees and property landlords (Andersson & Ehlers, 2020). Nevertheless, automated decisions should always be accompanied by a human element of review to avoid risks of algorithmic bias and human rights infringements.

There is evidence that also computational methods such as machine learning and neural networks might provide insights on migration. Simini et al. (2021) suggested a gravity model with deep neural network to predict flows of migrants and demonstrated that the model performed better than other models due to its geographic agnosticism. Moreover, convolutional neural networks might lead to new ways of fusing data and master high-frequency data (Pham et al., 2018).

18.4 The Way Forward

This chapter has demonstrated how the Digital Revolution has provided new data sources and opportunities to researchers. Timely data on migration are important not only for academics but also for policymakers and practitioners to design data-driven policies and programmes. The COVID-19 pandemic has stressed the importance of having timely and accurate mobility data for the study of the diffusion of the

---

[8] See https://immigrationlab.org/project/harnessing-big-data-to-improve-refugee-resettlement/.

virus (Alessandretti, 2022). However, data from digital traces often lacks a clear definition of what is being measured. Since such data are obtained from private companies, there may be no information available about the algorithms used to produce migration and mobility estimates, for example, about the specific criteria used to classify migrants. A clearer understanding of the construction of these measures would allow to include these data sources in models with more precision.

In the future, it would be important to create sustainable systems for safe and secure access to the data. At the moment, much of this research is dependent on application programming interfaces (API), which as attested by Freelon (2018) might be closed suddenly. When APIs are not available, web-scraping[9] might be a solution, but terms and conditions of the project as well as ethical implications should be taken into account. Initiatives such as the *Big Data for Migration Alliance* (BD4M),[10] convened by IOM's Global Migration Data Analysis Centre (GMDAC), the EU Commission Knowledge Centre on Migration and Demography (KCMD), and the Governance Lab (GovLab) at New York University, aim to provide a platform for cross-sectoral international dialogue and for guidance on ethical and responsible use of new data sources and methods. *Social Science One*[11] tries to create partnerships between academic researchers and businesses. At the moment, it has an active partnership with Facebook, established in April 2018. The initiative is led by Gary King (Harvard University) and Nathaniel Persily (Stanford University). The goal is to give researchers access to Facebook's micro-level data after having submitted a research proposal. There are significant privacy concerns from this, however, which has created delays in the process. On February 13, 2020, the first Facebook URLs dataset was made available; 'The dataset itself contains a total of more than 10 trillion numbers that summarize information about 38 million URLs shared more than 100 times publicly on Facebook (between 1/1/2017 and 7/31/2019)'.[12] A research proposal is needed to apply for access to such datasets; this is the first step in analysing large micro-level datasets from private social media companies. Companies also often control the analysis produced with their data. Researchers using companies' data have to follow strict contracts on its use and seek approval on the results before publication. The Social Science One initiative is interesting in this regard as it comes with pre-approval from Facebook. However, it also highlights challenges of relying on Facebook-internal teams to prepare the data in a non-transparent matter: recently, Facebook had to acknowledge that, accidentally, half of all of its US users were left out of the provided data.[13] This

---

[9] Web-scraping is defined as the process of automatically capturing online data from online websites (Marres & Weltevrede, 2013).

[10] https://data4migration.org

[11] https://socialscience.one

[12] https://socialscience.one/blog/unprecedented-facebook-urls-dataset-now-available-research-through-social-science-one

[13] https://www.washingtonpost.com/technology/2021/09/10/facebook-error-data-social-scientists/

essentially invalidated any work done with the data so far, including that of PhD students. To avoid such issues, ultimately caused by a lack of external oversight, researchers are increasingly calling for legally mandated corporate data-sharing programmes to enable outside, independent researchers to analyse and audit the platforms[14] (Guess et al., 2022).

Overall, the value of new data sources and new models cannot be underestimated. However, applications of these tools for research and public policy purposes should follow high ethical and data responsibility standards. New data sources and AI-based technologies could help researchers and policymakers improve prediction abilities and fill information gaps on migrants and migration, but the use of these technologies should be closely scrutinised and comprehensive risk assessments undertaken to ensure migrants' fundamental rights are safeguarded. The purposes of machine learning- and AI-based applications should be clearly communicated, and participatory approaches that empower migrant communities and 'data subjects' more generally should be promoted in research and policy domains, with a view to increasing transparency and public trust in these applications, but also provide guarantees for the protection of individual fundamental rights (Bircan & Korkmaz, 2021; Carammia et al., 2022). Many technologies come with a risk of being used to create 'digital fortresses'[15] in which these tools keep out migrants, rather than support them. Hence, social scientists and other researchers should carefully weigh the risks and potential repercussions when using digital traces.

References

Alessandretti, L. (2022) What human mobility data tell us about COVID-19 spread. *Nature Reviews Physics, 4,* 12–13.

Alexander, M., Polimis, K., & Zagheni, E. (2022). Combining social media and survey data to nowcast migrant stocks in the United States. *Population Research and Policy Review, 41,* 1–28. https://doi.org/10.1007/s11113-020-09599-3

Andersson, T., & Ehlers, L. (2020). Assigning refugees to landlords in Sweden: Efficient, stable, and maximum matchings. *The Scandinavian Journal of Economics, 122,* 937–965.

Avramescu, A., & Wiśniowski, A. (2021). Now-casting Romanian migration into the United Kingdom by using Google Search engine data. *Demographic Research, 45,* 1219–1254.

Azose, J. J., & Raftery, A. E. (2019) Estimation of emigration, return migration, and transit migration between all pairs of countries. *Proceedings of the National Academy of Sciences, 116,* 116–122.

Bansak, K., Ferwerda, J., Hainmueller, J., Dillon, A., Hangartner, D., Lawrence, D., & Weinstein, J. (2018) Improving refugee integration through data-driven algorithmic assignment. *Science, 359,* 325–329.

---

[14] https://www.brookings.edu/research/how-to-fix-social-media-start-with-independent-research/

[15] https://apnews.com/article/middle-east-europe-migration-technology-health-c23251bec65ba45205a0851fab07e9b6

Beduschi, A. (2017) The big data of international migration: Opportunities and challenges for states under international human rights law. *Georgetown Journal of International Law, 49*, 981–1018.

Bijak, J. (2010) *Forecasting international migration in Europe: A Bayesian view.* Springer Science & Business Media.

Billari, F. C. (2022). Demography: Fast and slow. *Population and Development Review, 48*, 9–30.

Billari, F. C., & Zagheni, E. (2017). Big data and population processes: A revolution. *Statistics and Data Science: New Challenges, New Generations,* In *Proceedings of the Conference of the Italian Statistical Society* (pp. 167–178). Firenze University Press, CC BY 4.0.

Bilsborrow, R. E., Hugo, G., Zlotnik, H., & Oberai, A. S. (1997). *International migration statistics: Guidelines for improving data collection systems.* International Labour Organization.

Bircan, T., & Korkmaz, E. E. (2021). Big data for whose sake? Governing migration through artificial intelligence. *Humanities and Social Sciences Communications, 8*, 1–5.

Böhme, M. H., Gröger, A., & Stöhr, T. (2020) Searching for a better life: Predicting international migration with online search keywords. *Journal of Development Economics, 142*, 102347.

Bosco, C., Grubanov-Boskovic, S., Iacus, S., Minora, U., Sermi, F., & Spyratos, S. (2022). Data innovation in demography, migration and human mobility. arXiv preprint arXiv:2209.05460.

Boswell, C., Geddes, A., & Scholten, P. (2011) The role of narratives in migration policy-making: A research framework. *The British Journal of Politics and International Relations, 13*, 1–11.

Carammia, M., Iacus, S. M., & Wilkin, T. (2022). Forecasting asylum-related migration flows with machine learning and data at scale. *Scientific Reports, 12*, 1–16.

Carling, J., Erdal, M. B., & Talleraas, C. (2021) Living in two countries: Transnational living as an alternative to migration. *Population, Space and Place, 27*, e2471.

Cesare, N., Lee, H., McCormick, T., Spiro, E., & Zagheni, E. (2018) Promises and pitfalls of using digital traces for demographic research. *Demography, 55*, 1979–1999.

Cooksey, B. (2014). *An Introduction to APIs.* https://zapier.com/learn/apis/

Dubois, A., Zagheni, E., Garimella, K., & Weber, I. (2018) Studying migrant assimilation through facebook interests. In *International Conference on Social Informatics* (pp. 51–60). Cham: Springer.

Fiorio, L., Abel, G., Cai, J., Zagheni, E., Weber, I., & Vinué, G. (2017) Using twitter data to estimate the relationship between short-term mobility and long-term migration. In *Proceedings of the 2017 ACM on Web Science Conference - WebSci '17* (pp. 103–110). Troy, New York, USA: ACM Press.

Fiorio, L., Zagheni, E., Abel, G., Hill, J., Pestre, G., Letouzé, E., & Cai, J. (2021) Analyzing the effect of time in migration measurement using georeferenced digital trace data. *Demography, 58*, 51–74.

Flores, R. D. (2017). Do anti-immigrant laws shape public sentiment? A study of Arizona's SB 1070 using Twitter data. *American Journal of Sociology, 123*, 333–384.

Freelon, D. (2018) Computational research in the post-API age. *Political Communication, 35*, 665–668.

Gendronneau, C., Wiśniowski, A., Yildiz, D., Zagheni, E., Fiorio, L., Hsiao, Y., Stepanek, M., Weber, I., Abel, G., & Hoorens, S. (2019) *Measuring labour mobility and migration using big data: Exploring the potential of social-media data for measuring EU mobility flows and stocks of EU movers.* Publications Office of the European Union.

Grow, A., Perrotta, D., Fava, E. D., Cimentada, J., Rampazzo, F., Gil-Clavel, S., & Zagheni, E. (2020) Addressing public health emergencies via facebook surveys: Advantages, challenges, and practical considerations. *Technical Report, SocArXiv.*

Guess, A., Aslett, K., Tucker, J., Bonneau, R. and Nagler, J. (2021) Cracking open the news feed: Exploring what us Facebook users see and share with large-scale platform data. *Journal of Quantitative Description: Digital Media, 1.* https://doi.org/10.51685/jqd.2021.006.

Guidotti, R., Nanni, M., Giannotti, F., Pedreschi, D., Bertoli, S., Speciale, B., & Rapoport, H. (2020). Measuring immigrants adoption of natives shopping consumption with machine learning. In *Joint European Conference on Machine Learning and Knowledge Discovery in Databases* (pp. 369–385). Springer.

Halford, S., Weal, M., Tinati, R., Carr, L., & Pope, C. (2018). Understanding the production and circulation of social media data: Towards methodological principles and praxis. *New Media & Society, 20*, 3341–3358.

Herdağdelen, A., State, B., Adamic, L., & Mason, W. (2016). The social ties of immigrant communities in the United States. In *Proceedings of the 8th ACM Conference on Web Science*, WebSci '16 (pp. 78–84). New York, NY, USA: Association for Computing Machinery.

Jones, W., & Teytelboym, A. (2018). The local refugee match: Aligning refugees' preferences with the capacities and priorities of localities. *Journal of Refugee Studies, 31*, 152–178.

Kashyap, R. (2021). Has demography witnessed a data revolution? Promises and pitfalls of a changing data ecosystem. *Population Studies, 75*, 47–75.

Kennan, J., & Walker, J. R. (2011). The effect of expected income on individual migration decisions. *Econometrica, 79*, 211–251.

Kim, J., Sîrbu, A., Giannotti, F., & Gabrielli, L. (2020) Digital footprints of international migration on twitter. In *International symposium on intelligent data analysis* (pp. 274–286). Springer.

Kühne, S., & Zindel, Z. (2020). Using facebook and instagram to recruit web survey participants: A step-by-step guide and application in Survey Methods: Insights from the Field (SMIF). Special issue: 'Advancements in Online and Mobile Survey Methods'. Retrieved from https://surveyinsights.org/?p=13558.

Kupiszewska, D., & Nowok, B. (2008). *Comparability of Statistics on International Migration Flows in the European Union* (pp. 41–71). Wiley.

Lee, E. S. (1966). A theory of migration. *Demography, 3*, 47–57.

Leetaru, K., Wang, S., Cao, G., Padmanabhan, A., & Shook, E. (2013) Mapping the global Twitter heartbeat: The geography of Twitter. *First Monday, 18*(5). https://doi.org/10.5210/fm.v18i5.4366.

Marres, N., & Weltevrede, E. (2013). Scraping the social? Issues in live social research. *Journal of Cultural Economy, 6*, 313–335.

Martín, Y., Cutter, S. L., Li, Z., Emrich, C. T., & Mitchell, J. T. (2020). Using geotagged tweets to track population movements to and from Puerto Rico after Hurricane Maria. *Population and Environment, 42*, 4–27.

Napierała, J., Hilton, J., Forster, J. J., Carammia, M., & Bijak, J. (2022). Toward an early warning system for monitoring asylum-related migration flows in Europe. *International Migration Review, 56*, 33–62.

Palotti, J., Adler, N., Morales-Guzman, A., Villaveces, J., Sekara, V., Herranz, M. G., Al-Asad, M., & Weber, I. (2020). Monitoring of the Venezuelan exodus through Facebook's advertising platform. *PLOS ONE, 15*, e0229175.

Petersen, W. (1958). A general typology of migration. *American Sociological Review, 23*, 256–266. http://www.jstor.org/stable/2089239.

Pham, K. H., Boy, J., & Luengo-Oroz, M. (2018). Data fusion to describe and quantify search and rescue operations in the Mediterranean sea. In *2018 IEEE 5th International Conference on Data Science and Advanced Analytics (DSAA)* (pp. 514–523). IEEE.

Pötzschke, S., & Braun, M. (2017). Migrant sampling using Facebook advertisements: A case study of Polish migrants in four European countries. *Social Science Computer Review, 35*, 633–653.

Pötzschke, S., & Weiß, B. (2021). Realizing a global survey of emigrants through Facebook and Instagram. https://doi.org/10.31219/osf.io/y36vr

Poulain, M., Perrin, N., & Singleton, A. (2006). *THESIM: Towards harmonised European statistics on international migration*. Presses universitaires de Louvain.

Rampazzo, F., Bijak, J., Vitali, A., Weber, I., & Zagheni, E. (2021). A framework for estimating migrant stocks using digital traces and survey data: An application in the united kingdom. *Demography, 58*, 2193–2218.

Rampazzo, F., & Weber, I. (2020). Facebook advertising data in Africa. *International Organization of Migration, Migration in West and North Africa and across the Mediterranean: Trends, Risks, Developments, Governance, 32*, 9.

Rowe, F., Mahony, M., Graells-Garrido, E., Rango, M., & Sievers, N. (2021). Using Twitter to track immigration sentiment during early stages of the COVID-19 pandemic. *Data & Policy, 3*, e36.

Salganik, M. J. (2019). *Bit by bit: Social research in the digital age.* Princeton University Press.

Simini, F., Barlacchi, G., Luca, M., & Pappalardo, L. (2021). A deep gravity model for mobility flows generation. *Nature Communications, 12*, 1–13.

Sîrbu, A., Andrienko, G., Andrienko, N., Boldrini, C., Conti, M., Giannotti, F., Guidotti, R., Bertoli, S., Kim, J., & Muntean, C. I. (2021). Human migration: The big data perspective. *International Journal of Data Science and Analytics, 11*, 341–360.

Sloan, L., & Quan-Haase, A. (2017) *The SAGE handbook of social media research methods.* SAGE.

Spyratos, S., Vespe, M., Natale, F., Weber, I., Zagheni, E., & Rango, M. (2018). *Migration data using social media: A European perspective.* EUR 29273 EN.

Stewart, I., Flores, R., Riffe, T., Weber, I., & Zagheni, E. (2019). Rock, Rap, or Reggaeton?: Assessing Mexican Immigrants' Cultural Assimilation Using Facebook Data. *arXiv:1902.09453 [cs].*

Stielike, L. (2022). Migration multiple? Big data, knowledge practices and the governability of migration. In *Research methodologies and ethical challenges in digital migration studies* (pp. 113–138). Cham: Palgrave Macmillan.

Sutherland, I. (1963). John Graunt: A tercentenary tribute. *Journal of the Royal Statistical Society: Series A (General), 126*, 537–556.

Taylor, L. (2023). Data justice, computational social science and policy. In *Handbook of computational social science for policy.* Springer.

UN (ed.) (1998). *Recommendations on statistics of international migration.* No. no. 58, rev. 1 in Statistical Papers. Series M. New York: United Nations.

US SEC Commision (2018). *Facebook Inc 2018 Annual Report 10-K.* https://www.sec.gov/Archives/edgar/data/1326801/000132680119000009/fb-12312018x10k.htm

US SEC Commision (2019). *Facebook Inc 2019 Annual Report 10-K.* https://sec.report/Document/0001326801-20-000013/fb-12312019x10k.htm

US SEC Commision (2020). *Facebook Inc 2020 Annual Report 10-K.* https://www.sec.gov/ix?doc=/Archives/edgar/data/1326801/000132680121000014/fb-20201231.htm

Wanner, P. (2021). How well can we estimate immigration trends using Google data? *Quality & Quantity, 55*, 1181–1202.

Willekens, F. (1994). Monitoring international migration flows in Europe: Towards a statistical data base combining data from different sources. *European Journal of Population, 10*, 1–42.

Willekens, F. (2019). Evidence-based monitoring of international migration flows in Europe. *Journal of Official Statistics, 35*, 231–277.

Zagheni, E., & Weber, I. (2015). Demographic research with non-representative internet data. *International Journal of Manpower, 36*, 13–25.

Zagheni, E., Weber, I., & Gummadi, K. (2017). Leveraging Facebook's advertising platform to monitor stocks of migrants. *Population and Development Review, 43*, 721–734.

Zlotnik, H. (1987). The concept of international migration as reflected in data collection systems. *The International Migration Review, 21*, 925–946.

Chapter 19
New Data and Computational Methods Opportunities to Enhance the Knowledge Base of Tourism

Gustavo Romanillos and Borja Moya-Gómez

Abstract Tourism is becoming increasingly relevant at different levels, intensifying its impact on the environmental, the economic and the social spheres. For this reason, the study of this rapidly evolving sector is important for many disciplines and requires to be quickly updated. This chapter provides an overview and general guidelines on the potential use of new data and computational methods to enhance tourism's knowledge base, encourage their institutional adoption and, ultimately, foster a more sustainable tourism.

First, the chapter delivers a brief review of the literature on new data sources and innovative computational methods that can significantly improve our understanding of tourism, addressing the big data revolution and the emergence of new analytic tools, such as artificial intelligence (AI) or machine learning (ML). Then, the chapter provides some guidelines and applications of these new datasets and methods, articulated around three topics: (1) measuring the environmental impacts of tourism, (2) assessing the socio-economic resilience of the tourism sector and (3) uncovering new tourists' preferences, facilitating the digital transition and fostering innovation in the tourism sector.

19.1 Introduction

Tourism is playing an increasingly important role at many levels, and its sector is evolving extraordinarily fast. Thus, the study of tourism, crucial for numerous disciplines, needs to be quickly updated. During the last few years, tourism research is starting to be renovated to keep pace with the ongoing transformations. Nowadays, new data sources and innovative quantitative and qualitative methods offer new possibilities for better analysing and planning tourism (Xu et al., 2020), overcoming many limitations of more conventional approaches.

G. Romanillos (✉) · B. Moya-Gómez
tGIS, Department of Geography, Universidad Complutense de Madrid, Madrid, Spain
e-mail: gustavro@ucm.es; bmoyagomez@ucm.es

© The Author(s) 2023   361
E. Bertoni et al. (eds.), *Handbook of Computational Social Science for Policy*,
https://doi.org/10.1007/978-3-031-16624-2_19

Although recent tourism research is exploring and taking advantage of new data sources and methods, there is still a long way to walk on innovation. This chapter aims to provide a general review and some guidelines on the potential use of new data and computational methods to enhance tourism's knowledge base and promote their institutional adoption and, ultimately, more sustainable tourism.

The chapter is articulated around three topics proposed by Barranco et al. and included in a publication of the Joint Research Centre that aimed at collecting the upcoming research needs in terms of policy questions (Bertoni et al., 2022). The first one measures the environmental impacts of tourism. Tourism is, directly and indirectly, consuming an increasing amount of global resources, including fossil fuel consumption with the associated CO_2 emissions, freshwater, land and food use (Gössling & Peeters, 2015). Therefore, assessing the impact of global tourism activity is one of the most relevant potential applications of new data sources and computational methods.

The second topic is assessing the socio-economic resilience of the tourism sector. Tourism economic weight and social impact have become even more evident in the context of the COVID-19 pandemic: the crisis has put between 100 and 120 million direct tourism jobs at risk, many of them in small- and medium-sized enterprises, according to the UNWTO (2021). Hence, it is relevant and urgent to explore how new data sources, analyses and models can contribute to planning a more resilient and balanced tourism sector in socio-economic terms.

Finally, the third topic is uncovering new tourists' preferences, facilitating the digital transition and fostering innovation in the tourism sector. How can we better analyse new tourist patterns? COVID-19 may have accelerated some existing changes in tourism trends, so there is an urgent need for quick analyses and predictions for the very near future, as the emergence of nowcasting techniques evidences it.

19.2   Existing Literature

Over the past years, new data sources and innovative computational methods emerged to significantly improve our understanding of tourism. A summary is provided next.

19.2.1   New Data Sources of Potential Interest for Tourism

Tourism is being transformed at an accelerated pace, and conventional data sources often do not reflect ongoing changes with enough velocity or spatiotemporal resolution to support the urgent studies to be carried out. In this scenario, new data sources emerge as the raw material to open further explorations in tourism.

New datasets can be grouped into different categories according to data sources. Next, a listing of relevant new datasets is provided, classified according to the nature of the data source and its potential interest for tourism studies, offering some specific examples.

First, we must point out big data from specific sources of the tourism sector, such as smart tourism cards or information systems in destinations. These sources provide data directly recorded in tourism points of interest, valid to monitor existing activity and analyse current or past trends. In this category, we can also include other data sources such as booking data from transportation companies (especially flight booking data from airline operators), which can help predict tourism activity quickly, feeding nowcasting models. Additionally, we can highlight online accommodation companies and apps, such as Tripadvisor[1] or Booking,[2] or new peer-to-peer accommodation online services such as Airbnb[3] (Calle Lamelas, 2017). These sources can be helpful not only for anticipating tourism demand but also because of the additional information collected from users, such as opinions, ratings, comments, etc. In addition, data with high spatiotemporal resolution allows us to analyse emergent spatial patterns, for instance, in the location of Airbnb accommodation in heritage cities (Gutiérrez et al., 2017).

Second, it is remarkable the potential use of GPS datasets. GPS data was actually ranked the top of big data in tourism research (accounting for 21%) and the first of device data (58%) according to the classification provided by Li et al. (2018). It is essential to use GPS tracks to study tourists' routes, with an unprecedented level of detail, thanks to the high spatiotemporal resolution of GPS records. In this group, we include the GPS routes recorded by vehicle navigation apps, such as TomTom,[4] Waze[5] or Google Maps,[6] and tracking apps such as Wikiloc[7] or Strava,[8] very useful when analysing tourism in natural areas, for instance (Barros et al., 2019), or GPS data collected through the emerging tourist mapping apps (Brilhante et al., 2013; Gupta & Dogra, 2017).

Third, it is also outstanding the interest of user-generated content (UGC), especially datasets obtained from social networks such as Twitter; photo-sharing social networks such as Instagram,[9] Flickr[10] or Panoramio[11]; or apps focused on the location of points of interest, such as Foursquare.[12] UGC allows us to explore

---

[1] Tripadvisor website, for more information: https://www.tripadvisor.com/

[2] Booking website, for more information: https://www.booking.com/

[3] Airbnb website, for more information: https://www.airbnb.com/

[4] TomTom website, for more information: https://www.tomtom.com/

[5] Waze website, for more information: https://www.waze.com/

[6] Google Maps website, for more information: https://maps.google.com/

[7] Wikiloc website, for more information: https://www.wikiloc.com/

[8] Strava website, for more information: https://www.strava.com/

[9] Instagram website, for more information: https://www.instagram.com/

[10] Flickr website, for more information: https://www.flickr.com/

[11] Panoramio website, for more information: https://www.panoramio.com/

[12] Foursquare website, for more information: https://foursquare.com/

different tourism dynamics. Semantic analysis of online textual data, such as tweets or travelling blog content, can uncover tourism preferences and trends (Ramanathan & Meyyappan, 2019). Spatial or temporal analyses can also be carried out because most users share data through mobile apps that register GPS coordinates. For instance, Flickr data can be the basis for different temporal analyses, such as estimating tourism demand over a day according to time slots or measuring tourism seasonality in national parks (Barros et al., 2019); also Twitter and Foursquare data can support spatial analyses, such as the identification of multifunction or specialised tourist spaces in cities (Salas-Olmedo et al., 2018) (Fig. 19.1).

Fourth, search engines' data constitute a precious data source, such as Google Trends records. Considering that search engines are a leading tool in planning vacations (Dergiades et al., 2018), these datasets provide information on tourists' interests and plans in advance and can feed models oriented to forecasting tourist arrivals (Havranek & Zeynalov, 2021).

Fifth, we must highlight the interest of datasets obtained from diverse information and communication technologies/devices. The rapid development of the Internet of Things (IoT) provides an increasing amount of Bluetooth data, RFID data and Wi-Fi data (Shoval & Ahas, 2016), which can be helpful to measure, for instance, tourist presence and consumer behaviour over time. Also, in this group, we must emphasise mobile phone data due to its potential use at different scales and for various purposes. The COVID-19 pandemic has accelerated the adoption of mobile phone data to monitor changes in tourism or general mobility trends with a high level of spatiotemporal resolution (Romanillos et al., 2021). This analysis may be extended beyond national borders. Nowadays, roaming services have become crucial for tourists, and roaming data allows us to track tourists globally. Lastly,

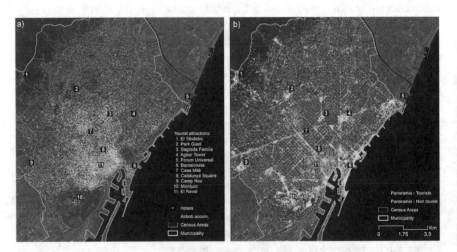

Fig. 19.1 Location of hotel and Airbnb offers (**a**) and density of photographs taken by tourists and residents (**b**) in Barcelona. Source: Gutiérrez et al. (2017)

credit card datasets should also be included here, given their potential for tourist consumption and behaviour analyses.

Finally, more conventional data sources can also provide "new" datasets and opportunities, due to improvements in the quality of data or the way data is shared, in real time, through mobile apps and online services. For example, it is the case of meteorological data. Given that weather is an essential factor in tourism demand, incorporating meteorological variables in tourism forecasting models can increase the predictability of tourist arrivals (Álvarez-Díaz & Rosselló-Nadal, 2010).

19.2.2 New Computational Methods with Application to Tourism Studies

In recent years, increased computational capacity, part of the *big data* revolution, has allowed for faster and cheaper analysis of massive databases by using new analytic tools, such as *artificial intelligence (AI)* or *machine learning (ML)*. Nowadays, tourism analysts may also access an enormous collection of methods for their studies (some are comprehensible; others are like "black boxes"). This section gives a brief and non-exhaustive list of computational methods used in tourism studies, applications and examples.

Unsupervised techniques can identify groups and relationships by analysing explanatory variables themselves: no already known responses exist. Outcomes must be validated – are they logic? – tagged and hypothesised. In tourism, *clustering* techniques were used for detecting the spatial patterns of new touristic accommodations (Carpio-Pinedo & Gutiérrez, 2020) or exploring topics of online tourists' reviews (Guo et al., 2017), *factor analysis* for uncovering latent motivational and satisfaction variables in tourist (Kau & Lim, 2005) and *association rules mining/learning* for discovering the most frequent and strong sets of visited places with Bluetooth data (Versichele et al., 2014).

Supervised techniques provide models to explain/predict responses. They need complete observations: explained (response) and explanatory variables. Outcomes must be compared to observed datasets. Some models investigate causalities and hypothetical "what-if" scenarios (key results are model's parameters): *linear regressions* for inferring causes on tourism industry employment and retention (Chen et al., 2021) or *structural equation models (SEM)* for modelling the quality of life in a tourist island (Ridderstaat et al., 2016). Other models, especially AI-based techniques, anticipate responses or classify observations (key results are responses): *autoregressive moving average (ARMAX) time series models* for forecasting weekly hotel occupation with online search engine queries and weather data (Pan & Yang, 2017) or *artificial neural networks (ANN)* for predicting tourist expenditures (Palmer et al., 2006).

Some datasets need to be treated before applying the above methods, especially for reusing datasets from other studies or online sources. Observations must be regrouped into another spatial or temporal unit. While aggregating is a straight-forward procedure, disaggregating data needs the use of other techniques; see estimating visitor data from regional to municipality scope (Batista e Silva et al., 2018).

Finally, data and models' outcomes need to be presented and stand out to the target public. They can be shown using innovative designs (word clouds, cartograms, etc.), such as the United Nations World Tourism Organization (UNWTO) tourism data dashboard (UNWTO, n.d.). Part of them should be used on digital social networks or in other analysis processes.

19.3   Guidelines

This section proposes some guidelines and potential applications of the described new data sources and computational methods to the three main topics mentioned in the introduction.

19.3.1   Assessing the Environmental Impacts of Tourism

To facilitate the green transition in the tourism sector, we need a concrete EU roadmap with a solid framework and measurable objectives. Working with key performance indicators (KPIs) can help guide and commit the tourism industry and destinations. This section aims to propose a set of KPIs related to central topics regarding the environmental impact of tourism, focusing on new data sources and computational methods.

The first topic concerns tourism mobility. Sustainable tourism should be linked to a concept of sustainable mobility, so we propose a set of KPIs that can reveal to what extent we are advancing in the transition to a more sustainable model (Table 19.1).

The second topic is tourism land consumption. As a consequence of the growth of tourism activity, land in tourist destinations is progressively occupied and degraded. Essential variables in this degradation process are land occupation, land fragmentation and changes in land-use patterns. We propose a set of KPIs that can improve the monitorisation of these variables, with the help of new data sources and methods (Table 19.2).

Finally, the third topic is tourism resources consumption and management. The increasing number of tourists leads to dramatic growth in the consumption of local resources, often leading to unsustainable scenarios. Next, a set of KPIs is proposed to help evaluate tourism resources consumption with the support of new data sources and methods (Table 19.3).

Table 19.1 KP Is for tourism mobility

Topic	KPIs' description [*units*] (from *low* (+) to *high feasibility* (+++))
Tourism mobility	*National/International tourist trips made by plane, according to distance ranges* [%] (+++) • The lowest the percentage, the lower the environmental impact of the tourism mobility model, since planes are generally the most pollutant transportation mode. Around 40% of tourism emissions are attributed to air travel (Higham et al., 2019) • Mobile phone data can be used to differentiate transportation modes by geolocating call detail records (CDR) and by map-matching estimated routes with transportation networks. Especially effective at the national level but increasingly feasible at the international one with the help of roaming databases (Ahas et al., 2018) • Data from transportation operators, especially airlines • Feasibility: high. Although monitorisation of different mobility and transport dynamics (including tourism) based on mobile phone data is still being adopted in other countries (at a very different pace), data from transportation operators are easily accessible
	National/international tourist trips made by collective transport modes [%] (+++) • Mobile phone data can be used to differentiate some collective transportation modes from private ones (train, planes or boats) but not correctly others (bus) • Data from transportation operators, car rental companies, etc. • Feasibility: high, for the same reasons explained above
	Proximity tourism, below a threshold to be defined [%] (++) • Mobile phone data can be effectively used to differentiate tourist trips, analyse trip distances and estimate tourist flow • Trends in the present and the near future can be estimated through data from online accommodation companies and apps, photo-sharing social networks, online accommodation companies and apps and search engines, such as Google Trends records • Feasibility: moderate–low. Monitorisation tourism based on mobile phone data is not yet adopted by all EU countries, or it does not have the necessary spatiotemporal resolution
	Tourism active mobility [%] (++) • Trekking and cycling trends can be estimated through data from tracking apps, such as Strava, and photo-sharing social networks (Barros et al., 2019) • Regarding urban tourism, cycling trends can be explored through bike share systems records and other tracking and navigation apps such as Google Maps. In addition, walking is starting to be explored through mobile phone data (Hunter et al., 2021) • Mobile phone data can be effectively used to differentiate trekking and cycling activity • Feasibility: moderate. Accessibility to the mentioned data sources depends on partnerships with private companies

(continued)

Table 19.1 (continued)

Topic	KPIs' description [*units*] (from *low* (+) to *high feasibility* (+++))
	CO_2*emissions associated with tourist trips [CO_2tonnes] (++)*
	• Mobile phone data can be used to differentiate transportation modes and estimate routes and trip distances, before estimating CO_2 emissions
	• Recently, specific methodologies to estimate the emissions from main touristic activities have been developed (Russo et al., 2020)
	• Emission inventory is the most reliable and comprehensive data source since it derives from many studies' compilation and analysis
	• Feasibility: moderate, depending on the accuracy of the estimation desired. Accurate estimations would rely on precise monitorisation of all tourist trips. Some useful data for this purpose (such as mobile phone data) have not been adopted by many EU Member States yet

19.3.2   Socio-Economic Resilience in the Tourism Sector

Tourism is an important sector in the EU economy. EU's tourists spent about $400 billion on trips across Europe before COVID-19 (Eurostat, 2021b). In 2016, tourism was 10% of the EU's GDP, and it employed 10% of workers in 3.2 million tourism-related enterprises (Eurostat, 2018). However, the tourism sector has high levels of temporal contracts and low retention rates (25%), women employment (~60%), younger workers 15–24 years old (~20%), lower educated workers (~20%) or foreign workers (~1/6) compared to other sectors.

The following KPIs can help key stakeholders assess their tourist offers and benchmark with competitors. These indicators could identify socio-economical relationships, vulnerabilities and weaknesses, undeveloped attractions and upcoming opportunities to make a more resilient sector. KPIs' spatiotemporal dimensions are essential, especially for regions characterised by stationarity. These KPIs should be calculated for several periods, for the whole touristic population in a location (descriptive) or the whole/specific touristic population in competitors (comparison).

This first group of KPIs points out socio-economic impacts of tourism in a region that can be used for comparing them with other industries and competitors (Table 19.4). Some of these KPIs measure tourism impacts directly, but others estimate effects through related activities.

The second set of KPIs concerns assessing tourist models' diversity for detecting excess dependencies on a few attractions and tourist profiles and their stationarity (Table 19.5). Less diverse territories might be very vulnerable to changes in the tourism demand, wildly unexpected events or incompatible weather, among other cases.

Table 19.2 KPIs for tourism land consumption

Topic	KPIs' description [*units*] (from *low* (+) to *high feasibility* (+++))
Tourism land consumption	*Land occupation related to tourism [%] (+++)* • In addition to conventional approaches, based on the analysis of satellite/aerial images, new methodologies based on new data sources can help differentiate tourist developments from residential ones and better assess tourism's particular impact • Mobile phone data and diverse online social networks can be analysed over time to discriminate tourism activity across the territory (Salas-Olmedo et al., 2018) • Feasibility: high. Satellite images are widely available, and characterisation of land occupation can be done based on several – more or less – accessible data sources
	Coast occupation related to tourism [%] (+++) • Defined as the land occupation in a distance range from the coast, generally 500 metres (Riitano et al., 2020) • Similar data sources and methods as in the percentage of land occupation KPI • Feasibility: high, for the same reasons mentioned above
	Land fragmentation indexes related to tourism (+++) • The European Environment Agency assesses the general land fragmentation pressure through the Effective Mesh Density (seff) Fragmentation Index • Land fragmentation indexes in tourist areas or the contribution of tourism to general land fragmentation could be estimated by differentiating tourist developments from residential ones using mobile phone and online social networks data • Feasibility: high. The datasets for calculating the most common fragmentation indexes are publicly accessible
	Tourism share in real estate market [%] (++) • New analyses can be based on data from companies monitoring its evolution daily – new opportunities for nowcasting • Tourist developments can be distinguished from residential ones using mobile phone and online social networks data • Feasibility: moderate. Accessibility to the mentioned datasets depends on partnerships with private companies
	Presence of short-term rentals platforms (e.g. Airbnb) (++) • At EU27, the share of international guest nights at short-stay accommodation offered via collaborative economy in 2019 was 67% (Eurostat, 2021a) • Analysis based on data from accommodation platforms or associated websites (e.g. Inside Airbnb) • Short-term tourism properties can be discriminated through estimations based on mobile phone data and diverse online social networks data • Feasibility: moderate. Accessibility to the mentioned datasets depends on partnerships with private companies

Table 19.3 KPIs for tourism resources consumption and management

Topic	KPIs' description [*units*] (from *low* (+) to *high feasibility* (+++))
Tourism resources consumption and management	*Local product consumption in tourist destinations [%] (++)* • A high percentage is related to the less environmental cost of product transportation and warehousing • Credit card datasets could be the basis for estimating the percentage of consumption in local businesses and markets related to local products • Google points of interest (POI) datasets allow estimating/monitoring people's presence in local businesses and markets, is an approximation of consumption • Feasibility: moderate. Accessibility to the mentioned datasets depends on partnerships with private companies
	Water consumption in tourist destinations [litres/person-year] (+++) • New water consumption datasets registered by water management companies, with high granularity • Innovative analyses and short-term forecasting of water consumption patterns can be done based on these new datasets (Candelieri et al., 2015) • New indicators are proposed considering direct and indirect water consumption, estimated from food consumption data, for instance (Gössling, 2015). In addition, food consumption patterns could be analysed based on credit card datasets • Feasibility: high, due to the increasing accessibility to the mentioned datasets
	Electric energy consumption per capita *(kWh/person-year) (+++)* • New datasets from electricity companies, which are recording consumption with unprecedented granularity • New electric energy consumer characterisation methodologies are proposed, based on data mining techniques (Pérez-Chacón et al., 2018) • Feasibility: high, for the same reasons explained above
	Solid waste generation [kg/person-year] and recycled solid waste generation [%] (+++) • New smart waste management systems are being proposed. For example, bins equipped with sensors can register geolocated data on total/recyclable waste (Folianto et al., 2015) • Feasibility: high, for the same reasons explained above

19.3.3 Uncovering New Tourists' Preferences, Digital Transition and Innovation in the Tourism Sector

New information technologies have revolutionised the tourism sector too. This section introduces how new technologies can be used to detect tourists' preferences and better manage touristic businesses and locations.

Table 19.4 KPIs for socio-economic impact of tourism in a region

Topic	KPIs' description [*units*] (from *low* (+) to *high feasibility* (+++))
Tourism employment	*Employment in tourism [%] (+++)* • Job portals include a large sample of available positions. The number of tourism positions can be compared with the whole set • Project Skills-OVATE is an example of getting data for this KPI (CEDEFOP, n.d.) • Feasibility: high, job portals usually have a good sample of positions everywhere
	Training/educational activities for tourism activities [%] (++) • Tourism-related post-compulsory educational offers within a location can point out the tourism industry employment demands • It is necessary to set some affinity between activities and tourism employment • Feasibility: moderate; educational offers usually are spread into many websites
Tourism economies	*New employee base salaries [€] (+++)* • Job positions on employment portals often include base salaries • This data can point out foreseeable income for employees and their relationship to any other sector's new position offer • Feasibility: high; position offers should include it
	Expenditure in destination [€] (+) • Aggregated credit card data can provide insight also regarding their consumptions • Additional data sources (accommodation websites, etc.) must be scanned to include them in the total expenditures • Feasibility: low; it highly depends on available aggregated bank data. Some payments were paid at home

19.3.4 Analysis of Preference Changes in the Tourism Sector

Businesses may use tourist demand data (accommodation booking, car renting) and users' responses (comments or reviews on products or services) to comprehend the needs of (new) customers to develop and/or to update their products and services and to improve their customer care. While the former may reveal tourist preferences based on their choices, the latter may also highlight some declared unsatisfied ones. Analyses of preference changes need benchmarking approaches; competitor performances provide insight into the strengths and weaknesses of the study location/business. Nevertheless, how may new data and methods aid in the detection of preferences and their changes? Some guidelines are provided next.

19.3.4.1 Searching for Holidays and Activities

Many trips or touristic activities begin with an online search. Potential tourists use either general online search engines or specific touristic planner services. Consequently, data on preferences may be extracted by using autocompletion to

Table 19.5 KPIs for assessing tourism diversity

Topic	KPIs' description [*units*] (from *low* (+) to *high feasibility* (+++))
Tourism offers diversity	*Type/category of tourism attractions [%] (++)* • Online travel agencies, trip advisor services and local tourist websites gather helpful information about tourist destinations (what to do, when, where, etc.). Businesses and other users feed them too • Data of those services for specific locations and nearby places and categorising them helps approximate the supplied tourist diversity • Feasibility: moderate; some attractions might not be online
	Affinity to be recommended [%] (++) • Previous services have also provided users with AI-based recommendations for choosing their destinations • These recommendations can show substitutive and complementarity options • These services can be used to know if any location is recommended for any specific tourism model or demand (search ranking, etc.) • Feasibility: moderate; it needs to create comprehensive groups before undertaking data collection
	Open attractions by weather [%] (+) • Some tourism attractions need particular short-/mid-term weather conditions • Attractions might post online messages regarding disruptions because of weather conditions • Feasibility: low; small attractions do not post it
Tourism profiles diversity	*Prominent group(s) of tourists per age/trip type [%] (++)* • Google Trends or similar can show the most frequent additional queries to some locations (main query) • The additional queries can include target words (children, family, party or elderly) that may describe a type of tourist • Longer queries might not retrieve results for having very few searches with that query • Feasibility: moderate; it needs to create comprehensive groups before undertaking data collection
	Prominent group(s) of tourists per activity [%] (++) • Some tourists post their experiences, opinions and revision on digital platforms. This usually has spatial data or metadata • Aggregating all this data can be used to get an approximation to identify which type of tourist usually visits any place • Feasibility: moderate; some groups might not be well represented (digital divide)
	Prominent group(s) of tourists per origin [%] (+) • Aggregated mobile phone data can be linked with the country of the user's operator (international and other territorial prefixes) and local antennas • It is possible to approximate which places are attractive to tourists, by origin, by counting their connections • This procedure can also determine national stationarities (Raun et al., 2016) • This KPI can also be obtained using credit card data • Feasibility: low–moderate; it highly depends on the availability of phone mobile/bank data

suggest current trending complete search queries, or using some services like Google Trends for a similar end to observe variations over time. These tools can use queries from specific countries to help segment tourist preferences per origin while planning their holidays. Search query data has been used in many academic studies; Dinis et al. (2019) gathered and summarised some of them into the following topic categories: forecasting, nowcasting, identifying interests and preferences, understanding relationships with official data and others.

19.3.4.2 Text Is a Mine

People use words to communicate, and they can publicly share their opinions, recommendations, suggestions and complaints towards touristic attractions in interactive platforms. An analyst can use *text mining* techniques, such as *natural language processing (NLP)*, to extract the sense of messages (including emojis) and undertake sentiment analyses (converting text into Likert scale values). However, this data may contain brief messages, with abbreviations, because of character restrictions. They must be translated into expanded statements. Also, fake/compulsive users should be dropped to avoid biases. Finally, text mining techniques have difficulties detecting ironic tones.

19.3.4.3 What a Beautiful Picture!

Some tourists also upload their pictures and videos on digital social networks. Unlike texts, images need to be described before automating processes to extract comprehensible data. Simple methods can summarise colours in pictures (they can explain weather conditions or infer day periods). More advanced ones, available in cloud computing services, can also identify locations, buildings and objects. Thus, pictures transformed into texts and previously mentioned *text mining* techniques can help determine preferences. In addition, images can include description text and comments that can be used to uncover revealed preferences. Finally, images' metadata include when and, sometimes, where they were taken. This data can be used for determining spatial preferences of what to take a picture of and from where (viewpoints).

19.3.4.4 Life Is Change

Tourists' preferences can evolve for many reasons (getting older, having children or new job positions or contextual reasons, among others). To detect these changes, it is required to have previous preferences to compare with the new ones and see significant changes. The above-mentioned methods can continuously process data, get further insights or update continuous datasets.

Notice that many tips and ideas use similar strategies for calculating the KPIs introduced in the previous section. Thus, they may also be reinterpreted to help analyse preference changes or warn regarding new successful tourist strategies in competitive locations by KPIs' variations.

19.3.5   Digital Transition and Innovation

We have seen that using new data sources and computational methods can improve our understanding of tourism dynamics and help plan and develop better tourism policies. However, institutions and companies still have a long way to go to use all these new resources. To accelerate what's been called the digital transition and foster innovation, we address several relevant questions in this section.

19.3.5.1   What Are the Main Challenges for Increasing Digitalisation and Innovation in the Tourism Sector? How Can Existing Difficulties Be Overcome?

Small and medium enterprises (SMEs) constitute the majority (around 90%) of Europe's tourism enterprises (UNWTO, 2020). These kinds of enterprises often do not keep pace regarding technological advances, and are behind large companies regarding the digital transition. Furthermore, it has been estimated that up to 25% of jobs in tourism need upskilling.

To maintain the competitiveness of the European travel destinations and satisfy the emerging interests of the travellers towards sustainable travelling options, we need to support the digital transition. Therefore, it is urgent to digitalise services and close the existing skills gap.

The private sector essentially provides this support, with most SMEs relying on a few private tech companies. Public institutions should provide similar platforms or foster new public-private partnerships (PPPs) to increase the accessibility to new technologies and facilitate the upskilling process.

19.3.5.2   What Are the Main Difficulties in Collecting New Data? What Strategies Towards Effective Data Collection Should Be Put in Place?

New datasets essentially come from digital data sources. Fostering digitalisation is, therefore, the first step in the way of increasing the collection of data. However, as previously mentioned, digitalisation is mainly led by a few private big tech companies. Consequently, most of the new datasets come from these companies. Two actions could be necessary, then: first, to foster new or better deals and partnerships with them as data providers and, second, to avoid an excessive dependence on big tech companies by developing public digital/online platforms and services for SMEs, where the whole ecosystem (companies, institutions and users/tourists) shares data.

19.3.5.3 How to Measure Innovation, Digital Transition and Digital Skills Needs in the Tourism Ecosystem?

Some indicators can reflect the advance in the digital transition or tourism. For example, quantifying the (1) number of public-private partnerships and the (2) budget allocated to these PPPs could be necessary, given the importance of big tech companies in the digital sphere. In addition, when providing license to new digital services, some authorities are pushing agreements in terms of data sharing, so that companies (in the fields of mobility, waste management, energy, etc.) have to make datasets public, which could be helpful for the mentioned analyses and models. Quantifying the (3) number of agreements on data sharing would then be another essential indicator.

19.3.5.4 How to Motivate and Monitor High-Quality Data Collection by the EU Member States?

The Member States must be aware of the usefulness of new data sources and computational methods. All campaigns and initiatives launched to incentivise/facilitate data collection should be supported by services provided in exchange. We need to strengthen the link between sharing data and getting benefits in better analyses and services. It could be a good strategy for incentivising bottom-up data collection initiatives, from users to companies, institutions and, eventually, Member States.

Monitoring the advances in data collection by the EU Member States is crucial and should be coordinated. Initiatives such as the Tourism Satellite Account (TSA)[13] are essential. As previously mentioned, this reflects the almost absence of indicators calculated based on new data sources, in the reports provided by the Member States. However, annual reports should be replaced by constantly open and updated online platforms that could also inform not only about results but also about Member States' progress, strategies, initiatives or agreements, regarding the digital transition.

Although recent tourism research is exploring and taking advantage of new data sources and methods, there is still a long way to walk on innovation in institutions at

---

[13] The Tourism Satellite Account (TSA) is a standard statistical framework and the main tool for the economic measurement of tourism. It has been developed by the World Tourism Organization (UNWTO), the Organisation for Economic Co-operation and Development (OECD), the Statistical Office of the European Commission (Eurostat) and the United Nations Statistics Division (UNSD). More information: https://www.oecd.org/cfe/tourism/tourismsatelliteaccountrecommendedmethodologicalframework.htm

the level of the European Union and national, regional or municipal levels. This fact is evidenced in the Tourism Satellite Account (TSA) 2019 (Eurostat, 2019) Annex II. All countries indicate the most relevant data sources used to calculate the related indicators for each TSA table. Annex II shows the almost absence of nontraditional or new data sources, such as "mobile positioning data" or "other Big Data sources".

19.4   The Way Forward

This chapter briefly discusses the potential of new data and computational methods to help stakeholders better understand and plan tourism.

The above KPIs might be measured almost everywhere in Europe and other regions of the world, in a wide range of periods and spatial scales, since they can be fed with similar data. If data sources are different, data must be reformatted to a common structure in comparative studies. Therefore, due to data's total/partial interoperability, KPIs can be measured for several locations or industries, including competitors, and undertake comparative studies.

Data, methods and KPIs proposed in this chapter have some limitations. They do not cover all the analyses needed regarding the complex tourism sector. Therefore, other traditional measurement techniques and data sources (surveys) are still required and used complementarily. Moreover, new techniques can create new problems. Some potential issues are:

- *Dependency on the digital footprint.* A significant number of tourists or tourist attractions may leave no or only a few digital footprints. The digital divide and digital infrastructure supply in the analysed locations must be considered while measuring and understanding KPIs' results. Traditional studies must help determine which tourism segments might be well represented by digital data and help fix biases.
- *User's privacy.* It must be guaranteed while estimating tourism KPIs and developing new products and strategies (Hall & Ram, 2020). KPIs should reveal the necessary information on a given topic without interfering with people's personal life and without explicitly fostering changes in their preferences. People can feel insecure if they feel constantly and unconsciously watched or worried that their data may be used against them. However, people may also agree to donate data to feed purely anonymous databases for good by an explicit agreement and fair and transparent methods.
- *Data availability/ownership.* Private companies own valuable data for analysing tourism. They usually restrict access to data since it might be a big part of their business. However, some have already released datasets worldwide or in events like hackathons and datathons. These actions have created new businesses/products and brought new insights into the tourism processes. It is necessary to explore win-win strategies among businesses and the public sector to access relevant data while keeping privacy and industrial know-how permanently (Robin et al., 2016).

Finally, the above KPIs are just values. Although some of those values seem to be easily interpretable (higher values are better than lower ones in some KPIs), they usually need some comparative or normative framework. These ranges must also be defined.

References

Ahas, R., Silm, S., & Tiru, M. (2018). Measuring transnational migration with roaming datasets. In *Adjunct Proceedings of the 14th International Conference on Location Based Services* (pp. 105–108).

Álvarez-Díaz, M., & Rosselló-Nadal, J. (2010). Forecasting British tourist arrivals in the Balearic Islands using meteorological variables. *Tourism Economics, 16*(1), 153–168.

Barros, C., Moya-Gómez, B., & Gutiérrez, J. (2019). Using geotagged photographs and GPS tracks from social networks to analyse visitor behaviour in national parks. *Current Issues in Tourism, 0.*(0, 1–20. https://doi.org/10.1080/13683500.2019.1619674

Batista e Silva, F., Marín Herrera, M. A., Rosina, K., Ribeiro Barranco, R., Freire, S., & Schiavina, M. (2018). Analysing spatiotemporal patterns of tourism in Europe at high-resolution with conventional and big data sources. *Tourism Management, 68*, 101–115. https://doi.org/10.1016/j.tourman.2018.02.020

Bertoni, E., Fontana, M., Gabrielli, L., Signorelli, S., & Vespe, M. (Eds). (2022). *Mapping the demand side of computational social science for policy.* EUR 31017 EN, Luxembourg, Publication Office of the European Union. ISBN 978-92-76-49358-7, https://doi.org/10.2760/901622

Brilhante, I., Macedo, J. A., Nardini, F. M., Perego, R., & Renso, C. (2013). Where shall we go today? Planning touristic tours with TripBuilder. In *Proceedings of the 22nd ACM International Conference on Information & Knowledge Management* (pp. 757–762).

Calle Lamelas, J. V. (2017). Revolución Big Data en el turismo: Análisis de las nuevas fuentes de datos para la creación de conocimiento en los Destinos Patrimonio de la Humanidad de España. *International Journal of Information Systems and Tourism (IJIST), 2*(2), 23–39.

Candelieri, A., Soldi, D., & Archetti, F. (2015). Short-term forecasting of hourly water consumption by using automatic metering readers data. *Procedia Engineering, 119*, 844–853.

Carpio-Pinedo, J., & Gutiérrez, J. (2020). Consumption and symbolic capital in the metropolitan space: Integrating 'old' retail data sources with social big data. *Cities, 106*(April), 102859. https://doi.org/10.1016/j.cities.2020.102859

CEDEFOP. (n.d.). *Skills-OVATE: Skills online vacancy analysis tool for Europe.*

Chen, T.-L., Shen, C. C., & Gosling, M. (2021). To stay or not to stay? The causal effect of interns' career intention on enhanced employability and retention in the hospitality and tourism industry. *Journal of Hospitality, Leisure, Sport and Tourism Education, 28*(1), 100305. https://doi.org/10.1016/j.jhlste.2021.100305

Dergiades, T., Mavragani, E., & Pan, B. (2018). Google trends and tourists' arrivals: Emerging biases and proposed corrections. *Tourism Management, 66*, 108–120. https://doi.org/10.1016/j.tourman.2017.10.014

Dinis, G., Breda, Z., Costa, C., & Pacheco, O. (2019). Google trends in tourism and hospitality research: A systematic literature review. *Journal of Hospitality and Tourism Technology, 10*(4), 747–763. https://doi.org/10.1108/JHTT-08-2018-0086

Eurostat. (2018). *Tourism industries—Employment.* Retrieved from https://ec.europa.eu/eurostat/statistics-explained/index.php?oldid=267583#Characteristics_of_jobs_in_tourism_industries

Eurostat. (2019). *Tourism satellite accounts in Europe—2019 edition.* Retrieved from https://doi.org/10.2785/78529

Eurostat. (2021a). *Share of international guest nights in total annual number of guest nights at short-stay accommodation offered via collaborative economy*. Retrieved from https://ec.europa.eu/eurostat/statistics-explained/index.php?title=File:Share_of_international_guest_nights_in_total_annual_number_of_guest_nights_at_short-stay_accommodation_offered_via_collaborative_economy,_2019_(%25).png

Eurostat. (2021b). *Tourism statistics—Expenditure*.

Folianto, F., Low, Y. S., & Yeow, W. L. (2015). Smartbin: Smart waste management system. In *2015 IEEE Tenth International Conference on Intelligent Sensors, Sensor Networks and Information Processing (ISSNIP)* (pp. 1–2).

Gössling, S. (2015). New performance indicators for water management in tourism. *Tourism Management, 46*, 233–244.

Gössling, S., & Peeters, P. (2015). Assessing tourism's global environmental impact 1900–2050. *Journal of Sustainable Tourism, 23*(5), 639–659. https://doi.org/10.1080/09669582.2015.1008500

Guo, Y., Barnes, S. J., & Jia, Q. (2017). Mining meaning from online ratings and reviews: Tourist satisfaction analysis using latent Dirichlet allocation. *Tourism Management, 59*, 467–483. https://doi.org/10.1016/j.tourman.2016.09.009

Gupta, A., & Dogra, N. (2017). Tourist adoption of mapping apps: A UTAUT2 perspective of smart travellers. *Tourism and Hospitality Management, 23*(2), 145–161.

Gutiérrez, J., García-Palomares, J. C., Romanillos, G., & Salas-Olmedo, M. H. (2017). The eruption of Airbnb in tourist cities: Comparing spatial patterns of hotels and peer-to-peer accommodation in Barcelona. *Tourism Management, 62*, 278–291. https://doi.org/10.1016/j.tourman.2017.05.003

Hall, C. M., & Ram, Y. (2020). Protecting privacy in tourism – A perspective article. *Tourism Review, 75*(1), 76–80. https://doi.org/10.1108/TR-09-2019-0398

Havranek, T., & Zeynalov, A. (2021). Forecasting tourist arrivals: Google trends meets mixed-frequency data. *Tourism Economics, 27*(1), 129–148.

Higham, J., Ellis, E., & Maclaurin, J. (2019). *Tourist aviation emissions: A problem of collective action*. Retrieved from https://doi.org/10.1177/0047287518769764.

Hunter, R. F., Garcia, L., de Sa, T. H., Zapata-Diomedi, B., Millett, C., Woodcock, J., & Moro, E. (2021). Effect of COVID-19 response policies on walking behavior in US cities. *Nature Communications, 12*(1), 1–9.

Kau, A. K., & Lim, P. S. (2005). Clustering of Chinese tourists to Singapore: An analysis of their motivations, values and satisfaction. *International Journal of Tourism Research, 7*(4–5), 231–248. https://doi.org/10.1002/jtr.537

Li, J., Xu, L., Tang, L., Wang, S., & Li, L. (2018). Big data in tourism research: A literature review. *Tourism Management, 68*, 301–323. https://doi.org/10.1016/j.tourman.2018.03.009

Palmer, A., José Montaño, J., & Sesé, A. (2006). Designing an artificial neural network for forecasting tourism time series. *Tourism Management, 27*(5), 781–790. https://doi.org/10.1016/j.tourman.2005.05.006

Pan, B., & Yang, Y. (2017). Forecasting destination weekly hotel occupancy with big data. *Journal of Travel Research, 56*(7), 957–970. https://doi.org/10.1177/0047287516669050

Pérez-Chacón, R., Luna-Romera, J. M., Troncoso, A., Martínez-Álvarez, F., & Riquelme, J. C. (2018). Big data analytics for discovering electricity consumption patterns in smart cities. *Energies, 11*(3), 683.

Ramanathan, V., & Meyyappan, T. (2019). Twitter text mining for sentiment analysis on people's feedback about Oman tourism. In *2019 4th MEC International Conference on Big Data and Smart City (ICBDSC)* (pp. 1–5).

Raun, J., Ahas, R., & Tiru, M. (2016). Measuring tourism destinations using mobile tracking data. *Tourism Management, 57*, 202–212.

Ridderstaat, J., Croes, R., & Nijkamp, P. (2016). A two-way causal chain between tourism development and quality of life in a small island destination: An empirical analysis. *Journal of Sustainable Tourism, 24*(10), 1461–1479. https://doi.org/10.1080/09669582.2015.1122016

Riitano, N., Dichicco, P., De Fioravante, P., Cavalli, A., Falanga, V., Giuliani, C., Mariani, L., Strollo, A., & Munafo, M. (2020). Land consumption in Italian coastal area. *Environmental Engineering & Management Journal (EEMJ), 19*(10), 1857.

Robin, N., Klein, T., & Jutting, J. (2016, February). Public-private partnerships for statistics: Lessons learned, future steps: A focus on the use of non-official data sources for national statistics. In *OECD development co-operation*.

Romanillos, G., García-Palomares, J. C., Moya-Gómez, B., Gutiérrez, J., Torres, J., López, M., Cantú-Ros, O. G., & Herranz, R. (2021). The city turned off: Urban dynamics during the COVID-19 pandemic based on mobile phone data. *Applied Geography, 134*, 102524. https://doi.org/10.1016/j.apgeog.2021.102524

Russo, M. A., Relvas, H., Gama, C., Lopes, M., Borrego, C., Rodrigues, V., Robaina, M., Madaleno, M., Carneiro, M. J., Eusébio, C., & Monteiro, A. (2020). Estimating emissions from tourism activities. *Atmospheric Environment, 220*(May 2019), 117048. https://doi.org/10.1016/j.atmosenv.2019.117048

Salas-Olmedo, M. H., Moya-Gómez, B., García-Palomares, J. C., & Gutiérrez, J. (2018). Tourists' digital footprint in cities: Comparing Big Data sources. *Tourism Management, 66*, 13–25. https://doi.org/10.1016/j.tourman.2017.11.001

Shoval, N., & Ahas, R. (2016). The use of tracking technologies in tourism research: The first decade. *Tourism Geographies, 18*(5), 587–606. https://doi.org/10.1080/14616688.2016.1214977

UNWTO. (2020). *European Union tourism trends*. World Tourism Organisation - UNWTO.

UNWTO. (2021). *2020: Worst year in tourism history with 1 billion fewer international arrivals*. Retrieved from https://www.unwto.org/news/2020-worst-year-in-tourism-history-with-1-billion-fewer-international-arrivals

UNWTO. (n.d.). *UNWTO tourism data dashboard*. Retrieved from https://www.unwto.org/unwto-tourism-dashboard

Versichele, M., de Groote, L., Claeys Bouuaert, M., Neutens, T., Moerman, I., & Van de Weghe, N. (2014). Pattern mining in tourist attraction visits through association rule learning on Bluetooth tracking data: A case study of Ghent, Belgium. *Tourism Management, 44*, 67–81. https://doi.org/10.1016/j.tourman.2014.02.009

Xu, F., Nash, N., & Whitmarsh, L. (2020). Big data or small data? A methodological review of sustainable tourism. *Journal of Sustainable Tourism, 28*(2), 147–166. https://doi.org/10.1080/09669582.2019.1631318

Chapter 20
Computational Social Science for Policy and Quality of Democracy: Public Opinion, Hate Speech, Misinformation, and Foreign Influence Campaigns

Joshua A. Tucker

Abstract The intersection of social media and politics is yet another realm in which Computational Social Science has a paramount role to play. In this review, I examine the questions that computational social scientists are attempting to answer – as well as the tools and methods they are developing to do so – in three areas where the rise of social media has led to concerns about the quality of democracy in the digital information era: online hate; misinformation; and foreign influence campaigns. I begin, however, by considering a precursor of these topics – and also a potential hope for social media to be able to positively impact the quality of democracy – by exploring attempts to measure public opinion online using Computational Social Science methods. In all four areas, computational social scientists have made great strides in providing information to policy makers and the public regarding the evolution of these very complex phenomena but in all cases could do more to inform public policy with better access to the necessary data; this point is discussed in more detail in the conclusion of the review.

20.1 Introduction

The advent of the digital information age – and, in particular, the stratospheric rise in popularity of social media platforms such as Facebook, Instagram, Twitter, YouTube, and TikTok – has led to unprecedented opportunities for people to share information and content with one another in a much less mediated fashion that was ever possible previously. These opportunities, however, have been accompanied by a myriad of new concerns and challenges at both the individual and societal levels, including threats to systems of democratic governance (Tucker et al., 2017). Chief among these are the rise of hateful and abusive forms of communication on these

J. A. Tucker (✉)
New York University, New York, NY, USA
e-mail: joshua.tucker@nyu.edu

platforms, the seemingly unchecked spread of mis- and disinformation,[1] and the ability of malicious political actors, including, and perhaps most notably, foreign adversaries, to launch coordinated influence attacks in an attempt to hijack public opinion.

Concurrently, the rise of computing power and the astonishing developments in the fields of information storage and retrieval, text-as-data, and machine learning have given rise to a whole new set of tools – collectively known as Computational Social Science – that have allowed scholars to study the digital trace data left behind by the new online activity of the digital information era in previously unimaginable ways. These Computational Social Science tools can enable scholars to characterize and describe the newly emerging phenomena of the digital information era but also, in the case of the more malicious of these new phenomena, to test ways to mitigate their prevalence and impact. Accordingly, this chapter of the handbook summarizes what we have learned about the potential for Computational Social Science tools to be used to address the three of these threats identified above: hate speech, mis-/disinformation, and foreign coordinated influence campaigns. As these topics are set against the backdrop of influencing public opinion, I begin with an overview of how Computational Social Science techniques can be harnessed to measure public opinion. Finally, the chapter concludes with a discussion of the paramount importance for any of these efforts of ensuring that independent researchers – that is, researchers not employed by the platforms themselves – have access to the data necessary to continue and build upon the research described in the chapter, as well as to inform, and ultimately facilitate, public regulatory policy.

All of these areas – using Computational Social Science to measure public opinion, and to detect, respond to, and possibly even remove hate speech, misinformation, and foreign influence campaigns – have important public policy connotations. Using social media to measure public opinion offers the possibility for policy makers to have additional tools at their disposal for gauging the opinions regarding, and salience of, issues among the general public, ideally helping to make governments more responsive to the public. Hate speech and misinformation together form the crux of the debate over "content moderation" on platforms, and Computational Social Science can provide the tools necessary to implement policy makers' proscriptions for addressing these potential harms but also, equally importantly, for understanding the actual nature of the problems that they are trying to address. Finally, foreign coordinated influence campaigns, regardless of the extent to which they actually influence politics in other countries, can rightly be conceived of as national security threats when foreign powers attempt to undermine the quality and functioning of democratic institutions. Here again, Computational Social Science has an important role to play in identifying such campaigns but also

---

[1] I follow Tucker et al. (2018) and Born and Edgington (2017) in defining misinformation online as information that is factually incorrect but is spread by people who are unaware that the information is incorrect; disinformation, by contrast, is knowingly spread false information.

in terms of attempting to measure the goals, strategies, reach, and ultimate impact of such campaigns.[2]

In the review that follows, I focus almost exclusively on publications and papers from the last 3–4 years. To be clear, this research all builds on very important prior work that will not be covered in the review.[3] In addition, in the time it has taken to bring this piece to publication, there have undoubtedly been many new and important contributions to the field that will not be addressed here. But hopefully the review is able to provide readers with a fairly up to date sense of the promises of – and challenges facing – new approaches from Computational Social Science to the study of democracy and its challenges.

20.2 Computational Social Science and Measuring Public Opinion

One of the great lures of social media was that it would lead to new ways to analyse and measure public opinion (Barberá & Steinert-Threlkeld, 2020; Klašnja et al., 2017). Traditional survey-based methods of measuring public opinion of course have all sort of important advantages, to say nothing of a 70-year pedigree of developing appropriate methods around sampling and estimation. There are, however, drawbacks too: surveys are expensive; there are limits to how many anyone can run; they are dependent on appropriate sampling frames; they rely on an "artificial" environment for measuring opinion and are correspondingly subject to social desirability bias; and, perhaps most importantly, they can only measure opinions for the questions pollsters decide to ask. Social media, on the other hand, holds open the promise of inexpensive, real-time, finely grained time-series measurement of people's opinions in a non-artificial environment where there is no sense of being observed for a study or needing to respond to a pollster (Beauchamp, 2017). Moreover, analysis can also be retrospective, going back in time to study the evolution of opinion on a topic for which one might not thought to have previously asked questions in public opinion surveys.[4]

The way the field has developed has not, however, been in a way that uses social media to mimic the traditional public opinion polling approach of an omnibus survey

---

[2] See as well Chap. 22, "Political Analysis, Misinformation, and Democracy" of *Mapping the Demand Side of Computational Social Science for Public Policy* (Bertoni et al., 2022).

[3] For longer, prior reviews of related literature, see Tucker et al. (2018) and Persily and Tucker (2020b).

[4] More generally, if we can extract public opinion data from existing stores of social media data, we can retrospectively examine public opinion on *any* topic, which is of course impossible in traditional studies of public opinion via survey questionnaires, which are by definition limited to the questions asked in the past. Of course, social media data vary in the extent to which past data are available for retrospective analysis, but platforms where most posts are public (e.g. Twitter, Reddit) offer important opportunities in this regard.

that presents attitudes among the public across a large number of topics on a regular basis. Instead, we have seen two types of Computational Social Science studies take centre stage: studies that examine attitudes over time related to one particular issue or topic and studies that attempt to use social media data to assess the popularity of political parties and politicians, often in an attempt to predict election outcomes.[5]

The issue-based studies generally involve a corpus of social media posts (usually tweets) being collected around a series of keywords related to the issue in question and then sentiment analysis (usually positive or negative sentiment towards the issue) being measured over a period of time. Studies of this nature have examined attitudes towards topics such as Brexit (Georgiadou et al., 2020), immigration (Freire-Vidal & Graells-Garrido, 2019), refugees (Barisione et al., 2019), austerity (Barisione & Ceron, 2017), COVID-19 (Dai et al., 2021; Gilardi et al., 2021; Lu et al., 2021), the police (Oh et al., 2021), gay rights (Adams-Cohen, 2020), and climate change (Chen et al., 2021b). Studies of political parties and candidates follow similar patterns although sometimes using engagement such as "likes" to measure popularity instead of sentiment analysis. Recent examples include studies that have been conducted in countries including Finland (Vepsäläinen et al., 2017), Spain (Bansal & Srivastava, 2019; Grimaldi et al., 2020), and Greece (Tsakalidis et al., 2018).[6]

Of course, studying public opinion using computational social methods and social media data is not without its challenges. First and foremost is the question of representativeness: whose opinions are being measured when we analyse social media data? There are two layers of concern here: whether the people whose posts are being analysed are representative of the overall users of the platform but also whether the overall users of the platform are representative of the population of interest (Klašnja et al., 2017). If the goal is simply to ascertain the opinions of those using the platform, then the latter question is less problematic. Of course, the "people" part of the question can also be problematic, as social media accounts can also be "bots", accounts that are automated to produce content based on algorithms as opposed to having a one-to-one relationship to a human being, although this varies by platform (Grimaldi et al., 2020; Sanovich et al., 2018; Yang et al., 2020). Another problem for representativeness can arise when significant portions of the population lack internet access, or when people are afraid to voice their opinions online due to fear of state repression (Isani, 2021).

Even if the question of representativeness can be solved and/or an appropriate population of interest identified, the original question of how to extract opinions out of unstructured text data still remains. Here, however, we have seen great strides by computational social scientists in developing innovative methods. Loosely speaking,

---

[5] The one exception has been a few studies that attempt to use the discussion of issues as a way of teasing out who is leading the public conversation on important policy issues, elites or the mass public (Barberá et al., 2019; Gilardi et al., 2021, 2022). These studies, however, tend to measure attention to multiple topics and issues, but not opinions in regard to these issues.

[6] The Greek case involved a referendum, as opposed to a parliamentary election. For a meta-review of 74 related studies, see Skoric et al. (2020).

we can identify two basic approaches. The first set of methods are characterized by a priori identifying text that is positively or negatively associated with a certain topic and then simply tracking the prevalence (e.g. counts, ratios) of these words over time (Barisione et al., 2019; Georgiadou et al., 2020; Gilardi et al., 2022). For example, in Siegel and Tucker (2018), we took advantage of the fact that when discussing ISIS in Arabic, the term "Islamic State" suggests support for the organization, while the derogatory term "Daesh" is used by those opposed to ISIS. Slight variations on this approach can involve including emojis as well as words (Bansal & Srivastava, 2019) or focusing on likes instead of text (Vepsäläinen et al., 2017).

The more popular approach, however, is to rely on one of the many different machine learning approaches to try to classify sentiment. These approaches include nonnegative matrix factorization (Freire-Vidal & Graells-Garrido, 2019), deep learning (Dai et al., 2021), convolutional and recurrent neural nets (Wood-Doughty et al., 2018), and pre-trained language transformer models (Lu et al., 2021; Terechshenko et al., 2020); many papers also compare a number of different supervised machine learning models and select the one that performs best (Adams-Cohen, 2020; Grimaldi et al., 2020; Tsakalidis et al., 2018). While less common, some studies use unsupervised approaches for stance relying on networks and activity to cluster accounts (Darwish et al., 2019). Closely related to these latter approaches are network-based models that are not focused on positive or negative sentiment towards a particular topic, but rather attempt to place different users along a latent dimension of opinion, such as partisanship (Barberá, 2015; Barberá et al., 2015) or attitudes towards climate change (Chen et al., 2021b).

With this basic background on the ways in which Computational Social Science can be utilized to measure public opinion using social media data, in the remainder of this chapter, I examine the potential of Computational Social Science to address three pernicious forms of online behaviour that have been identified as threats to the quality of democracy: hate speech, misinformation, and foreign influence campaigns.

20.3   Computational Social Science and Hate Speech

The rise of Web 2.0 brought with it the promise of a more interactive internet, where ordinary users could be contributing content in near real time (Ackland, 2013). Social media in many ways represented the apex of this trend, with the most dominant tech companies becoming those that did not actually produce content, but instead provided platforms on which everyone could create content. While removing the gatekeepers from the content production process has many attractive features from the perspective of democratic participation and accountability, it also has its downsides – perhaps no more obvious than the fact that gatekeepers could also play a role in policing online hate. As that observation became increasingly obvious, a wave of scholarship has developed utilizing Computational Social Science tools to attempt to characterize the extent of the problem, measure its impact, and assess the effectiveness of various countermeasures (Siegel, 2020).

Attempts to measure the prevalence and diffusion of hate speech have been at the forefront of this work, including studies that take place on single platforms (Gallacher & Bright, 2021; He et al., 2021; Mathew et al., 2018) and those on multiple platforms (Gallacher, 2021; Velásquez et al., 2021) with the latter including studies of what happens to user's hate speech on one platform when they are banned from another one (Ali et al., 2021; Mitts, 2021). Other studies have focused on more specific topics, such as the amount of hate speech produced by bots as opposed to humans (Albadi et al., 2019), examining whether there are serial producers of hate in Italy (Cinelli et al., 2021) or hate speech targeted at elected officials and politicians (Greenwood et al., 2019; Rheault et al., 2019; Theocharis et al., 2020).

A second line of research has involved attempting to ascertain both the causes and effects of hate speech and in particular the relationship between offline violence, including hate crimes, and online hate speech. For example, a number of papers have examined the rise in online anti-Muslim hate speech on Twitter and Reddit following terrorist attacks in Paris (Fischer-Preßler et al., 2019; Olteanu et al., 2018) and Berlin (Kaakinen et al., 2018). Conversely, other studies have examined the relationship between hate speech on social media and hate crimes (Müller & Schwarz, 2021; Williams et al., 2020). Other work examines the relationship between political developments and the rise of hate speech, such as the arrival of a boat of refugees in Spain (Arcila-Calderón et al., 2021). Closely related are studies, primarily of an experimental nature, that attempt to measure the impact of being exposed to incivility (Kosmidis & Theocharis, 2020) or hate speech on outcomes such as prejudice (Soral et al., 2018) or fear (Oksanen et al., 2020).

A third line of research has focused on attempts to not just detect but also to counter hate speech online. The main approach here has been field experiments, where researchers detect users of hate speech on Twitter, use "sock puppet" accounts to deliver some sort of message designed to reduce the use of hate speech using an experimental research design, and then monitor users' future behaviour. Stimuli tested have involved varying the popularity, race, and partisanship of the account delivering the message (Munger, 2017, 2021), embedding the exhortation in religious (Islamic) references (Siegel & Badaan, 2020), and threats of suspension from the platform (Yildirim et al., 2021). Researchers have also employed survey experiments to measure the impact of counter-hate speech (Sim et al., 2020) as well as observational studies, such as Garland et al. (2022)'s study of 180,000 conversations on German political Twitter.

Computational Social Science sits squarely at the root of all of this research, as any study that involves detecting hate speech at scale needs to rely on automated methods.[7] There are essentially two different research strategies employed by researchers. The first is to utilize dictionary methods – identifying hateful words that

---

[7] Some studies of hate speech do avoid the need to identify hate speech at scale by the use of surveys and survey experiments (Kaakinen et al., 2018; Kunst et al., 2021; Oksanen et al., 2020;

are either available in existing databases or identified by the researchers conducting the study and then collecting posts that contain those particular terms (Arcila-Calderón et al., 2021; Greenwood et al., 2019; Mathew et al., 2018; Mitts, 2021; Olteanu et al., 2018).

The second option is to rely on supervised machine learning. As with the study of opinions and sentiment generally, we can see a wide range of supervised ML methods employed, including pre-trained language models based on the BERT architecture (Cinelli et al., 2021; Gallacher, 2021; Gallacher & Bright, 2021; He et al., 2021), SVM models (Rheault et al., 2019; Williams et al., 2019), random forest (Albadi et al., 2019), doc2vec (Garland et al., 2022), and logistic regression with L1 regularization (Theocharis et al., 2020). Siegel et al. (2021) combine dictionary methods with supervised machine learning to screen out false positives from the dictionary methods using a naive Bayes classifier and, signaling a potential warning for the dictionary methods, find that large numbers (in many cases approximately half) of the tweets identified by the dictionary methods are removed by the supervised machine learning approach as false positives.

Unsupervised machine learning is less prevalent in this research – other than for identifying subtopics in a general area in which to look for the relative prevalence of hate speech (e.g. Arcila-Calderón et al. 2021, (refugees), Velásquez et al. 2021 (COVID-19), Fischer-Preßler et al. 2019 (terrorist attacks)) – although Rasmussen et al. (2021) propose what they call a "super-unsupervised" method for hate speech detection that relies on word embeddings and does not require human-coded training data.

One important development of note is that in recent years it is becoming more and more possible to find studies of hate speech involving language other than English, including Spanish (Arcila-Calderón et al., 2021), Italian (Cinelli et al., 2021), German (Garland et al., 2022), and Arabic (Albadi et al., 2019; Siegel & Badaan, 2020). Other important Computational Social Science innovations in the field include matching accounts across multiple platforms to observe how the same people behave on multiple platforms, including how content moderation actions on one platform can impact hate speech on another (Mitts, 2021) and network analyses of the spread of hateful content (Velásquez et al., 2021). Finally, it is important to remember that any form of identification of hate speech that relies on humans to classify speech as hateful or not is subject to whatever biases underlie human coding (Ross et al., 2017), which includes all supervised machine learning methods. One warning here can be found in Davidson et al. (2019), who demonstrate that a number of hate speech classifiers are more likely to classify tweets written in what the authors call "African-American English" as hate speech than tweets written in standard English.

Sim et al., 2020; Soral et al., 2018), or creating one's own platform in which to observe participant behavior (Álvarez-Benjumea & Winter, 2018, 2020).

20.4   Computational Social Science and Misinformation

In the past 6 years or so, we have witnessed a very significant increase in research related to misinformation online.[8] One can conceive of this field as attempting to answer six closely related questions, roughly in order of time sequence:

1. Who produces misinformation?
2. Who is exposed to misinformation?
3. Conditional on exposure, who believes misinformation?
4. Conditional on belief, is it possible to correct misinformation?
5. Conditional on exposure, who shares misinformation?
6. Through production and sharing, how much misinformation exists online/on platforms?

Computational Social Science can be used to shed light on any of these questions but is particularly important for questions 2, 5, and 6: who is exposed, who shares, and how much misinformation exists online?[9]

To answer these questions, Computational Social Science is employed in one of two ways: to trace the spread of misinformation or to identify misinformation. The former of these is a generally easier task than the latter, and studies that employ Computational Social Science in this way generally follow the following pattern. First, a set of domains or news articles are identified as being false. In the case of news articles, researchers generally turn to fact checking organizations for lists of articles that have been previously identified as being false such as Snopes or PolitiFact (Allcott et al., 2019; Allcott & Gentzkow, 2017; Shao et al., 2018). Two points are worth noting here. First, this means that such studies are limited to countries in which fact checking organizations exist. Second, such studies are also limited to articles that fact checking organizations have chosen to check (which might be subject to their own organizational biases).[10] For news domains, researchers generally rely either on outside organization that ranks the quality of news domains, such as NewsGuard (Aslett et al., 2022), or else lists of suspect news sites published by journalists or other scholars (Grinberg et al., 2019; Guess et al., 2019). Scholars have also found other creative ways to find sources of suspect information, such as public pages on Facebook associated with conspiracy theories (Del Vicario et al., 2016) or videos that were removed from YouTube

---

[8] For reviews, see Guess and Lyons (2020), Tucker et al. (2018), and Van Bavel et al. (2021).

[9] The questions of who believes misinformation and how to correct misinformation are of course crucially important but are generally addressed using survey methodology (Aslett et al., 2022). For a review of the literature on correcting misinformation, see Wittenberg and Berinsky (2020); for more recent research on the value of "accuracy nudges" and games designed to inoculate users against believing false news, see Pennycook et al. (2021) and Maertens et al. (2021), respectively.

[10] For an exception to this approach, however, see Godel et al. (2021) which relies on an automated method to select popular articles from five news streams (three of which are low-quality news streams) in real time and then send those articles to professional fact checkers for evaluation as part of the research pipeline.

(Knuutila et al., 2020). Once the list of suspect domains or articles are identified, the Computational Social Science component of researching the spread comes from interacting with and/or scraping online information to track where these links are found. This can be as simple as querying an API, and as complicated as developing methods to track the spread of information.[11]

The second – and primary – use of Computational Social Science techniques in the study of misinformation is the arguably more difficult task of using Computational Social Science to identify content as misinformation. As might be expected, using dictionary methods to do so is much more difficult than for tasks such as identifying hate speech or finding posts about a particular topic or issue. Accordingly, when we do see dictionary methods in the study of misinformation, they are generally employed in order to identify posts about a specific topic (e.g. Facebook ads related to a Spanish general election in Cano-Orón et al., 2021) that are then coded by hand; Gorwa (2017) and Oehmichen et al. (2019) follow similar procedures of hand labelling small numbers of posts/accounts as examples of misinformation in Poland and the United States, respectively.

Although still a very challenging computational task, recent research has begun to attempt to use machine learning to build supervised classifiers to identify misinformation on Twitter using SVMs (Bojjireddy et al., 2021), BERT embeddings (Micallef et al., 2020), and ensemble methods (Al-Rakhami & Al-Amri, 2020). Jagtap et al. (2021) comparatively test a variety of different supervised classifiers to identify misinformation in YouTube comments. Jachim et al. (2021) have built a tool based on unsupervised machine learning called "Troll Hunter" that while not identifying misinformation per se can be used to surface narratives across multiple posts online that might form the basis of disinformation campaign. Karduni et al. (2019) also incorporate images into their classifier.

Closely related, other studies have sought to harness network analysis to identify misinformation online. For example, working with leaked documents that identify actors paid by the South Korean government, Keller et al. (2020) show how retweet and co-tweet networks can be used to identify possible purveyors of misinformation. Zhu et al. (2020) utilize a "heuristic greedy algorithm" to attempt to identify nodes in networks that, if removed, would greatly reduce the spread of misinformation. Sharma et al. (2021) train a network-based model on data from the Russian Internet Research Agency (IRA) troll datasets released by Twitter and use it to identify coordinated groups spreading anti-vaccination and anti-masks conspiracies.

A different use of machine learning to identify misinformation – in this case, false news articles – can be found in Godel et al. (2021). Here we assess the possibility of crowdsourcing fact checking of news articles by testing a wide range of different possible rules for how decisions could possibly be made by crowds. Compared with intuitively simple rules such as "take the mode of the crowd", we find that machine learning methods that draw upon a richer set of features – and in particular when analysed using convolutional neural nets – far outperform simple aggregation rules

---

[11] See, for example, https://informationtracer.com/, which is presented in Z. Chen et al. (2021).

in having the judgment of the crowd match the assessment of a set of professional fact checkers.

Given the scale at which misinformation spreads, it is clear that any content moderation policy related to misinformation will need to rely on machine learning to at least some extent. From this vantage point, the progress the field has made in recent years must be seen as encouraging; still, important challenges remain. First, the necessary data to train models is not always available, either because platforms do not make it available to researchers due to privacy or commercial concerns or because it has, ironically, been deleted as part of the process of content moderation.[12] In some cases, platforms have released data of deleted accounts for scholarly research, but even here the method by which these accounts were identified generally remains a black box. Second, for any supervised learning method, the question of the robustness of a classifier designed to identify misinformation in one context to detect it in another context (different language, different country, different context even in the same country and language) remains paramount. While this is a problem for measuring sentiment on policy issues or hate speech as well, we have reason to suspect that the contextual nature of misinformation might make this even more challenging and suggests the potential value of unsupervised and/or network-based models. Third, so many of the methods to date rely on training classifiers based on news that has existed in the information ecosystem for extended periods of time, while the challenge for content moderation is to be able to identify misinformation in near real time before it spreads widely (Godel et al., 2021). Finally, false positives can have negative consequences as well, if the reaction to identifying misinformation is to suppress its spread. While reducing the spread of misinformation receives the most attention, it is important to remember that reducing true news in circulation is also costly, so future studies should try to explicitly address this trade-off, perhaps by attempting to assess the impact of methods of identifying misinformation for the overall makeup of the information ecosystem.

20.5 Computational Social Science and Coordinated Foreign Influence Operations

A third area in which Computational Social Science plays an important role in protecting democratic integrity is in the study of foreign influence operations. Here, I define foreign influence operations as coordinated attempts online by one state to influence the attitudes and behaviours of citizens of another state.[13] While foreign

---

[12] Tools such as the Wayback Machine have been creatively applied in some instances to get around this issue of deletion (Bastos & Farkas, 2019; Knuutila et al., 2020).

[13] For reviews of media reports of foreign influence operations globally, see Bradshaw et al. (2021), O'Connor et al. (2020), and Martin and Shapiro (2019).

propaganda efforts of course precede the advent of the modern digital information age, the cost of mounting coordinated foreign influence operations has significantly dropped in the digital information era, especially due to the rise of social media platforms.[14]

Research on coordinated foreign influence operations (hereafter CFIOs) can loosely be described as falling into one of two categories: attempts to describe what actually happened as part of previously identified CFIOs and attempts to develop methods to identify new CFIOs. Notably, the scholarly literature on the former is much larger (although one would guess that research on the latter is being conducted by social media platforms). Crucially, almost all of this literature, though, is dependent on having a list of identified accounts and/or posts that are part of CFIOs – by definition if the goal is to describe what happened in a CFIO and for use as training data if the goal is to develop methods to identify CFIOs. Accordingly, the primary source of data for the studies described in the remainder of this section are collections of posts from (or list of accounts involved with) CFIOs released by social media platforms. After having turned over lists of CFIO accounts to the US government as part of congressional testimony, Twitter has emerged as a leader in this regard; however other platforms including Reddit and Facebook have made CFIO data available for external research as well.[15]

By far the most studied subject of CFIOs is the activities of the Russian IRA in the United States (Bail et al., 2020; Bastos & Farkas, 2019) and in particular in the period of time surrounding the 2016 US presidential election (Arif et al., 2018; Boyd et al., 2018; DiResta et al., 2022; Golovchenko et al., 2020; Kim et al., 2018; Linvill & Warren, 2020; Lukito, 2020; Yin et al., 2018; Zannettou et al., 2020).

Studies of CFIOs in other countries include Russian influence attempts in Germany (Dawson & Innes, 2019), across 12 European countries (Innes et al., 2021), Syria (Metzger & Siegel, 2019), Libya, Sudan, Madagascar, Central African Republic, and Mozambique (Grossman et al., 2019, 2020); Chinese influence attempts in the United Kingdom (Schliebs et al., 2021), Hong Kong, and Taiwan

---

[14] In a way, coordinated foreign influence operations that rely on disguised social media accounts – that is, accounts pretending to be actors that they are not – could be considered another form of misinformation, with the identity of the online actors here being the misinformation. It is important to note, though, that coordinated foreign influence operations are not used solely to spread misinformation. Foreign influence operations can, and do, rely on true information in addition to misinformation; indeed, Yin et al. (2018) found that Russian foreign influence accounts on Twitter were actually much more likely to share links to legitimate news sources – and in particular to local news sources – than they were to low-quality news sources.

[15] Two other potential sources of data included leaked data and data from actors that researchers can identify – or at least speculate – as being involved in foreign influence activities, such as Chinese ambassadors (Schliebs et al., 2021), the FB pages of Chinese state media (Molter & DiResta, 2020), or the Twitter accounts of Russian state media actors (Metzger & Siegel, 2019). While there have been a series of very interesting papers published based on leaked data to identify coordinated domestic propaganda efforts (Keller et al., 2020; King et al., 2017; Sobolev, 2019), I am not aware of any CFIO studies at this time based on leaked data.

(Wallis et al., 2020); and the US (Molter & DiResta, 2020) and Iranian influence attempts in the Middle East (Elswah et al., 2019).

The methods employed in these studies vary, but many involve a role for Computational Social Science. In Yin et al. (2018) and Golovchenko et al. (2020), we extract hyperlinks shared by Russian IRA trolls using a custom-built Computational Social Science tool; in the latter study, we also utilize methods described earlier in this review in the measuring public opinion section to automate the estimation of the ideological placement of the shared links. Zannettou et al. (2020) extract and analyse the images shared by Russian IRA accounts. Innes et al. (2021), Dawson and Innes (2019), and Arif et al. (2018) all rely on various forms of network analysis to track the spread of IRA content in Germany, Europe, and the United States, respectively. Two studies of Chinese influence operations use sentiment analysis – again, in a manner similar to the one described earlier in the measuring public option section – to measure whether influence operations are relying on positive or negative messages (Molter & DiResta, 2020; Wallis et al., 2020). In a similar vein, Boyd et al. (2018) use NLP tools to chart the stylistic evolution of Russian IRA posts over time. DiResta et al. (2022) and Metzger and Siegel (2019) use structural topic models to dig deeper into the topics discussed by Russian influence operations in the United States and tweets by Russian state media about Syria, respectively. Lukito (2020) employs a similar method to the one discussed earlier in the measuring public opinion section regarding whether elites or masses drive the discussion of political topics to argue that the Russian IRA was trying out topics on Reddit before purchasing ads on those subjects on Facebook. Other papers combine digital trace data from social media platforms such as Facebook ads (Kim et al., 2018) or exposure to IRA tweets (Bail et al., 2020; Eady et al., 2022) with survey data.

A number of studies rely on qualitative analyses based on human annotation of CFIO account activity (e.g. Innes et al. (2021) include a case study of a Russian influence in Estonia to supplement a network-based study of Russian influence in 12 European countries; see also Bastos and Farkas, 2019; Dawson and Innes, 2019; DiResta et al., 2022; and Linvill and Warren, 2020), but even in these cases, Computational Social Science plays a role in allowing scholars to extract the relevant posts and accounts for analysis.

What there is much less of, though, are studies of the actual influence of exposure to CFIOs, which is a direction in which the literature should try to expand in the future. Two exceptions are Bail et al. (2020) and Eady et al. (2022), both of which rely on panel survey data combined with data on exposure to tweets by Russian trolls that took place between waves of the panel survey.

A second strand of the Computational Social Science literature involves trying to use machine learning to identify CFIOs.[16] One approach has been to use the releases of posts from CFIOs by social media platforms as training data for supervised

---

[16] Note that there is also a much larger literature on detecting automated social media accounts or bots (Ferrara et al., 2016; Stukal et al., 2017) which is beyond the subject of this review. Bots come

models to identify new CFIOs (or at least new CFIOs that are unknown to the models); both Alizadeh et al. (2020) and Marcellino et al. (2020) report promising findings using this approach. Innes et al. (2021) filter on keywords and then attempt to identify influence campaigns through network analysis; this approach has the advantage of not needing to use training data, although the ultimate findings will of course be a function of the original keyword search. Schliebs et al. (2021) use NLP techniques to look for common phrases or patterns across the posts from Chinese diplomats, thus suggesting evidence of a coordinated campaign. This method also does not require training data, but, unlike either of the previous approaches, does require identifying the potential actors involved in the CFIO as a precursor to the analysis.

Taken together, it is clear that a great deal about the ways in which CFIOs operate in the modern digital era has been learned in a short period of time. That being said, a strikingly large proportion of recent research has focused on the activities of Russian CFIOs around the 2016 US elections; future research should continue to look at influence operations run by other countries with other targets.[17] There is also clearly a lot more work to be done in terms of understanding the impact of CFIOs, as well as in developing methods for identifying these campaigns. This latter point reflects a fundamental reality of the field, which is that its development has occurred largely because the platforms chose (or were compelled) to release data, and it is to this topic that I turn in some brief concluding remarks in the following section.

20.6   The Importance of External Data Access

Online hate, disinformation, and online coordinated influence operations all pose potential threats to the quality of democracy, to say nothing of the threats to people whose personal lives may be impacted by being attacked online or being exposed to dangerous misinformation. Computational Social Science – and in particular tools that facilitate working with large collections of (digital trace) data and innovations in machine learning – have important roles to play in helping society understand the nature of these threats, as well as potential mitigation strategies. Indeed, social scientists are getting better and better at incorporating the newest developments in machine learning (e.g. neural networks, pre-trained transformer models) into their research. So many of the results laid out in the previous sections are incredibly impressive and represent research we would not have even conceived of being able to do a decade ago.

---

up a lot in discussion for CFIOs, as bots can be a useful vehicle for such campaigns. Suffice it to say, Computational Social Science methods play a very important role in the detection of bots.

[17] This picture looks a lot more troubling if one takes out the numerous excellent reports produced by the Stanford Internet Observatory on CFIOs targeting Africa and the Middle East.

That being said, the field as a whole remains dependent on the availability of data. And here, social scientists find themselves in a different position than in years past. Previously, most quantitative social research was conducted either with administrative data (e.g. election results, unemployment data, test scores) or with data – usually survey or experimental – that we could collect ourselves. As Nathaniel Persily and I have noted in much greater detail elsewhere (Persily & Tucker, 2020a, b, 2021), we now find ourselves in a world where the data which we need to do our research on the kinds of topics surveyed in this handbook chapter are "owned" by a handful of very large private companies. Thus, the key to advancing our knowledge of all of the topics discussed in this review, as well as the continued development of related methods and tools, is a legal and regulatory framework that ensures that outside researchers that are not employees of the platforms, and who are committed to sharing the results of their research with both the mass public and policy makers, are able to continue to access the data necessary for this research.[18]

Let me give just two examples. First, almost none of the work surveyed in the previous section on CFIOs would have been possible had Twitter not decided to release its collections of tweets produced by CFIOs after they were taken off the platform. Yes, it is fantastic that Twitter did (and has continued to) release these data, but we as a society do not want to be at the mercy of decisions by platforms to release data for matters as crucial as understanding whether foreign countries are interfering in democratic processes. And just because Twitter has chosen to do this in the past, it does not mean that it will continue to do so in the future. Second, even with all the data that Twitter releases publicly through its researcher API, external researchers still do not have access to impressions data (e.g. how many times tweets were seen and by whom). While some have come up with creative ways to try to estimate impressions, this means that any research that is built around impressions is carrying out studies with unnecessary noise in our estimates; a decision by Twitter tomorrow could change this reality. For all of the topics in this review – hate speech, misinformation, foreign influence campaigns – impressions are crucially important pieces of the puzzle that we are currently missing.

As of the final editing of this essay, though, important steps are being taken on both sides of the Atlantic to try to address this question of data access for external academic researchers. In the United States, a number of bills have recently been introduced in the national legislature that include components aimed

---

[18] Of course, issues surrounding data access raise very important issues in terms of obligations to users of social media to both protect their privacy and to make sure their voices are heard. The myriad of trade-offs in this regard are far beyond the purview of this chapter, but I invite interested readers to see the discussion of trade-offs between data privacy and data access for public-facing research to inform public policy in Persily and Tucker (2020b, pp. 321–324), the chapter by Taylor (2023) in the present handbook, as well as the proposal for a "Researcher Code of Conduct" – as laid out in Article 40 of the General Data Protection Regulation (GDRP) – by the European Digital Media Observatory multi-stakeholder Working Group on Platform-to-Researcher Data Access: https://edmo.eu/wp-content/uploads/2022/02/Report-of-the-European-Digital-Media-Observatorys-Working-Group-on-Platform-to-Researcher-Data-Access-2022.pdf.

at making social media data available to external researchers for public-facing analysis.[19] While such bills are a still a long way from being made into law, the fact that multiple lawmakers are taking the matter seriously is a positive step forwards. Perhaps more importantly in terms of immediate impact, the European Union's Digital Services Act (DSA) has provisions allowing data access to "vetted researchers" of key platforms, in order for researchers to evaluate how platforms work and how online risk evolves and to support transparency, accountability, and compliance with the new laws and regulations.[20]

Computational Social Science has a huge role to play in helping us understand some of the most important challenges faced by democratic societies today. The scholarship that is being produced is incredibly inspiring, and the methodological leaps that are occurring in such short periods of time were perhaps previously unimaginable. But at the end of the day, the ultimate quality of the work we are able to do will depend on the data to which we have access. Thus data access needs to be a fundamental part of any forward-facing research plan for improving what Computational Social Science can teach us about threats to democracy.

Acknowledgements I am extremely grateful to Sophie Xiangqian Yi and Trellace Lawrimore for their incredible research assistance in helping to locate almost all of the literature cited in this review, as well as for providing excellent summaries of what they had found. I would also like to thank Roxanne Rahnama for last-minute research assistance with the current status of EU efforts regarding data access (which included writing most of the text of footnote 20); Rebekah Tromble, Brandon Silverman, and Nate Persily provided helpful suggestions on this topic as well. I would

---

[19] See, for example, https://www.coons.senate.gov/news/press-releases/coons-portman-klobuchar-announce-legislation-to-ensure-transparency-at-social-media-platforms; https://www.bennet.senate.gov/public/index.cfm/2022/5/bennet-introduces-landmark-legislation-to-establish-federal-commission-to-oversee-digital-platforms; and https://trahan.house.gov/news/documentsingle.aspx?DocumentID=2112.

[20] The Act defines "vetted researchers" as individuals "with an affiliation with an academic institution, independence from commercial interests, proven subject or methodological expertise, and the ability to comply with data security and confidentiality requirements" (Nonnecke & Carlton, 2022). The Act requires platforms to make three categories of data available with online databases or APIs: data needed to assess systemic risks (dissemination of illegal content, impacts on fundamental rights, coordinated manipulation of the platform's services), "data on the accuracy, functioning, and testing of algorithmic systems for content moderation, recommender systems or advertising systems, and data on processes and outputs of content moderation or internal complaint-handling systems" (Nonnecke & Carlton, 2022), Moreover, VLOPs (very large online platforms) are required, by Article 63, to create a public digital ad repository with information on ad content, those behind ads, whether it was targeted, parameters for targeting, and number of recipients (Nonnecke & Carlton, 2022). Member states will be required to designate independent "Digital Service Coordinators", who will supervise compliance with the new rules on their territory (https://ec.europa.eu/commission/presscorner/detail/en/QANDA_20_2348). The EU Parliament and Council and Commission reached a compromise regarding the text for the DSA on April 23, 2022. The final text is expected to be confirmed soon, and once formally approved, it will apply after 15 months or from January 1, 2024 (https://ec.europa.eu/commission/presscorner/detail/en/QANDA_20_2348). See as well the discussion of data altruism, and the possibility of donating data for research, in https://www.consilium.europa.eu/en/press/press-releases/2022/05/16/le-conseil-approuve-l-acte-sur-la-gouvernance-des-donnees/.

like to thank Matteo Fontana for his very helpful feedback on the first draft of this chapter, as well as the rest of the CSS4P team (Eleonora Bertoni, Lorenzo Gabrielli, Serena Signorelli, and Michele Vespe) for inviting me to contribute the chapter, their patience with my schedule, and their helpful comments and suggestions along the way.

References

Ackland, R. (2013). *Web social science: Concepts, data and tools for social scientists in the digital age*. Sage. https://doi.org/10.4135/9781446270011

Adams-Cohen, N. J. (2020). Policy change and public opinion: Measuring shifting political sentiment with social media data. *American Politics Research, 48*(5), 612–621. https://doi.org/10.1177/1532673X20920263

Albadi, N., Kurdi, M., & Mishra, S. (2019). Hateful people or hateful bots? In *Detection and characterization of bots spreading religious hatred in Arabic social media*. Retrieved from https://doi.org/10.48550/ARXIV.1908.00153

Ali, S., Saeed, M. H., Aldreabi, E., Blackburn, J., De Cristofaro, E., Zannettou, S., & Stringhini, G. (2021). Understanding the effect of deplatforming on social networks. In *13th ACM web science conference 2021* (pp. 187–195). Retrieved from https://doi.org/10.1145/3447535.3462637

Alizadeh, M., Shapiro, J. N., Buntain, C., & Tucker, J. A. (2020). Content-based features predict social media influence operations. *Science Advances, 6*(30), eabb5824. https://doi.org/10.1126/sciadv.abb5824

Allcott, H., & Gentzkow, M. (2017). Social media and fake news in the 2016 election. *Journal of Economic Perspectives, 31*(2), 211–236. https://doi.org/10.1257/jep.31.2.211

Allcott, H., Gentzkow, M., & Yu, C. (2019). Trends in the diffusion of misinformation on social media. *Research & Politics, 6*(2), 205316801984855. https://doi.org/10.1177/2053168019848554

Al-Rakhami, M. S., & Al-Amri, A. M. (2020). Lies kill, facts save: Detecting COVID-19 misinformation in twitter. *IEEE Access, 8*, 155961–155970. https://doi.org/10.1109/ACCESS.2020.3019600

Álvarez-Benjumea, A., & Winter, F. (2018). Normative change and culture of hate: An experiment in online environments. *European Sociological Review, 34*(3), 223–237. https://doi.org/10.1093/esr/jcy005

Álvarez-Benjumea, A., & Winter, F. (2020). The breakdown of antiracist norms: A natural experiment on hate speech after terrorist attacks. *Proceedings of the National Academy of Sciences, 117*(37), 22800–22804. https://doi.org/10.1073/pnas.2007977117

Arcila-Calderón, C., Blanco-Herrero, D., Frías-Vázquez, M., & Seoane-Pérez, F. (2021). Refugees welcome? Online hate speech and sentiments in twitter in Spain during the reception of the boat Aquarius. *Sustainability, 13*(5), 2728. https://doi.org/10.3390/su13052728

Arif, A., Stewart, L. G., & Starbird, K. (2018). Acting the part: Examining information operations within #BlackLivesMatter discourse. *Proceedings of the ACM on Human-Computer Interaction, 2*, 1–27. https://doi.org/10.1145/3274289

Aslett, K., Guess, A. M., Bonneau, R., Nagler, J., & Tucker, J. A. (2022). News credibility labels have limited average effects on news diet quality and fail to reduce misperceptions. *Science Advances, 8*(18), eabl3844. https://doi.org/10.1126/sciadv.abl3844

Bail, C. A., Guay, B., Maloney, E., Combs, A., Hillygus, D. S., Merhout, F., Freelon, D., & Volfovsky, A. (2020). Assessing the Russian Internet Research Agency's impact on the political attitudes and behaviors of American Twitter users in late 2017. *Proceedings of the National Academy of Sciences, 117*(1), 243–250. https://doi.org/10.1073/pnas.1906420116

Bansal, B., & Srivastava, S. (2019). Lexicon-based Twitter sentiment analysis for vote share prediction using emoji and N-gram features. *International Journal of Web Based Communities, 15*(1), 85. https://doi.org/10.1504/IJWBC.2019.098693

Barberá, P. (2015). Birds of the same feather tweet together: Bayesian ideal point estimation using Twitter data. *Political Analysis, 23*(1), 76–91. https://doi.org/10.1093/pan/mpu011

Barberá, P., & Steinert-Threlkeld, Z. C. (2020). How to use social media data for political science research. In I. L. Curini & R. Franzese (Eds.), *The Sage handbook of research methods in political science and international relations* (pp. 404–423). SAGE Publications Ltd.. https://doi.org/10.4135/9781526486387.n26

Barberá, P., Jost, J. T., Nagler, J., Tucker, J. A., & Bonneau, R. (2015). Tweeting from left to right: Is online political communication more than an echo chamber? *Psychological Science, 26*(10), 1531–1542. https://doi.org/10.1177/0956797615594620

Barberá, P., Casas, A., Nagler, J., Egan, P. J., Bonneau, R., Jost, J. T., & Tucker, J. A. (2019). Who leads? Who follows? Measuring issue attention and agenda setting by legislators and the mass public using social media data. *American Political Science Review, 113*(4), 883–901. https://doi.org/10.1017/S0003055419000352

Barisione, M., & Ceron, A. (2017). A digital movement of opinion? Contesting austerity through social media. In M. Barisione & A. Michailidou (Eds.), *Social media and European politics* (pp. 77–104). Palgrave Macmillan UK. https://doi.org/10.1057/978-1-137-59890-5_4

Barisione, M., Michailidou, A., & Airoldi, M. (2019). Understanding a digital movement of opinion: The case of #RefugeesWelcome. *Information, Communication & Society, 22*(8), 1145–1164. https://doi.org/10.1080/1369118X.2017.1410204

Bastos, M., & Farkas, J. (2019). "Donald Trump is my president!": The internet research agency propaganda machine. *Social Media + Society, 5*(3), 205630511986546. https://doi.org/10.1177/2056305119865466

Beauchamp, N. (2017). Predicting and interpolating state-level polls using Twitter textual data. *American Journal of Political Science, 61*(2), 490–503.

Bertoni, E., Fontana, M., Gabrielli, L., Signorelli, S., & Vespe, M. (Eds). (2022). *Mapping the demand side of computational social science for policy.* EUR 31017 EN, Luxembourg, Publication Office of the European Union. ISBN 978-92-76-49358-7, https://doi.org/10.2760/901622

Bojjireddy, S., Chun, S. A., & Geller, J. (2021). Machine learning approach to detect fake news, misinformation in COVID-19 pandemic. In *DG.O2021: The 22nd Annual International Conference on Digital Government Research* (pp. 575–578). https://doi.org/10.1145/3463677.3463762

Born, K., & Edgington, N. (2017). *Analysis of philanthropic opportunities to mitigate the disinformation/propaganda problem.*

Boyd, R. L., Spangher, A., Fourney, A., Nushi, B., Ranade, G., Pennebaker, J., & Horvitz, E. (2018). *Characterizing the Internet research Agency's social media operations during the 2016 U.S. presidential election using linguistic analyses* [preprint]. *PsyArXiv.* https://doi.org/10.31234/osf.io/ajh2q

Bradshaw, S., Bailey, H., & Howard, P. (2021). Industrialized disinformation: 2020 global inventory of organized social media manipulation. *Computational Propaganda Research Project.*

Cano-Orón, L., Calvo, D., López García, G., & Baviera, T. (2021). Disinformation in Facebook ads in the 2019 Spanish General Election Campaigns. *Media and Communication, 9*(1), 217–228. https://doi.org/10.17645/mac.v9i1.3335

Chen, Z., Aslett, K., Reynolds, J., Freire, J., Nagler, J., Tucker, J. A., & Bonneau, R. (2021a). *An automatic framework to continuously monitor multi-platform information spread.*

Chen, T. H. Y., Salloum, A., Gronow, A., Ylä-Anttila, T., & Kivelä, M. (2021b). Polarization of climate politics results from partisan sorting: Evidence from Finnish Twittersphere. *Global Environmental Change, 71*, 102348. https://doi.org/10.1016/j.gloenvcha.2021.102348

Cinelli, M., Pelicon, A., Mozetič, I., Quattrociocchi, W., Novak, P. K., & Zollo, F. (2021). Online Hate. *Behavioural Dynamics and Relationship with Misinformation.* https://doi.org/10.48550/ARXIV.2105.14005

Dai, Y., Li, Y., Cheng, C.-Y., Zhao, H., & Meng, T. (2021). Government-led or public-led? Chinese policy agenda setting during the COVID-19 pandemic. *Journal of Comparative Policy Analysis: Research and Practice, 23*(2), 157–175. https://doi.org/10.1080/13876988.2021.1878887

Darwish, K., Stefanov, P., Aupetit, M., & Nakov, P. (2019). *Unsupervised user stance detection on Twitter*. Retrieved from https://doi.org/10.48550/ARXIV.1904.02000.

Davidson, T., Bhattacharya, D., & Weber, I. (2019). Racial bias in hate speech and abusive language detection datasets. *Proceedings of the Third Workshop on Abusive Language Online, 2019*, 25–35. https://doi.org/10.18653/v1/W19-3504

Dawson, A., & Innes, M. (2019). How Russia's internet research agency built its disinformation campaign. *The Political Quarterly, 90*(2), 245–256. https://doi.org/10.1111/1467-923X.12690

Del Vicario, M., Bessi, A., Zollo, F., Petroni, F., Scala, A., Caldarelli, G., Stanley, H. E., & Quattrociocchi, W. (2016). The spreading of misinformation online. *Proceedings of the National Academy of Sciences, 113*(3), 554–559. https://doi.org/10.1073/pnas.1517441113

DiResta, R., Grossman, S., & Siegel, A. (2022). In-house vs. outsourced trolls: How digital mercenaries shape state influence strategies. *Political Communication, 39*(2), 222–253. https://doi.org/10.1080/10584609.2021.1994065

Eady, G., Paskhalis, T., Zilinsky, J., Stukal, D., Bonneau, R., Nagler, J., & Tucker, J. A. (2022). *Exposure to the Russian foreign influence campaign on Twitter in the 2016 US election and its relationship to political attitudes and voting behavior.*

Elswah, M., Howard, P., & Narayanan, V. (2019). *Iranian digital interference in the Arab World. Data memo.* Project on Computational Propaganda.

Ferrara, E., Varol, O., Davis, C., Menczer, F., & Flammini, A. (2016). The rise of social bots. *Communications of the ACM, 59*(7), 96–104. https://doi.org/10.1145/2818717

Fischer-Preßler, D., Schwemmer, C., & Fischbach, K. (2019). Collective sense-making in times of crisis: Connecting terror management theory with Twitter user reactions to the Berlin terrorist attack. *Computers in Human Behavior, 100*, 138–151. https://doi.org/10.1016/j.chb.2019.05.012

Freire-Vidal, Y., & Graells-Garrido, E. (2019). *Characterization of local attitudes toward immigration using social media.* Retrieved from https://doi.org/10.48550/ARXIV.1903.05072

Gallacher, J. D. (2021). *Leveraging cross-platform data to improve automated hate speech detection.* Retrieved from https://doi.org/10.48550/ARXIV.2102.04895

Gallacher, J. D., & Bright, J. (2021). *Hate contagion: Measuring the spread and trajectory of hate on social media* [preprint]. *PsyArXiv.* https://doi.org/10.31234/osf.io/b9qhd

Garland, J., Ghazi-Zahedi, K., Young, J.-G., Hébert-Dufresne, L., & Galesic, M. (2022). Impact and dynamics of hate and counter speech online. *EPJ Data Science, 11*(1), 3. https://doi.org/10.1140/epjds/s13688-021-00314-6

Georgiadou, E., Angelopoulos, S., & Drake, H. (2020). Big data analytics and international negotiations: Sentiment analysis of Brexit negotiating outcomes. *International Journal of Information Management, 51*, 102048. https://doi.org/10.1016/j.ijinfomgt.2019.102048

Gilardi, F., Gessler, T., Kubli, M., & Müller, S. (2021). Social media and policy responses to the COVID-19 pandemic in Switzerland. *Swiss Political Science Review, 27*(2), 243–256. https://doi.org/10.1111/spsr.12458

Gilardi, F., Gessler, T., Kubli, M., & Müller, S. (2022). Social media and political agenda setting. *Political Communication, 39*(1), 39–60. https://doi.org/10.1080/10584609.2021.1910390

Godel, W., Sanderson, Z., Aslett, K., Nagler, J., Bonneau, R., Persily, N., & Tucker, J. A. (2021). Moderating with the mob: Evaluating the efficacy of real-time crowdsourced fact-checking. *Journal of Online Trust and Safety, 1*(1), 10.54501/jots.v1i1.15.

Golovchenko, Y., Buntain, C., Eady, G., Brown, M. A., & Tucker, J. A. (2020). Cross-platform state propaganda: Russian trolls on twitter and YouTube during the 2016 U.S. Presidential Election. *The International Journal of Press/Politics, 25*(3), 357–389. https://doi.org/10.1177/1940161220912682

Gorwa, R. (2017). *Computational propaganda in Poland: False amplifiers and the digital public sphere* (2017.4; Computational Propaganda Research Project). University of Oxford.

Greenwood, M. A., Bakir, M. E., Gorrell, G., Song, X., Roberts, I., & Bontcheva, K. (2019). *Online abuse of UK MPs from 2015 to 2019: Working paper.* Retrieved from https://doi.org/10.48550/ARXIV.1904.11230

Grimaldi, D., Cely, J. D., & Arboleda, H. (2020). Inferring the votes in a new political landscape: The case of the 2019 Spanish Presidential elections. *Journal of Big Data, 7*(1), 58. https://doi.org/10.1186/s40537-020-00334-5

Grinberg, N., Joseph, K., Friedland, L., Swire-Thompson, B., & Lazer, D. (2019). Fake news on Twitter during the 2016 U.S. Presidential election. *Science, 363*(6425), 374–378. https://doi.org/10.1126/science.aau2706

Grossman, S., Bush, D., & DiResta, R. (2019). Evidence of Russia-linked influence operations in Africa. *Technical Report Stanford Internet Observatory.*

Grossman, S., Ramali, K., DiResta, R., Beissner, L., Bradshaw, S., Healzer, W., & Hubert, I. (2020). *Stoking conflict by keystroke: An operation run by IRA-linked individuals targeting Libya, Sudan, and Syria [Technical report].* Stanford Internet Observatory.

Guess, A. M., & Lyons, B. A. (2020). Misinformation, disinformation, and online propaganda. In J. A. Tucker & N. Persily (Eds.), *Social media and democracy: The state of the field, prospects for reform* (pp. 10–33). Cambridge University Press. Retrieved from https://www.cambridge.org/core/books/social-media-and-democracy/misinformation-disinformation-and-online-propaganda/D14406A631AA181839ED896916598500

Guess, A. M., Nagler, J., & Tucker, J. A. (2019). Less than you think: Prevalence and predictors of fake news dissemination on Facebook. *Science Advances, 5*(1), eaau4586. https://doi.org/10.1126/sciadv.aau4586

He, B., Ziems, C., Soni, S., Ramakrishnan, N., Yang, D., & Kumar, S. (2021). Racism is a virus: Anti-Asian hate and counterspeech in social media during the COVID-19 crisis. *ArXiv:2005.12423 [Physics].* Retrieved from http://arxiv.org/abs/2005.12423

Innes, M., Innes, H., Roberts, C., Harmston, D., & Grinnell, D. (2021). The normalisation and domestication of digital disinformation: On the alignment and consequences of far-right and Russian state (dis)information operations and campaigns in Europe. *Journal of Cyber Policy, 6*(1), 31–49. https://doi.org/10.1080/23738871.2021.1937252

Isani, M. A. (2021). Methodological problems of using Arabic-language Twitter as a gauge for Arab attitudes toward politics and society. *Contemporary Review of the Middle East, 8*(1), 22–35. https://doi.org/10.1177/2347798920976283

Jachim, P., Sharevski, F., & Pieroni, E. (2021). TrollHunter2020: Real-time detection of trolling narratives on Twitter during the 2020 U.S. elections. In *Proceedings of the 2021 ACM Workshop on Security and Privacy Analytics* (pp. 55–65). https://doi.org/10.1145/3445970.3451158.

Jagtap, R., Kumar, A., Goel, R., Sharma, S., Sharma, R., & George, C. P. (2021). Misinformation detection on YouTube using video captions. *ArXiv:2107.00941 [Cs].* Retrieved from http://arxiv.org/abs/2107.00941

Kaakinen, M., Oksanen, A., & Räsänen, P. (2018). Did the risk of exposure to online hate increase after the November 2015 Paris attacks? A group relations approach. *Computers in Human Behavior, 78*, 90–97. https://doi.org/10.1016/j.chb.2017.09.022

Karduni, A., Cho, I., Wesslen, R., Santhanam, S., Volkova, S., Arendt, D. L., Shaikh, S., & Dou, W. (2019). Vulnerable to misinformation?: Verifi! In *Proceedings of the 24th International Conference on Intelligent User Interfaces* (pp. 312–323). https://doi.org/10.1145/3301275.3302320.

Keller, F. B., Schoch, D., Stier, S., & Yang, J. (2020). Political astroturfing on twitter: How to coordinate a disinformation campaign. *Political Communication, 37*(2), 256–280. https://doi.org/10.1080/10584609.2019.1661888

Kim, Y. M., Hsu, J., Neiman, D., Kou, C., Bankston, L., Kim, S. Y., Heinrich, R., Baragwanath, R., & Raskutti, G. (2018). The stealth media? Groups and targets behind divisive issue campaigns on Facebook. *Political Communication, 35*(4), 515–541. https://doi.org/10.1080/10584609.2018.1476425

King, G., Pan, J., & Roberts, M. E. (2017). How the Chinese government fabricates social media posts for strategic distraction, not engaged argument. *American Political Science Review, 111*(3), 484–501. https://doi.org/10.1017/S0003055417000144

Klašnja, M., Barberá, P., Beauchamp, N., Nagler, J., & Tucker, J. A. (2017). *Measuring public opinion with social media data* (Vol. 1). Oxford University Press. https://doi.org/10.1093/oxfordhb/9780190213299.013.3

Knuutila, A., Herasimenka, A., Au, H., Bright, J., Nielsen, R., & Howard, P. N. (2020). COVID-related misinformation on YouTube: The spread of misinformation videos on social media and the effectiveness of platform policies. *COMPROP Data Memo, 6.*

Kosmidis, S., & Theocharis, Y. (2020). Can social media incivility induce enthusiasm? *Public Opinion Quarterly, 84*(S1), 284–308. https://doi.org/10.1093/poq/nfaa014

Kunst, M., Porten-Cheé, P., Emmer, M., & Eilders, C. (2021). Do "good citizens" fight hate speech online? Effects of solidarity citizenship norms on user responses to hate comments. *Journal of Information Technology & Politics, 18*(3), 258–273. https://doi.org/10.1080/19331681.2020.1871149

Linvill, D. L., & Warren, P. L. (2020). Troll factories: Manufacturing specialized disinformation on twitter. *Political Communication, 37*(4), 447–467. https://doi.org/10.1080/10584609.2020.1718257

Lu, Y., Pan, J., & Xu, Y. (2021). Public sentiment on Chinese social media during the emergence of COVID19. *Journal of quantitative description: Digital Media, 1,* 10.51685/jqd.2021.013.

Lukito, J. (2020). Coordinating a multi-platform disinformation campaign: Internet Research Agency Activity on three U.S. Social Media Platforms, 2015 to 2017. *Political Communication, 37*(2), 238–255. https://doi.org/10.1080/10584609.2019.1661889

Maertens, R., Roozenbeek, J., Basol, M., & van der Linden, S. (2021). Long-term effectiveness of inoculation against misinformation: Three longitudinal experiments. *Journal of Experimental Psychology: Applied, 27*(1), 1–16. https://doi.org/10.1037/xap0000315

Marcellino, W., Johnson, C., Posard, M., & Helmus, T. (2020). *Foreign interference in the 2020 election: Tools for detecting online election interference.* RAND Corporation. https://doi.org/10.7249/RRA704-2

Martin, D. A., & Shapiro, J. N. (2019). *Trends in online foreign influence efforts.* Princeton University.

Mathew, B., Dutt, R., Goyal, P., & Mukherjee, A. (2018). *Spread of hate speech in online social media.* Retrieved from https://doi.org/10.48550/ARXIV.1812.01693.

Metzger, M. M., & Siegel, A. A. (2019). *When state-sponsored media goes viral: Russia's use of RT to shape global discourse on Syria. Working Paper.*

Micallef, N., He, B., Kumar, S., Ahamad, M., & Memon, N. (2020). *The role of the crowd in countering misinformation: A case study of the COVID-19 infodemic.* Retrieved from https://doi.org/10.48550/ARXIV.2011.05773.

Mitts, T. (2021). *Banned: How Deplatforming extremists mobilizes hate in the dark corners of the Internet.*

Molter, V., & DiResta, R. (2020). Pandemics & propaganda: How Chinese state media creates and propagates CCP coronavirus narratives. *Harvard Kennedy School Misinformation Review.*https://doi.org/10.37016/mr-2020-025

Müller, K., & Schwarz, C. (2021). Fanning the flames of hate: Social media and hate crime. *Journal of the European Economic Association, 19*(4), 2131–2167. https://doi.org/10.1093/jeea/jvaa045

Munger, K. (2017). Tweetment effects on the tweeted: Experimentally reducing racist harassment. *Political Behavior, 39*(3), 629–649. https://doi.org/10.1007/s11109-016-9373-5

Munger, K. (2021). Don't @ Me: Experimentally reducing partisan incivility on Twitter. *Journal of Experimental Political Science, 8*(2), 102–116. https://doi.org/10.1017/XPS.2020.14

Nonnecke, B., & Carlton, C. (2022). EU and US legislation seek to open up digital platform data. *Science, 375*(6581), 610–612. https://doi.org/10.1126/science.abl8537

O'Connor, S., Hanson, F., Currey, E., & Beattie, T. (2020). *Cyber-enabled foreign interference in elections and referendums.* Australian Strategic Policy Institute Canberra.

Oehmichen, A., Hua, K., Amador Diaz Lopez, J., Molina-Solana, M., Gomez-Romero, J., & Guo, Y. (2019). Not all lies are equal. A study into the engineering of political misinformation in the 2016 US Presidential Election. *IEEE Access, 7,* 126305–126314. https://doi.org/10.1109/ACCESS.2019.2938389

Oh, G., Zhang, Y., & Greenleaf, R. G. (2021). Measuring geographic sentiment toward police using social media data. *American Journal of Criminal Justice*. https://doi.org/10.1007/s12103-021-09614-z

Oksanen, A., Kaakinen, M., Minkkinen, J., Räsänen, P., Enjolras, B., & Steen-Johnsen, K. (2020). Perceived societal fear and cyberhate after the November 2015 Paris terrorist attacks. *Terrorism and Political Violence, 32*(5), 1047–1066. https://doi.org/10.1080/09546553.2018.1442329

Olteanu, A., Castillo, C., Boy, J., & Varshney, K. R. (2018). *The effect of extremist violence on hateful speech online*. Retrieved from https://doi.org/10.48550/ARXIV.1804.05704

Pennycook, G., Epstein, Z., Mosleh, M., Arechar, A. A., Eckles, D., & Rand, D. G. (2021). Shifting attention to accuracy can reduce misinformation online. *Nature, 592*(7855), 590–595. https://doi.org/10.1038/s41586-021-03344-2

Persily, N., & Tucker, J. A. (2020a). Conclusion: The challenges and opportunities for social media research. In J. A. Tucker & N. Persily (Eds.), *Social media and democracy: The state of the field, prospects for reform* (pp. 313–331). Cambridge University Press. Retrieved from https://www.cambridge.org/core/books/social-media-and-democracy/conclusion-the-challenges-and-opportunities-for-social-media-research/232F88C00A1694FA25110A318E9CF300

Persily, N., & Tucker, J. A. (Eds.). (2020b). *Social media and democracy: The state of the field, prospects for reform* (1st ed.). Cambridge University Press. https://doi.org/10.1017/9781108890960

Persily, N., & Tucker, J. A. (2021). *How to fix social media? Start with independent research. (Brookings Series on The Economics and Regulation of Artificial Intelligence and Emerging Technologies)*. Brookings Institution. Retrieved from https://www.brookings.edu/research/how-to-fix-social-media-start-with-independent-research/

Rasmussen, S. H. R., Bor, A., Osmundsen, M., & Petersen, M. B. (2021). Super-unsupervised text classification for labeling online political hate [Preprint]. *PsyArXiv*. https://doi.org/10.31234/osf.io/8m5dc

Rheault, L., Rayment, E., & Musulan, A. (2019). Politicians in the line of fire: Incivility and the treatment of women on social media. *Research & Politics, 6*(1), 205316801881622. https://doi.org/10.1177/2053168018816228

Ross, B., Rist, M., Carbonell, G., Cabrera, B., Kurowsky, N., & Wojatzki, M. (2017). *Measuring the reliability of hate speech annotations: The case of the European refugee crisis*. Retrieved from https://doi.org/10.48550/ARXIV.1701.08118

Sanovich, S., Stukal, D., & Tucker, J. A. (2018). Turning the virtual tables: Government strategies for addressing online opposition with an application to Russia. *Comparative Politics, 50*(3), 435–482. https://doi.org/10.5129/001041518822704890

Schliebs, M., Bailey, H., Bright, J., & Howard, P. N. (2021). *China's public diplomacy operations: Understanding engagement and inauthentic amplifications of PRC diplomats on Facebook and Twitter*.

Shao, C., Ciampaglia, G. L., Varol, O., Yang, K.-C., Flammini, A., & Menczer, F. (2018). The spread of low-credibility content by social bots. *Nature Communications, 9*(1), 4787. https://doi.org/10.1038/s41467-018-06930-7

Sharma, K., Zhang, Y., Ferrara, E., & Liu, Y. (2021). Identifying coordinated accounts on social media through hidden influence and group behaviours. *ArXiv:2008.11308 [Cs]*. Retrieved from http://arxiv.org/abs/2008.11308

Siegel, A. A. (2020). Online Hate Speech. In J. A. Tucker & N. Persily (Eds.), *Social media and democracy: The state of the field, prospects for reform* (pp. 56–88). Cambridge University Press. Retrieved from https://www.cambridge.org/core/books/social-media-and-democracy/online-hate-speech/28D1CF2E6D81712A6F1409ED32808BF1

Siegel, A. A., & Badaan, V. (2020). #No2Sectarianism: Experimental approaches to reducing sectarian hate speech online. *American Political Science Review, 114*(3), 837–855. https://doi.org/10.1017/S0003055420000283

Siegel, A. A., & Tucker, J. A. (2018). The Islamic State's information warfare: Measuring the success of ISIS's online strategy. *Journal of Language and Politics, 17*(2), 258–280. https://doi.org/10.1075/jlp.17005.sie

Siegel, A. A., Nikitin, E., Barberá, P., Sterling, J., Pullen, B., Bonneau, R., Nagler, J., & Tucker, J. A. (2021). Trumping hate on Twitter? Online hate speech in the 2016 U.S. election campaign and its aftermath. *Quarterly Journal of Political Science, 16*(1), 71–104. https://doi.org/10.1561/100.00019045

Sim, J., Kim, J. Y., & Cho, D. (2020). *Countering sexist hate speech on YouTube: The role of popularity and gender.* Bright Internet Global Summit. Retrieved from http://brightinternet.org/wp-content/uploads/2020/11/Countering-Sexist-Hate-Speech-on-YouTube-The-Role-of-Popularity-and-Gender.pdf

Skoric, M. M., Liu, J., & Jaidka, K. (2020). Electoral and public opinion forecasts with social media data: A meta-analysis. *Information, 11*(4), 187. https://doi.org/10.3390/info11040187

Sobolev, A. (2019). *How pro-government "trolls" influence online conversations in Russia.*

Soral, W., Bilewicz, M., & Winiewski, M. (2018). Exposure to hate speech increases prejudice through desensitization. *Aggressive Behavior, 44*(2), 136–146. https://doi.org/10.1002/ab.21737

Stukal, D., Sanovich, S., Bonneau, R., & Tucker, J. A. (2017). Detecting bots on Russian political Twitter. *Big Data, 5*(4), 310–324. https://doi.org/10.1089/big.2017.0038

Taylor, L. (2023). Data justice, computational social science and policy. In *Handbook of computational social science for policy.* Springer.

Terechshenko, Z., Linder, F., Padmakumar, V., Liu, F., Nagler, J., Tucker, J. A., & Bonneau, R. (2020). A comparison of methods in political science text classification: Transfer learning language models for politics. *SSRN Electronic Journal.* https://doi.org/10.2139/ssrn.3724644

Theocharis, Y., Barberá, P., Fazekas, Z., & Popa, S. A. (2020). The dynamics of political incivility on twitter. *SAGE Open, 10*(2), 215824402091944. https://doi.org/10.1177/2158244020919447

Tsakalidis, A., Aletras, N., Cristea, A. I., & Liakata, M. (2018). Nowcasting the stance of social media users in a sudden vote: The case of the Greek referendum. In *Proceedings of the 27th ACM International Conference on Information and Knowledge Management* (pp. 367–376). https://doi.org/10.1145/3269206.3271783

Tucker, J. A., Theocharis, Y., Roberts, M. E., & Barberá, P. (2018). From liberation to turmoil: Social media and democracy. *Journal of Democracy, 28*(4), 46–59. https://doi.org/10.1353/jod.2017.0064

Tucker, J. A., Guess, A., Barbera, P., Vaccari, C., Siegel, A., Sanovich, S., Stukal, D., & Nyhan, B. (2018). Social media, political polarization, and political disinformation: A review of the scientific literature. *SSRN Electronic Journal.* https://doi.org/10.2139/ssrn.3144139

Van Bavel, J. J., Harris, E. A., Pärnamets, P., Rathje, S., Doell, K. C., & Tucker, J. A. (2021). Political psychology in the digital (mis)information age: A model of news belief and sharing. *Social Issues and Policy Review, 15*(1), 84–113. https://doi.org/10.1111/sipr.12077

Velásquez, N., Leahy, R., Restrepo, N. J., Lupu, Y., Sear, R., Gabriel, N., Jha, O. K., Goldberg, B., & Johnson, N. F. (2021). Online hate network spreads malicious COVID-19 content outside the control of individual social media platforms. *Scientific Reports, 11*(1), 11549. https://doi.org/10.1038/s41598-021-89467-y

Vepsäläinen, T., Li, H., & Suomi, R. (2017). Facebook likes and public opinion: Predicting the 2015 Finnish parliamentary elections. *Government Information Quarterly, 34*(3), 524–532. https://doi.org/10.1016/j.giq.2017.05.004

Wallis, J., Uren, T., Thomas, E., Zhang, A., Hoffman, S., Li, L., Pascoe, A., & Cave, D. (2020). *Retweeting through the great firewall.*

Williams, M. L., Burnap, P., Javed, A., Liu, H., & Ozalp, S. (2019). Hate in the machine: Anti-Black and Anti-Muslim social media posts as predictors of offline racially and religiously aggravated crime. *The British Journal of Criminology, azz049.* https://doi.org/10.1093/bjc/azz049

Williams, M. L., Burnap, P., Javed, A., Liu, H., & Ozalp, S. (2020). Hate in the machine: Anti-Black and Anti-Muslim social media posts as predictors of offline racially and religiously aggravated crime. *The British Journal of Criminology, azz049,* 242. https://doi.org/10.1093/bjc/azz049

Wittenberg, C., & Berinsky, A. J. (2020). Misinformation and its correction. In J. A. Tucker & N. Persily (Eds.), *Social media and democracy: The state of the field, prospects for reform* (pp. 163–198). Cambridge University Press. Retrieved from https://www.cambridge.org/core/books/social-media-and-democracy/misinformation-and-its-correction/61FA7FD743784A723BA234533012E810

Wood-Doughty, Z., Andrews, N., Marvin, R., & Dredze, M. (2018). Predicting Twitter user demographics from names alone. *Proceedings of the Second Workshop on Computational Modeling of People's Opinions, Personality, and Emotions in Social Media, 2018*, 105–111. https://doi.org/10.18653/v1/W18-1114

Yang, K.-C., Hui, P.-M., & Menczer, F. (2020). How Twitter data sampling biases U.S. voter behavior characterizations. *ArXiv:2006.01447 [Cs]*. Retrieved from http://arxiv.org/abs/2006.01447

Yildirim, M. M., Nagler, J., Bonneau, R., & Tucker, J. A. (2021). Short of suspension: How suspension warnings can reduce hate speech on twitter. *Perspectives on Politics, 1–13*, 1. https://doi.org/10.1017/S1537592721002589

Yin, L., Roscher, F., Bonneau, R., Nagler, J., & Tucker, J. A. (2018). *Your friendly neighborhood troll: The Internet Research Agency's use of local and fake news in the 2016 US presidential campaign*. SMaPP Data Report, Social Media and Political Participation Lab, New York University.

Zannettou, S., Caulfield, T., Bradlyn, B., De Cristofaro, E., Stringhini, G., & Blackburn, J. (2020). Characterizing the use of images in state-sponsored information warfare operations by Russian trolls on Twitter. *Proceedings of the International AAAI Conference on Web and Social Media, 14*, 774–785.

Zhu, J., Ni, P., & Wang, G. (2020). Activity minimization of misinformation influence in online social networks. *IEEE Transactions on Computational Social Systems, 7*(4), 897–906. https://doi.org/10.1109/TCSS.2020.2997188

Chapter 21
Social Interactions, Resilience, and Access to Economic Opportunity: A Research Agenda for the Field of Computational Social Science

Theresa Kuchler and Johannes Stroebel

Abstract We argue that the increasing availability of digital trace data presents substantial opportunities for researchers and policy makers to better understand the importance of social networks and social interactions in fostering economic opportunity and resilience. We review recent research efforts that have studied these questions using data from a wide range of sources, including online social networking platform such as Facebook, call detail record data, and network data from payment systems. We also describe opportunities for expanding these research agendas by using other digital trace data, and discuss various promising paths to increase researcher access to the required data, which is often collected and owned by private corporations.

21.1 Introduction

Social networks facilitate much of modern economic activity. Workers use them to find jobs and investors to learn about new investment opportunities. Social networks can serve to spread information, enforce social norms, and sustain collaboration, trade, and lending. The tangible and intangible resources that individuals can access through their social networks—that is, the social capital available to them—are central to fostering their resilience to a range of economic shocks, from recessions to health emergencies to environmental disasters.

Understanding the relationships between social interactions and economic outcomes is therefore of central importance to policy makers. For example, in 2020,

T. Kuchler · J. Stroebel (✉)
NYU Stern, NBER, CEPR, CESifo, New York, NY, USA
e-mail: theresa.kuchler@nyu.edu; johannes.stroebel@nyu.edu

© The Author(s) 2023
E. Bertoni et al. (eds.), *Handbook of Computational Social Science for Policy*,
https://doi.org/10.1007/978-3-031-16624-2_21

the European Commission's first annual Strategic Foresight Report[1] prominently identified the concept of resilience as a central compass for EU policy making. Increasing the resilience of communities involves strengthening their abilities not only to "withstand and cope with challenges" but also to "undergo transitions in a sustainable, fair, and democratic manner". Investing in social capital is crucial to achieving these objectives. However, policy makers who hope to increase resilience and economic opportunity by fostering social networks face challenges, in part due to a number of important gaps in the academic literature that studies the economic effects of social networks. To close some of these gaps, researchers need to better understand which features of social networks—for example, their size, connectedness, homogeneity, or geographic spread—contribute to the resilience of communities and their access to economic opportunities. Similarly, it is unclear how resilience and economic opportunity are affected by different types of social connections, such as connections among family members, friends, neighbours, or colleagues. As we discuss in this chapter, computational social scientists are in a strong position to answer such questions about the role of networks and social capital in fostering community resilience and economic opportunity.

While policy makers are naturally interested in the economic effects of social networks, fostering strong networks might also be a direct policy objective. For example, following the large increase in immigration to Europe from refugees fleeing violence in Syria and Afghanistan, many European governments are highly concerned with the question of how to best achieve the social integration of these refugees (European Commission, 2020). While social integration has multiple aspects, including labour market attachment and language acquisition, a central aspect is the formation of ties of camaraderie between immigrants and natives. Such ties are desirable in themselves, independent of their positive economic effects, and researchers are increasingly interested in measuring and explaining the formation of such links between different groups (Bailey et al., 2022).

The primary objective of this chapter is to discuss a number of approaches and data sources that hold promise for computational social science research studying the economic effects of social networks. In particular, we focus on opportunities to use the recent explosion in digital trace data—the footprints produced by users' interactions with information systems such as websites and smartphone apps—to make progress on questions of policy interest (see also Lazer et al., 2009, 2020). Compared to traditional survey instruments to measure social networks, digital trace data offers a host of advantages: with it, researchers can observe social interactions as they organically occur, circumvent response biases, and measure social networks at unprecedented scales. But while some sources of digital trace data have recently become more accessible to researchers, others have not. In this sense, and in the spirit of this volume, our discussion of future research avenues will be partly aspirational.

---

[1] https://ec.europa.eu/info/strategy/strategic-planning/strategic-foresight/2020-strategic-foresight-report_en

21.2 Current Progress

The role of social networks and social capital in creating economic resilience, exchange, and opportunity has been the focus of research across several fields, including sociology and economics. While it is impossible to do justice to this wide-ranging literature in the few short paragraphs available here, we next describe several research papers that have worked with particularly new and promising datasets. We encourage readers who are interested in obtaining more comprehensive overviews to start with recent review articles. Readers interested in more theoretical treatments could start with Jackson (2011) and Jackson et al. (2017), who discuss various economic applications of social networks, and Jackson (2020), who provides a formal typology of measures of social capital and their interactions with network measures. Readers who are looking for an overview of empirical work on the economic effects of social networks might start with Kuchler and Stroebel (2021), who review the role of social interactions in household financial decision-making, and Jackson (2021), who summarizes the evidence on the interaction between social capital and economic inequality. In addition, several chapters of the *Handbook of Social Economics* (edited by Benhabib, 2011) summarize the evidence on peer effects across a wide range of settings. Finally, for discussions of identification challenges in the peer effects literature, see Bramoullé et al. (2020) and Kuchler and Stroebel (2021).

One takeaway from these reviews is that much of the existing empirical research into the economic effects of social networks has measured networks either by using data from a few relatively small surveys (e.g. the National Longitudinal Study of Adolescent Health) or by defining networks according to individuals' memberships in observable associations (e.g. groups of neighbours or work colleagues). However, in recent years, the increased availability of digital trace data has led to a surge of interest in using tools from the computational social science to better understand economic activity. (An earlier literature has studied the topological structure of social graphs across a variety of online social networking services but without explicitly linking the structure of networks to economic outcome variables of interest; see Magno et al. (2012) and Ugander et al. (2011).) We next discuss some data sources that have the ability to push forward the frontier of this field of research.

21.2.1 Online Social Networking Services

The appeal of working with data from online social networking services is clear: these widely adopted services record social links between many individuals and even, in some cases, the strength of these ties. The scale of the most successful online social networks is astonishing. As of the second quarter of 2021, Facebook had 2.9 billion monthly active users—nearly 40 per cent of the world's population— and as of their last reports, Twitter and LinkedIn each had over 300 million active users. WeChat, a China-based online platform that includes a substantial social networking element, had 1.25 billion users. The enormous user bases of these platforms dwarf the sample sizes traditionally studied by economists and social scientists and provide researchers not only with sufficient statistical power to detect granular patterns but also with data that is difficult or expensive to obtain directly via surveys.

Already, a number of researchers have worked with anonymized (individual-level) microdata from Facebook to study a broad range of economic and social outcomes. For example, Gee et al. (2017) explore the extent to which weak and strong ties might help individuals find new jobs. Similarly, Bailey et al. (2018a, 2019a, b) study the role of social interactions in driving optimism in housing and mortgage markets. Bailey et al. (2019a, b) use data from Facebook to study the role of peer effects in product adoption, and Bailey et al. (2020a) study the role of information obtained through friends on individuals' social distancing behaviours during the COVID-19 pandemic. Bailey et al. (2022) use data from Facebook to explore the determinants of the social integration of Syrian migrants in Germany.

Data from online social networking platforms can also be a rich record of cross-country and cross-regional connections. Using more aggregated data from Facebook—data we describe in more detail in Sect. 21.3.2—researchers have explored the historical and cultural drivers of social connectedness across European regions (Bailey et al., 2020b), as well as the relationship between social connections and international trade flows (Bailey et al., 2021), migration (Bailey et al., 2018b), investment (Kuchler et al., 2021), bank lending (Rehbein et al., 2020), and the spread of COVID-19 (Kuchler et al., 2020).

While Facebook is the largest online social networking platform in the world, other platforms—in particular those that offer different services and therefore measure different types of networks—are also valuable data sources for researchers. Jeffers (2017) uses LinkedIn data on professional networks to study the role of labour mobility frictions in reducing entrepreneurship. Bakshy et al. (2011) quantify the influence of Twitter users by studying the diffusion of information that they post, and Bollen et al. (2011) measure the sentiment of Tweets to predict stock market movements. In a similar vein, Vosoughi et al. (2018) examine the network structure of sharing behaviour on Twitter to document that false news often spreads faster and more widely than true news.

As illustrated by these studies, social networking platforms have information on a large set of variables. Besides the connections between pairs of individuals, these services collect data on the personal characteristics that users choose to share—for example, education, employment, and relationship status—as well as the content they produce or engage with (such as posts, messages, and "likes"). With advances in natural language processing (NLP) methods, which extract meaning from text, the latter type of data provides increasing opportunities for researchers to measure opinions and beliefs that are otherwise hard to capture at scale. A recent example is Bailey et al. (2020a), who use Facebook posts to measure attitudes towards social distancing policies during the COVID-19 pandemic. (For a review of text mining and NLP research with Facebook and Twitter data, see Salloum et al. (2017)).

Moreover, many of these services record a rich set of metadata, including users' log-in times and geographic locations. Several recent studies have exploited location data from Facebook to study social distancing behaviour during the COVID-19 pandemic (Ananyev et al., 2021; Bailey et al., 2020a; Tian et al., 2022).

Similarly, most apps record information on the phone type used to log into the apps. Combined with other information, this can provide a proxy of a users' income or socio-economic status (see Chetty et al., 2022a, b). Such data can be very helpful to researchers hoping to study the effects of social capital on outcomes such as social mobility. Indeed, many measures of social capital that the literature associates with beneficial outcomes relate to the extent to which relatively poor individuals are connected with relatively rich individuals—see, for example, the work of Loury (1976), and Bourdieu (1986), and the discussion in Chetty et al. (2022a, b). Measuring the variation of such "bridging capital" across regions or other groups requires information not only on networks but also on the income or socio-economic status of each individual node.

21.2.2 Other Communication Networks

The widespread adoption of smartphones has generated a trove of data capturing various aspects of economic and social behaviours. A large body of research has used smartphone location data—available from companies such as SafeGraph, Veraset, and Unacast—to study a range of topics, from the effect of partisanship on family ties (Chen & Rohla, 2018) to the role of staff networks in spreading COVID-19 in nursing homes (Chen et al., 2021) to racial segregation and other racial disparities (Athey et al., 2020; Chen et al., 2020).

Another set of research has used call detail record (CDR) data to understand the economic effects of social networks. This literature includes Björkegren (2019), who uses CDR data from Rwanda to study the spread of network goods (goods whose benefits to a user depend on the network of other users), as well as Büchel and Ehrlich (2020) and Büchel et al. (2020), who use CDR data to analyse how geographic distance impacts interpersonal exchange and how social networks affect residential mobility decisions, respectively.

Other sources of digital trace data suggest further avenues for advancing research on social networks and resilience. For example, researchers who wish to study the relationship between segregation and resilience might follow Davis et al. (2019) in using data from services such as Yelp—a platform that allows users to review local businesses—to test whether people of different racial or socio-economic backgrounds visit the same parks, restaurants, hotels, stores, or other public places. Email and direct messaging networks can also offer insights into the structure of networks. For example, data on who communicates with whom within a corporation or community can allow researchers to establish how hierarchical organizations are, or how quickly information spreads within a community—both of which can be related to economic resilience and opportunity. For example, the analysis by Diesner et al. (2005) of the Enron email corpus illustrates the patterns of communication within a collapsing organization. Data from other professional communication tools, such as Slack, Skype, or Bloomberg chat, might also offer insights into how the communications of traders and other finance professionals shape trading behaviour and asset prices.

21.2.3 Financial or Business Transaction Networks

One crucial way through which social networks bolster economic resilience is by providing a foundation for the flow of credit and insurance, and a long line of sociological research illustrates this phenomenon in myriad communities. An early example is Geertz's (1962) description of the rotating credit associations of small communities in Asia and Africa, where members periodically contribute money to a fund that can be claimed by each member on a schedule. More recently, Banerjee et al. (2013) document how well-connected individuals in Indian villages—for instance, shopkeepers and teachers—play an essential role in spreading information about a microfinance programme.

But the importance of social networks in fostering access to financial resources is not limited to less-developed countries. In Europe, crowdfunding platforms such as GoFundMe and Kickstarter have hosted campaigns to help refugees, rescue small businesses during the COVID-19 recession, and finance individuals' medical needs, educational expenses, or creative ventures. Data from such crowdfunding platforms is thus an interesting and valuable source of information for researchers hoping to measure the strength of social capital across communities. Social networks can also provide essential resources to small businesses. Two classic discussions in the literature are provided by Light (1984), who attributes the entrepreneurial success of Korean immigrants in Los Angeles to social solidarity, nepotistic hiring, mutual support groups, and political connections, and by Coleman (1988), who describes Jewish diamond merchants in New York City exchanging stones with each other for inspection, relying on close ethnic ties, rather than expensive formal contracts, as insurance against theft.

Furthermore, with the growth of online payment platforms (e.g. PayPal, Venmo, WeChat Pay, and Wise) and peer-to-peer lending websites (e.g. Zopa and LendingClub), it is increasingly possible to observe networks of financial transactions among friends and family as well as strangers. An example of work benefiting from such data is by Sheridan (2020) who uses data from MobilePay, a Danish mobile payment platform, to measure social networks. Sheridan (2020) shows that individuals' spending responds to their friends' unemployment shocks, thereby documenting that spending and consumption are linked across social networks. In an international context, remittances by immigrants to their home countries are an important economic force in many countries with substantial expat communities. Increasingly, such remittances are sent electronically, allowing for systematic measurement. We view the use of these types of data sources as highly promising directions for researchers interested in studying the contribution of various types of social capital to the resilience of communities.

21.2.4 Civic Networks

Although sociologists have characterized a central product of social networks—social capital—in various different ways (see the discussion in Chetty et al., 2022a), one influential description by Putnam (2000) emphasizes citizens' participation in civic and community life, their respect for moral norms and obligations, and their trust in institutions and in one another. Digital trace data can be used to provide new ways of measuring these aspects of civic social capital.

A growing body of literature has used digital trace data to analyse the relationship between social networks and political trends, especially polarization. Employing innovative text, content, and sentiment analysis techniques, researchers have quantified patterns in political news and discourse on Facebook and Twitter (e.g. Alashri et al., 2016; Engesser et al., 2017; Moody-Ramirez & Church, 2019). Other work has found that individuals' socio-economic backgrounds can predict their civic engagement on social media (e.g. Hopp and Vargo, 2017; Lane et al., 2017) and that social media can drive their real-life political opinions and behaviours (e.g. Amador Diaz Lopez et al., 2017; Bond et al., 2017; Gil de Zúñiga et al., 2012; Groshek and Koc-Michalska, 2017; Kosinski et al., 2013). In particular, there has been enormous interest in researching the causes and consequences of "fake news" on social media (e.g. Allcott and Gentzkow, 2017; Guess et al., 2019; Lazer et al., 2018).

Besides Facebook and Twitter, other sources of digital trace data provide further opportunities to measure civic beliefs and behaviours and to construct measures of civic social capital. An emerging strand of research uses data from e-petition platforms—including governmental sites established by the White House (Dumas et al., 2015) and the Bundestag (Puschmann et al., 2017), as well as commercial sites such as Change.org (Halpin et al., 2018)—to study the forces that motivate citizens' political engagement. Elnoshokaty et al. (2016), for instance, have found that the success of petitions is more strongly driven by emotional elements than by moral or cognitive ones. Combined with records of online and offline social connections, this data offers the opportunity to study attitudes not only towards governmental policies and programmes but also towards those of communities such as universities and neighbourhood associations.

21.3 The Way Forward

Despite the economic and political importance of better understanding the effects of various types of social networks, research has long been hindered by the lack of large-scale data on individuals' social interactions. Moreover, to study how individuals' networks affect their economic outcomes, economists must not only measure connections between individuals but also match these measurements to data on income, savings, consumption, health, or other variables of interest. The difficulty of obtaining such complex data can pose a serious roadblock to researchers.

21.3.1 Increasing Access to Microdata

As illustrated in our discussion in the previous section, the richest datasets on social networks are usually not in the public domain, but are instead held by corporations. The digital trace data created on platforms such as Facebook, Instagram, WhatsApp, Twitter, Snapchat, YouTube, WeChat, TikTok (Douyin), Meetup, and Nextdoor hold immense promise for empirical research on which types of people form connections, how and where they meet, and whether their acquaintances and friends shape their future behaviours. As for research on professional networks, LinkedIn, along with its European competitors XING and Viadeo, possesses records that can shed light on important labour market patterns.

While these microdata hold much promise for conducting research of substantial value to policy makers and the academic community, there are obvious challenges to facilitating large-scale data access to researchers. Most importantly, the firms holding the data are responsible for safeguarding the privacy of their users and have to trade off the benefits of research to the broader public against potential reputational and legal risks from collaborating with researchers on these projects.

There are a number of paths that researchers have followed in navigating the challenge of accessing microdata owned by corporations. On the one hand, some researchers have gained access to proprietary data by working directly with companies as employees, contractors, or consultants. These agreements often involve signing nondisclosure agreements, and companies usually retain the right to veto publication if they are concerned, for example, that their users' anonymity is compromised by the results. Because of the potential for various conflicts of interest in such relationships, some members of the research community have expressed concerns about bias in the questions asked or the results generated by researchers with such arrangements.

On the other hand, researchers may attempt to work independently of the companies whose data they analyse. For example, they might be able to use data that companies publicize through application programming interfaces (APIs) or data purchased from market research firms. However, the former source of data may be unstructured or incomplete, while the latter, collected through methods that are sometimes opaque, might be unrepresentative or prohibitively expensive. (For a longer discussion of the tradeoffs researchers face in accessing proprietary microdata, see Lazer et al. (2020)).

To help navigate these challenges, there have been recent advances in developing models of industry-academic data-sharing collaborations that seek to facilitate researchers' access to anonymized microdata held by firms while guaranteeing their ability to publish findings independent of a final review by the company. Most prominent is Facebook's relationship with Social Science One, launched after the 2016 US elections (see King and Persily (2020) for details).

We believe that policy makers have the opportunity—and even the responsibility—to play a key role in advancing the various attempts by firms and academic researchers to collaborate on producing publicly accessible research on questions of high social importance. A key aspect of this is to create legal certainty about how academic research would be treated within various privacy frameworks, ideally carving out exemptions for public good research to the frameworks' most restrictive provisions. For example, the US Federal Trade Commission recently highlighted[2] that its consent decree with Facebook "does not bar Facebook from creating exceptions for good-faith research in the public interest". Increasing support from policy makers to facilitate public interest research within the frameworks of other privacy regulations, such as the European Union's GDPR, would be hugely beneficial to the academic research community and broader society.

21.3.2   Increasing Access to Aggregated Data

While working with individual-level data offers several important advantages for researchers, these collaborations are often hard to scale, in part due to the substantial resources that companies must invest to provide privacy-protected access to their data. In addition, many outcome variables of interest cannot be merged to individual-level data in a privacy-preserving way. On the flipside, there are many opportunities for better understanding the role of social networks and social interactions by using more aggregated data on social networks, social capital, and mobility.

One prominent example of such aggregated data is the Social Connectedness Index (SCI), which was introduced by Bailey et al. (2018a, b). The SCI is based on the universe of friendship links on Facebook and measures the relative probability that a random pair of Facebook users across two locations are friends with each other on Facebook. For example, Fig. 21.1 shows a heat map of the social connectedness to Düsseldorf in Germany to all European NUTS2 regions.

Importantly, the SCI data is publicly available[3] to researchers through the Humanitarian Data Exchange (HDX). As of February 2022, this data has been downloaded more than 16,000 times, demonstrating that the research and policy communities are highly interested in accessing aggregated data sets, even at such

---

[2] https://www.ftc.gov/blog-posts/2021/08/letter-acting-director-bureau-consumer-protection-samuel-levine-facebook

[3] https://data.humdata.org/dataset/social-connectedness-index

Düsseldorf (DEA1)

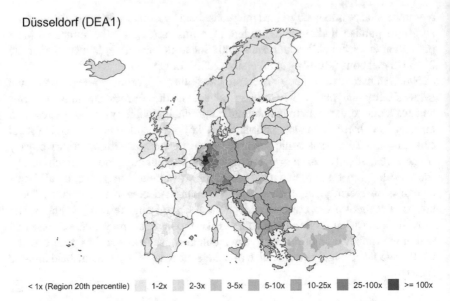

< 1x (Region 20th percentile)   1-2x   2-3x   3-5x   5-10x   10-25x   25-100x   >= 100x

Fig. 21.1 Heat map shows the strength of social connections of European NUTS2 regions to Düsseldorf. Darker colours correspond to stronger social ties. The data source is the Social Connectedness Index described in Bailey et al. (2018b)

relatively coarse levels of aggregation. We believe that there are many opportunities to deepen the insights from such data sets by further disaggregation—for example, by demographics or by setting. This finer level of information would allow researchers to study questions about how social connectedness varies across individuals of different ages, ethnicities, nationalities, genders, or educational and professional backgrounds.

Other sources of aggregated data also offer opportunities to understand how social networks and social capital affect economic outcomes. For example, LinkedIn could provide aggregated measures of connectedness across geographic locations, allowing researchers to study similarities and differences between the structure of professional networks and friendship networks. Similarly, measures of the connectedness between firms could be useful to study the determinants of labour flows.

We believe that policy makers should communicate to firms that such data efforts are perceived as valuable by both the academic and the policy communities, thereby encouraging more firms to engage in similar efforts.

21.4 Summary

There are many interesting opportunities to work with nontraditional data sources to understand the role of social networks in fostering resilience and access to economic opportunities. Indeed, many of the data sources required to further study these questions already exist or could be collected in a relatively straightforward way. Many of these data sets are owned by private companies. An important question, then, is what can be done to facilitate more broad-based access to such data.

It is critical that the private companies collecting digital trace data, aware of their unique positions to advance important research agendas, continue and expand their engagement with researchers to find paths to improve our understanding of the economic effects of social networks. Our hope is that, over time, we reach an equilibrium where such efforts to engage with academic researchers become the expectation of companies holding unique and important data assets. We are encouraged by the creation of "Data for Good" efforts across a variety of firms such as Meta and Acxiom, as well as by the creation of formal research institutes within many corporations, such as the JPMorgan Chase Institute and the ADP Research Institute. The further expansion of such efforts holds much promise for the future of the computational social sciences.

Policy makers can help this process by creating frameworks that incentivize firms to collaborate with researchers. For firms, collaborations with researchers involve substantial financial costs and can carry reputational and legal risks. In the decision of whether to engage in collaborations that are not directly related to the core business of the firm (as is the case with many of the research questions reviewed in this chapter), these costs and risks are then weighed against potential benefits to the firm, such as positive press and public goodwill.

Policy makers can alter both the perceived costs and benefits to firms from such collaborations. On the cost side, as highlighted above, an important element is the provision of legal certainty about how research for the social good will be treated under data privacy regulations such as GDPR. Policy makers interested in encouraging firms to collaborate with researchers on social good questions should also consider providing explicit carve-outs for these research activities in various privacy regulations. Similarly, policy makers can increase the perceived benefits for firms from academic collaborations, for example, by publicly recognizing that firms' facilitation of such collaborations contributes to the public good.

References

Alashri, S., Kandala, S. S., Bajaj, V., Ravi, R., Smith, K. L., & Desouza, K. C. (2016). An analysis of sentiments on Facebook during the 2016 U.S. presidential election. In *2016 IEEE/ACM International Conference on Advances in Social Networks Analysis and Mining (ASONAM)* (pp. 795–802). https://doi.org/10.1109/ASONAM.2016.7752329

Allcott, H., & Gentzkow, M. (2017). Social media and fake news in the 2016 election. *Journal of Economic Perspectives, 31*(2), 211–236. https://doi.org/10.1257/jep.31.2.211

Amador Diaz Lopez, J. C., Collignon-Delmar, S., Benoit, K., & Matsuo, A. (2017). Predicting the Brexit vote by tracking and classifying public opinion using twitter data. *Statistics, Politics and Policy, 8*(1). https://doi.org/10.1515/spp-2017-0006

Ananyev, M., Poyker, M., & Tian, Y. (2021). The safest time to fly: Pandemic response in the era of Fox News. *Journal of Population Economics, 34*(3), 775–802. https://doi.org/10.1007/s00148-021-00847-0

Athey, S., Ferguson, B., Gentzkow, M., & Schmidt, T. (2020). *Experienced segregation* (no. w27572, p. w27572). National Bureau of Economic Research. https://doi.org/10.3386/w27572

Bailey, M., Cao, R., Kuchler, T., & Stroebel, J. (2018a). The economic effects of social networks: Evidence from the housing market. *Journal of Political Economy, 126*(6), 2224–2276. https://doi.org/10.1086/700073

Bailey, M., Cao, R., Kuchler, T., Stroebel, J., & Wong, A. (2018b). Social connectedness: Measurement, determinants, and effects. *Journal of Economic Perspectives, 32*(3), 259–280. https://doi.org/10.1257/jep.32.3.259

Bailey, M., Dávila, E., Kuchler, T., & Stroebel, J. (2019a). House price beliefs and mortgage leverage choice. *The Review of Economic Studies, 86*(6), 2403–2452. https://doi.org/10.1093/restud/rdy068

Bailey, M., Johnston, D. M., Kuchler, T., Stroebel, J., & Wong, A. (2019b). *Peer effects in product adoption*. National Bureau of Economic Research.

Bailey, M., Johnston, D., Koenen, M., Kuchler, T., Russel, D., & Stroebel, J. (2020a). *Social networks shape beliefs and behavior: Evidence from social distancing during the COVID-19 pandemic* (no. w28234; p. w28234). National Bureau of Economic Research. https://doi.org/10.3386/w28234

Bailey, M., Johnston, D., Kuchler, T., Russel, D., State, B., & Stroebel, J. (2020b). The determinants of social connectedness in Europe. In S. Aref, K. Bontcheva, M. Braghieri, F. Dignum, F. Giannotti, F. Grisolia, & D. Pedreschi (Eds.), *Social informatics* (Vol. 12467, pp. 1–14). Springer. https://doi.org/10.1007/978-3-030-60975-7_1

Bailey, M., Gupta, A., Hillenbrand, S., Kuchler, T., Richmond, R., & Stroebel, J. (2021). International trade and social connectedness. *Journal of International Economics, 129*, 103418. https://doi.org/10.1016/j.jinteco.2020.103418

Bailey, M., Johnston, D., Koenen, M., Kuchler, T., Russel, D., & Stroebel, J. (2022). *The social integration of international migrants: Evidence from the networks of Syrians in Germany.*

Bakshy, E., Hofman, J. M., Mason, W. A., & Watts, D. J. (2011). Everyone's an influencer: Quantifying influence on twitter. In *Proceedings of the fourth ACM international conference on web search and data mining – WSDM'11*, p. 65. https://doi.org/10.1145/1935826.1935845.

Banerjee, A., Chandrasekhar, A. G., Duflo, E., & Jackson, M. O. (2013). The diffusion of microfinance. *Science, 341*(6144), 1236498. https://doi.org/10.1126/science.1236498

Benhabib, J. (Ed.). (2011). *Handbook of social economics* (Vol. 1A, 1st ed.). Elsevier.

Björkegren, D. (2019). The adoption of network goods: Evidence from the spread of mobile phones in Rwanda. *The Review of Economic Studies, 86*(3), 1033–1060. https://doi.org/10.1093/restud/rdy024

Bollen, J., Mao, H., & Zeng, X.-J. (2011). Twitter mood predicts the stock market. *Journal of Computational Science, 2*(1), 1–8. https://doi.org/10.1016/j.jocs.2010.12.007

Bond, R. M., Settle, J. E., Fariss, C. J., Jones, J. J., & Fowler, J. H. (2017). Social endorsement cues and political participation. *Political Communication, 34*(2), 261–281. https://doi.org/10.1080/10584609.2016.1226223

Bourdieu, P. (1986). The forms of capital. In J. G. Richardson (Ed.), *Handbook of theory and research for the sociology of education* (p. 19). Greenwood Press.

Bramoullé, Y., Djebbari, H., & Fortin, B. (2020). Peer effects in networks: A survey. *Annual Review of Economics, 12*(1), 603–629. https://doi.org/10.1146/annurev-economics-020320-033926

Büchel, K., & Ehrlich, M. V. (2020). Cities and the structure of social interactions: Evidence from mobile phone data. *Journal of Urban Economics, 119*, 103276. https://doi.org/10.1016/j.jue.2020.103276

Büchel, K., Ehrlich, M. V., Puga, D., & Viladecans-Marsal, E. (2020). Calling from the outside: The role of networks in residential mobility. *Journal of Urban Economics, 119*, 103277. https://doi.org/10.1016/j.jue.2020.103277

Chen, M. K., & Rohla, R. (2018). The effect of partisanship and political advertising on close family ties. *Science, 360*(6392), 1020–1024. https://doi.org/10.1126/science.aaq1433

Chen, M. K., Haggag, K., Pope, D. G., & Rohla, R. (2020). Racial disparities in voting wait times: Evidence from smartphone data. *ArXiv: 1909.00024*. Retrieved from http://arxiv.org/abs/1909.00024

Chen, M. K., Chevalier, J. A., & Long, E. F. (2021). Nursing home staff networks and COVID-19. *Proceedings of the National Academy of Sciences, 118*(1), e2015455118. https://doi.org/10.1073/pnas.2015455118

Chetty, R., Hendren, N., Jackson, M. O., Kuchler, T., Stroebel, J., Fluegge, R., Gonzalez, F., Jacob, M., Koenen, M., Laguna-Muggenburg, E., et al. (2022a). *Social capital in the United States I: Measurement and associations with economic mobility*. Harvard University.

Chetty, R., Hendren, N., Jackson, M. O., Kuchler, T., Stroebel, J., Fluegge, R., Gonzalez, F., Jacob, M., Koenen, M., Laguna-Muggenburg, E., et al. (2022b). *Social Capital in the United States II: Exposure, friending bias, and the determinants of economic connectedness*. Harvard University.

Coleman, J. S. (1988). Social capital in the creation of human capital. *American Journal of Sociology, 94*, S95–S120.

Davis, D. R., Dingel, J. I., Monras, J., & Morales, E. (2019). How segregated is urban consumption? *Journal of Political Economy, 127*(4), 1684–1738. https://doi.org/10.1086/701680

Diesner, J., Frantz, T. L., & Carley, K. M. (2005). Communication networks from the Enron email corpus "It's always about the people. Enron is no different". *Computational and Mathematical Organization Theory, 11*(3), 201–228. https://doi.org/10.1007/s10588-005-5377-0

Dumas, C. L., LaManna, D., Harrison, T. M., Ravi, S., Kotfila, C., Gervais, N., Hagen, L., & Chen, F. (2015). Examining political mobilization of online communities through e-petitioning behavior in *We the People*. *Big Data & Society, 2*(2), 205395171559817. https://doi.org/10.1177/2053951715598170

Elnoshokaty, A. S., Deng, S., & Kwak, D.-H. (2016). Success factors of online petitions: Evidence from Change.org. In *2016 49th Hawaii International Conference on System Sciences (HICSS)* (pp. 1979–1985). https://doi.org/10.1109/HICSS.2016.249.

Engesser, S., Ernst, N., Esser, F., & Büchel, F. (2017). Populism and social media: How politicians spread a fragmented ideology. *Information, Communication & Society, 20*(8), 1109–1126. https://doi.org/10.1080/1369118X.2016.1207697

European Commission. (2020). *Communication from the Commission to the European Parliament, the council, the European economic and social committee and the Committee of the Regions: Action plan on integration and inclusion 2021-2027*. Retrieved from https://eur-lex.europa.eu/legal-content/EN/ALL/?uri=COM%3A2020%3A758%3AFIN.

Gee, L. K., Jones, J., & Burke, M. (2017). Social networks and labor markets: How strong ties relate to job finding on Facebook's social network. *Journal of Labor Economics, 35*(2), 485–518. https://doi.org/10.1086/686225

Geertz, C. (1962). The rotating credit association: A 'middle rung' in development. *Economic Development and Cultural Change, 10*(3), 241–263. https://doi.org/10.1086/449960

Gil de Zúñiga, H., Jung, N., & Valenzuela, S. (2012). Social media use for news and individuals' social capital, civic engagement and political participation. *Journal of Computer-Mediated Communication, 17*(3), 319–336. https://doi.org/10.1111/j.1083-6101.2012.01574.x

Groshek, J., & Koc-Michalska, K. (2017). Helping populism win? Social media use, filter bubbles, and support for populist presidential candidates in the 2016 US election campaign. *Information, Communication & Society, 20*(9), 1389–1407. https://doi.org/10.1080/1369118X.2017.1329334

Guess, A., Nagler, J., & Tucker, J. (2019). Less than you think: Prevalence and predictors of fake news dissemination on Facebook. *Science Advances, 5*(1), eaau4586. https://doi.org/10.1126/sciadv.aau4586

Halpin, D., Vromen, A., Vaughan, M., & Raissi, M. (2018). Online petitioning and politics: The development of Change.org in Australia. *Australian Journal of Political Science, 53*(4), 428–445. https://doi.org/10.1080/10361146.2018.1499010

Hopp, T., & Vargo, C. J. (2017). Does negative campaign advertising stimulate uncivil communication on social media? Measuring audience response using big data. *Computers in Human Behavior, 68*, 368–377. https://doi.org/10.1016/j.chb.2016.11.034

Jackson, M. O. (2011). An overview of social networks and economic applications. In *Handbook of social economics* (Vol. 1, pp. 511–585). Elsevier. https://doi.org/10.1016/B978-0-444-53187-2.00012-7

Jackson, M. O. (2020). A typology of social capital and associated network measures. *Social choice and welfare, 54*(2–3), 311–336. https://doi.org/10.1007/s00355-019-01189-3

Jackson, M. O. (2021). Inequality's economic and social roots: The role of social networks and homophily. *SSRN Electronic Journal*. https://doi.org/10.2139/ssrn.3795626

Jackson, M. O., Rogers, B. W., & Zenou, Y. (2017). The economic consequences of social-network structure. *Journal of Economic Literature, 55*(1), 49–95. https://doi.org/10.1257/jel.20150694

Jeffers, J. (2017). The impact of restricting labor mobility on corporate investment and entrepreneurship. *SSRN Electronic Journal.*https://doi.org/10.2139/ssrn.3040393

King, G., & Persily, N. (2020). A new model for industry–academic partnerships. *PS: Political Science & Politics, 53*(4), 703–709. https://doi.org/10.1017/S1049096519001021

Kosinski, M., Stillwell, D., & Graepel, T. (2013). Private traits and attributes are predictable from digital records of human behavior. *Proceedings of the National Academy of Sciences, 110*(15), 5802–5805. https://doi.org/10.1073/pnas.1218772110

Kuchler, T., & Stroebel, J. (2021). Social finance. *Annual Review of Financial Economics, 13*(1), 37–55. https://doi.org/10.1146/annurev-financial-101320-062446

Kuchler, T., Russel, D., & Stroebel, J. (2020). *The geographic spread of COVID-19 correlates with the structure of social networks as measured by Facebook* (no. w26990, p. w26990). National Bureau of Economic Research. https://doi.org/10.3386/w26990

Kuchler, T., Li, Y., Peng, L., Stroebel, J., & Zhou, D. (2021). Social proximity to capital: Implications for investors and firms. *The Review of Financial Studies, hhab111*. https://doi.org/10.1093/rfs/hhab111

Lane, D. S., Kim, D. H., Lee, S. S., Weeks, B. E., & Kwak, N. (2017). From online disagreement to offline action: How diverse motivations for using social media can increase political information sharing and catalyze offline political participation. *Social Media + Society, 3*(3), 205630511771627. https://doi.org/10.1177/2056305117716274

Lazer, D., Pentland, A., Adamic, L., Aral, S., Barabasi, A. L., Brewer, D., Christakis, N., Contractor, N., Fowler, J., Gutmann, M., Jebara, T., King, G., Macy, M., Roy, D., & Van Alstyne, M. (2009). Life in the network: The coming age of computational social science. *Science (New York, N.Y.), 323*(5915), 721–723. https://doi.org/10.1126/science.1167742

Lazer, D. M. J., Baum, M. A., Benkler, Y., Berinsky, A. J., Greenhill, K. M., Menczer, F., Metzger, M. J., Nyhan, B., Pennycook, G., Rothschild, D., Schudson, M., Sloman, S. A., Sunstein, C. R., Thorson, E. A., Watts, D. J., & Zittrain, J. L. (2018). The science of fake news. *Science, 359*(6380), 1094–1096. https://doi.org/10.1126/science.aao2998

Lazer, D. M. J., Pentland, A., Watts, D. J., Aral, S., Athey, S., Contractor, N., Freelon, D., Gonzalez-Bailon, S., King, G., Margetts, H., Nelson, A., Salganik, M. J., Strohmaier, M., Vespignani, A., & Wagner, C. (2020). Computational social science: Obstacles and opportunities. *Science, 369*(6507), 1060–1062. https://doi.org/10.1126/science.aaz8170

Light, I. (1984). Immigrant and ethnic enterprise in North America*. *Ethnic and Racial Studies, 7*(2), 195–216. https://doi.org/10.1080/01419870.1984.9993441

Loury, G. C. (1976). *A dynamic theory of racial income differences*. Discussion paper.

Magno, G., Comarela, G., Saez-Trumper, D., Cha, M., & Almeida, V. (2012). New kid on the block: Exploring the Google+ social graph. In *Proceedings of the 2012 ACM conference on internet measurement conference – IMC'12*, p. 159. https://doi.org/10.1145/2398776.2398794

Moody-Ramirez, M., & Church, A. B. (2019). Analysis of Facebook meme groups used during the 2016 US presidential election. *Social Media + Society, 5*(1), 205630511880879. https://doi.org/10.1177/2056305118808799

Puschmann, C., Bastos, M. T., & Schmidt, J.-H. (2017). Birds of a feather petition together? Characterizing e-petitioning through the lens of platform data. *Information, Communication & Society, 20*(2), 203–220. https://doi.org/10.1080/1369118X.2016.1162828

Putnam, R. D. (2000). Bowling alone: America's declining social capital: Originally published in Journal of Democracy 6 (1), 1995. In L. Crothers & C. Lockhart (Eds.), *Culture and politics* (pp. 223–234). Palgrave Macmillan US. https://doi.org/10.1007/978-1-349-62965-7_12

Rehbein, O., Rother, S., et al. (2020). *Distance in bank lending: The role of social networks.* University of Bonn and University of Mannheim.

Salloum, S. A., Al-Emran, M., Monem, A. A., & Shaalan, K. (2017). A survey of text mining in social media: Facebook and Twitter perspectives. *Advances in Science, Technology and Engineering Systems Journal, 2*(1), 127–133. https://doi.org/10.25046/aj020115

Sheridan, A. (2020). *Learning about social networks from mobile money transfer.*

Tian, Y., Caballero, M. E., & Kovak, B. K. (2022). Social learning along international migrant networks. *Journal of Economic Behavior & Organization, 195*, 103–121.

Ugander, J., Karrer, B., Backstrom, L., & Marlow, C. (2011). *The anatomy of the Facebook social graph.* ArXiv:1111.4503 [Physics]. Retrieved from http://arxiv.org/abs/1111.4503

Vosoughi, S., Roy, D., & Aral, S. (2018). The spread of true and false news online. *Science, 359*(6380), 1146–1151. https://doi.org/10.1126/science.aap9559

Chapter 22
Social Media Contribution to the Crisis Management Processes: Towards a More Accurate Response Integrating Citizen-Generated Content and Citizen-Led Activities

Caroline Rizza

Abstract The two policy questions addressed in this chapter cover the whole crisis management cycle from the response and recovery to prevention and preparedness. They consider both the benefit of using citizen-generated content and the challenges of integrating citizen-led initiatives in the response. On the one hand, focusing on data allows interrogating the IT methods available to collect, process and deliver relevant information to support decision-making and response engagement. On the other hand, considering citizens' contribution and initiatives to the crisis management processes and response requires working on organizational and collaborative processes from local, regional, national or transnational levels. This chapter frames an up-do-date state of the art on the questions of citizens' generated content and led initiatives for crisis management and response, and it proposes directions to policy makers to that respect. It places the question of mutual trust between institutions and citizens as a key problematic in a hybrid world where mediated communication and interactions with citizens required new and adapted practices from professionals of crisis management.

22.1 Introduction

The current global context, characterized by climate change and the COVID-19 pandemic, poses new challenges in terms of crisis management and collaboration between world regions, countries and actors (policy makers, emergency management services, citizens and private sector actors) (European Commission Joint Research Centre, 2021). Despite their more or less long-term consequences on

C. Rizza (✉)
Information and Communication Sciences, I3-Telecom Paris (UMR 9217), Institut Polytechnique de Paris, Paris, France
e-mail: caroline.rizza@telecom-paris.fr

E. Bertoni et al. (eds.), *Handbook of Computational Social Science for Policy*,
https://doi.org/10.1007/978-3-031-16624-2_22

the environment and human activities, the manifestations of these phenomena are increasingly violent and frequent in a short-term perspective. Thus, so-called civil security crises such as natural disasters, technological events or urban crises are characterized, among other things, by rapid kinetics (with a crisis peak and a return to "normal"), uncertainty, tension, victims, etc.

The field of crisis informatics studies how networked digital technologies, for instance, social media from the 2000s onwards, interact with crisis management, from both social sciences and computational sciences sensibility, notably through data science (Palen et al., 2007, 2020; Palen & Anderson, 2016). More specifically, scientific literature in this field has highlighted the presence and simultaneous manifestation of citizen initiatives to respond to a crisis. During a disaster, people immediately react and help each other providing first aid to victims and very often organize themselves for helping with cleaning and rebuilding during the recovery phase. As illustrations, during the Nice attacks in July 2016 in France, taxis immediately organized themselves to evacuate people from the Promenade des Anglais; a few months earlier, during the Bataclan attacks in November 2015, Parisians opened their doors to welcome those who were unable to return home; Genoa in 1976 and in 2011, having experienced two exceptionally violent flash floods, twice saw its young city dwellers volunteering to clean up the streets and to help shopkeepers and residents for days on end (Rizza & Guimarães Pereira, 2014).

The use of social media in daily life has enriched this range of initiatives by allowing them to be manifested and organized online, in addition to the actions that usually arise spontaneously on the ground. In the examples given above, the hashtag #parisportesouvertes was used to publicize and organize the reception of Parisians during the attacks; in Genoa in 2011, a Facebook page "Gli angeli col fango sulle magliette" became the hub of communication and organization during and after the flooding, involving institutional bodies in particular.

Social media, as a virtual public space, allow the emergence and organization of citizen initiatives and make available new data supporting to build a more accurate situational picture of the event and its consequences on the ground. Nevertheless, they also complexify, in ways we will develop below, crisis management and constitute a challenge in the response provided by its managers. Their integration into the crisis management process requires mutual trust that the proximity of institutional and citizen actors may facilitate.

22.2 State of the Art

22.2.1 Social Media and Crisis Management: New Perspectives

Social media are Web 2.0 platforms or applications that allow their users to create content online, exchange it, consume it and interact with other users or their environment in real time (e.g., Kaplan & Haenlein, 2010; Luna & Pennock, 2018;

Reuter et al., 2020). In 2019 Facebook, Twitter and WhatsApp had 5.7 billion users worldwide (Statista, 2019) in Reuter et al. (2020). Thus, in recent years, the use of social media has increased considerably, and its nature has changed by becoming more collaborative, especially during crises or emergencies (Reuter et al., 2020).

In general, social media allow users to communicate and interact in different and often combined ways: information creation and dissemination, relationship management, communication and self-expression. Based on these activities, we can distinguish (Reuter et al., 2011):

- Wikis: for information gathering and knowledge creation according to a collaborative logic
- Blogs and microblogs: for publishing information and/or self-expression
- Social networks: for relationship management, self-expression and communication and information gathering
- Content sharing and indexing systems: for the creation and exchange of multimedia information (photos, videos)

Last but not least, platforms specialized in crisis management also exist: they are run by communities of volunteers and allow, for instance, collaborative mapping (e.g., Crisis mapper, a variation of OpenStreetMap), on-site and remote contribution (e.g., Ushahidi) and public-private-citizen partnership (Wendling et al., 2013).

Based on this categorization, social media are differentiated according to their main functions: Twitter as a microblog is used for the dissemination or collection of information; Facebook as a social network allows interaction between "friends" or within a "group" community; Wikipedia, as a collaborative encyclopaedia, supports the creation of collaborative knowledge and sense-making (Bubendorff & Rizza, 2021; Kaplan & Haenlein, 2010). Literature in the field highlights that during a major event, specific uses of social media rise such as a combination of the main function of a platform with other functions needed in the moment. As an illustration, to make sense to an ongoing event and face its uncertainty, the discussion pages of Wikipedia become the place for exchanges within the community of contributors in the same way as a group on a social network (Bubendorff & Rizza, 2020).

"Sharing and obtaining factual information is the primary function of social media usage consistently across all disaster types" (Eismann et al., 2016). Much of the literature in the field of crisis informatics has focused on "microblogging" activities, i.e., the use of social media by citizens to report on what is happening on site during a major event. These microblogging activities have been documented based on real events: they cover both the creation and distribution of information as well as the communication and response to requests for help (Palen et al., 2009; Palen & Vieweg, 2008; Reuter et al., 2011; Tapia et al., 2013). Due to their ubiquity in citizens' life, their speed as a relay of information and communication and their accessibility through different platforms, microblogging activities have been considered very early as an opportunity for crisis management and communication. They constitute a place where real-time information about an event is being collected (Palen et al., 2010; Reuter et al., 2011; Vieweg et al., 2010). Interestingly, Reuter et al. (2020) distinguish two reasons for harnessing information from social media: to establish a more complete picture of the situation, also known as "situational awareness", and to engage a response on the ground, also mentioned as "actionable information" (see also Coche et al., 2019).

To this respect, several challenges related to the quality of the data and its relevance for the crisis management processes have been underlined such as issues of format, reliability, quantity, attention required, effective interpretation and contextualization (Grant et al., 2013; Ludwig et al., 2015; Moore et al., 2013; Tapia et al., 2011, 2013). From an organizational perspective, other challenges exist: issues of verification, accountability, credibility, information overload, dedicated resource allocation as well as lack of time and experienced and trained staff (Castagnino, 2019; Hiltz et al., 2014; Hughes & Palen, 2014; Kaufhol et al., 2019; San et al., 2013).

Lately, the literature on crisis management and institutional practices has been emphasizing new challenges. In order to effectively benefit from the multiple sources of available data, Munkvold et al. (2019), Pilemalm et al. (2021), and Steen-Tveit and Erik Munkvold (2021) show how "situational awareness and understanding" and "common operational picture" both require more effective collaboration between the engaged stakeholders based on specific organizational processes to be established. Technical and organizational challenges have been rising in terms of combination of different information sources (e.g., video and images, social media, sensor data, body-worn devices, UAVs and open data).

22.2.2 From Citizen-Generated Content to Citizen-Led Activities: Opportunities and Challenges

As mentioned in the introduction, social media have been supporting emergence and organization of online citizens' initiatives at the occasion of a major event. Literature commonly distinguishes "real volunteers", who act on site to respond to the crisis from "virtual volunteers", who, located anywhere, provide help and support by organizing action and processing information on social media (Reuter et al., 2013). This distinction helps to understand how social media has become a place for expressing and organizing solidarity (Batard et al., 2018; Rizza & Guimarães Pereira, 2014). Whether these citizen initiatives take place on site or online, they are mostly spontaneous: spontaneous volunteers are people who act in response to or in anticipation of a disaster and who may or may not have the required skills (Drabek & McEntire, 2003). The notion of "affiliation" ('affiliated'/'unaffiliated volunteer') with a crisis management organization allows refining this characterization (Batard et al., 2019; Stallings & Quarantelli, 1985; Zettl et al., 2017). Some volunteers have signed agreements with public institutions and their actions are coordinated. This is the case of the VOST (Virtual Operations Support Team) in Europe, for example, but other user communities such as the Waze community can also be mentioned.

Consequently, social media enable citizens to build a collective and coherent approach to the event (Stieglitz et al., 2018). The generated content can be understood as a key element in the achievement of social resilience (Jurgens & Helsloot, 2018) where resilience is the ability of social groups and communities to recover from or respond positively to crises (Maguire & Hagan, 2007; Reuter & Spielhofer, 2017).

As described above, there is a need for collaboration and reliable models between heterogeneous actors (such as police, firefighters, infrastructure providers, public administration and citizens) in order to improve collaborative resilience, i.e., the ability of a community to prepare for, respond to and recover from a crisis (Board on Earth Sciences and Resources, 2011; Goldstein, 2011). Therefore, the opportunities of organization and collaboration with social media offer in response to the crisis concretized their organizational dimension.

However, this aspect should not minimize the challenges raised by citizens' initiatives or engagement during a major event and the added complexity, time and organization they require from crisis management institutions. Citizen activism can have negative effects (Reuter et al., 2020). Three examples illustrate this view. During the 2011 attack in Norway, citizens' action to save people from the attack and expression of public opinion on social media made the management of the crisis more complex for the rescue teams and crisis managers who had to respond to these citizen dynamics at some point of the crisis (Perng et al., 2013). During the 2015 Bataclan attacks, the use of the hashtag #parisportesouvertes associated with personal addresses of Parisians who were offering places to victims or people stranded outside has also required regulation from the authorities to protect citizens who were putting themselves in danger. The manhunt against the rioters of the 2011 Vancouver riots also underlines the negative side of this activism and necessity from public institutions to be fully prepared when mobilizing it (Rizza et al., 2014).

22.3 Computational Guidelines

This section is articulated around two policy questions proposed by De Groeve et al. and included in a publication of the Joint Research Centre that aimed at collecting the upcoming research needs in terms of policy questions around different topics, including emergency response and disaster risk management (Bertoni et al., 2022).

22.3.1 Which Contribution to the Crisis Management Cycle?

The two proposed policy questions in Bertoni et al. (2022; Chap. 15) cover the whole crisis management cycle from the response and recovery to prevention and preparedness. They consider both the benefit of using citizen-generated content and the challenges of integrating citizen-led initiatives in the response. Focusing on data allows interrogating the IT methods available to collect, process and deliver relevant information to support decision-making and response engagement. Social network analysis can also play a relevant role in the context of disinformation campaign (Starbird, 2020; Starbird et al., 2019). Considering citizens' contribution and initiatives to the crisis management processes and response requires working on organizational and collaborative processes from local, regional, national or transnational levels.

22.3.2   Towards an Actionable Information for Practitioners

In this section, we aim to address the first set of policy questions from Bertoni et al. (2022; Chap. 15) related to the optimization of crisis response and computational methods supporting to both harness and process multiple sources of citizen-generated content. The multiplication of such data sources brings to crisis managers several visions of an event and may support settling a more accurate situational awareness based on more information, geo-localization of the data collected and cross-verification through several platforms or formats (e.g., text, images, sensors). In that respect, in the field of crisis informatics, several systems have been developed to process emerging sources of data from social media, sensors in smart cities, UAVs as well as external data such as open data and multidisciplinary data archives.

Nevertheless, as pointed out by Coche et al. (2021a), the adoption of such systems in crisis management practices is low and may be understood or interpreted by a gap between practitioners' expectations and what these systems provide: actionable information vs. situational awareness. The key element here relies on the fact that systems aim at improving the situational awareness by addressing practitioners' information needs about the event while practitioners expect an "actionable information", that is to say, a complementary piece of information allowing them to take a decision and engage concretely a response on the ground.

Once settled, systems supporting EMS should collect, process and match multiple data sources and formats in order to both establish a relevant situational awareness of the event and to aggregate data in order to build actionable information supporting the engagement of a response. In this context, actionable information is relevant, timely, precise and reliable. In their research works on social media contribution to crisis management and response, Coche et al. (2021a) demonstrate that actionable information can be identified by systems only if a situational awareness is established first. Underlining the issue that information management and filtering systems for actionable information detection remain mostly unexplored in the field, they propose to design and build new systems based on a four-step architecture where the two last steps focus on actionable information: (1) data collection and management; (2) what they call "information creation" to establish a sufficient perception of the situation; (3) "information management" to understand the situation and be able to take a decision; and (4) "information filtering" to anticipate the evolution of the situation.

22.3.2.1   Designing Automatic Emergency Systems to Support Local EMS and EU Supervision: Directions

Once settled the objective of data processing systems in terms of situational awareness and actionable information, recommendations about design and implementation of such systems into practices can be addressed.

Designing crisis situation models is based on the data available at the time of the event, and, for this purpose, heterogeneous data sources such as phone calls, the information provided by the rescue teams on the ground, sensors and UAVs or news media exist. Nevertheless, these channels do not allow automated implementation and therefore neither implementation of viable crisis models.

Interestingly, social media data are already in a digital format and can be processed by a computer with minimal human interaction to input the data (Coche et al., 2021b). About automatic social media processing systems, three main types of systems to provide information to decision-makers can also be identified:

- Automatic data filtering systems
- Real-time systems providing semantic enrichment (based on spatial, social and/or emotional context of the content)
- Automatic clustering of tweets based on machine learning to filter messages from social media and successfully classify them in a category or another depending on their content and the context of the crisis, or to extract and cluster the information according to different parameters and similarities

Open data and multidisciplinary open archives also constitute relevant sources of data both to contextualize an event and to analyse its impact. Chasseray et al. (2021) and Lorini et al. (2020) propose to use meta-modelling and ontology to structure this available knowledge and feed decision support systems. Relevantly, they insist on the necessity to mobilize experts to validate the information extracted through this processing. While computational method supports data extraction and processing, expert intervention can specifically focus on decision-making and response engagement.

Decision support systems in crisis management and response should also facilitate collaboration between stakeholders. Based on Fogli and Guida (2013), Fertier et al. (2020) assign three properties to these systems related to collaborative dimensions: sharing information with citizens, interacting with other information systems and coordinating heterogeneous and independent stakeholders while anticipating the effects of decisions made. To that respect, it is important that support decision systems based on the heterogeneous data available ("decision support environment") provide to each crisis cell an up-to-date common operational picture allowing them to take decision, coordinate and collaborate. Fertier et al. (2020) also assign four key capabilities to these systems: improving the situation awareness through automatic collection and interpretation of raw data; processing data, in real time, by means of easy subscription to new sources; managing the issues related to big data; and processing heterogeneous data to update the model of a complex situation in real time.

However, systems of systems enabling multiagency crisis management strengthen the issue of making mass surveillance possible and require a specific attention. Interoperability combined with systems of systems and big data processing can foster the development of a technological and bureaucratic apparatus for all, encompassing surveillance and eroding civil liberties (Büscher et al., 2014; Rizza et al., 2017). The potentiality of collecting and processing data from participatory sensing makes fuzzy the boundary between decision support and control or surveillance. For instance, the knowledge database created through such system could contain pervasive information revealing individuals' habits, routines or decisions and, consequently, constitutes a privacy infringement.

To address these issues, a human practice focused approach is particularly useful when designing crisis management information systems: it allows designing and developing tools in close collaboration with EMS to and supporting them in restructuring their services in integrating these tools in their practices. It also prevents from misuse by closely working with stakeholders, understanding and framing their needs at the multiple level of the command chain. Indeed, in this context, crisis management system processing data, providing decision support and collaboration between stakeholders are more likely to be integrated into practices in respect of the rules and processes of each institution.

Box 1: In Summary

- Multiple formats and sources of citizen-generated content raised opportunities in terms of crisis management and response: some of these data support to contextualize and understand the event (e.g., open data, multidisciplinary data archives) and others to make decision and engage response on the ground (sensors, UAVs, social media).
- Computational modelling methods can support establishing a more accurate and timely situational awareness and providing pieces of data constitutive of actionable information: they move expert intervention from the collection and processing of the data to the analysis and decision-making phases.
- In the case of disinformation campaign, social network analysis constitutes a relevant tool to reveal and understand the ongoing dynamics.
- Such systems constitute challenges for both practitioners and researchers and IT designers. Practitioners need to integrate new information and their processing in their decision-making processes and to coordinate and collaborate more closely. Researchers and IT designers have to develop data processing and decision-making systems embedding these new properties, answering to practitioners' needs and giving a specific attention to legal and ethical frameworks.

22.3.3 Integrating Citizen-Led Activities in the Crisis Management Processes

This section addresses the second round of policy questions from Bertoni et al. (2022; Chap. 15) more focused on citizen-led activities and their possible integration to the crisis management processes. Social media have made possible most of these activities through the organizational dimension they propose. Beyond the benefits of using citizen-generated content by means of new computational methods, how making the most from these grassroots initiatives in the crisis management cycle?

22.3.3.1 Social Media as a Communication and Organizational Infrastructure

We usually think of social media as a means of communication used by institutions (e.g., ministries, municipalities, fire and emergency services) to communicate with citizens top-down and improve the situational analysis of the event through the information conveyed bottom-up from citizens (Zaglia, 2021). The literature has demonstrated the changes brought by social media, how citizens have used them to communicate in the course of an event, provide information or organize to self-help.

There are therefore an informational dimension and an organizational dimension to the contribution of social media to crisis management (Batard, 2021; Rizza, 2020):

– Informational in that the published content constitutes a source of relevant information to assess what is happening on site. For example, fire officers can use data from social media, or photos and videos during a fire outbreak, to readjust the means they need to deploy.
– Organizational in that aim is to work together to respond to the crisis: for example, creating a Wikipedia page about an ongoing event (and clearing up uncertainties), communicating pending an institutional response (as it has been the case during Hurricane Irma in Cuba in 2017), helping to evacuate a place (Gard, France, July 2019), taking in victims (Paris, 2015; Var, November 2019) or helping to rebuild or to clean a city (Genoa, November 2011).

22.3.3.2 Citizens: First Links of the Crisis Management Chain?

While institutions according to a top-down perspective use citizen-generated content and more largely social media to assess and communicate with citizens, citizen initiatives affect institutions horizontally in their professional practices. There is indeed a significant difference between harnessing and using online published data to understand an ongoing event or its consequences on the ground from supporting organized grassroots initiatives or engaging citizens on site to face an event. There is still a prevailing idea that citizens need to be protected, even if the COVID-19 crisis has been showing that the public also wants to play an active role in protecting themselves and others. In that respect, during the first 2020 lockdown, panels of citizen-led initiatives have emerged to support states facing the first peak of the crisis: sewing masks or making them with 3D printers to public hospitals, turning soap production into hand sanitizer production, proposing to translate information on preventative measures into different languages and sharing it to reach as many citizens as possible, etc. Despite this experience, according to some institutions, doing so would be recognizing that, somehow, crisis managers are failing (Batard, 2021).

Consequently, another dimension is delaying the integration of these initiatives in the common or virtual public space. It implies placing the public on the same level as the institution; in other words, citizen-led initiatives do not just have an "impact" horizontally on professional practices and their internal rules and processes (doctrines), but their integration requires citizens to be recognized as full participants, as actors, of the crisis management and response processes. Then, the main question concerns the required conditions supporting to both recognize citizens as actors of the crisis management processes and response and integrate their initiatives in these processes.

22.3.3.3 Building Specific Partnerships and Collaborations with Existing Online Communities

As underlined in the introduction of this chapter, the integration of citizen-led initiatives into the crisis management processes requires mutual trust from both sides. Again, the COVID-19 crisis illustrates the existing distrust against institutions, which has taken the form of misinformation campaign and required specific actions to counter this phenomenon. Among them, Wikipedia community has been working to making sense to major events, and, in that respect, the discussion pages associated to each article related to the crisis reveal the specific work done by Wikipedian contributors (Bubendorff & Rizza, 2020, 2021). Even if its contribution has not been yet fully recognized in crisis management, Wikipedia is a notorious community. Other online communities not affiliated to crisis management, such as forecast or road traffic groups, have also been playing an increasing role at the time of an event and need specific attention by publishing, for instance, prevention messages. Consequently, working closely with these communities in order to be able to mobilize them at the time of an event from the prevention to the recovery phases would be an asset. Affiliated communities of volunteers such as the VOST play already fully their part by providing online support to crisis managers. Their collaboration has been recognized and formalized through institutional agreements. They constitute today a trustful and reliable network to be mobilized before, during or even after a crisis. Establishing such agreements with other citizens, communities (as already mentioned, forecast or road traffic group but also related to air or water quality, earthquake monitoring, etc.) would allow crisis managers and decision-makers to rely on complementary and reliable raw data sources easily mobilizing in case of need.

The geographical proximity of actors in the same area enables them to get to know each other better and therefore encourages mutual trust – this trust constitutes a key component to success. In order to build such partnership with online communities, it is necessary working locally on specific areas to understand the composition of the network of local actors and initiate collaborations at each level of the national territory. At the European level, mobilizing and animating these communities by topics (air, water, fire, etc.), types of crises (floods, earthquake, technological event, urban crisis, etc.), type of data (social media, sensors, etc.), etc. would allow an EU monitoring of data and an EU kind of virtual taskforce.

Box 2: In Summary

- Social media constitute an infrastructure and a public place allowing citizens to communicate, coordinate and provide self-help initiatives.
- From an organizational perspective, integrating citizen-led initiatives into the crisis management processes and the response implies recognizing citizens as the first link of the crisis management chain.
- Building collaboration and partnership with online affiliated and (non-)affiliated communities will support the constitution of relevant networks of experts easily mobilized from a local to a European level to provide data and monitor citizen-led initiatives.

22.4 The Way Forward

Crisis management institutions are increasingly challenged by citizen-generated content and citizen-led initiatives. The first one constitutes new opportunity to better assess, monitor the situation and to engage a more accurate response on the ground but, at the same time, requires time, human resources and competences to aggregate, analyse and integrate these data in the usual processes. While responding specifically to the effects of an event on site, the second ones make the usual processes more complex in several ways: they require a specific and additional attention; they can disturb the institutional response. Nevertheless, affiliated and non-affiliated volunteers and their initiatives have been demonstrating their relevant contributions to crisis management at the time of an event.

What we argue in this chapter is the necessity from both types of actors (crisis managers and citizens) to both getting used of each other's online practices. Bridging the gap between such initiatives and their integration in the crisis management processes and response relies on (re-)establishing a mutual trust between institutions and citizens. Citizens have to get used to online official communication in their daily life in order to be able to get the message and understand it when it is published at the time of an event. Crisis management institutions need to understand and adapt to online rules of communication to be able to be heard on online public sphere. This communication goes beyond the usual diffusion of prevention or behavioural messages and requires real interaction with citizens.

In this context, computational methods and decision support systems have been improved: they tend to collect and analyse more and more multiple data sources and rely on different methodologies. They still require to be designed closely with practitioners to guarantee an answer to their needs and to ensure their integration into practices. Despite their contribution, such systems constitute only a decision support; in other words, they allow the expert intervention moving from the data extraction and processing to the analysis and decision-making phases. As an illustration, computational methods and systems may allow the VOST to focus on the analysis of the situation to be reported to crisis management institutions instead of manually screening and collecting information on social media.

Despite the benefit of such methods and systems, what we would like to underline is the necessary reorganization of institution internal processes in order to be able to fully and relevantly integrate citizen-generated content, as well as citizen-led initiatives in the crisis management processes.

References

Batard, R. (2021). *Intégrer les contributions citoyennes aux dispositifs de gestion de crise: L'apport des médias sociaux.* I3-Telecom Paris, Institut Polytechnique de Paris et IMT Mines Albi-Carmaux.

Batard, R., Rizza, C., Montarnal, A., & Benaben, F. (2018, March). Ethical, legal and social considerations surrounding the use of social media by citizens during hurricane Irma in Cuba. In *Proceedings of the 15th International Conference on Information Systems for Crisis Response and Management.*

Batard, R., Rizza, C., Montarnal, A., Benaben, F., & Prieur, C. (2019). Taxonomy of post-impact volunteerism types to improve citizen integration into crisis response. In *Proceedings of the 16th International Conference on Information Systems for Crisis Response and Management* (pp. 1114–1125).

Bertoni, E., Fontana, M., Gabrielli, L., Signorelli, S., & Vespe, M. (Eds). (2022). *Mapping the demand side of computational social science for policy.* EUR 31017 EN, Luxembourg, Publication Office of the European Union. ISBN 978-92-76-49358-7, https://doi.org/10.2760/901622

Board on Earth Sciences and Resources. (2011). *Building community disaster resilience through private-public collaboration.*

Bubendorff, S., & Rizza, C. (2020). The Wikipedia contribution to social resilience during terrorist attacks. In A. Hughes, F. McNeill, & C. Zobel (Éds.), *ISCRAM 2020 Conference Proceedings – 17th International Conference on Information Systems for Crisis Response and Management* (pp. 790–801). Virginia Tech. Retrieved from http://idl.iscram.org/files/sandrinebubendorff/2020/2271_SandrineBubendorff+CarolineRizza2020.pdf

Bubendorff, S., & Rizza, C. (2021, paraître). Faire collectivement sens en temps de crise: L'utilisation de Wikipédia lors de la pandémie de Covid-19. *Communiquer: Revue de communication sociale et politique.*

Büscher, M., Lieg, M., Rizza, C., & Watson, H. (Éds.). (2014). *How to do IT more carefully?: Ethical, Legal and Social Issues (ELSI) in IT supported crisis response and management* (Vol. 6).

Castagnino, F. (2019). What can we learn from a crisis management exercise ? Trusting social media in a French firefighters' department. In Z. Franco, J. J. González, & J. H. Canós (Éds.), *Proceedings of the 16th International Conference on Information Systems for Crisis Response and Management*. Iscram. Retrieved from http://idl.iscram.org/files/florentcastagnino/2019/1967_FlorentCastagnino2019.pdf

Chasseray, Y., Négny, S., & Barthe-Delanoë, A.-M. (2021). Automated unsupervised ontology population system applied to crisis management domain. In A. Adrot, R. Grace, K. Moore, & C. Zobel (Éds.), *ISCRAM 2021—18th International conference on Information Systems for Crisis Response and Management* (p. 968–981).

Coche, J., Montarnal, A., Tapia, A., & Benaben, F. (2019). Actionable collaborative common operational picture in crisis situation: A comprehensive architecture powered with social media data. In L. M. Camarinha-Matos, H. Afsarmanesh, & D. Antonelli (Éds.), Collaborative networks and digital transformation (p. 151–162). Springer. doi:https://doi.org/10.1007/978-3-030-28464-0_14

Coche, J., Kropczynski, J., Montarnal, A., Tapia, A., & Benaben, F. (2021a). Actionability in a situation awareness world: Implications for social media processing system design. In A. Adrot, R. Grace, K. Moore, & C. Zobel (Éds.), *ISCRAM 2021—18th International conference on Information Systems for Crisis Response and Management* (pp. 994–1001).

Coche, J., Rodriguez, G., Montarnal, A., Tapia, A., & Bénaben, F. (2021b, janvier 1). *Social media processing in crisis response: An attempt to shift from data to information exploitation.* Hawaii International Conference on System Sciences. https://doi.org/10.24251/HICSS.2021.279

Drabek, T. E., & McEntire, D. A. (2003). Emergent phenomena and the sociology of disaster: Lessons, trends and opportunities from the research literature. *Disaster Prevention and Management, 12*, 97–112.

Eismann, K., Posegga, O., & Fishbach, K. (2016). Collective behaviour, social media, and disasters: A systematic literature review. In *Twenty-Fourth European Conference on Information Systems (ECIS)*.

European Commission. Joint Research Centre. (2021). *1st workshop on social media for disaster risk management: Researchers meet practitioners.* Publications Office. Retrieved from https://data.europa.eu/doi/10.2760/477815

Fertier, A., Barthe-Delanoë, A.-M., Montarnal, A., Truptil, S., & Bénaben, F. (2020). A new emergency decision support system: The automatic interpretation and contextualisation of events to model a crisis situation in real-time. *Decision Support Systems, 133*, 113260. https://doi.org/10.1016/j.dss.2020.113260

Fogli, D., & Guida, G. (2013). Knowledge-centered design of decision support systems for emergency management. *Decision Support Systems, 55*(1), 336–347. https://doi.org/10.1016/j.dss.2013.01.022

Goldstein, B. E. (2011). *Collaborative resilience—Moving through crisis to opportunity.* MIT Press.

Grant, T. J., Geugies, F. L. E., & Jongejan, P. A. (2013). Social media in command & control: A proof-of principle experiment. In *Proceedings of the 10th International ISCRAM Conference* (pp. 52–61).

Hiltz, S. R., Kushma, J. Ann., & Plotnick, L. (2014). Use of social media by US public sector emergency managers: Barriers and wish lists. In *Proceedings of 11th International Conference on Information Systems for Crisis Response and Management* (pp. 600–609).

Hughes, A., & Palen, L. (2014). Social media in emergency management: Academic perspective. In J. E. Trainor & T. Subbio (Éds.), *Critical issues in disaster science and management: A dialogue between scientists and emergency managers.* University of Delaware.

Jurgens, M., & Helsloot, I. (2018). The effect of social media on the dynamics of (self) resilience during disasters: A literature review. *Contingencies and Crisis Management, 26*, 79–88.

Kaplan, A., & Haenlein, M. (2010). Users of the world, unite! The challenges and opportunities of social media. *Business Horizons, 53*, 59–68.

Kaufhol, M.-A., Rupp, N., Reuter, C., & Habdank, M. (2019). Mitigating information overload in social media during conflicts and crises: Design and evaluation of a cross-platform alerting system. *Behaviour and Information Technology*, 319–342.

Lorini, V., Rando, J., Saez-Trumper, D., & Castillo, C. (2020). Uneven coverage of natural disasters in Wikipedia: The case of flood. In A. Hughes, F. McNeill, & C. Zobel (Éds.), *Proceedings of the 17th ISCRAM Conference*.

Ludwig, T., Reuter, C., & Pipek, V. (2015). Social haystack: Dynamic quality assessment of citizen-generated content during emergencies. *ACM Transactions on Computer-Human Interaction, 22*, 1.

Luna, S., & Pennock, M. J. (2018). Social media applications and emergency management: A literature review and research agenda. *International Journal of Disaster Risk Reduction, 28*, 565–577.

Maguire, B., & Hagan, P. (2007). Disasters and communities: Understanding social resilience. *The Australian Journal of Emergency Management*, 16–20.

Moore, K., Tapia, A., & Griffin, C. (2013). Research in progress: Understanding how emergency managers evaluate crowdsourced data: A trust game-based approach. *Proceedings of the 10th International ISCRAM Conference, 2013*, 272–277.

Munkvold, B. E., Rød, J. K., Snaprud, M., Radianti, J., Opach, T., Pilemalm, S., & Bunker, D. (2019). Sharing incident and threat information for common situational understanding. In Z. Franco, J. J. González, & J. H. Canós (Eds.), *Proceedings of the 16th ISCRAM Conference*. ISCRAM.

Palen, L., & Anderson, K. M. (2016). Crisis informatics—New data for extraordinary times. *Science, 353*(6296), 224–225. https://doi.org/10.1126/science.aag2579

Palen, L., & Vieweg, S. (2008). The emergence of on line widescale interaction in unexpected events: Assistance, alliance. *Proceedings of CSCW 2008*.

Palen, L., Vieweg, S., Sutton, J., Liu, S. B., & Hughes, A. (2007). *Crisis informatics: Studying crisis in a networked world*.

Palen, L., Vieweg, S., Liu, S. B., & Hughes, A. (2009). Crisis in a networked world: Features of computer-mediated communication in the April 16, 2007, Virginia Tech event. *Social Science Computer Review, 27*, 467–480.

Palen, L., Anderson, K. M., Mark, G., Martin, J., Sicker, D., Palmer, M., & Grunwald, D. (2010). A vision for technology-mediated support for public participation & assistance in mass emergencies & disasters. *Proceedings of ACM-BCS Visions of Computer Science*.

Palen, L., Anderson, J., Bica, M., Castillos, C., Crowley, J., Díaz, P., Finn, M., Grace, R., Hughes, A., Imran, M., Kogan, M., LaLone, N., Mitra, P., Norris, W., Pine, K., Purohit, H., Reuter, C., Rizza, C., St Denis, L., ... Wilson, T. (2020). *Crisis informatics: Human-centered research on tech & crises*. Retrieved from https://hal.archives-ouvertes.fr/hal-02781763

Perng, S.-Y., Büscher, M., Wood, L., Halvorsrud, R., Stiso, M., Ramirez, L., & Al-Akkad, A. (2013). Peripheral response: Microblogging during the 22/7/2011 Norway attacks. *International Journal of Information Systems for Crisis Response and Management, 5*(1), 41–57. https://doi.org/10.4018/jiscrm.2013010103

Pilemalm, S., Radianti, J., Munkvold, B. E., Majchrzak, T. A., & Steen-Tveit, K. (2021). *Turning common operational picture data into double-loop learning from crises – Can vision meet reality?* 14.

Reuter, C., & Spielhofer, T. (2017). Towards social resilience: A quantitative and qualitative survey on citizens' perception of social media in emergencies in Europe. *Technological Forecasting and Social Change, 121*, 168–180.

Reuter, C., Marx, A., & Pipek, V. (2011, Social software as an infrastructure for crisis management – A case study about current practice and potential usage. In *Proceedings of the 8th International ISCRAM Conference*.

Reuter, C., Heger, O., & Pipek, V. (2013, Combining real and virtual volunteers through social media. In *Proceedings of the 10th International ISCRAM Conference*.

Reuter, C., Kaufhol, M.-A., Spahr, F., Spielhofer, T., & Hahne, A. S. (2020). Emergency service staff and social media – A comparative empirical study of the attitude by emergency services staff in Europe in 2014 and 2017. *International Journal of Disaster Risk Reduction, 46*, 101516.

Rizza, C. (2020, juillet 20). Gestion de crise: Mieux intégrer la réponse des citoyens. *The Conversation*. Retrieved from http://theconversation.com/gestion-de-crise-mieux-integrer-la-reponse-des-citoyens-141741

Rizza, C., & Guimarães Pereira, Â. (2014). Building a resilient community through social network: Ethical considerations about the 2011 Genoa floods. In *Proceedings of the 11th International Conference on Information Systems for Crisis Response and Management*, pp. 294–298.

Rizza, C., Pereira, Â. G., & Curvelo, P. (2014). "Do-it-yourself justice": Considerations of social media use in a crisis situation: The case of the 2011 Vancouver riots. *International Journal of Information Systems for Crisis Response and Management, 6*(4), 42–59. https://doi.org/10.4018/IJISCRAM.2014100104

Rizza, C., Büscher, M., & Watson, H. (Éds.). (2017). *Working with data: Ethical legal and social considerations surrounding the use of crisis data and information sharing during a crisis* (Vol. 1, p. 2).

San, S. Y., Wardell, III, C., & Thorkildsen, Z. (2013). *Social media in the emergency management field 2012 survey results* (IPP-2013-U-004984/final; p. 78). CNA. Retrieved from http://cdrmaguire.com/emkey/Resources/Social/SocialMedia_EmergencyManagement.pdf

Stallings, R. A., & Quarantelli, E. L. (1985). Emergent citizen groups and emergency management. *Public Administration Review, 45*, 93–100.

Starbird, K. (2020, mars 19). *How a crisis researcher makes sense of Covid-19 misinformation*. Medium. Retrieved from https://onezero.medium.com/reflecting-on-the-covid-19-infodemic-as-a-crisis-informatics-researcher-ce0656fa4d0a

Starbird, K., Arif, A., & Wilson, T. (2019). Disinformation as collaborative work: Surfacing the participatory nature of strategic information operations. *Proceedings of the ACM on Human-Computer Interaction, 3*, 1–26. https://doi.org/10.1145/3359229

Statista, I. (2019). Ranking der größten sozialen Netzwerke und Messenger nach der Anzahl der monatlich aktive Nutzer (MAU) im Januar 2019 (in Millionen). *Online im Internet*. Retrieved from https://de-1statista-1com-1001347iz0073.emedia1.bsb-muenchen.de/statistik/daten/studie/181086/umfrage/die-weltweit-groesste n-social-networks-nach-anzahl-der-user/

Steen-Tveit, K., & Erik Munkvold, B. (2021). From common operational picture to common situational understanding: An analysis based on practitioner perspectives. *Safety Science, 142*, 105381. https://doi.org/10.1016/j.ssci.2021.105381

Stieglitz, S., Bunker, D., Mirmabaie, M., & Ehnis, C. (2018). Sense-making in social media during extreme events. *Contingencies and Crisis Management, 26*, 4–15.

Tapia, A., Bajpai, K., Jansen, B. J., & Yen, J. (2011). Seeking the trustworthy Tweet: Can microblogged data fit the information needs of disaster response and humanitarian relief organizations. In *Proceedings of the 8th International ISCRAM Conference*.

Tapia, A., Moore, K., & Johnson, N. (2013). Beyond the trustworthy Tweet: A deeper understanding of microblogged data use by disaster response and humanitarian relief organizations. In *Proceedings of the 10th International ISCRAM Conference*, pp. 770–779.

Vieweg, S., Hughes, A., Starbird, K., & Palen, L. (2010). Microblogging during two natural hazards events: What twitter may contribute to situational awareness. *Proceedings CHI, 2010*, 1079–1088.

Wendling, C., Radish, J., & Jacobzone, S. (2013). *The use of social media in risk and crisis communication (working paper No 24; OECD working papers on public governance)*. OECD Publishing. https://doi.org/10.1787/5k3v01fskp9s-en

Zaglia, C. (2021). *Communication opérationnelle: Communiquer avec les médias traditionnels et les médias sociaux (Carlo Zaglia)*.

Zettl, V., Ludwig, T., Kotthaus, C., & Skudelny, S. (2017). Embedding unaffiliated volunteers in crisis management systems: Deploying and Supporting the concept of intermediary organizations. *Proceedings of the 14th ISCRAM Conference, 2017*, 421–431.

Chapter 23
The Empirical Study of Human Mobility: Potentials and Pitfalls of Using Traditional and Digital Data

Ettore Recchi and Katharina Tittel

Abstract The digitization of human mobility research data and methods can temper some shortcomings of traditional approaches, particularly when more detailed or timelier data is needed to better address policy issues. We critically review the capacity of non-traditional data sources in terms of accessibility, availability, populations covered, geographical scope, representativeness bias and sensitivity, with special regard to policy purposes. We highlight how digital traces about human mobility can assist policy-making in relation to issues such as health or the environment differently to migration policy, where digital data can lead to stereotyped categorizations, unless analysis is carefully tailored to account for people's real needs. In a world where people move for myriad reasons and these reasons may vary quickly without being incorporated in digital traces, we encourage researchers to constantly assess if what is being measured reflects the social phenomenon that the measurement is intended to capture and avoids rendering people visible in ways that are damaging to their rights and freedoms.

23.1 Introduction

Besides the shock of the human lives lost to the disease, when the Covid-19 pandemic broke out, the global public opinion was in awe by the sight of the main non-pharmaceutical intervention that governments put in place almost everywhere – lockdowns. The images of an immobile world created an unexpected dystopic landscape. The prohibition to step out of home emptied cities, highways, stations and airports. Like in a postatomic fantasy, media spread out pictures and videos of the usually most crowded venues of the planet – from Shibuya Station in Tokyo to

E. Recchi (✉) · K. Tittel
Sciences Po, Centre de Recherche sur les Inégalités Sociales (CRIS), CNRS, Paris, France

Migration Policy Centre (MPC), EUI, Florence, Italy Institut Convergences Migrations, Paris, France
e-mail: ettore.recchi@sciencespo.fr; katharina.tittel@sciencespo.fr

© The Author(s) 2023
E. Bertoni et al. (eds.), *Handbook of Computational Social Science for Policy*,
https://doi.org/10.1007/978-3-031-16624-2_23

Heathrow Airport in London – without a living soul in the middle of the day. After decades of reckless increase in the number of airline travellers, for instance, in April 2020 these were no more than 3 per cent their number a year earlier (Recchi et al., 2022). No other nonviolent event could have looked more antinomic – and thus revealing – of the nature of social life in late modernity.

In the pre-Covid-19 era, the number of international trips had been on a steady and uninterrupted rise since at least 1960 – from 69 million to almost 3 billion border crossings per year (Recchi, 2015, 2016; Recchi et al., 2019). Never in history have human beings had such an ease to move out of their usual residences – whether on daily commutes, weekend trips, holiday travel or (although not across the board) long-term migration spells.[1] This is clearly more the case for the rich, whose lifestyle is often patterned after frequent journeys. Nonetheless, the drop in the cost of travel has also facilitated the (shorter-haul and less exotic) mobility of the less privileged, at least in high-income countries (e.g. Demoli and Subtil, 2019). Spatial mobility has thus progressively become a hallmark of our age, as social theorists Zygmunt Bauman and John Urry happened to remark already by the turn of the millennium (Bauman, 1998; Urry, 2000). In the second half of the twentieth century, the absence of large-scale wars, economic growth, progress in transportation and ICT developments – that is, the major keys to globalization – paved the way to a more mobile world. While the aftermath of the Covid-19 pandemic and a possibly higher sensitivity to the climate impact of fuel-propelled mobility may styme this pre-existing trend, human mobility will hardly cease to be part and parcel of what it means to live in the twenty-first century – not the least because global inequality and climate change also spur migration (Barnett & McMichael, 2018; Milanović, 2019; Rigaud et al., 2018).

For a comprehensive take of the topic, we must acknowledge its different mani-festations – first of all, in spatial terms. Human mobility is a multiscale phenomenon, spanning from the micro (local) to the meso (national) and macro (international) level. Repeated national surveys show that all of these levels saw an increase in the last decades of the twentieth century (for instance, in Germany: Zumkeller, 2009). People spend more time in mobility and cover a larger distance per time/unit than they were used to do. The second dimension that thus needs to be acknowledged is the temporal manifestation of mobility. Movements can be temporary, seasonal or longer term/permanent. The third dimension that fundamentally shapes individuals' mobility experiences as well as legal and policy responses to it is the reason for mobility, including voluntary reasons (such as tourism, work, education or family reasons) or forced displacement (as a result of conflict or natural disasters), with or

---

[1] While short-term mobility has skyrocketed in comparative terms, international migration only modestly increased in recent decades: in 2019, 3.5% of the global population qualified as migrants compared to 2.9% in 1990 (UNDESA, 2019). This proportion is obtained applying the UN definition, according to which an international migrant is somebody who has moved into a different country for 12 months or more (see the Glossary on Migration in IOM 2019). Issues of definition are discussed in the draft *Handbook on Measuring International Migration through Population Censuses* (UNDESA, 2017).

without documentation. In practice demarcations are not always clear, since reasons are often mixed and people are often motivated to move by a multiplicity of factors (Mixed Migration Centre, 2020), and intersections between migration and other forms of circular mobility are growing (Skeldon, 2018). Think, for instance, about 'gap years' and 'sabbaticals', not only in academia.

While mobility is a fundamental element of human freedom with real and perceived value for all groups affected by it, the historical intensification of mobility – and its likely spatial-temporal clustering in certain sites (for instance, global cities) or certain periods (for instance, during end-of-year festivities) – is a key concern from a policy perspective. We mention here just two issues that have come to the fore in recent years, partly in conjunction with the Covid-19 crisis – beyond the epidemiological risk that is openly associated with the movement of virus-carrying population.

First is the 'evasiveness of remote workers' through mobility. Since the outbreak of the pandemic, firms increasingly operate partly or completely remote, asking or allowing workers to 'work from home'. Such a shift, which was already present in some industries like IT, may spur white collars' relocation even at long distance – including abroad – as 'digital nomads'. Some countries introduced so-called digital nomad visas (Bloom, 2020; Hughes, 2018) that can be used to track such mobility. Mobility in free movement zones, or if workers relocate short term on tourist visas, often goes unrecorded yet may be of increasing interest for policies related to such mobilities, including tax issues.

Second is the 'environmental and epidemiological risks' linked to human mobility. The bulk of human mobility is fed by fossil fuel burned for road, rail, air and maritime transportation. Almost all (95 per cent) of the world's transportation energy comes from petroleum-based fuels, largely gasoline and diesel. Globally, transportation accounts for no more than 14 per cent of the global greenhouse emissions – less than electricity and heating (25 per cent), agriculture (24 per cent) and industry (21 per cent) (Pachauri et al., 2014). These proportions vary significantly by level of economic development though, and transportation takes the lion's share of greenhouse gas emission in richer countries (e.g. the US: EPA, 2019). Therefore, the impact of mobility on the environment is bound to become more severe as economic development advances, unless major changes in fuel emission take place. In parallel, travel spreads diseases, and increased travel may have made the world more vulnerable to epidemics, although the intensity of long-distance mobility does not necessarily entail a stronger incidence of epidemics (Clemens & Ginn, 2020; Recchi et al., 2022).

The importance of effective policy responses to mobility has been amplified during the Covid-19 pandemic. A major puzzle is whether human mobility will change in size, scope and form in the coming years. At the macroscale, the appetite of human beings for travel does not seem shaken, but travel limitations and travellers' biometric and health controls are likely to be enhanced as 'new normal' ways of restricting (even surreptitiously) access to undesirable travellers (Favell & Recchi, 2020). At the meso and micro level, it is unlikely that the economic and cultural attractiveness of cities as poles of mobility will be disrupted, although the take-off of telework – as complement or substitute to office spaces – may incentivize some sort of 'flight to the suburbs' (Florida et al., 2021). Clearly, the pandemic moment and its aftermath expose the importance of monitoring human mobility with adequate measurement tools to improve response capacities. At the same time, the pandemic experience has also brought more attention to the ethical components of tracking mobility among the general public.

23.2 Monitoring Human Mobility: Traditional and New Data

Tracking human movements in space has always been a challenge for population statistics, but new data open up new opportunities and challenges. Previous studies (European Commission, 2016; Bosco et al., 2022) offer a systematic review of the literature about measuring migration with traditional and new data sources, which we complement with up-to-date information and critical consideration from a policy-related angle. As suggested also by Taylor in this volume (Taylor, 2023), we pay particular attention to what digital data reflects and how it can be used for policy-relevant analysis, foregrounding the policy issue to be solved and what evidence is needed in support, discarding the 'panopticon illusion' (and danger) of making everything visible through mobility data.

23.2.1 *Traditional Data: Pros and Cons*

Traditionally, mobility has been measured with censuses, population registers, administrative sources and household surveys. Data from these sources are cleaned, edited, imputed, aggregated and used to produce official statistics, including the datasets documenting international migration flows and migrant stocks released by agencies of the United Nations.

The major advantages of traditional data sources are that they are transparent, frequently curated and stored in public databases (with varied degrees of accessibility), allowing comparability over time and across countries. However, there are some important limitations. First, they are not reliably available in many parts of the world. Estimates on in- and out-migration flows by country of origin and destination are only reported by 45 countries to the United Nations (UN DESA, 2015). Second, these aggregate statistics have a poor time and spatial resolution. Moreover, with these standard approaches, the category of internally displaced persons is often overlooked despite its policy relevance. In general, inconsistent definitions of a *migrant* make it difficult to compare data across different countries (Sîrbu et al., 2021). Third, and as an extension of the previous point, traditional sources typically do not capture circular, short-term, seasonal or temporary mobility (Hannam et al., 2006). Fourth, surveys that include large enough samples of people with different migratory backgrounds and socioeconomic profiles, particularly the most vulnerable, in different contexts and over time are not at all or not systematically available despite the importance to understand inequalities in terms of education, housing, employment, discrimination, well-being, access to services and protection, etc. Finally, censuses and surveys have data publishing lags of several years.[2] This is particularly problematic in a context in which migratory flows become increasingly complex and dynamic, and in emergency situations, including environmental or health crisis situations.

Both academic and nonacademic actors have tried to improve comparability and availability of traditional data sources and reconcile measurement problems, such as undercount, varying duration of stay criteria and coverage (de Beer et al., 2010; European Commission, 2016; Raymer et al., 2013). To respond to these shortcomings, and with the purpose of informing the humanitarian community and government partners, different international organizations have established data collection and dissemination mechanisms on specific aspects of human mobility, such as UNHCR's refugee statistics, ILO's labour migration statistics, the World bank remittances database or IOM's data on various migrations matters including internal displacement and their 'missing migrants' project. These organizations are increasingly aiming to incorporate more digital, non-traditional data sources as part of their migration data strategies.

23.2.2 Non-traditional Data Usages: An Overview

The mass use of digital devices across the globe has generated large repositories of spatiotemporal 'trace data' (Chi et al., 2020), some of which provide new opportunities for ad hoc measurements and modelling of human mobility. While new technologies are capturing mobility rather than migration data (McAuliffe & Sawyer, 2021), some can also be used to better understand certain aspects of migration. As outlined in Table 23.1, different non-traditional data sources differ significantly in terms of the information available, the populations covered, geographical availability, the data level (individual or grouped), representativeness bias issues, sensitivity and in consequence in terms of who they reflect, the mobility events they capture (micro, meso and macro level), ethical issues and their usefulness to provide information relevant for policy purposes. In the policy sphere, categories are used to define 'groups of people who are assumed to share particular qualities that make it reasonable to subject them to the same outcomes of policy' (Bakewell, 2008: 436). While in relation to issues such as health or the environment, information about mobility events (how many individuals move, where, when and how) provide key information and the characteristics of who moves may be secondary, in the context of migration, analytical or administrative categories, such

---

[2] For example, in the case of the International Migration Database of the OECD, the lag is between 2 and 3 years.

Table 23.1 Characteristics of some traditional and non-traditional data sources for the empirical study of human mobility

	Data source	Information[a]	Populations covered				Accessibility and costs			
			Geographical availability	Data level[b]	Time invariance[c]	Representativeness bias issues[d]	Data costs	Accessibility	Frequency of collection	Release time[e]
Traditional data sources	Censuses	Country of birth, citizenship, year of arrival in the country for foreign-born persons, place of residence (or change of)	Absent or less reliable in low- and middle-income countries	Country level	Yes	Biased toward documented citizens, often failing to cover undocumented migrants	Expensive, time consuming	Usually public	Every 5 or 10 years	Delayed
	Population registers	Inventory of the population living in a particular area	Absent or less reliable in low- and middle-income countries	Local or country level	Yes	Emigrants tend to be under-counted lacking incentives to deregister	Cheap, research data is by-product	Usually made accessible for research purposes	In real time (updated following vital events)	Delayed

	Administrative data	Absent or less reliable in low- and middle-income countries	Local or country level	Yes	Not capturing free or irregular mobility	Cheap, research data is by-product	Usually made accessible for research purposes	In real time (updated following vital events)	Delayed
	Data retrieved from issuance of residence permits, border statistics and health insurance registrations								
	Surveys	Exceptional in low- and middle-income countries	Country level, some coordinated internationally	In theory, but few longitudinal surveys	Inferential problems with small minorities in sample surveys	Expensive, time consuming	Usually public	Sporadically	Delayed
	Targeted information, also retrospective panel-like								
Non-traditional data sources	Geotagged CDR	Global, but use and ownership vary across countries and over time	Country level, some coordinated internationally	No, but population relatively constant	Likely and hard to assess, depending on phone use patterns	Cheap, research data is by-product	Proprietary, requires purchase agreements	In real time	Depends on purchase agreement
	Individual ID, timestamps and geo-coordinates from each activity (e.g. calling)								

(continued)

Table 23.1 (continued)

Data source	Information[a]	Populations covered				Accessibility and costs			
		Geographical availability	Data level[b]	Time invariance[c]	Representativeness bias issues[d]	Data costs	Accessibility	Frequency of collection	Release time[e]
Twitter	Individual ID, timestamps and geo-coordinates from each activity (e.g. posting)	Global, but adoption rate very different by country	Global	No, population changes rather quickly	Likely and hard to assess, biased toward educated, young, politically active, exacerbated by bots	Cheap, research data is by-product	Through API	In real time	Instantly
Facebook ads	Stock estimates and profiles of specific populations also for non-probability sampling	Global, but adoption rate very different by country	Global	No, population changes rather quickly	Biased toward younger and middle-aged individuals	Cheap, research data is by-product	Through API	In real time	Instantly
Satellite data	Images	Global, but can detect mobility only in certain instances	Depends on local context					In real time	Instantly

Google/Apple mobility datasets	Pre-calculated indexes that show intensity by different transport types	Global, but adoption rate very different by country	Global	No, population tends to change rather quickly	Likely and hard to assess as companies do not share the characteristics of users	Cheap, research data is by-product	Limited by company	In real time	Depends on company choices
Flickr, Instagram	Geolocation of photos and text	Mostly in high- and middle-income countries, but adoption rate differs by country	Global	No, population tends to change rather quickly	Biased toward younger, female users	Cheap, research data is by-product	Through API	In real time	Instantly
LinkedIn	Workplace, occupational sector, languages spoken, education	Mostly in high- and middle-income countries, but adoption rate differs by country	Global	No, population tends to change rather quickly	Biased toward highly educated, younger workers	Cheap, research data is by-product	Through API	In real time	Instantly
Flight data	Numbers of passengers and routes	Global	Global	Potentially longitudinal data but unreleased	No bias, captures all travellers	Cheap, research data is by-product	Subject to (purchase) agreements with firms	In real time	Depends on purchase agreement
Ticket reservation data in bus, railway or sea lines	Numbers of passengers and routes	Mostly available in high- and middle-income countries	Local, national and international, but not integrated	Potentially longitudinal data but unreleased	No bias, captures all travellers	Cheap, research data is by-product	Subject to (purchase) agreements with firms	In real time	Depends on purchase agreement

(continued)

Table 23.1 (continued)

	Data source	Mobility events captured			Types of mobility captured		Risk for privacy and confidentiality			
		Micro	Meso	Macro	Information on migrant status[f]	Information on short-term/circular migration[g]	Nature of data[h]	Informed consent	Confidentiality risk[i]	Vulnerable population
Traditional data sources	Censuses	No	Imprecise (can be inferred from residence change)	Imprecise (can be inferred from residence change)	Yes (imperfect, based on, e.g. place of birth)	No	Individual	Yes	Low (informed consent)	No
	Population registers	Yes (if involving change of residence)	Partly (in countries where individuals have to register their residence)	Incoming migration captured; outgoing often is not	Yes (imperfect, based on, e.g. place of birth)	No	Individual	Yes	Low (informed consent)	No
	Administrative data	No	Partly (in countries where individuals have to register their residence)	Partly (e.g. for visas)	Yes	No	Individual	Yes	Low (informed consent)	No
	Surveys	No	Partly	Partly	Yes (depends on the survey)	No	Individual	Yes	Low (informed consent)	No

Non-traditional data sources									
Geotagged CDR	Yes	Yes	Depends on context	No (exception: Data4 Development challenge)	Depends on context	Individual	No	Very high (re-identification and linkage to sensitive places)	Yes
Twitter	Yes (limited)	Yes (limited)	Yes (limited)	Possibly (imperfect, based on language or network properties)	Yes (in theory)	Individual	No	High (unobtrusively collected on the level of the individual)	Depends
Facebook ads	No	No	No	Estimated by Facebook, methodology not transparent	No	Aggregate	No	Low (either just aggregated data or further surveys with informed consent)	Depends
Satellite data	Yes	Yes	No	Yes (possibly to detect displaced populations)	Depends on context	Aggregate	No	Occasionally (i.e. population movements in contested areas)	Yes

(continued)

Table 23.1 (continued)

Data source	Mobility events captured			Types of mobility captured		Risk for privacy and confidentiality			
	Micro	Meso	Macro	Information on migrant status[f]	Information on short-term/circular migration[g]	Nature of data[h]	Informed consent	Confidentiality risk[i]	Vulnerable population
Google/Apple mobility datasets	Yes (limited)	No	No	No	No	Aggregate	No	Low (unobtrusively collected but only shared in aggregated form)	No
Flickr, Instagram	Yes (limited)	Limited	Limited	Possibly (imperfect, based on language or network properties)	Yes (in theory)	Individual	No	Medium (unobtrusively collected at individual level but information relatively less sensitive)	Depends

LinkedIn	No	Yes (limited)	Yes (limited)	Possibly (imperfect, based on language or education)	No	Individual	No	Medium (unobtrusively collected at individual level but information relatively less sensitive)	No
Flight data	Depends on type of transport	In case of internal flights	Yes	No	No	Aggregate	No	Low (unobtrusively collected but low risk of identification)	No
Ticket reservation data in bus, railway or sea lines	Depends on type of transport	Depends on type of transport	Depends on type of transport	No	No	Aggregate	No	Low (unobtrusively collected but low risk of identification)	No

[a] Which information is usually used to detect mobility events or migrant identity?
[b] At which level is data usually collected?
[c] Is the population relatively constant over time?
[d] Is there a systematic bias in the sample?
[e] When is the data usually made available?
[f] Can data be used to infer migrant status?
[g] Can short-term or circular migration be measured?
[h] For research purposes, is data usually shared at the level of individuals or as aggregate?
[i] Is there a risk of re-identification of the individual?

as 'migrant', 'foreign worker', 'internally displaced person' or 'refugee', fundamentally shape the interactions between individuals and bureaucratic organizations.
As Taylor stressed in this volume (Taylor, 2023), that connection is often obscured
when computational methods and new data sources are used.

23.2.2.1   A Review of the Usefulness of Non-traditional Data to Study Different Types of Mobility

Local and National Mobility

At micro and meso level, geotagged digital trace data from call detail records (CDR)
(e.g. Song et al., 2010), GPS technology (e.g. Bachir et al., 2019; Cui et al., 2018;
Huang et al., 2018) or social media data (e.g. Bao et al., 2016) can be used to
study individual (Giannotti et al., 2011; González et al., 2008; Pappalardo et al.,
2015; Wang et al., 2011) as well as group mobility (Hiir et al., 2019; Lulli et al.,
2017; Tosi, 2017). Because of their wide coverage[3] and ad hoc availability, these
data allow studying population movements in emergency situations, such as during
natural disasters (Bengtsson et al., 2011) or events like the Covid-19 pandemic (e.g.
Xiong et al., 2020). In other contexts, satellite data have been used to estimate
the effect of extreme climate events, such as flooding, on migration (Chen et al.,
2017). Compared to self-reporting on causes of migration in surveys, they offer the
advantage of not being affected by subjective factors such as recall bias.

While individual characteristics, such as gender or age or motivations for mobility, that are key variables to consider for policy responses, are usually unavailable,
researchers started to collect or link survey data with geotagged digital trace data
to alleviate this limitation and get more information on demographic characteristics
of the populations covered (Blumenstock & Fratamico, 2013). Other sources used
for mobility research include Twitter (e.g. Fiorio et al., 2017; Zagheni et al., 2014),
Skype (e.g. Kikas et al., 2015), LinkedIn (e.g. Li et al., 2019) or Flickr (Bojic et al.,
2016) and could include any other platform that provides geotagged data of their
users. Their usefulness for policy purposes fundamentally depends on how well
represented the population of interest is on the specific platform.

Large platform companies like Apple or Google also possess vast repositories
of human movement data that could be used to understand local mobility patterns.
While these companies do not normally publish their data for research purposes,
they offered ad hoc data products and visualizations of aggregated mobility of
customers, including the use of travel modes (public transport, driving, walking),
during the Covid-19 pandemic (Apple, 2021; Google, 2021). Notably, however,
omitted information on methods and on the underlying population that is captured

---

[3] As of 2018, mobile phone penetration is around 100 per cent in high- and middle-income and 55
per cent in low-income countries. This has raised from 12 per cent of the world population in 2000
(Worldbank, 2019).

leads to a lack of clarity of these data and their biases, limiting their usefulness for policy purposes.

International Mobility

Social media advertising platforms can help estimate stocks and sociodemographic profiles of certain populations and facilitate non-probability sampling, since the platforms support showing ads exclusively to certain audiences. This information has also been used to target specific populations in order to invite them through paid Facebook advertisement to participate in a survey, such as Polish migrants in European countries (Pötzschke & Braun, 2017). Compared to traditional surveys, this approach offers the advantage of targeting demographic characteristics to reach a larger sample size at a global scale quickly and at lower cost (Rampazzo et al., 2021).

Böhme et al. (2020) used georeferenced online search data from Google Trends (looking for the combination of migration- and target country-related keywords as a proxy for migration intentions) in origin countries to improve the predictive power of international migration models. While there are promising examples in different areas employing Google Trends data, such as to forecast private consumption (Vosen & Schmidt, 2011), the precision and goodness of fit of such models can also rapidly change (Lazer et al., 2014).

A Special Case: Airline Mobility

A major source of big data on travel are airline reservation systems (ARS). A handful of private companies dominate this market. They handle such information omitting not only personal information but also categorical groupings about sociodemographic characteristics of passengers. One of these companies, Sabre, sells an air travel dataset that reports monthly data on the numbers of air travellers between all world airports and regular airline routes. Capitalizing on this source, in combination with the more traditional statistical reports of the United Nations World Tourism Organization, researchers have created a Global Transnational Mobility Dataset which details cross-border trips between all sovereign states worldwide from 2011 to 2016 (Recchi et al., 2019). Other studies have used Sabre data to infer types of transnational mobility (Gabrielli et al., 2019) the economic impact of reduced mobility due to Covid-19 (Iacus et al., 2020), and the global spread of the pandemic in 2020 (Recchi et al., 2022).

Potentially, similar data could be collected for other ticket reservation systems in bus, railway or sea lines, but such forms of transportations tend to be highly national or regional, rather than global, and thus there is possibly an issue of integration of different sources. At any rate, this is an evolving area of data collection that has proven fruitful for macro analyses of international flows. Its major limit is that a travel is an event, not a person, thus leaving uncharted the characteristics of human populations that experience cross-border travel, which survey research describes as mostly – albeit not exclusively – drawn from among the middle-upper classes (Demoli & Subtil, 2019).

Along these lines, Chareyron et al. (2021) used data scraped from the digital platform Tripadvisor to examine privileged mobility patterns. Other platforms for evaluating tourism consumption (accommodation, places, activities) that might be leveraged for research on this issue in certain contexts include Booking, Airbnb, Hotels.com or Weibo.

Difficulties to Infer Policy-Relevant Categories from Digital Trace Data

While digital devices trace the geolocalizations of their users, there are no standards or commonly respected methodological frameworks for how to produce estimates of policy-relevant information from granular geo-located data points (Bell et al., 2015), and the analysis of such data by data scientists without context-specific knowledge and understanding of the social phenomena underlying human mobility creates new risks (McAuliffe & Sawyer, 2021). Unlike survey data about respondents' residential history, georeferenced digital trace data only record locations at a specific moment in time. Blondel et al. (2015) and Chi et al. (2020) introduce different estimation techniques to infer patterns of human mobility from observational geo-tagged data. Without further context-specific information, it is not straightforward to determine what the location of a given individual corresponds to (Fiorio et al., 2021). For this reason, how researchers choose to define features of trips for the ambiguous distinction between migration and other kinds of movements, and how they group geo-located data points together based on their temporality, greatly affects the consistency of human mobility estimates generated from digital trace data (Ahas et al., 2018; Fiorio et al., 2021). This points to the challenge of how to discern policy-relevant categories from inferred mobility patterns. A risk that is linked to this labelling process is called *delinkage*, which refers to the replacement of an individual identity by a 'stereotyped identity with a categorical prescription of assumed needs' (Zetter, 1991: 44).

In theory, the possibilities opened by the new data sources suggest revisiting some presuppositions of such labelling and categorization processes and question the labels researchers apply to people and the functions those categories fulfil to design policies that effectively cater for real and not stereotyped human needs (see Turton, 2005; Bakewell, 2008). In practice, however, there are several challenges to correctly understand and interpret the underlying meaning of data variables in different contexts. Certain data, such as those scraped from LinkedIn, may offer relatively straightforward ways to identify a 'foreign worker' on the platform, whereas such classification is more difficult and more sensitive to contextual changes when using CDR data. Ahas et al. (2018), for example, in their roaming dataset operationalize a 'foreign worker' as someone who did 1 to 52 trips to a certain country in a certain time period. Such an identification strategy would not have worked, however, during the Covid-19 pandemic, when remote work was widespread. Geolocated messages or posts are often the key variable in estimating the geo-coordinates of users, while other studies use the language used on social networks; friend or follower networks; profile pictures; names; or other textual information available (e.g. Huang et al., 2014; Kim et al., 2020) to infer users' sociodemographic characteristics. In the case of Facebook marketing data, researchers have to rely on the categories provided by the platform, even though, as Zagheni et al. (2017) highlight, categories are not documented according to scientific research standards. This may introduce biases that are hard to disentangle from biases related to selection and non-representativeness, or other inconsistencies. 'Naming' mobile individuals is often based on legal definitions and should be carefully considered, particularly in a context where inaccurate estimates can cause confusion and be fuel for heavily contested public and political discourses (McAuliffe & Sawyer, 2021).

Without relevant content knowledge of migration and technology use, errors or wrong assumptions can lead to misspecification and misinterpretation (McAuliffe & Sawyer, 2021), exemplified by Pew Research's 2019 estimates of irregular migrants in Europe (Connor & Passel, 2019). The authors wrongly included asylum seekers whose applications were being processed in the category of irregular migrants, leading to inaccurate and inflated estimates. Drawing on examples like this one, McAuliffe and Sawyer (2021) highlight that, in reality, the application of so-called new data science in the study of migration often fails to take into account the most basic understanding of the topic.

23.2.2.2 Limitations and Caveats in the Use of Non-traditional Data on Human Mobility

Despite the opportunities offered by non-traditional data sources, their use comes with important limitations and caveats that add to the potentials we outlined so far. In this final section, we list and discuss four of them. Importantly, the different non-traditional data sources differ significantly with regard to their properties and hence related concerns.

Proprietariness

A first, and rather mundane, problem with some of the above-mentioned data is difficult access. Some data sources require appropriate technical skills (e.g. Facebook and LinkedIn marketing API); some data can be purchased; some sources lack formalized purchasing mechanisms (e.g. mobile phone providers), and others do not share their data at all. Moreover, by employing terms of service (TOS)-compliant methods, a researcher may respect the business prerogatives of the company that created the platform studied, but this may or may not respect the dignity and privacy of the platform users (Freelon, 2018). This is particularly sensitive in a context of radical power asymmetries with the platform/service providers, as users often have far less understanding of who can access their data and under which circumstances, as well as of the functioning of the tools they use online (Broeders & Dijstelbloem, 2015; Taylor, 2023).

Non-representativeness

Second, another key and too often overlooked issue with digital trace data – like in many other social science data – is selection bias: users of a particular social media platform or mobile phone provider are not representative of the underlying general population. In the analysis of CDR, selection bias regarding mobile phone ownership and usage must be considered when extrapolating from the number of moving SIM cards to the number of moving persons (Blumenstock, 2012; Blumenstock & Fratamico, 2013). For instance, in some sub-Saharan African countries, men are more likely to be mobile phone owners, while phone sharing is common among rural women, and there is considerable cross-country variation: while mobile phone records in Kenya are an excellent proxy for mobility, regardless of socioeconomic factors, mobile phone data in Rwanda are a good proxy only for the mobility of wealthy and educated men (Luca et al., 2021). While existing studies showed that approaches using CDR data work well in one-off emergencies, such as the earthquake in Haiti (Bengtsson et al., 2011) and other disaster events (Chen et al., 2017), for estimating general population displacement, ad hoc knowledge is needed about who is using phones or services. Otherwise, such approaches cannot identify vulnerabilities of specific populations, a key aspect of targeting social protection and relief (Lu et al., 2016). Similarly, Facebook and Twitter adoption rates differ between countries and depending on user characteristics, such as age or gender (Zagheni et al., 2017). By relying on data from highly specialized online services, users' self-selection into these services hence limits the generalization of these results (Böhme et al., 2020). For instance, LinkedIn may be useful to study the labour mobility of highly educated individuals in rich countries and allows researchers to link this to career choices and industry-specific patterns. However, it cannot yield mobility estimates for the global population. This is problematic because, as Sîrbu et al. (2021) highlight, being unable to track specific groups of users can steer migration policies in directions that unwillingly perpetuate discriminations or neglect the needs of invisible groups.

Different statistical approaches help to correct for selection bias. Zagheni and Weber (2015) propose a method that relies on calibration of the digital trace data against reliable official statistics. When the data also contains demographic information about users of a given platform, that information can be leveraged to de-bias non-representative results by adjusting the responses via multilevel regression prediction models and post-stratification (Wang et al., 2015). Importantly, statistical calibration models require datasets containing enough variables for the use of post-stratification techniques, as well as knowledge about specific functional relationship between estimates of migration and how this relationship varies by geography and population characteristics as well as how it changes over time – information that is often not available, and that requires systematic and hence costly on-the-ground research. Since the composition of the user bases of new data sources may change rapidly, predicting over time variation is usually more difficult than understanding cross-spatial variation in human mobility.

Beyond challenges that are common to any survey, such as selection bias and nonresponse, some pitfalls are specific to non-probability sampling on social media. Not only does non-probability inclusion lead to non-representative data, but the sampling error is further enhanced by the self-selection of users, which may be affected by issues of trust and incentives, and by the platform's algorithm.

To alleviate some of these shortcomings and generate more reliable and comprehensive estimates, it is key to borrow from a number of different data sources and develop methods to analyse them that are robust to the lack of a specific data source (European Commission, 2016). For example, Huang et al. (2021) used Twitter, Google, Apple and Descartes Labs data to disentangle the disparities in mobility dynamics from lower- and upper-income US counties during Covid-19. They found that mobility from each source presented unique and even contrasting characteristics. Their (optimistic) conclusion is that hierarchical Bayesian methods can be used effectively to combine different mobility data in a consistent way. However, this requires the availability of different mobility datasets as proxies for the same phenomenon, which at the global level – given that the data showed contrasting characteristics even for the USA – seems extremely hard to achieve.

No Gold Standard

Third, it must be acknowledged that a proper gold standard does not exist since precise current and past mobility patterns are unknown. Therefore, validation of nowcasting models of human mobility is not straightforward. While traditional data sources have a number of limitations and caveats, without a benchmark it is difficult to trust new data sources and innovative approaches and assess their validity (European Commission, 2016). Therefore, a combination of traditional and new data might yield more accurate estimates and predictions than solely relying on non-representative sources (Lazer et al., 2014; Zagheni et al., 2017).

Ethical Concerns

Finally, we deem appropriate to underscore some ethical caveats and raise frequently ignored data justice-related questions, which Taylor discusses in this issue (Taylor, 2023). While statistical techniques may alleviate the shortcomings of new data sources, the use of some of this information, notably individual-level data (CDR, social media data), raises severe ethical issues. Anonymization – i.e. removing personal identifiers – is a commonly used method to protect users' privacy, but it is not sufficient to shield privacy nor address issues related to informed consent, since in large mobility datasets, individuals can be reidentified with as little as four spatial-temporal data points, even if they do not contain identifiable information like names or email addresses (de Montjoye et al., 2013). Having a precise, always-on tracking of individuals, with a spatiotemporal history of their trajectories, and drawing a picture of how people use city space or move across borders and how they break rules and create informal ways to support themselves are a sensitive matter in any context, especially when data are unobtrusively collected without informed consent. Risks are aggravated in an environment where geolocations might be mapped to addresses such as religious places, abortion clinics and other sensitive areas. In the context of migration, where many individuals are vulnerable, and political freedoms cannot be taken for granted, these concerns are particularly important. It is key to consider what it means if mobile populations become more legible and, thereby, more amenable to control from above (Scott, 2008). Individual invisibility may sometimes be life-saving or, at least, grant a basic right to personal freedom. As Polzer and Hammond (2008) insist, 'researchers who lift this veil [of invisibility] in the name of illuminating 'creative livelihood strategies' or 'flexible identities' may inadvertently be alerting powerful states, the UN or NGOs to the ways in which their rules are circumvented, and thereby reduce the space for life-saving creativity and flexibility in remaining invisible'. While visibility to institutions that are seen as potential allies might increase access to resources, defence of rights and legitimacy and can hence be seen as an ethical imperative, invisibility may serve as a protective shield in the absence of true legal, political and social protection, and in contexts of xenophobic and majoritarian violence. As governments and public agencies are increasingly using digital technologies for a more efficient, neutral and disembodied migration management and border control (Latonero & Kift, 2018; Trimikliniotis et al., 2015), Leurs and Smets (2018) remind us that it is important to ponder how approaches, methodologies, tools and findings may be coopted or used in unintended and undesirable ways. Actors interested in better understanding human mobility may include organizations like the United Nations and aid agencies, but also private sector subcontractors, as well as actors in the 'migration industry of connectivity services' (Gordano Peile, 2014), such as money transfer services, mobile phone companies targeting refugees or even illegal organizations exploiting irregular migration. In a context where Western states direct much attention and investment to monitoring and combatting irregular migration in some geographical areas (Andersson, 2016; Słomczyńska & Frankowski, 2016; Triandafyllidou & McAuliffe, 2018), journalistic coverage (BBC, 2021) of the terrifying final hours of a fatal attempt to cross the English channel exemplifies not only the centrality of technology use in life-saving efforts but also the risks digital traces pose to individuals, reflected in them tossing their phone into the waves to protect people traffickers' identities or to hide details that may prevent their asylum claims being accepted.

The key here is to try to minimize people's vulnerability in the face of unequal power relations. This may entail very different decisions when trying to better understand labour mobility of highly educated intra-European migrants, or analysing intra-city mobility patterns by vehicle type, rather than when dealing with marginalized and vulnerable groups. Since we know about certain minority populations' reluctance to participate in routine demographic exercises after experiences of marginalization and stigmatization (Weitzberg, 2015), any choice to make populations visible without informed consent should be carefully considered.

23.3 Concluding Remarks

Based on the above-described potentials and pitfalls of different sources, we recommend that policy-makers use digital data to temper the shortcomings of traditional mobility data – namely, their poor space-time resolution, the limited availability of data disaggregated by sociodemographic characteristics, their delayed availability – when more detailed data is needed to better address policy issues related to inequalities, such as regarding housing, education, health, employment and non-discrimination. Ultimately, the usefulness of digital data and arising methodological challenges depend on research goals. For example, longitudinal mobility estimations using digital data are rendered difficult by changing user bases. No single non-traditional data source captures all types of mobility, but the different sources discussed here capture related but partly different phenomena, including urban and international transport, tourism, population displacements, labour mobility of the highly educated and large-scale mobility data from different data providers, where usability depends on coverage and accessibility. Because of their higher granularity, digital data can monitor and evaluate human mobility and population presence at a higher scale, resolution and detail – in real time – spanning from the micro (local) to the meso (national) and macro (international) level. This is particularly important for policy responses in emergency situations (such as humanitarian or public health emergencies). Here, estimations based on digital data sources can help make faster and more informed decisions. However, in other contexts, it should be carefully considered who benefits if individuals on the move and their practices are made visible.

For models to be correctly specified and for estimations to be reliable, in-depth context-specific knowledge about the ways human mobility occurs on the ground as well as knowledge about quickly changing technology use among the populations of interest is fundamental, although difficult and costly to obtain, and has to be constantly updated. Traditional statistics remain important to evaluate and complement estimations as baselines, especially in a context in which migration is the focus of significant political and media attention and is all too frequently misunderstood or misinterpreted. Moreover, as regards migration, it is crucial that legal policy definitions and normative frameworks are respected. The bottom line is that any model of human movement should be carefully tailored to the specific local context. In an unpredictable world, where people move or not move for myriad reasons and these reasons may vary quickly, as the case of Covid-19 exemplifies, we encourage researchers to constantly reassess if what is being measured reflects the social phenomenon that the measurement is intended to assess and to ensure that their analysis does not generate injustice by rendering people visible in ways that are damaging to their rights and freedoms. This makes the data collection and analysis process more expensive and less universal than sometimes suggested in relation to new data sources and their usefulness for policy-relevant analysis.

Beyond the realm of migration, digital data on human mobility can assist evidence-based policies on transportation – a primary concern in the field of environmental policies. A well-informed understanding of human mobility and its forms is particularly urgent in a context in which a reduction of fossil fuel-propelled transports in rich countries (flights, cruises and car use) is needed to mitigate global warming (Holden et al., 2019; Peeters & Dubois, 2010). A primary instance is the design of incentives to shift passengers to less polluting travel means – e.g. from airplanes to trains.

Whatever the domain of interest, researchers must be aware that precise, always-on data about individuals, often unobtrusively collected without informed consent, raise several issues concerning privacy and security. Associated risks largely depend on the legal, political and social situations of the individuals or groups eventually covered by it, the actors handling this data and their interests. Operational guidelines on data responsibility are provided, for example, by the Inter-Agency Standing Committee (IASC) of the United Nations system.[4] Although far from frontline operations, migration research analysing big data can have a quick impact on policies (for instance, border management) and, thus, on human lives – something that traditional studies of migrants rarely had (McAuliffe & Sawyer, 2021). Just because research and policy-making on human mobility have an unprecedented potential to go hand in hand – a good news in itself – we can only urge any actor collecting, using, storing and sharing human mobility data to commit to 'do no harm while maximizing the benefits' principles (IASC, 2021), always prioritizing the safe, ethical and effective management of personal and nonpersonal data.

References

Ahas, R., Silm, S., & Tiru, M. (2018). *Measuring transnational migration with roaming datasets [application/pdf]*. https://doi.org/10.3929/ETHZ-B-000225599

Andersson, R. (2016). Europe's failed 'fight' against irregular migration: Ethnographic notes on a counterproductive industry. *Journal of Ethnic and Migration Studies, 42*(7), 1055–1075. https://doi.org/10.1080/1369183X.2016.1139446

Apple. (2021). COVID-19 – Mobility trends reports. *Apple*. Retrieved from https://www.apple.com/covid19/mobility

Bachir, D., Khodabandelou, G., Gauthier, V., El Yacoubi, M., & Puchinger, J. (2019). Inferring dynamic origin-destination flows by transport mode using mobile phone data. *Transportation Research Part C: Emerging Technologies, 101*, 254–275. https://doi.org/10.1016/j.trc.2019.02.013

[4] They highlight the importance of accountability, confidentiality, coordination and collaboration, data security, necessity and proportionality, fairness and legitimacy, personal data protection, quality, retention and destruction and transparency, within a human rights-based, people-centered and inclusive approach to handling data.

Bakewell, O. (2008). Research beyond the categories: The importance of policy irrelevant research into forced migration. *Journal of Refugee Studies, 21*(4), 432–453. https://doi.org/10.1093/jrs/fen042

Bao, J., Lian, D., Zhang, F., & Yuan, N. J. (2016). Geo-social media data analytic for user modeling and location-based services. *SIGSPATIAL Special, 7*(3), 11–18. https://doi.org/10.1145/2876480.2876484

Barnett, J., & McMichael, C. (2018). The effects of climate change on the geography and timing of human mobility. *Population and Environment, 39*(4), 339–356. https://doi.org/10.1007/s11111-018-0295-5

Bauman, Z. (1998). *Globalization: The human consequences (Repr.)*. Columbia University Press.

BBC. (2021, December 21). *Channel migrants tragedy: Terrifying final hours of their fatal journey - BBC News*. News. Retrieved from https://www.bbc.co.uk/news/resources/idt-b7bd2274-88b1-4ef9-a459-be22e180b52c

Bell, M., Charles-Edwards, E., Ueffing, P., Stillwell, J., Kupiszewski, M., & Kupiszewska, D. (2015). Internal migration and development: Comparing migration intensities around the world. *Population and Development Review, 41*(1), 33–58. https://doi.org/10.1111/j.1728-4457.2015.00025.x

Bengtsson, L., Lu, X., Thorson, A., Garfield, R., & von Schreeb, J. (2011). Improved response to disasters and outbreaks by tracking population movements with mobile phone network data: A post-earthquake geospatial study in Haiti. *PLoS Medicine, 8*(8), e1001083. https://doi.org/10.1371/journal.pmed.1001083

Blondel, V. D., Decuyper, A., & Krings, G. (2015). A survey of results on mobile phone datasets analysis. *EPJ Data Science, 4*(1), 1–55. https://doi.org/10.1140/epjds/s13688-015-0046-0

Bloom, L. B. (2020, July 30). Want to live and work in paradise? 7 countries inviting Americans to move abroad. *Forbes*. Retrieved from https://www.forbes.com/sites/laurabegleybloom/2020/07/30/live-work-remote-move-abroad-coronavirus/

Blumenstock, J. E. (2012). Inferring patterns of internal migration from mobile phone call records: Evidence from Rwanda. *Information Technology for Development, 18*(2), 107–125. https://doi.org/10.1080/02681102.2011.643209

Blumenstock, J., & Fratamico, L. (2013). Social and spatial ethnic segregation: A framework for analyzing segregation with large-scale spatial network data. In *Proceedings of the 4th Annual Symposium on Computing for Development - ACM DEV-4'13*, pp. 1–10. https://doi.org/10.1145/2537052.2537061

Böhme, M. H., Gröger, A., & Stöhr, T. (2020). Searching for a better life: Predicting international migration with online search keywords. *Journal of Development Economics, 142*, 102347. https://doi.org/10.1016/j.jdeveco.2019.04.002

Bojic, I., Sobolevsky, S., Nizetic-Kosovic, I., Podobnik, V., Belyi, A., & Ratti, C. (2016). Sublinear scaling of country attractiveness observed from Flickr dataset. *ArXiv*. Retrieved from https://dspace.mit.edu/handle/1721.1/109842

Bosco, C., Grubanov-Boskovic, S., Iacus, S., Minora, U., Sermi, F., & Spyratos, S. (2022). *Data Innovation in Demography, Migration and Human Mobility*. EUR 30907 EN, Publications Office of the European Union: Luxembourg.

Broeders, D., & Dijstelbloem, H. (2015). The Datafication of mobility and migration management: The mediating state and its consequences. In *Digitizing identities*. Routledge.

Chareyron, G., Cousin, S., & Jacquot, S. (2021, July 13). L'Europe rythmée par ses visiteurs. 20 ans de commentaires géolocalisés et chronoréférencés. *«Migrer sans entraves», De facto [En ligne], 27*. Retrieved from https://www.icmigrations.cnrs.fr/2021/06/16/defacto-027-06/

Chen, J. J., Mueller, V., Jia, Y., & Tseng, S. K.-H. (2017). Validating migration responses to flooding using satellite and vital registration data. *American Economic Review, 107*(5), 441–445. https://doi.org/10.1257/aer.p20171052

Chi, G., Lin, F., Chi, G., & Blumenstock, J. (2020). A general approach to detecting migration events in digital trace data. *PLoS One, 15*(10), e0239408. https://doi.org/10.1371/journal.pone.0239408

Clemens, M. A., & Ginn, T. (2020). Global mobility and the threat of pandemics: Evidence from three centuries. In *Institute of Labor Economics (IZA), IZA Discussion Papers 13947*.

Connor, P., & Passel, J. S. (2019). *Europe's unauthorized immigrant population peaks in 2016, then levels off* (p. 53). Pew Research. Retrieved from https://www.pewresearch.org/global/wp-content/uploads/sites/2/2019/11/PG_2019.11.13_EU-Unauthorized_FINAL.pdf

Cui, Y., Meng, C., He, Q., & Gao, J. (2018). Forecasting current and next trip purpose with social media data and Google places. *Transportation Research Part C: Emerging Technologies, 97*, 159–174. https://doi.org/10.1016/j.trc.2018.10.017

de Beer, J., Raymer, J., van der Erf, R., & van Wissen, L. (2010). Overcoming the problems of inconsistent international migration data: A new method applied to flows in Europe. *European Journal of Population / Revue Européenne de Démographie, 26*(4), 459–481. https://doi.org/10.1007/s10680-010-9220-z

de Montjoye, Y.-A., Hidalgo, C. A., Verleysen, M., & Blondel, V. D. (2013). Unique in the crowd: The privacy bounds of human mobility. *Scientific Reports, 3*(1), 1376. https://doi.org/10.1038/srep01376

Demoli, Y., & Subtil, J. (2019). Boarding Classes. Mesurer la démocratisation du transport aérien en France (1974-2008). *Sociologie, 10*, 2. Retrieved from https://journals.openedition.org/sociologie/5295

EPA. (2019). *Fast facts on transportation greenhouse gas emissions | US EPA*. Retrieved from https://www.epa.gov/greenvehicles/fast-facts-transportation-greenhouse-gas-emissions

European Commission. (2016). *Inferring migrations, traditional methods and new approaches based on mobile phone, social media, and other big data: Feasibility study on inferring (labour) mobility and migration in the European Union from big data and social media data*. Publications Office. Retrieved from https://data.europa.eu/doi/10.2767/61617

Favell, A., & Recchi, E. (2020). Mobilities, neo-nationalism and the lockdown of Europe: Will the European Union survive? In *COMPAS*. Retrieved from https://www.compas.ox.ac.uk/2020/mobilities-and-the-lockdown-of-europe-will-the-european-union-survive/

Fiorio, L., Abel, G., Cai, J., Zagheni, E., Weber, I., & Vinué, G. (2017). Using twitter data to estimate the relationship between short-term mobility and long-term migration. In *Proceedings of the 2017 ACM on Web Science Conference*, pp. 103–110. https://doi.org/10.1145/3091478.3091496

Fiorio, L., Zagheni, E., Abel, G., Hill, J., Pestre, G., Letouzé, E., & Cai, J. (2021). Analyzing the effect of time in migration measurement using georeferenced digital trace data. *Demography, 58*(1), 51–74. https://doi.org/10.1215/00703370-8917630

Florida, R., Rodríguez-Pose, A., & Storper, M. (2021). Cities in a post-COVID world. *Urban Studies, 004209802110180*, 004209802110180. https://doi.org/10.1177/00420980211018072

Freelon, D. (2018). Computational research in the post-API age. *Political Communication, 35*(4), 665–668. https://doi.org/10.1080/10584609.2018.1477506

Gabrielli, L., Deutschmann, E., Natale, F., Recchi, E., & Vespe, M. (2019). Dissecting global air traffic data to discern different types and trends of transnational human mobility. *EPJ Data Science, 8*(1), 26. https://doi.org/10.1140/epjds/s13688-019-0204-x

Giannotti, F., Nanni, M., Pedreschi, D., Pinelli, F., Renso, C., Rinzivillo, S., & Trasarti, R. (2011). Unveiling the complexity of human mobility by querying and mining massive trajectory data. *The VLDB Journal, 20*(5), 695. https://doi.org/10.1007/s00778-011-0244-8

González, M. C., Hidalgo, C. A., & Barabási, A.-L. (2008). Understanding individual human mobility patterns. *Nature, 453*(7196), 779–782. https://doi.org/10.1038/nature06958

Google. (2021). *COVID-19 community mobility reports*. Retrieved October 03, 2021, from https://www.google.com/covid19/mobility/

Gordano Peile, C. (2014). The migration industry of connectivity services: A critical discourse approach to the Spanish case in a European perspective. *Crossings: Journal of Migration & Culture, 5*(1), 57–71. https://doi.org/10.1386/cjmc.5.1.57_1

Hannam, K., Sheller, M., & Urry, J. (2006). Editorial: Mobilities, immobilities and moorings. *Mobilities, 1*(1), 1–22. https://doi.org/10.1080/17450100500489189

Hiir, H., Sharma, R., Aasa, A., & Saluveer, E. (2019). Impact of natural and social events on Mobile call data records – An Estonian case study. *Complex Networks and Their Applications VIII Studies in Computational Intelligence, 882*, 415–426. https://doi.org/10.1007/978-3-030-36683-4_34

Holden, E., Gilpin, G., & Banister, D. (2019). Sustainable mobility at thirty. *Sustainability, 11*(7), 1965. https://doi.org/10.3390/su11071965

Huang, W., Weber, I., & Vieweg, S. (2014). Inferring nationalities of twitter users and studying inter-national linking. In *Proceedings of the 25th ACM Conference on Hypertext and Social Media*, pp. 237–242. https://doi.org/10.1145/2631775.2631825

Huang, Z., Ling, X., Wang, P., Zhang, F., Mao, Y., Lin, T., & Wang, F.-Y. (2018). Modeling real-time human mobility based on mobile phone and transportation data fusion. *Transportation Research Part C: Emerging Technologies, 96*, 251–269. https://doi.org/10.1016/j.trc.2018.09.016

Huang, X., Li, Z., Jiang, Y., Ye, X., Deng, C., Zhang, J., & Li, X. (2021). The characteristics of multi-source mobility datasets and how they reveal the luxury nature of social distancing in the U.S. during the COVID-19 pandemic. *International Journal of Digital Earth, 14*(4), 424–442. https://doi.org/10.1080/17538947.2021.1886358

Hughes, N. (2018). 'Tourists go home': Anti-tourism industry protest in Barcelona. *Social Movement Studies, 17*(4), 471–477. https://doi.org/10.1080/14742837.2018.1468244

Iacus, S. M., Natale, F., Santamaria, C., Spyratos, S., & Vespe, M. (2020). Estimating and projecting air passenger traffic during the COVID-19 coronavirus outbreak and its socio-economic impact. *Safety Science, 129*, 104791. https://doi.org/10.1016/j.ssci.2020.104791

IASC (Inter-Agency Standing Committee). (2021). Operational guidance on data responsibility in humanitarian action. https://interagencystandingcommittee.org/system/files/2021-02/IASC%20Operational%20Guidance%20on%20Data%20Responsibility%20in%20Humanitarian%20Action-%20February%202021.pdf

Kikas, R., Dumas, M., & Saabas, A. (2015). Explaining international migration in the Skype network: The role of social network features. In *Proceedings of the 1st ACM Workshop on Social Media World Sensors*, pp. 17–22. https://doi.org/10.1145/2806655.2806658

Kim, J., Sîrbu, A., Giannotti, F., & Gabrielli, L. (2020). Digital footprints of international migration on Twitter. In M. R. Berthold, A. Feelders, & G. Krempl (Eds.), *Advances in intelligent data analysis XVIII* (pp. 274–286). Springer. https://doi.org/10.1007/978-3-030-44584-3_22

Latonero, M., & Kift, P. (2018). On digital passages and Borders: Refugees and the new infrastructure for movement and control. *Social Media + Society, 4*(1), 205630511876443. https://doi.org/10.1177/2056305118764432

Lazer, D., Kennedy, R., King, G., & Vespignani, A. (2014). The parable of Google flu: Traps in big data analysis. *Science, 343*(6176), 1203–1205.

Leurs, K., & Smets, K. (2018). Five questions for digital migration studies: Learning from digital connectivity and forced migration in(to) Europe. *Social Media + Society, 4*(1), 2056305118764425. https://doi.org/10.1177/2056305118764425

Li, L., Yang, J., Jing, H., He, Q., Tong, H., & Chen, B. C. (2019). NEMO: Next career move prediction with contextual embedding. In *26th International World Wide Web Conference 2017, WWW 2017 Companion*, pp. 505–513. https://doi.org/10.1145/3041021.3054200

Lu, X., Wrathall, D. J., Sundsøy, P. R., Nadiruzzaman, M., Wetter, E., Iqbal, A., Qureshi, T., Tatem, A., Canright, G., Engø-Monsen, K., & Bengtsson, L. (2016). Unveiling hidden migration and mobility patterns in climate stressed regions: A longitudinal study of six million anonymous mobile phone users in Bangladesh. *Global Environmental Change, 38*, 1–7. https://doi.org/10.1016/j.gloenvcha.2016.02.002

Luca, M., Barlacchi, G., Oliver, N., & Lepri, B. (2021). Leveraging mobile phone data for migration flows. *ArXiv:2105.14956 [Cs]*. Retrieved from http://arxiv.org/abs/2105.14956

Lulli, A., Gabrielli, L., Dazzi, P., Dell'Amico, M., Michiardi, P., Nanni, M., & Ricci, L. (2017). Scalable and flexible clustering solutions for mobile phone-based population indicators. *International Journal of Data Science and Analytics, 4*(4), 285–299. https://doi.org/10.1007/s41060-017-0065-y

McAuliffe, M., & Sawyer, A. (2021). The roles and limitations of data science in understanding international migration flows and human mobility. In *Research handbook on international migration and digital technology*. Retrieved from https://www.elgaronline.com/view/edcoll/9781839100604/9781839100604.00012.xml

Milanović, B. (2019). *Capitalism, alone: The future of the system that rules the world*. The Belknap Press of Harvard University Press.

Mixed Migration Centre. (2020). *The mixed migration Centre in 2020*. Mixed Migration Centre. Retrieved from https://mixedmigration.org/wp-content/uploads/2021/01/157_annual_catalogue_2020.pdf

Pachauri, R. K., Allen, M. R., & Barros, V. R., Broome, J., Cramer, W., Christ, R., Church, J. A., Clarke, L., Dahe, Q., Dasgupta, P., Dubash, N. K., Edenhofer, O., Elgizouli, I., Field, C. B., Forster, P., Friedlingstein, P., Fuglestvedt, J., Gomez-Echeverri, L., Hallegatte, S., Hegerl, G., Howden, M., et al. (Eds.). (2014). *Climate change 2014: Synthesis report. Contribution of working groups I, II and III to the Fifth assessment report of the intergovernmental panel on Climate change*. Intergovernmental Panel on Climate Change.

Pappalardo, L., Simini, F., Rinzivillo, S., Pedreschi, D., Giannotti, F., & Barabási, A.-L. (2015). Returners and explorers dichotomy in human mobility. *Nature Communications, 6*, 8166. https://doi.org/10.1038/ncomms9166

Peeters, P., & Dubois, G. (2010). Tourism travel under climate change mitigation constraints. *Journal of Transport Geography, 18*(3), 447–457. https://doi.org/10.1016/j.jtrangeo.2009.09.003

Polzer, T., & Hammond, L. (2008). Invisible displacement. *Journal of Refugee Studies, 21*(4), 417–431. https://doi.org/10.1093/jrs/fen045

Pötzschke, S., & Braun, M. (2017). Migrant sampling using Facebook advertisements: A case study of polish migrants in four European countries. *Social Science Computer Review, 35*(5), 633–653. https://doi.org/10.1177/0894439316666262

Rampazzo, F., Bijak, J., Vitali, A., Weber, I., & Zagheni, E. (2021). A framework for estimating migrant stocks using digital traces and survey data: An application in the United Kingdom. *Demography*. Retrieved from https://eprints.soton.ac.uk/448283/

Raymer, J., Wiśniowski, A., Forster, J. J., Smith, P. W. F., & Bijak, J. (2013). Integrated modeling of European migration. *Journal of the American Statistical Association, 108*(503), 801–819.

Recchi, E. (2015). *Mobile Europe*. Palgrave Macmillan UK. https://doi.org/10.1057/9781137316028

Recchi, E. (2016). Space, mobility and legitimacy. In *Oxford Research Encyclopedia of Politics*. https://oxfordre.com/politics/view/10.1093/acrefore/9780190228637.001.0001/acrefore-9780190228637-e-11?rskey=W5XrDb&result=1

Recchi, E., Deutschmann, E., & Vespe, M. (2019). *Estimating transnational human mobility on a global scale*. Robert Schuman Centre for Advanced Studies Research Paper, No 30(RSCAS).

Recchi, E., Ferrara, A., Rodriguez Sanchez, A., Deutschmann, E., Gabrielli, L., Iacus, S., Bastiani, L. Spyratos, S. & Vespe, M. (2022). The impact of air travel on the precocity and severity of COVID-19 deaths in sub-national areas across 45 countries. *Scientific reports, 12*(1), 1–13.

Rigaud, K. K., de Sherbinin, A., Jones, B., Bergmann, J., Clement, V., Ober, K., Schewe, J., Adamo, S., McCusker, B., Heuser, S., & Midgley, A. (2018). *Groundswell: Preparing for internal climate migration* (p. 222). World Bank.

Scott, J. C. (2008). *Seeing like a state: How certain schemes to improve the human condition have failed (Nachdr.)*. Yale University Press.

Sîrbu, A., Andrienko, G., Andrienko, N., Boldrini, C., Conti, M., Giannotti, F., Guidotti, R., Bertoli, S., Kim, J., Muntean, C. I., Pappalardo, L., Passarella, A., Pedreschi, D., Pollacci, L., Pratesi, F., & Sharma, R. (2021). Human migration: The big data perspective. *International Journal of Data Science and Analytics, 11*(4), 341–360. https://doi.org/10.1007/s41060-020-00213-5

Skeldon, R. (2018). *International migration, internal migration, mobility and urbanization: Towards more integrated approaches* (p. 15). *Migration Research Series N° 53*. International Organization for Migration (IOM).

Słomczyńska, I., & Frankowski, P. (2016). Patrolling power Europe: The role of satellite observation in EU border management. In R. Bossong & H. Carrapico (Eds.), *EU Borders and shifting internal security: Technology, externalization and accountability* (pp. 65–80). Springer. https://doi.org/10.1007/978-3-319-17560-7_4

Song, C., Koren, T., Wang, P., & Barabási, A.-L. (2010). Modelling the scaling properties of human mobility. *Nature Physics, 6*(10), 818–823. https://doi.org/10.1038/nphys1760

Taylor, L. (2023). Data justice, computational social science and policy. In Bertoni, E., Fontana, M., Gabrielli, L., Signorelli, S., (Eds.), *Handbook of computational social science for policy.* Springer.

Tosi, D. (2017). Cell phone big data to compute mobility scenarios for future smart cities. *International Journal of Data Science and Analytics, 4*(4), 265–284. https://doi.org/10.1007/s41060-017-0061-2

Triandafyllidou, A., & McAuliffe, M. (2018). *Migrant smuggling data and research: A global review of the emerging evidence base* (Vol. 2). Retrieved from https://cadmus.eui.eu//handle/1814/57084

Trimikliniotis, N., Parsanoglou, D., & Tsianos, V. (2015). *Mobile commons, migrant digitalities and the right to the city.* Palgrave Macmillan.

Turton, D. (2005). The meaning of place in a world of movement: Lessons from long-term field research in southern Ethiopia. *Journal of Refugee Studies, 18*(3), 258–280.

UN DESA. (2015). *International Migration Flows to and from selected countries: The 2015 Revision* [POP/DB/MIG/Flow/Rev.2015]. United Nations, Department of Economic and Social Affairs, Population Division. Retrieved from https://www.un.org/en/development/desa/population/migration/data/empirical2/docs/migflows2015documentation.pdf

UNDESA (2017). *Handbook on Measuring International Migration through Population Censuses.* Retrieved from https://unstats.un.org/unsd/demographic-social/Standards-and-Methods/files/Handbooks/international-migration/2017-draft-E.pdf

UNDESA. (2019). *International migrant stock 2019.* United Nations. Retrieved from https://www.un.org/en/development/desa/population/migration/data/estimates2/estimates19.asp

Urry, J. (2000). Mobile sociology. *The British Journal of Sociology, 51*(1), 185–203. https://doi.org/10.1111/j.1468-4446.2000.00185.x

Vosen, S., & Schmidt, T. (2011). Forecasting private consumption: Survey-based indicators vs. Google trends. *Journal of Forecasting, 30*(6), 565–578. https://doi.org/10.1002/for.1213

Wang, D., Pedreschi, D., Song, C., Giannotti, F., & Barabasi, A.-L. (2011). Human mobility, social ties, and link prediction. In *Proceedings of the 17th ACM SIGKDD international conference on knowledge discovery and data mining*, pp. 1100–1108. https://doi.org/10.1145/2020408.2020581

Wang, W., Rothschild, D., Goel, S., & Gelman, A. (2015). Forecasting elections with non-representative polls. *International Journal of Forecasting, 31*(3), 980–991. https://doi.org/10.1016/j.ijforecast.2014.06.001

Weitzberg, K. (2015). The unaccountable census: Colonial enumeration and its implications for the Somali people of Kenya. *The Journal of African History, 56*(3), 409–428. https://doi.org/10.1017/S002185371500033X

Worldbank. (2019). *Mobile cellular subscriptions (per 100 people)—Low & middle income | Data.* Retrieved from https://data.worldbank.org/indicator/IT.CEL.SETS.P2?locations=XO&most_recent_value_desc=false

Xiong, C., Hu, S., Yang, M., Younes, H., Luo, W., Ghader, S., & Zhang, L. (2020). Mobile device location data reveal human mobility response to state-level stay-at-home orders during the COVID-19 pandemic in the USA. *Journal of the Royal Society Interface, 17*(173), 20200344. https://doi.org/10.1098/rsif.2020.0344

Zagheni, E., & Weber, I. (2015). Demographic research with non-representative internet data. *International Journal of Manpower, 36*(1), 13–25. https://doi.org/10.1108/IJM-12-2014-0261

Zagheni, E., Garimella, V. R. K., Weber, I., & State, B. (2014). Inferring international and internal migration patterns from Twitter data. In *Proceedings of the 23rd International Conference on World Wide Web*, pp. 439–444. https://doi.org/10.1145/2567948.2576930

Zagheni, E., Weber, I., & Gummadi, K. (2017). Leveraging Facebook's advertising platform to monitor stocks of migrants: Leveraging Facebook's advertising platform. *Population and Development Review, 43*(4), 721–734. https://doi.org/10.1111/padr.12102

Zetter, R. (1991). Labelling refugees: Forming and transforming a bureaucratic identity. *Journal of Refugee Studies, 4*(1), 39–62. https://doi.org/10.1093/jrs/4.1.39

Zumkeller, D. (2009). The dynamics of change: Latest results from the German mobility panel. In *12th International Conference on Travel Behaviour Research.*

Chapter 24
Towards a More Sustainable Mobility

Fabiano Pallonetto

Abstract The transport sector is the second most important source of emissions in the EU. It is paramount to act now towards the decarbonisation of our transport system to mitigate climate change effects. Waiting for future technological advancements to minimise the existing anthropogenic emissions and dramatically boost its sustainability is risky for human survival. The current chapter highlights how the path towards a sustainable transport system is a whole stakeholders' effort involving the mass deployment of available technology, changing user behaviours, data-driven legislation, and researching and developing future disruptive technologies. The author analyses and classifies the available data on various transport modals and assesses the impact of the technologies and policy measures in terms of potential reduction of carbon emissions, challenges, and opportunities. It also exemplifies outstanding test settings across the world on how already available technologies have contributed to the development of a lower-carbon transport setting. The chapter considers developing countries' economic and infrastructural challenges in upgrading to a low-carbon transport system and the lack of data-driven decisions and stakeholders' engagement measures in addressing the sector sustainability challenges. It also emphasised how a sustainable transport system should lay the foundation on data harmonisation and interoperability to accelerate innovation and promote a fast route for deploying new and more effective policies.

24.1 Introduction

The transport sector has been a critical economic area for the world from the industrialisation era to the present. It is an essential financial sector as it employs more than 11 million people, enabling international trade both in Europe and developing countries (Maparu & Mazumder, 2017). The trade-off of advanced

F. Pallonetto (✉)
Maynooth University, Maynooth, Ireland
e-mail: Fabiano.Pallonetto@mu.ie

© The Author(s) 2023
E. Bertoni et al. (eds.), *Handbook of Computational Social Science for Policy*,
https://doi.org/10.1007/978-3-031-16624-2_24

transport infrastructures is their environmental impact. Although the greenhouse emissions in EU decreased by 22.4% between 1990 and 2014, the greenhouse gas (GHG) contribution of the transport sector has considerably increased, amounting to more than 20.8% and rated as the second most important source of emissions in the EU (Andrés & Padilla, 2018). The pollutants emitted by endothermic engines powered by fossil fuels can also elicit harmful health effects like heart disease, asthma, and cancer. In Europe, the transport sector is the second sector for greenhouse gas (GHG) emissions after the power system sector. In the USA, the emissions caused by the transport sector have overtaken the environmental impact of electricity generation (Fan et al., 2018). Road transport is responsible for 72.9% of emissions within the transport sector, followed by aviation and maritime, which account for 13.3% and 12.8%, respectively. The growing demand for transport services could determine increased air pollution, reducing the sustainability of the whole sector. However, the pandemic has disrupted the status quo offering a cause for reflection on how transportation needs can evolve in the coming future. Significantly during times of restricting travel and activity measures, the travel behaviour has radically changed, showing a slight increase in shares of cycling and walking while public transport usage dropped significantly.

At the same time, private cars remained the preferred travel mode across countries (Eisenmann et al., 2021). It is also interesting to evaluate how the pandemic has provided alternative solutions to reduce the transport demand, such as blended, flexible, and hybrid working, and it has established a new essential travel baseline. At the same time, technology is presenting imminent breakthroughs in the transportation sector. Electric vehicles and scooters have already been used through-out Europe but are not yet perceived as a potential replacement for endothermic cars because of their range or safety limitations (Kopplin et al., 2021). Autonomous vehicles and drones are the future technologies announced as having a high potential to reduce energy consumption and emissions, especially in the last mile operations (Figliozzi, 2020; Staat, 2018). The deployment of these technologies is imminent, but social, economic, and technological barriers and untested negative implications are slowing down the adoption. Among the concerns of these technologies, there is a potential increase in congestion, unanswered ethical questions on the control of the vehicle, and excessive travel demand (EU Directorate-General for Communication, 2020). Therefore, it is essential to assess these innovations' potential impact and evaluate possible already available alternatives if the technology will not deliver what has been envisioned. The scalability of analysed advanced mobility solutions is probably one of the leading global challenges, especially for developing countries lacking infrastructure with less structured and competent governmental bodies. Another possible alternative tested during the pandemic could be to establish policies that could, in some ways, limit the mobility of the population. However, such an approach could become incompatible with the concept of democracy and freedom of movement. The relevance of these issues is underlined by the related policy questions in a recently published European Commission policy report (Bertoni et al., 2022, pp. 136–140).

The current chapter tries to reflect on the points mentioned above, and it is structured into six main sections. Section 24.2 provides a detailed overview of the future and already present technologies that could positively impact the sector, including the integration perspective. Section 24.3 highlights the potential of the illustrated technologies to reduce emissions and improve the sector's sustainability. Section 24.4 will give an overview of the impact of the pandemic on the transport sector and highlights some findings. Section 24.5 will assess the applicability of technologies to developing countries and their challenges. Section 24.6 shows how policies can further contribute to transport sustainability, while Sect. 24.7 will summarise the chapter and provide some recommendations.

24.2 Background: Computational, Environmental, and Data Aspects of Sustainable Mobility Technologies

This section analyses the main characteristics of the more imminent and high-potential technologies in the mobility sector that will lead to the decarbonisation of the transport system, trying to estimate the adoption rate, the economic and environmental impact, and the data dimension. All the low emission technologies identified are compared across different characteristics and summarised in Table 24.1. The table compares each technology's advantages, barriers, and travel range and highlights its scalability and the economic impact. A set of labels for each technology analysed is assigned to assess the solution's effects and define an uptaking timeline. The labels identified are leverage, application timeline and potential risks. *Leverage* identifies a technology/set of technologies that can solve one or more mobility challenges and it has been divided as *high leverage* and *low leverage*. *High leverage* identifies technologies that could significantly solve one or more challenges identified as critical or bottlenecks to the uptake. The *low leverage* label identifies technologies that could solve a limited subset of broader mobility challenges. *Application timeline* evaluates the temporal applicability of the technology, and it has been identified as long term, medium or short term. When a solution is tagged long term, the technology will have its immediate impact after 2030. In the medium-term, it could already impact, but the adoption phase could be delayed after 2030. In short term, the technology has already impacted the mobility sector, and it is in the adoption phase. Short- and long-term solutions are necessary to improve the sector's sustainability. In *potential risk* assesses the uncertainty of the impact of the technology on the emissions reduction and side effects of the full-scale deployment. It is divided in *high risk*, when the technology is risky because of its uncertain emission reduction, or it is not yet at commercialisation maturity or could lead to adverse side effects. If is tagged *medium risk*, the emission reduction impact at the full scale has been modelled, and a contingency plan for adverse risks is outlined. In case of *low risk* the technology could positively impact the emission reductions, and there are no relevant adverse risks on a full-scale deployment. The labels represent an evaluation based on the literature, and they should not be

Table 24.1 Recent and upcoming low emission technologies for a more sustainable mobility

Technology	Hybrid and plug-in EV	Connected autonomous EV	Compressed natural gas V	Hydrogen fuel cell vehicles	Unmanned aerial vehicles	Carsharing	Cycling electric bike	Electric scooter	MaaS
Environmental	30–40% fewer emissions than conventional ICEs	40–60% fewer emissions than conventional ICEs[a]	25–60% fewer emissions than traditional ICEs	100% fewer emissions than traditional ICEs	25–50% fewer emissions than traditional ICEs[a]	13–18% fewer emissions than traditional ICEs	40–70% fewer emissions than traditional ICEs	40–70% fewer emissions than traditional ICEs	40–70% fewer emissions than traditional ICEs
Economic	Limited – high-impact power system/battery still expensive	High – high impact on the transport system model	Limited – high prices of CNG reduce the savings	Limited – high prices of manufacturing	High – high impact on the transport system model	High – cost reduced by up to 60% of ownership	High – low capital and maintenance cost	High – low capital and maintenance cost	High – low cost per km and no maintenance
Advantages	Decarbonisation is shifted to the power system/high efficiency	Decarbonisation to power system/control/high efficiency	Reduced carbon emissions/cost-effective	Maximum reduction of carbon emissions/hydrogen	Decarbonisation to power system/reduced travel time, congestion	Reduced vehicles on the road, carbon emissions/increased cost-effectiveness	Reduced congestion and emissions/cost-effective	Reduced congestion and emissions/cost-effective	Reduced congestion, carbon emissions, changed behaviour
Barriers	Range/raw material for batteries/GDP dependent	Ethical/range/raw material for batteries/safety/tech barriers/law	Range/CNG import/reduced power train	Range/green hydrogen limited/tech early stage	Ethical/range/acceptance/tech barriers/battery capacity density	Public acceptance/safety/legislation gaps	Cycle infrastructure/integration with other modals	Safety/no infrastructure/regulations	Public acceptance/safety/legislation/road infrastructure/LCA

Adoption	Positive trend (+20.5%)[b]	Undefined – future technology	Negative trend (−0.04%)[a]	Positive trend (+0.01%)[a]	Undefined – future technology	Negative trend (−32%)	Positive trend (+3.73%)[a]	Positive trend (+27.6%)	Negative trend (−20%)
Scalability	In progress – charging infrastructure not ready	Requires charging infrastructure, connection	CNG infrastructure is lacking in capillarity	Hydrogen charging infrastructure is not existent	Requires landing/takeoff infrastructure, weight dependent	High scalable – IT infrastructures already present	High scalable and available	High scalable and available	High scalable – already available on the market
Travel distance	0 to >100 km	0 to >100 km	0 to >100 km	0 to >100 km	0 to 10 km	0 to >100 km	0 to 20 km	0 to 20 km	0 to 20 km
Notes	43–69% more efficient, commercial vehicles are at an early stage of adoption	Challenges due to lack of infrastructures in developing countries	Natural gas reserves dependency could promote shale gas	Only for passenger vehicles, not available for commercial	Suitable for short-distance trips because of battery capacity limitation	The pandemic has reverted the adoption rate, and it will be a few years delayed	The pandemic has promoted e-bikes and cycling more investment in the sector needed	Safety concerns and regulations are different for each country	The pandemic has reverted the adoption trend, and it will be a few years delayed

Impact

Leverage	High leverage	High leverage	Low leverage	High leverage	High leverage	Low leverage	Low leverage	Low leverage	Low leverage
Application timeline	Short term	Long term	Short term	Medium term	Long term	Short term	Short term	Short term	Short term
Potential risk	Medium risk	High risk	High risk	High risk	High risk	Low risk	Low risk	Low risk	Low risk

[a] Estimate based on preliminary studies and literature review – https://www.mordorintelligence.com/industry-reports/europe-bicycle-market
[b] ACEA, E., 2021. Making the Transition to Zero-emission Mobility – trend data from 2014 to 2020

considered definitive or specific. However, such a classification will provide a quick overview of what is coming, the timeline, and the potential impact. The following section adds further details to Table 24.1 on data and computational perspectives for the upcoming and future mobility technologies, starting from the imminent uptaking of electric vehicles and extending to multimodal sharing mobility concepts.

24.2.1 Hybrid and Plug-In Vehicles

Electric vehicles provide an alternative to meet the needs for a green and clean source of transportation with fewer emissions and better fuel economy. There are three main categories of EVs: fully electric vehicles (EVs), hybrid electric vehicles (HEVs), and plug-in hybrid electric vehicles (PHEV). As illustrated in Table 24.1, the technology is low risk and commercially available across Europe, with Norway leading the uptake of EVs with a share of 75% (Wangsness et al., 2020). In the EU, HEVs and PHEV reached the penetration of 1.25% over the 3.46% share of the whole car sector (European Environmental Agency, n.d.). The EV's lifetime energy consumption costs are significantly lower than conventional, between 45% and 70%. EVs are between 60% and 70% more efficient than gas vehicles; however, the benefit is offset by the high capital cost of the EVs' battery technology (Habib et al., 2018). One of the adoption barriers is the battery capacity which is not enough to provide a comparable driving range to internal combustion engine (ICE) vehicles and requires a lengthy charging duration (Capuder et al., 2020). Another major obstacle to the mass deployment of EVs is the slow implementation of charging infrastructures such as fast-charging stations and the challenges of integrating the power system. One of the primary data challenges is data interoperability between power systems, mobility providers, parking data, and charging infrastructure (Karpenko et al., 2018). Since cars are parked approximately 90% of the time, interoperability can support grid services by modulating/injecting/absorbing electricity based on grid operators' needs and market opportunities. At the same time, operators can deliver local benefits via behind-the-meter optimisation leading to a maximised energy efficiency and local use of renewables, fostering customers' involvement through new services and tools. From the life cycle analysis of critical components such as EV batteries, the charging and travel historical data could lead to a life extension of the battery or innovative business models focused on second-life battery applications (Shahjalal et al., 2022).

24.2.2 Connected Autonomous Vehicles (CAV)

Autonomous vehicles (AVs) have been identified as a possible solution to various modern transport issues. The adoption of autonomous cars can provide environmental benefits of up to 60% and economic and social advantages (Kopelias et al., 2020).

Table 24.1 shows that the benefits of connected electric autonomous vehicles involve reducing emissions and energy consumption through their ability to implement eco-driving, which continuously optimises the engine to run consistently at the most efficient operating points. As a result, it will also reduce emissions (Wadud et al., 2016). Additionally, the environmental advantages begin from the reduced demand for vehicles to the car's standard maintenance and optimal operation. AVs can also provide more significant economic benefits by offering ridesharing services (Bahamonde-Birke et al., 2018). The ridesharing economy allows greater efficiencies by reallocating underutilised resources for more productive purposes, such as achieving new sources of supply at a lower cost. It requires integrated datasets and intense computational resources. Legislations for all road traffic aim to ensure the best road safety; therefore, autonomous vehicles must meet their predecessors' complex and new strict requirements. The legal challenges include public policies, traffic codes, technological standards, and ethical dilemmas (Barabás et al., 2017).

Furthermore, AVs pose a significant threat to the job of professional drivers as they would change the required skills for workers whose careers are linked to mobility systems. It also may impact taxi drivers and other on-demand driver services as corporations have already begun experimenting with offering driverless experiences (Sousa et al., 2018). One of AVs' risks is cybersecurity, which could lead to terrorist attacks and privacy intrusion (Ahangar et al., 2021). Therefore, in Table 24.1, the technology has been identified as high risk and high leverage. However, its impact on society is still projected in the long term. From the data perspective, AV requires a different approach to mobility data, such as seamless integration across data providers and social media to forecast trajectories, optimise routes, and understand common mobility patterns (Giannotti et al., 2016).

24.2.3 Compressed Natural Gas (CNG) Vehicles

Over 23.5 million natural gas vehicles (NGV) are on roads worldwide. The leading countries in natural gas are the Asian countries with 15.7 million natural gas vehicles, closely followed by the Latin American countries with 5.4 million natural gas vehicles (Khan et al., 2015). NGV have been identified as leading candidates for green transportation among sustainable fuel alternatives. CNG is a clean energy fuel when used as motor fuel, and there are relatively low particulate emissions and toxicity of exhaust gasses (Agarwal et al., 2018). However, there are high costs with developing the refuelling infrastructure, such as pipelines and filling stations, which are the more significant disadvantage of the technology (Imran Khan, 2017). The considerable challenge of natural gas vehicles is the lower efficiency compared to gasoline vehicles and longer refilling time. Other environmental challenges to the adoption of CNG vehicles concern fuel treatment and natural gas distribution (Chala et al., 2018). From the data perspective, geographical data can reduce waiting time at gas stations for refilling, and satellite data analysis can support the identification of leakages. Despite advances, the technology is highly dependent on gas imports, and it is unlikely to scale, so in Table 24.1, it has been classified as low leverage.

24.2.4 Hydrogen Fuel Cell Vehicles

Fuel cell vehicles result in nearly zero tailpipe emissions during vehicle operations (Sharma & Strezov, 2017). The implementation and use of hydrogen fuel cell vehicles were found to have a positive impact and result in economic savings over internal combustion engine vehicles (ICEVs) (Watabe & Leaver, 2021). The study found that hydrogen fuel cell vehicles using hydrogen from solar and wind electrolysis will have positive economic benefits beyond 2050. The first significant barrier concerns the safety of hydrogen vehicles and linked awareness campaigns. The concern for safety arises as hydrogen can burn in lower concentrations, and a possible spark or fire may occur if there is a mixture of hydrogen and air (Manoharan et al., 2019). The second barrier involves the storage of hydrogen. A sizeable onboard storage tank is required to transport the fuel. The barrier to adoption is concerning finding the appropriate material for the storage container. As described in Table 24.1, another barrier is a lack of hydrogen infrastructure that could lead to slow adoption, the inability to charge from home, and cost-related issues to the adoption. From the social data perspective, the technology requires strong awareness campaigns to limit the focus on security concerns, and integrated data on the infrastructure could support adoption in the long term.

24.2.5 Unmanned Aerial Vehicles (UAVs)

Drones, also known as unmanned aerial vehicles (UAVs), combine three critical principles of technology: data processing, autonomy, and boundless mobility. They enable new access to new spaces and analysis with data collection aid (Kellermann et al., 2020). UAVs have the potential to reduce energy consumption and emissions in some scenarios significantly. Current UAVs are approximately 47 times more CO_2 efficient than US delivery vehicles in terms of energy consumption and approximately over 1000 times concerning emissions. Drone delivery will also significantly shift energy and greenhouse gas consumption (Figliozzi, 2020). For instance, drones will shift energy usage and greenhouse gas emissions from vehicle fuels such as diesel and gasoline to varying regional sources of electricity to be charged. The wide-scale implementation of drones will lead to economic and commercial benefits. Drones can be deployed in several contexts and for varying purposes; however, drones for parcel delivery services are still in infancy, along with their "air taxi" services to transport passengers between cities. As a result of its ability to serve multiple needs, the European Commission estimates that drones will have an economic impact of 10 billion euros annually by 2035 and expects approximately 250,000 to 450,000 jobs to be created (de Miguel Molina & Santamarina Campos, 2018). Despite being identified as high leverage, there are several barriers to the public adoption of drones (Table 24.1). The most significant anticipated obstacles to adopting drones are concerning the technical, legal, and public acceptance of drones (Kellermann et al., 2020). The technical concerns refer to autonomous flying, airspace integration, and questions about battery capacity and data communication. UAV trips can flood the suburb and city airspace, providing traffic and safety concerns. Therefore, prioritising accurate, centralised data acquisition and control of airspace traffic is required to fully deploy the technology. The second biggest potential barrier is ethical aspects, which are heavily related to privacy threats. Drones may threaten privacy because of their ability to capture imagery and collect sensitive data (Merkert & Bushell, 2020).

24.2.6   Carsharing

Carsharing significantly impacts car usage and ownership, enabling a reduction in environmental impacts. The annual environmental benefit per capita is between 240 and 390 kilograms (Nijland & van Meerkerk, 2017). The same study found that the total impact of carsharing versus ownership leads to an annual emission reduction between 13% and 18%. Similar findings have been highlighted in other studies, which have found an emission reduction between 35% when hybrid vehicles are utilised and 65% when utilising electric vehicles (Baptista et al., 2014; Te & Lianghua, 2020). Carsharing is a short-term technology (Table 24.1), and it enables users to gain economic benefits such as reducing travel costs associated with travel style and car ownership. Car owners saved approximately 74%, and public transit car owners saved around 60% by adopting carsharing in Ireland (Rabbitt & Ghosh, 2016). However, non-car owners may have to adapt to a multimodal active traveller lifestyle, which was found to have incurred additional costs. Safety was highlighted as an area of concern as respondents had commented that security is one of the most significant inhibitors of carsharing. Carsharing requires a data-driven approach to learning the mobility habits of users, providing flexibility in case of delays or route changes. Additionally, carsharing requires reassurance to users on the reliability and safety of the trips and drivers. Such a reassurance and safety layer can be entrusted by social network data and previous users' feedback.

24.2.7   Micromobility

Micromobility aims at providing short-distance, flexible, sustainable, and cost-effective on-demand short-distance transport (between 3 and 20 km). Micromobility involves a range of small vehicles that operate at approximately 20 to 25 km/h, such as bicycles, scooters, skateboards, and electric bikes. These vehicles encourage a shift towards low-carbon and sustainable modes of transport that can reduce carbon emissions from 40 to 70 per cent compared to an ICE (Abduljabbar et al., 2021).

24.2.7.1 Cycling and Electric Bikes

Cycling with traditional bicycles is environmentally friendly as it does not emit emissions and is economically viable to produce (Pucher & Buehler, 2017). E-bikes are substantially more efficient, with an average CO_2 emission for km of 22 g, which is significantly lower than ICE vehicles (Elliot et al., 2018; Philips et al., 2020). However, the benefits of electric bikes are also varied depending on the mode of transport they are replacing (Edge et al., 2018). The wide-scale implementation and encouragement of cycling as a sustainable mode of transportation is not without its drawbacks. Cycling has been marginalised in many cities' transports planning, and significant barriers to adopting and implementing pro-cycling policies are caused by a lack of infrastructure, funding, and leadership (Wang, 2018). There are often compact urban structure and a lack of street space in European cities, especially inner-city areas, therefore making it challenging to implement cycling infrastructure. Concerning electric bikes, the disposal of their batteries and their manufacturing emissions is the most significant environmental concern (Liu et al., 2021). Besides the lack of cycling infrastructure, other relevant barriers to cycling are the limited feasible trip range, personal safety concerns, the safety of bike storage, and lack of flexibility for sudden route extension beyond a particular length or passenger transport. However, as identified in Table 24.1, a low-risk technology could significantly contribute to a sustainable transport system if the ecosystem is enriched with data-driven technologies for charging, sharing, and improving the infrastructure.

24.2.7.2 Electric Scooter

During the last few years, electric scooters' uptake has been soaring. The transport solution has been identified as economic, clean, and sustainable (Table 24.1). However, introducing electric scooters as a transportation mode has caused several conflicts, such as problems with space, speed, and safety (Gössling, 2020; O'Keeffe, 2019). Some researchers found that the barriers varied greatly on whether respondents had used an e-scooter and how often they used it in the last month. In the survey, 46% of non-riders were satisfied with the current modes of transport and were not interested in e-scooters (Sanders et al., 2020). Issues with e-scooter equipment, such as being hard to find or easy to break, were a significant barrier among e-scooter users. Safety-related barriers were found to be more even between both groups. From the data perspective, localisation and the computation of optimal collection routes of dead scooters require accurate GPS data. Additional parking verification through advanced computer vision techniques requires heavy computational capabilities.

24.2.7.3 Mobility as a Service (MaaS)

Shared mobility refers to the shared use of a vehicle. These vehicles can range from scooters to bikes or electric bikes. It is a modern and innovative transportation strategy that enables users to have short-term access to a mode of transport when required. Thus, it may increase multimodality, minimise vehicle ownership and distance travelled, and provide new ways to access goods and services. Shared mobility has an extensive and wide range of modalities; however, the development of newer mobility options, alongside the development of new technology, led to the development of the service concept known as mobility as a service (MaaS) (Machado et al., 2018). Such a concept is often described as a one-stop management platform that unifies and links the purchase and delivery of mobility services such as bike sharers, share riders, and car sharers (Wong et al., 2020).

Additionally, the subscription to MaaS enables tailoring and developing mobility services around an individual's preferences, which may be beneficial to both transport users and providers. Therefore, the seamless and affordable travel experience MaaS provides may play a significant role in pursuing sustainable transport. Its goals are to create an integrated multimodal system and substitution private vehicles with alternative options (Jittrapirom et al., 2017). The efficient running of MaaS platforms requires seamless data interoperability between operators and mobility data providers such as navigation systems to forecast demand and dynamically allocate resources such as vehicles or public transport routes.

24.3 Questions and Challenges: Decarbonisation of the Transport Sector with the Currently Available Technology

As illustrated in the previous section, three leading technologies could disrupt the personal transport sector: unmanned aerial vehicles (drones) and connected autonomous and hydrogen cars. Although these technologies have been classified as high leverage technologies with a potentially disruptive impact on the industry, critical technological and social barriers exist and rely on research and development progress and political and financial factors. Additionally, without an environmental analysis of mobility behaviour and clear directives on how to shift towards more sustainable mobility solutions, it is impossible to outline a feasible roadmap to the decarbonisation of the transport system. For example, it is essential to understand if low leverage technologies can achieve the same benefit as the leading future technologies illustrated. For this purpose, the aggregated impact of low leverage technologies on the annual per capita carbon footprint measured in tons of CO_2 emissions is considered and computed with an open mobility dataset. It is further compared with the potential impact of the appraised high leverage technologies to assess the viability of the combined solutions.

It should be noted that the low leverage and low-risk solutions such as cycling, electric bike, and electric scooters have a constraint on the trip length. At the same time, carsharing and electric vehicles can cover virtually any distance, as demonstrated by numerous examples of successful carsharing initiatives such as BlaBlaCar (Quirós et al., 2021).

The first step in the analysis was to evaluate if short trips within the 20 km range represent a significant percentage of the overall total of car trips. Users would switch to micromobility transports such as bicycles, e-scooters, electric bikes, and so on for trips between 3 and 20 km (Fiorello et al., 2016). The distance below 3 km can easily be covered by walking, so the uptake is much lower. As per literature, above the 20 km range, micromobility is not suitable and more comfortable transport modes are required.

Therefore, we have analysed the share of domestic trips with private vehicles that can be replaced with micromobility transport services. The data for the analysis were extracted from the US national travel survey in 2009 and 2017 (Federal Highway Administration, 2020), which details more than 1.9 million private trips from participants across the USA, and similar results have been found throughout Europe.[1] The participants, during the trial, have logged detailed data for each trip, such as distance, type of vehicle, starting and end time, duration, and destination. As illustrated in Fig. 24.1, the cumulative percentage of car trips versus distance reveals that 47.9% of trips are within the 3 km to 20 km range. The low leverage micromobility solutions could uptake a significant percentage of the transport needs in such a range. Above the 20 km range, electric vehicles and carsharing can increase their market share until reaching their full potential.

The impact of each low leverage technology has been evaluated through the literature and in terms of annual per capita carbon footprint reduction. Four scenarios for each technology have been considered and their carbon emission reduction compared to the average EU carbon emissions per capita, as illustrated in Fig. 24.2. The four scenarios illustrate a stepwise increased share of a single technology from 10% (Scenario A) to 75% (Scenario D). In the graph, carsharing does not assume any vehicle upgrading. Still, it calculates the reduction of emissions caused by fewer vehicles on the road and shared mobility, and it includes the whole range of trips above 3 km. The remaining low leverage technologies can significantly impact the share of trips between 3 and 20 km, while between 20 and 30 km, the number of trips affected by the modal switch was reduced by 50%.

Interestingly, there is a marginal positive impact of station-based bike-sharing uptake compared to dockless bike-sharing. The potential emission reduction of these technology spans from 13% associated with a carsharing penetration of 75% in Scenario D up to 33% of station-based bike-sharing in the same scenario. The scenarios do not separate the electric versus not electric bike-sharing because the two results averaged. Each scenario also considers a 5% uncertainty derived from

---

[1] https://ec.europa.eu/eurostat/statistics-explained/index.php?title=Passenger_mobility_statistics#Distance_covered

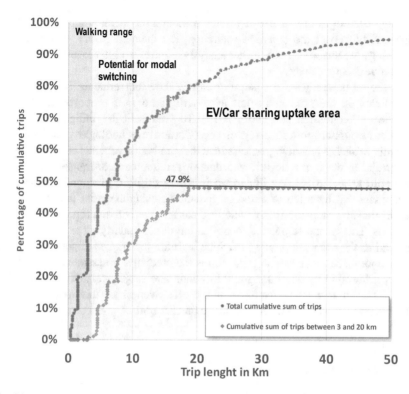

Fig. 24.1 Percentage of domestic trips within the range for modal shifting. The blue line is the total cumulative percentage of the trips, while the orange line is the cumulative percentage of trips between 3 and 20 km that a low leverage/low-risk technology could replace

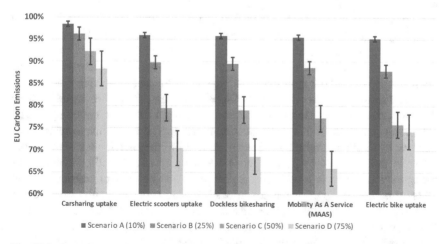

Fig. 24.2 Annual per capita potential EU emission reduction for each low leverage technology where 100% is the average EU emissions per capita baseline for transport

slightly different results in the literature. Although the technologies analysed can significantly impact the emission reduction, the interaction and uptake of each technology are uncertain, and further analysis requires a higher level of complexity and different perspectives.

Moreover, the study and the literature clearly show that reducing the total number of vehicles on the road could not necessarily lead to a proportional emission reduction (Commission & Centre, 2019). To systematically mitigate the carbon emissions from transport, focusing on a few disruptive technologies or assessing the currently available technologies to reduce the number of vehicles is not enough. It is essential to develop a holistic and data-driven approach to transport emissions (Giannotti et al., 2016). As illustrated in Wang et al., a profound discovery of phenomena that have led to emission reductions using essential temporal-spatial data is the first step towards developing sustainable and interoperable mobility solutions. The second step is the identification of the underlying explanations that have led to the events exploiting machine learning techniques of social-economic and temporal-spatial data. The third step is the predictive assessment of solutions' impact achievable by combining mobility data and subject-related data (Wang et al., 2021). An interesting example of the effectiveness of this methodology is represented by the impact of the pandemic on transport.

24.4 Impact of the Pandemic

The Covid-19 pandemic has had an enormous impact on social life and transport. The lockdowns and restrictions imposed by governments worldwide to reduce transmissions have drastically affected the transport sector. Such measure was evident as global road transport had fallen below 50% compared to 2019, and commercial flight transport had dropped below 75% by mid-April of 2020 (Abu-Rayash & Dincer, 2020). The pandemic has altered people's social life routines, travel, and working behaviours. The government-imposed restrictions caused a surge of immediate change towards remote working. Remote working has dramatically affected our mobility, reducing congestion and improving productivity in several sectors (Philips et al., 2020). Global road transport has decreased by more than 50% compared to March 2019. April 2020 flight activity dropped by almost 75% compared to 2019 due to a reduction in transport demand due to restrictions imposed by the government. By the end of April 2020, the total number of passenger transport had declined by 77% compared to January 2020. Lastly, air passenger transport was the least used mode of transportation because of restrictions. On utilising public transport, when conditions began to ease, the public had become more cautious with their transportation choice due to anxiety and fear of infection of Covid-19 (Campisi et al., 2020). As a result, the preference for travelling employing private vehicles increased as they felt public transport was unsafe. As reported in a survey from Jenelius et al., 25% of respondents have entirely resigned from using public transport. These results suggest that people's perception of their well-being in public transport is essential in determining their willingness to use it. After several months, the general perception identified it as risky (Jenelius & Cebecauer, 2020; Przybylowski et al., 2021). The pandemic has resulted in developing a preference for private vehicles as their mode of transport rather than public transport. In parallel, different private transport modes such as e-scooter and bikes had peak sales trends (Eisenmann et al., 2021; Nundy et al., 2021). It has been deemed the most appropriate mode of road transport in several countries. Berlin has expanded its yellow tapes on its roads to encourage and allow more room for cyclists. In Budapest, cycles were implemented, and there has been a 300% reduction in tariffs for bikes. Lastly, in the UK, the government supported bicycles as a mode of transport.

24.5 Developing Countries

Developing countries tend to suffer from a variety of issues. These countries face significant environmental challenges due to rapid urbanisation, population growth, climate and environmental issues, and inefficient governance and environmental management, therefore making it extremely difficult for these countries to pursue sustainable transportation (Ameen & Mourshed, 2017). However, an interesting perspective is that developing countries could implement measures to avoid the same path as developed countries. For instance, during the pandemic, the municipality of Bogota transformed a car road lane of 100 km into a bike lane to facilitate citizens to commute by bicycle (Rodriguez-Valencia et al., 2021). In most developing countries, urban areas are affected by prevailing global megatrends such as population growth and urbanisation. The main problem in achieving sustainable transportation in developing countries is a lack of quality infrastructure (Gordon, 2012). Poor infrastructure contributes to a high quantity of accidents and mortality rates. For instance, Bangladesh's fatality rates are the highest globally at 85.6 per 10,000 vehicles in 2004, which was double the South Asian average of 40.56. Secondly, pollution is exponentially increasing and affecting population health. Lack of necessary transport infrastructure and planning leads to high traffic congestion. Therefore, it is challenging to design the infrastructure to match the current needs (Kyriacou et al., 2019). However, digitalisation and the availability without strong privacy concerns of large mobility datasets could open up to test innovative solutions. For instance, the city of Manila (Philippines) has made an effort to propel digital transition and sustainable transport modes to respond to the pandemic measures. The government has pushed towards systematic data collection and establishing an open database system for all governmental transport agencies to adopt other MaaS solutions (Hasselwander et al., 2022).

24.6 Policy Restrictions

Covid-19 has forced society to partially renounce its freedom of movement, especially during the early stage of the pandemic when limited studies on the virus and reckless citizens' behaviour threatened public health. From the economic perspective, the externalities cost of the restrictions posed a significant burden on society in unequal shares (Zivin & Sanders, 2020). In such a context, personal and individual decisions no longer match the greater benefit of society. Therefore, policy interventions such as the shutdown of businesses and limitations to mobility and personal freedom have caused significant backlashes. These periods have further stressed the evidence that subsidies and incentives for sustainable and virtuous behaviours are more effective than restrictions on personal freedom. Because the climate crisis is reaching an emergency level and reducing the burden for citizens of such externality, policymakers could develop a subsidy infrastructure for environmentally friendly behaviours exploiting big data such as mobile data, metering, and location data. Privacy concerns, cybersecurity, and reliability of the data sources are still open challenges to reaching such ambitious objectives. If the policy framework is accurately planned, some technologies such as edge computing, anonymisation, gamification, and distributed ledgers reduce the security risks and mitigate the consequences of opening the data to the public.

There is often a lack of adequate planning and regulation in developing countries, leading to problems such as congestion problems and high costs and travel times. In some of these developing countries, this is caused by a poor public transport system, sense of community, and education. In contrast, mobility bottlenecks could be caused by high motorisation rates and private car use in other countries, leading to economic, social, and environmental problems (Sánchez-Atondo et al., 2020).

24.7 Conclusions and Recommendations

This chapter analysed a set of future/imminent transport technologies, and they were classified based on their suitability to solve a specific transport issue (high/low leverage), deployment time (short/long term), and associated risk (high/low risk). Among the high leverage and long-term technologies, connected autonomous vehicles, hydrogen fuel cells, and unmanned aerial vehicles have high disruptive potential and the necessary features to reduce carbon emissions from transport globally. Carsharing, shared electric mobility (MaaS), electric scooter, and cycling have been classified as low leverage and short-term technologies that can improve the transport sector's sustainability. One of the main questions addressed was determining if low leverage sustainable transport modes could replace future high leverage solutions if the technology advancements do not deliver what they have promised. The study indicated that the expected average emission reduction of

9.3% can be associated with the micromobility shared technologies identified if adequately promoted at the European level.

It should be noted that the combination of micromobility solutions, EV adoption, and carsharing could reach a similar level of decarbonisation for passenger transport expected by high leverage technologies. However, the path towards full decarbonisation of the transport sector is still long. Thus, waiting for future technologies will not bring any additional benefit.

As a first recommendation, we can state that a sustainable transport system requires all stakeholders to work together towards the common objective of adopting low leverage technologies and creating a new data-driven infrastructure to reward sustainable mobility behaviours. As described above, it is fundamental to collect data to analyse patterns, establish a baseline, and test and verify new technologies and measures. A shared and open data repository could support the analysis of positive and negative phenomena that impact the system. The security risks of such an open data repository are well known and should be carefully considered; however, nowadays, several data distributed infrastructures based on semantic interoperability are rising and could be good candidates to be scaled across the EU.

The second recommendation is to implement a reward system to promote sustainable mobility behaviours. EV owners have been rewarded with free motorway tolls, reduced parking tariffs and taxes in several cities. Such a reward mechanism could be extended to exploiting recurrent mobility patterns. A reward for using a planning tool for short/long trips or utilising a MaaS infrastructure instead of a personal car can be awarded, and shared data can be utilised for further optimisation.[2] The technology to implement a flexible transport system is low leverage; Google's Matrix APIs already provide timing, traffic, and distance for different transport systems such as bikes, public transport, and cars. In the USA, Google Maps started to embed public transport tickets. The Uber CEO clearly stated they wanted to become the leader in a safe, electric, shared, and connected transportation system for cities. Private efforts should be backed up by policies to reduce mobility needs, foster remote working, and pave the way to fast, steady, and effective adoption of technologies.

The third recommendation for a sustainable transport system is to exploit the heterogeneous public and big data to forecast demand and optimise road capacity, reduce peak hours' traffic, and integrate with the rewards mechanisms. Using data collected through different sources and gamification can promote sustainable behaviours, especially if combined with social media and linked to a reward mechanism. Data integration and harmonisation are essential for the mass adoption of existing low leverage and future technologies. One of the limits of the low leverage technologies and EV adoption is to rely on the decarbonisation of the power system to deliver an essential contribution to the sector emissions. Therefore, data interoperability is necessary also for sector coupling and integration. The

---

[2] https://www.forbes.com/sites/gusalexiou/2021/05/23/mobility-as-a-service-concept-promises-to-revolutionize-transport-accessibility/?sh=7c0524df7fe6

situation is more complex in developing countries because of the lack of regulation and infrastructure. In this case, probably the most appropriate solution to reduce congestion and emissions is adopting a combination of low and high leverage technologies that do not require massive investment in infrastructures such as MaaS, UAVs and waterborne or air transport. The decarbonisation of the sector for these countries will be undoubtedly further delayed in time compared to the more economically developed countries because of a slower orchestration between public and private interests.

As a final recommendation, the pandemic has also highlighted the importance of remote working and provided an estimated baseline for essential travel requirements. These data should be seen as a reference scenario and used to develop subsidies and incentives towards more sustainable mobility. Although government personal freedom restrictions are not compatible with democracy unless there is a tangible health and safety risk, the associated risks of climate change emergency could justify implementing more decisive policies and actions to reduce anthropogenic carbon emissions.

References

Abduljabbar, R. L., Liyanage, S., & Dia, H. (2021). The role of micro-mobility in shaping sustainable cities: A systematic literature review. *Transportation Research Part D: Transport and Environment, 92*, 102734–102734. https://doi.org/10.1016/j.trd.2021.102734

Abu-Rayash, A., & Dincer, I. (2020). Analysis of mobility trends during the COVID-19 coronavirus pandemic: Exploring the impacts on global aviation and travel in selected cities. *Energy Research and Social Science, 68*, 101693. https://doi.org/10.1016/j.erss.2020.101693

Agarwal, A. K., Ateeq, B., Gupta, T., Singh, A. P., Pandey, S. K., Sharma, N., Agarwal, R. A., Gupta, N. K., Sharma, H., Jain, A., & Shukla, P. C. (2018). Toxicity and mutagenicity of exhaust from compressed natural gas: Could this be a clean solution for megacities with mixed-traffic conditions? *Environmental Pollution, 239*, 499–511. https://doi.org/10.1016/j.envpol.2018.04.028

Ahangar, M. N., Ahmed, Q. Z., Khan, F. A., & Hafeez, M. (2021). A survey of autonomous vehicles: Enabling communication technologies and challenges. *Sensors, 21*(3). https://doi.org/10.3390/s21030706

Ameen, R. F. M., & Mourshed, M. (2017). Urban environmental challenges in developing countries—A stakeholder perspective. *Habitat International, 64*, 1–10. https://doi.org/10.1016/j.habitatint.2017.04.002

Andrés, L., & Padilla, E. (2018). Driving factors of GHG emissions in the EU transport activity. *Transport Policy, 61*, 60–74. https://doi.org/10.1016/j.tranpol.2017.10.008

Bahamonde-Birke, F. J., Kickhöfer, B., Heinrichs, D., & Kuhnimhof, T. (2018). A systemic view on autonomous vehicles: Policy aspects for a sustainable transportation planning. *DISP, 54*(3), 12. https://doi.org/10.1080/02513625.2018.1525197

Baptista, P., Melo, S., & Rolim, C. (2014). Energy, environmental and mobility impacts of car-sharing systems. Empirical results from Lisbon, Portugal. *Procedia - Social and Behavioral Sciences, 111*, 28–37. https://doi.org/10.1016/j.sbspro.2014.01.035

Barabás, I., Todoruct, A., Cordocs, N., & Molea, A. (2017). *Current challenges in autonomous driving* (Vol. 252, p. 12096). https://doi.org/10.1088/1757-899x/252/1/012096

Bertoni, E., Fontana, M., Gabrielli, L., Signorelli, S., & Vespe, M. (Eds). (2022). *Mapping the demand side of computational social science for policy*. EUR 31017 EN, Luxembourg,

Publication Office of the European Union. ISBN 978-92-76-49358-7, https://doi.org/10.2760/901622

Campisi, T., Basbas, S., Skoufas, A., Akgün, N., Ticali, D., & Tesoriere, G. (2020). The impact of Covid-19 pandemic on the resilience of sustainable mobility in Sicily. *Sustainability (Switzerland), 12*(21). https://doi.org/10.3390/su12218829

Capuder, T., Sprčić, D. M., Zoričić, D., & Pandžić, H. (2020). Review of challenges and assessment of electric vehicles integration policy goals: Integrated risk analysis approach. *International Journal of Electrical Power & Energy Systems, 119*, 105894–105894.

Chala, G. T., Abd Aziz, A. R., & Hagos, F. Y. (2018). Natural gas engine technologies: Challenges and energy sustainability issue. *Energies, 11*(11). https://doi.org/10.3390/en11112934

Commission, E., & Centre, J. R. (2019). *The future of road transport: Implications of automated, connected, low-carbon and shared mobility*. Publications Office. https://doi.org/10.2760/524662

de Miguel Molina, M., & Santamarina Campos, V. (2018). *Ethics and civil drones: European policies and proposals for the industry*. Springer Nature.

Edge, S., Dean, J., Cuomo, M., & Keshav, S. (2018). Exploring e-bikes as a mode of sustainable transport: A temporal qualitative study of the perspectives of a sample of novice riders in a Canadian city. *The Canadian Geographer/Le Géographe Canadien, 62*(3), 384–397. https://doi.org/10.1111/cag.12456

Eisenmann, C., Nobis, C., Kolarova, V., Lenz, B., & Winkler, C. (2021). Transport mode use during the COVID-19 lockdown period in Germany: The car became more important, public transport lost ground. *Transport Policy, 103*, 60–67. https://doi.org/10.1016/j.tranpol.2021.01.012

Elliot, T., McLaren, S. J., & Sims, R. (2018). Potential environmental impacts of electric bicycles replacing other transport modes in Wellington, New Zealand. *Sustainable Production and Consumption, 16*, 227–236. https://doi.org/10.1016/j.spc.2018.08.007

EU Directorate-General for Communication. (2020). *Special Eurobarometer 496: Expectations and concerns from a connected and automated mobility*.

European Environmental Agency. (n.d.). *Monitoring of CO_2 emissions from passenger cars. New Registrations of Electric Vehicles in Europe*. Retrieved from https://www.eea.europa.eu/data-and-maps/data/co2-cars-emission-20

Fan, Y. V., Perry, S., Klemeš, J. J., & Lee, C. T. (2018). A review on air emissions assessment: Transportation. *Journal of Cleaner Production, 194*, 673. https://doi.org/10.1016/j.jclepro.2018.05.151

Federal Highway Administration. (2020). *2020 NextGen NHTS National Passenger OD Data*. U.S. Department of Transportation, Washington, DC. Available online: https://nhts.ornl.gov/

Figliozzi, M. A. (2020). Carbon emissions reductions in last mile and grocery deliveries utilizing air and ground autonomous vehicles. *Transportation Research Part D: Transport and Environment, 85*, 102443–102443. https://doi.org/10.1016/j.trd.2020.102443

Fiorello, D., Martino, A., Zani, L., Christidis, P., & Navajas-Cawood, E. (2016). Mobility data across the EU 28 member states: Results from an extensive CAWI survey. *Transportation Research Procedia, 14*, 1104–1113. https://doi.org/10.1016/j.trpro.2016.05.181

Giannotti, F., Gabrielli, L., Pedreschi, D., & Rinzivillo, S. (2016). Understanding human mobility with big data. In *Solving large scale learning tasks. Challenges and algorithms* (p. 208220). Springer. https://doi.org/10.1007/978-3-319-41706-6_10

Gordon, C. (2012). The challenges of transport PPP's in low-income developing countries: A case study of Bangladesh. *Transport Policy, 24*, 296–301. https://doi.org/10.1016/j.tranpol.2012.06.014

Gössling, S. (2020). Integrating e-scooters in urban transportation: Problems, policies, and the prospect of system change. *Transportation Research Part D: Transport and Environment, 79*, 102230–102230. https://doi.org/10.1016/j.trd.2020.102230

Habib, S., Khan, M. M., Abbas, F., Sang, L., Shahid, M. U., & Tang, H. (2018). A comprehensive study of implemented international standards, technical challenges, impacts and prospects for electric vehicles. *IEEE Access, 6*, 13866–13890.

Hasselwander, M., Bigotte, J. F., Antunes, A. P., & Sigua, R. G. (2022). Towards sustainable transport in developing countries: Preliminary findings on the demand for mobility-as-a-service (MaaS) in metro Manila. *Transportation Research Part A: Policy and Practice, 155*, 501–518. https://doi.org/10.1016/j.tra.2021.11.024

Imran Khan, M. (2017). Policy options for the sustainable development of natural gas as transportation fuel. *Energy Policy, 110*, 126–136. https://doi.org/10.1016/j.enpol.2017.08.017

Jenelius, E., & Cebecauer, M. (2020). Impacts of COVID-19 on public transport ridership in Sweden: Analysis of ticket validations, sales and passenger counts. *Transportation Research Interdisciplinary Perspectives, 8*, 100242. https://doi.org/10.1016/j.trip.2020.100242

Jittrapirom, P., Caiati, V., Feneri, A.-M., Ebrahimigharehbaghi, S., Alonso González, M. J., & Narayan, J. (2017). Mobility as a service: A critical review of definitions, assessments of schemes, and key challenges.

Karpenko, A., Kinnunen, T., Madhikermi, M., Robert, J., Främling, K., Dave, B., & Nurminen, A. (2018). Data exchange interoperability in IoT ecosystem for smart parking and EV charging. *Sensors, 18*(12). https://doi.org/10.3390/s18124404

Kellermann, R., Biehle, T., & Fischer, L. (2020). Drones for parcel and passenger transportation: A literature review. *Transportation Research Interdisciplinary Perspectives, 4*, 100088–100088. https://doi.org/10.1016/j.trip.2019.100088

Khan, M. I., Yasmin, T., & Shakoor, A. (2015). Technical overview of compressed natural gas (CNG) as a transportation fuel. *Renewable and Sustainable Energy Reviews, 51*, 785. https://doi.org/10.1016/j.rser.2015.06.053

Kopelias, P., Demiridi, E., Vogiatzis, K., Skabardonis, A., & Zafiropoulou, V. (2020). Connected & autonomous vehicles – Environmental impacts – A review. *Science of the Total Environment, 712*, 135237–135237. https://doi.org/10.1016/j.scitotenv.2019.135237

Kopplin, C. S., Brand, B. M., & Reichenberger, Y. (2021). Consumer acceptance of shared e-scooters for urban and short-distance mobility. *Transportation Research Part D: Transport and Environment, 91*, 102680.

Kyriacou, A. P., Muinelo-Gallo, L., & Roca-Sagalés, O. (2019). The efficiency of transport infrastructure investment and the role of government quality: An empirical analysis. *Transport Policy, 74*, 93–102. https://doi.org/10.1016/j.tranpol.2018.11.017

Liu, W., Liu, H., Liu, W., & Cui, Z. (2021). Life cycle assessment of power batteries used in electric bicycles in China. *Renewable and Sustainable Energy Reviews, 139*, 110596–110596. https://doi.org/10.1016/j.rser.2020.110596

Machado, C. A. S., De Salles Hue, N. P. M., Berssaneti, F. T., & Quintanilha, J. A. (2018). An overview of shared mobility. *Sustainability, 10*(12). https://doi.org/10.3390/su10124342

Manoharan, Y., Hosseini, S. E., Butler, B., Alzhahrani, H., Senior, B. T. F., Ashuri, T., & Krohn, J. (2019). Hydrogen fuel cell vehicles; current status and future prospect. *Applied Sciences, 9*(11). https://doi.org/10.3390/app9112296

Maparu, T. S., & Mazumder, T. N. (2017). Transport infrastructure, economic development and urbanization in India (1990–2011): Is there any causal relationship? *Transportation Research Part A: Policy and Practice, 100*, 319. https://doi.org/10.1016/j.tra.2017.04.033

Merkert, R., & Bushell, J. (2020). Managing the drone revolution: A systematic literature review into the current use of airborne drones and future strategic directions for their effective control. *Journal of Air Transport Management, 89*, 101929–101929. https://doi.org/10.1016/j.jairtraman.2020.101929

Nijland, H., & van Meerkerk, J. (2017). Mobility and environmental impacts of car sharing in the Netherlands. *Environmental Innovation and Societal Transitions, 23*, 84–91. https://doi.org/10.1016/j.eist.2017.02.001

Nundy, S., Ghosh, A., Mesloub, A., Albaqawy, G. A., & Alnaim, M. M. (2021). Impact of COVID-19 pandemic on socio-economic, energy-environment and transport sector globally and sustainable development goal (SDG). *Journal of Cleaner Production, 312*, 127705–127705. https://doi.org/10.1016/j.jclepro.2021.127705

O'Keeffe, B. (2019). *Regulatory challenges arising from disruptive transport technologies—The case of e-scooters.* Australasian Transport Research Forum (ATRF), 41st, 2019, Canberra, ACT, Australia.

Philips, I., Anable, J., & Chatterton, T. (2020). *E-bike carbon savings—How much and where.* Centre for Research into energy demand solutions. Retrieved from https://www.creds.ac.uk/wp-content/pdfs/creds-e-bikes-briefing-may2020.pdf

Przybylowski, A., Stelmak, S., & Suchanek, M. (2021). Mobility behaviour in view of the impact of the COVID-19 pandemic-public transport users in Gdansk case study. *Sustainability (Switzerland), 13*(1). https://doi.org/10.3390/su13010364

Pucher, J., & Buehler, R. (2017). Cycling towards a more sustainable transport future. *Transport Reviews, 37*(6), 689. https://doi.org/10.1080/01441647.2017.1340234

Quirós, C., Portela, J., & Marín, R. (2021). Differentiated models in the collaborative transport economy: A mixture analysis for BlaBlacar and Uber. *Technology in Society, 67*, 101727. https://doi.org/10.1016/j.techsoc.2021.101727

Rabbitt, N., & Ghosh, B. (2016). Economic and environmental impacts of organised car sharing services: A case study of Ireland. *Research in Transportation Economics, 57*, 3–12. https://doi.org/10.1016/j.retrec.2016.10.001

Rodriguez-Valencia, A., Rosas-Satizabal, D., Unda, R., & Handy, S. (2021). The decision to start commuting by bicycle in Bogotá, Colombia: Motivations and influences. *Travel Behaviour and Society, 24*, 57–67. https://doi.org/10.1016/j.tbs.2021.02.003

Sánchez-Atondo, A., García, L., Calderón-Ramírez, J., Gutiérrez-Moreno, J. M., & Mungaray-Moctezuma, A. (2020). Understanding public transport ridership in developing countries to promote sustainable urban mobility: A case study of Mexicali, Mexico. *Sustainability, 12*(8). https://doi.org/10.3390/su12083266

Sanders, R. L., Branion-Calles, M., & Nelson, T. A. (2020). To scoot or not to scoot: Findings from a recent survey about the benefits and barriers of using E-scooters for riders and non-riders. *Transportation Research Part A: Policy and Practice, 139*, 217–227. https://doi.org/10.1016/j.tra.2020.07.009

Shahjalal, M., Roy, P. K., Shams, T., Fly, A., Chowdhury, J. I., Ahmed, M. R., & Liu, K. (2022). A review on second-life of Li-ion batteries: Prospects, challenges, and issues. *Energy, 241*, 122881–122881. https://doi.org/10.1016/j.energy.2021.122881

Sharma, A., & Strezov, V. (2017). Life cycle environmental and economic impact assessment of alternative transport fuels and power-train technologies. *Energy, 133*, 1132–1141. https://doi.org/10.1016/j.energy.2017.04.160

Sousa, N., Almeida, A., Coutinho-Rodrigues, J., & Natividade-Jesus, E. (2018). Dawn of autonomous vehicles: Review and challenges ahead. *Proceedings of the Institution of Civil Engineers - Municipal Engineer, 171*(1), 3–14. https://doi.org/10.1680/jmuen.16.00063

Staat, D. W. (2018). *Facing an exponential future: Technology and the community college.* Rowman & Littlefield.

Te, Q., & Lianghua, C. (2020). Carsharing: Mitigation strategy for transport-related carbon footprint. *Mitigation and Adaptation Strategies for Global Change, 25*(5), 791–818. https://doi.org/10.1007/s11027-019-09893-2

Wadud, Z., MacKenzie, D., & Leiby, P. (2016). Help or hindrance? The travel, energy and carbon impacts of highly automated vehicles. *Transportation Research Part A: Policy and Practice, 86*, 1. https://doi.org/10.1016/j.tra.2015.12.001

Wang, L. (2018). Barriers to implementing pro-cycling policies: A case study of Hamburg. *Sustainability, 10*(11). https://doi.org/10.3390/su10114196

Wang, A., Zhang, A., Chan, E. H. W., Shi, W., Zhou, X., & Liu, Z. (2021). A review of human mobility research based on big data and its implication for smart city development. *ISPRS International Journal of Geo-Information, 10*(1). https://doi.org/10.3390/ijgi10010013

Wangsness, P. B., Proost, S., & Rødseth, K. L. (2020). Vehicle choices and urban transport externalities. Are Norwegian policy makers getting it right? *Transportation Research Part D: Transport and Environment, 86*, 102384. https://doi.org/10.1016/j.trd.2020.102384

Watabe, A., & Leaver, J. (2021). Comparative economic and environmental benefits of ownership of both new and used light duty hydrogen fuel cell vehicles in Japan. *International Journal of Hydrogen Energy, 46*(52), 26582–26593. https://doi.org/10.1016/j.ijhydene.2021.05.141

Wong, Y. Z., Hensher, D. A., & Mulley, C. (2020). Mobility as a service (MaaS): Charting a future context. *Transportation Research Part A: Policy and Practice, 131*, 5–19. https://doi.org/10.1016/j.tra.2019.09.030

Zivin, J. G., & Sanders, N. (2020). The spread of COVID-19 shows the importance of policy coordination. *Proceedings of the National Academy of Sciences, 117*(52), 32842–32844. https://doi.org/10.1073/PNAS.2022897117

Conclusions: Status and a Way Forward for Computational Social Science in Policymaking[1]

The number and diversity of contributions in the present handbook (24 chapters, divided in 3 sections and authored by 40 different leading experts in their field) underline the breadth and depth of topics of interest when thinking about the role and potential of Computational Social Science in a policy context. Despite the intrinsic horizontal nature of Computational Social Science, the present contribution provides a first refined picture of the scientific and practical landscape of the discipline and its applications to policy.

The role of this book, both for the policymaking world and the different research communities interested in exploring the intersection between policymaking and Computational Social Science (statisticians, econometricians, data scientists, machine learners, legal scholars, qualitative and quantitative political scientists, philosophers, etc.), is twofold. As a first goal, we wish this book to be both a reference for terminology, definitions and concepts relative to Computational Social Science for Policy and a picture of the status of the different possible fields of application. At the same time, this book should not to be seen as an end point for the research in Computational Social Science in policymaking, but a stimulus for scientists to advance knowledge in this very ebullient field of research. We additionally hope policymakers to be interested and inspired in implementing Computational Social Science solutions in the policy cycle, as well as to engage in a co-creation exercise with scientists and practitioners to better steer research in this very important applied field. The set of foundational aspects described and analysed in the chapters of the first section is fundamental in setting up the ethical, moral, legal and political context which scientists or policymakers should put themselves into when talking about Computational Social Science for Policy. This may appear not immediately obvious to many quantitative researchers who are not trained to take into consideration these aspects when performing Computational

---

[1] The views expressed are purely those of the authors and may not in any circumstances be regarded as stating an official position of the European Commission.

Social Science research. Indeed, it may also be a stimulus (already advocated by several academic institutions across the world) to include data ethics or data justice courses when training data analytics professionals (statisticians, data scientists, data engineers, machine learning specialists, etc.) (Saltz et al., 2018).

The main take-home message of the first section of this handbook is that Computational Social Science research and applications, especially if applied to policy, must be planned and performed considering all the possible stakeholders in an inclusive and open fashion and in an environment that is able to foster cross-disciplinarity and cross-fertilisation. While this is not necessarily a new concept in the academic research devoted to Computational Social Science (see, e.g. Lazer et al., 2020), the series of chapters in the first section of the handbook describe effectively how the environment changes when policymaking comes into the picture.

Computational Social Science research has to be performed ethically (Chap. 4) and taking into account a strong social justice perspective with respect to those whose actions are being studied and analysed, as well as those affected by policy decisions triggered by Computational Social Science research (Chap. 3). Moreover, it is fundamental to consider the ecosystem in which Computational Social Science is developed (Chap. 2), also with the use of specific professional figures such as the "Data Steward". The organisational dimension is also stressed in Chap. 1 which provides insights into the functions of public sector bodies that can be helped by Computational Social Science (detection, measurement, prediction, explanation and simulation), setting the scene for the second section.

The second section reviews and presents methodologies aimed at performing those "functions of government" described in Chap. 1. Apart from the sheer power of some of the techniques presented (which are, e.g. the ability to infer political and social sentiment from unstructured text or to map the spread of a fake news on a social network), a direct connection between the phases of the policy cycle and some sets of techniques emerges, especially when talking about impact assessment and impact evaluation of policies. This focus is mentioned also fairly explicitly in some of the applied chapters (Chaps. 14 and 15 among all). Namely, a subset of the computational techniques can be used for the formulation/ex ante phases of the policy cycle, when a decision is still being formulated, and thus data about its impact (or about the impact of similar phenomena) is absent. We are referring to simulation techniques, among which we may encounter agent-based models, microsimulation, dynamic stochastic general equilibrium models in macroeconomics, computable general equilibrium ones or integrated assessment models for micro and climate economics issues. For the ex-post evaluation of policies, reliable structured or unstructured data sources – where available and accessible and characterised – allow for the use of more empirical techniques, such as statistical or machine learning ones.

Two other common lines of reasoning can be deducted from the methodological section: the first one is that policies designed or assessed using Computational Social Science methods should be made available for public scrutiny, the issue of openness of data processing and of modelling techniques and the replicability of the findings used for policy purposes. The second one relates to modelling and communicating uncertainty in policymaking, which has proven to be of key importance especially during the COVID-19 pandemic (as described, e.g. in Chap. 6). In fact, we have learned what the effective replication number R_t is, and many policymakers have based decisions on confinement measures based on statistical estimates of this parameter, usually without much consideration to the uncertainty connected to the estimates, or to the robustness of the estimation procedure. We believe that further research in the perception of uncertainty in decision-making and methodological research in flexible forecasting and causal inference methods are in order.

As the reader may have observed, the first two sections set up the methodological and foundational scene for CSS4P to be performed. At this point, we can observe how to develop a connection between the "offer" of Computational Social Science methods and the demand of pressing societal questions coming from policymakers; the role of "science-policy bilinguals" starts to be fundamental. This necessity of competences that cross different domains poses additional challenges to governments and supranational organisations aiming and innovate policymaking with scientific and computationally driven insights, as well as research institutions and universities who need to train established data scientists as well as students with this new paradigm of competences.

The third section of the handbook (14 chapters) provides a critical review of the state of the art of the use of Computational Social Science methods in specific disciplines. In some fields the use of Computational Social Science methods has nearly reached the production level, meaning that insights from Computational Social Science are already mainstreamed into the policy cycle. Notable examples are the field of macroeconomic forecasting (Chap. 12), where advanced forecasting models (using also nontraditional data) have been used to inform economic policy during the COVID-19 pandemic (see, e.g. Barbaglia et al., 2022), the use of integrated assessment models for the fit for 55 package (Chap. 14) or the labour market intelligence through text mining on job advertisement data (Chap. 13). Other fields instead, despite showing a great deal of potential for policy, are still in their infancy (notably mobility, with respect to both sustainability aspects in Chap. 24 and the direct analysis of human mobility patterns in Chap. 23).

In terms of data sources and methodologies, we can observe again the clear partition between policy fields for which the focus is on the ex ante evaluation and others for which the main interest is on ex-post assessments. Among the first, one can observe, e.g. Chap. 11 or 15, while the focus on ex-post modelling is typical of Chaps. 12 or 20. Some fields interestingly propose a promising fusion of these modelling approaches (mainly Chaps. 14 and 17), via the use of advanced calibration techniques and/or advanced post-processing for computational models. This partition is also reflected by the type of data that are currently used to perform analytical work. Some disciplines are still exploring administrative data sources (e.g. Chap. 11), some others are starting to exploit less-traditional data sources (Chaps. 12 and 14), while others have fully embraced their full potential (Chaps. 18 and 15).

To conclude, this book aims at presenting an important contribution in establishing the context, theoretical and methodological underpinnings as well as the

Printed in the United States
by Baker & Taylor Publisher Services